Universal Algebra

Fundamentals and Selected Topics

Clifford Bergman
Iowa State University
Ames, USA

CRC Press
Taylor & Francis Group
Boca Raton London New York

CRC Press is an imprint of the
Taylor & Francis Group, an **informa** business
A CHAPMAN & HALL BOOK

PURE AND APPLIED MATHEMATICS

A Program of Monographs, Textbooks, and Lecture Notes

MONOGRAPHS AND TEXTBOOKS IN PURE AND APPLIED MATHEMATICS

Recent Titles

Sergio Macías, Topics on Continua (2005)

Mircea Sofonea, Weimin Han, and Meir Shillor, Analysis and Approximation of Contact Problems with Adhesion or Damage (2006)

Marwan Moubachir and Jean-Paul Zolésio, Moving Shape Analysis and Control: Applications to Fluid Structure Interactions (2006)

Alfred Geroldinger and Franz Halter-Koch, Non-Unique Factorizations: Algebraic, Combinatorial and Analytic Theory (2006)

Kevin J. Hastings, Introduction to the Mathematics of Operations Research with *Mathematica*®, Second Edition (2006)

Robert Carlson, A Concrete Introduction to Real Analysis (2006)

John Dauns and Yiqiang Zhou, Classes of Modules (2006)

N. K. Govil, H. N. Mhaskar, Ram N. Mohapatra, Zuhair Nashed, and J. Szabados, Frontiers in Interpolation and Approximation (2006)

Luca Lorenzi and Marcello Bertoldi, Analytical Methods for Markov Semigroups (2006)

M. A. Al-Gwaiz and S. A. Elsanousi, Elements of Real Analysis (2006)

Theodore G. Faticoni, Direct Sum Decompositions of Torsion-Free Finite Rank Groups (2007)

R. Sivaramakrishnan, Certain Number-Theoretic Episodes in Algebra (2006)

Aderemi Kuku, Representation Theory and Higher Algebraic K-Theory (2006)

Robert Piziak and P. L. Odell, Matrix Theory: From Generalized Inverses to Jordan Form (2007)

Norman L. Johnson, Vikram Jha, and Mauro Biliotti, Handbook of Finite Translation Planes (2007)

Lieven Le Bruyn, Noncommutative Geometry and Cayley-smooth Orders (2008)

Fritz Schwarz, Algorithmic Lie Theory for Solving Ordinary Differential Equations (2008)

Jane Cronin, Ordinary Differential Equations: Introduction and Qualitative Theory, Third Edition (2008)

Su Gao, Invariant Descriptive Set Theory (2009)

Christopher Apelian and Steve Surace, Real and Complex Analysis (2010)

Norman L. Johnson, Combinatorics of Spreads and Parallelisms (2010)

Lawrence Narici and Edward Beckenstein, Topological Vector Spaces, Second Edition (2010)

Moshe Sniedovich, Dynamic Programming: Foundations and Principles, Second Edition (2010)

Drumi D. Bainov and Snezhana G. Hristova, Differential Equations with Maxima (2011)

Willi Freeden, Metaharmonic Lattice Point Theory (2011)

Murray R. Bremner, Lattice Basis Reduction: An Introduction to the LLL Algorithm and Its Applications (2011)

Clifford Bergman, Universal Algebra: Fundamentals and Selected Topics (2011)

First published in paperback 2024

First published 2012
by CRC Press
2385 NW Executive Center Drive, Suite 320, Boca Raton FL 33431

and by CRC Press
4 Park Square, Milton Park, Abingdon, Oxon, OX14 4RN

CRC Press is an imprint of Taylor & Francis Group, LLC

Publisher's Note
The publisher has gone to great lengths to ensure the quality of this reprint but points out that some imperfections in the original copies may be apparent.

ISBN: 978-1-4398-5129-6 (hbk)
ISBN: 978-1-03-292218-8 (pbk)
ISBN: 978-0-429-10808-2 (ebk)

DOI: 10.1201/9781439851302

Visit the Taylor & Francis Web site at
http://www.taylorandfrancis.com

and the CRC Press Web site at
http://www.crcpress.com

Dedicated to Ralph McKenzie
on the occasion of his 70th *birthday*

Contents

Preface

This text is based on the two-semester course that I have taught over the years at Iowa State University. In the writing, as in my course, I attempt to convey my enthusiasm for the subject and my feelings that it is a worthy object of study for both graduate students and professional mathematicians.

In choosing the level of detail, I have taken my inspiration more from the tradition of first-year algebra texts such as van der Waerden, Lang, and Dummit and Foote, than from a typical research monograph. The book is addressed to newcomers to the field, whom I do not wish to overwhelm, more than to veterans seeking an encyclopedic reference work.

It is the job of the author to decide what to omit. One rule of thumb that I have always used in my classes is to introduce a tool only if it will be applied later in the course. As a teacher, I have always found it frustrating to expend a lot of effort and class time developing some construction and then not be able to demonstrate its importance. Thus, for example, in Chapter 7, the basics of commutator theory are developed in the context of congruence-permutable varieties and applied to the characterization of directly representable varieties. The more involved development in the congruence-modular case is omitted since it isn't needed for this application.

As I have matured as a teacher, I have come to incorporate many more examples into all of my classes. I have applied that philosophy to the writing of this book. Throughout the text a series of examples is developed that can be used repeatedly to illustrate new concepts as they are introduced. Whenever possible I work with objects that are already familiar to students, such as Abelian groups and commutative rings. Of course, part of the fun of the subject is the introduction of new sorts of structures (quasigroups, pseudocomplemented distributive lattices, implication algebras) but especially by working with familiar structures, students will understand how this new subject of universal algebra is related to topics they have already studied.

I have also put a great deal of effort into the exercises. Working problems is still the best way to reinforce new concepts in a student's mind. The number of exercises is not huge, and most of them are not hard. Rather, they have been carefully selected to complement the text and push the reader to a deeper understanding of the theorems and techniques presented in each section. I encourage the ambitious student to do every single one.

The book is divided into two parts. Part I consists of the core components that I think every student of the subject must master. There is a brief first chapter that introduces the central concepts of operation, algebra and homomorphism. This should be an easy conceptual step for anyone with a modest background in abstract algebra. The discussion of the set of subalgebras and congruences provides natural motivation for the second chapter on lattices.

The emphasis in Chapter 2 is on complete and algebraic lattices. We do include a brief diversion into distributivity and modularity, but the primary focus is on lattices that arise from closure operators, including Galois connections.

Chapter 3 covers direct and subdirect products and the isomorphism theorems. It is at this point that we begin to develop some of our primary examples. We characterize subdirectly irreducible Abelian groups, commutative rings that satisfy $x^n \approx 1$, distributive lattices, and pseudocomplemented distributive lattices. We then use these characterizations to study the lattices of subvarieties of each of these classes of algebras.

In Chapter 4 we introduce the definition of a clone of operations. From this point on we take every opportunity to discuss concepts from a clone-theoretic point of view. Terms and term algebras are covered as the natural analog of term operations and clones. I know from experience that free algebras are difficult for students to grasp, so we devote several pages to examples. Of course, the whole point is to prove Birkhoff's result that the notions of variety and equational class coincide. Part I concludes with a derivation of the most familiar Maltsev conditions, which seems to fit nicely with our discussion of terms and free algebras.

Part II is, as billed, a selection of topics that I feel illustrate the power and breadth of the subject. These seemed to divide up naturally according to the properties of congruence lattices, but the chapter titles shouldn't be taken too religiously. Chapter 5 covers the consequences of Jónsson's lemma and Baker's theorem on finitely based algebras. While we are at it, we provide an example of a nonfinitely based algebra and a discussion of definable principal congruences.

Chapter 6 treats the work of Foster and Pixley on primal and quasiprimal algebras. It also includes a proof of Murskiĭ's theorem that almost every finite algebra is primal. Many instructors will probably choose to skip this proof because of its length and nature, relying as it does, on asymptotics. I was anxious to include this result for several reasons. First, because everyone finds the theorem to be counterintuitive upon first learning it. Second, because this particular proof, largely due to Bob Quackenbush, is really quite nice; and third, because the only published proof in English is a translation from the original Russian that is very hard to follow.

The goal in Chapter 7 is Ralph McKenzie's characterization of directly representable varieties. To my mind, this theorem is the single best illustration of the power of the universal algebraic toolbox. In order to give a complete proof, it is necessary to develop the commutator operation on congruences as well as provide a complete characterization of congruence-permutable Abelian varieties.

Finally, Chapter 8 covers the rudiments of tame congruence theory. It is certainly not a complete treatment. But I hope it goes far enough to introduce the reader to these new ideas, and also, despite how different these constructions seem, that they grow naturally out of the considerations in Chapters 1–7. As always, I have included several illustrations of the use of the new tools, including the characterization of idempotent varieties omitting type 1 via "Taylor terms."

Since the book largely parallels the course that I have taught, one can simply start at the beginning and progress as far as is possible in the time allotted. In one semester it should be possible to cover almost all of Part I, at least through Birkhoff's theorem in Section 4.4. If the instructor feels the need to omit something, the "case studies" are a likely candidate, as are the examples that compose Section 4.4. A one-semester course need not cover the section on interpretations.

For a two-semester course, I would encourage the instructor to cover everything in Part I and then choose some of the sections in Part II. There are several very long proofs that one might choose to skip. These are Baker's Theorem (5.39), Murskiĭ's Theorem (all of Section 6.2, although one might omit only the proof of Lemma 6.22), and two theorems used in the characterization of directly representable varieties, 7.55 and 7.56.

A very recent development is the discovery that the tools of universal algebra can be brought to bear on constraint satisfaction problems. A reader specifically interested in this line of research could, after reading most of Part I, concentrate on Sections 7.2–7.4 and then move on to Chapter 8.

I cringe a bit at the amount of specialized notation that appears in these pages. I have done what I can to help the reader by providing an index of symbols, alphabetized to the degree it is possible. In mathematical writing, there is always a trade-off between precision and readability. While one never wants to leave a reader guessing about the exact meaning of any assertion, it is equally true that an author should not become a slave to his notation. Thus, when I feel that the context allows, I omit superscript or subscript qualifiers and closure under isomorphism.

I am greatly indebted to my students and colleagues in the universal algebraic community for all of the ideas, suggestions and criticisms that they supplied. Libor Barto, Keith Kearnes, Ralph McKenzie, George McNulty, Bob Quackenbush, Jonathan Smith and Ross Willard all very patiently answered my many questions on the fine points of the subject. The treatment in Chapter 8 relies very heavily on some notes of Matt Valeriote. And I very gratefully acknowledge both Joel Berman and David Failing for their painstaking reading of portions of the manuscript.

This text is obviously influenced by the books that have come before, especially those of Burris and Sankappanavar [BS81] and McKenzie, McNulty and Taylor [MMT87], both of which are undeservedly out of print. My copies of each are dog-eared and marked up. I fondly hope that the book you are now holding will meet the same fate.

Part I

Fundamentals of Universal Algebra

Chapter 1

Algebras

This first, short, chapter sets the stage for the entire text. Many of the fundamental notions and several central examples are introduced. This will allow us to provide examples and motivation for the deeper ideas we encounter in the later chapters.

1.1 Operations

Let A be a set and n a positive integer. We define A^n to be the set of all n-tuples of elements of A, and $A^0 = \{\varnothing\}$.

Remark. More generally, for any two sets A and B, A^B denotes the set of all functions from B to A. How do you reconcile this with the definition of A^n just given? In particular, what about A^0? What does the definition tell us about the case $A = \varnothing$?

For any A and n as above, we call a function $A^n \to A$ an *n-ary operation on A*. The natural number n is called the *rank* of the operation. Operations of rank 1 and 2 are usually called unary and binary operations, respectively. Notice that an operation of rank 0 is a function $c\colon \{\varnothing\} \to A$. Such a function is completely determined by the value $c(\varnothing) \in A$. Thus, for all intents and purposes, nullary operations (as those of rank 0 are often called) are just the elements of A. They are frequently called constants.

A bit more notation before we proceed. We denote by ω the set of *natural numbers,* $\{0, 1, 2, \ldots\}$. For any set X, $\mathrm{Sb}(X)$ denotes the set of all subsets of X (i.e., the power set). We write $Y \subseteq_\omega X$ to indicate that Y is a finite subset of X. Finally $\mathrm{Sb}_\omega(X) = \{\, Y : Y \subseteq_\omega X \,\}$. The symbols $\mathbb{Z}$, $\mathbb{Q}$ and $\mathbb{R}$ are used to represent the sets of integers, rational numbers and real numbers, respectively.

Definition 1.1. An *algebra* is a pair $\langle A, F\rangle$ in which A is a nonempty set and $F = \langle f_i : i \in I\rangle$ is a family of operations on A, indexed by some set I. The set A is called the *universe* of the algebra, and the f_i are the *fundamental* or *basic* operations.

Let $\mathbf{A} = \langle A, F\rangle$ be an algebra, with $F = \langle f_i : i \in I\rangle$. The function $\rho\colon I \to \omega$ which assigns to each $i \in I$ the rank of f_i is called the *similarity type* of $\mathbf{A}$. Two algebras are called *similar* if they have the same similarity type. It is common to write "type" in lieu of "similarity type" when we feel we can get away with it.

Definition 1.2. Let $\mathbf{A} = \langle A, F\rangle$ and $\mathbf{B} = \langle B, G\rangle$ be algebras of the same similarity type $\rho\colon I \to \omega$.

(1) We call $\mathbf{B}$ a *subalgebra* of $\mathbf{A}$ if $B \subseteq A$ and for every $i \in I$, $g_i = f_i{\restriction}_{B^{\rho(i)}}$.

(2) A function $h\colon B \to A$ is called a *homomorphism* if for every $i \in I$, and every $b_1, b_2, \ldots, b_n \in B$, $h(g_i(b_1, \ldots, b_n)) = f_i(h(b_1), \ldots, h(b_n))$. (Here, $n = \rho(i)$.)

Algebras in which every basic operation is unary are called *unary algebras.* An algebra with a single basic operation which is binary is called a *binar.*

1.2 Examples

Here are a few examples just to whet your appetite and give you an idea of how we use the notation and terminology. We will introduce more examples as we go.

Definition 1.3. A *semigroup* is an algebra $\langle S, \cdot\rangle$ satisfying the associative law

$$x \cdot (y \cdot z) \approx (x \cdot y) \cdot z. \tag{1–1}$$

A *monoid* is an algebra $\langle M, \cdot, e\rangle$ satisfying the associative law and the identities

$$x \cdot e \approx x \text{ and } e \cdot x \approx x. \tag{1–2}$$

Of course, we should really be specifying the ranks of the basic operations of our algebras. It is hard to imagine a scenario in which a multiplication-like symbol such as "$\cdot$" could be anything but binary. The operation "e" here is nullary. Monoids are often referred to as "semigroups with identity."

Examples of semigroups and monoids abound in mathematics. $\langle \mathbb{Z}, +\rangle$ and $\langle \omega, +\rangle$ are semigroups, with the latter being a subalgebra of the former. $\langle \omega, \cdot\rangle$ is a semigroup, although not a subalgebra of $\langle \mathbb{Z}, +\rangle$. Each of these three can be made into a monoid by including the appropriate nullary operation. For any nonempty set A, A^A forms a semigroup (in fact, a monoid) under function composition, called the semigroup of *self-maps* of A.

Why are we using "squiggly equals" instead of the usual equals sign in Definition 1.3? The short answer is that (1–1) and (1–2) are not equalities but *identities,* since they assert the equality of the left- and right-hand sides for each choice of the variables (in this case x, y and z). The study of the identities satisfied by an algebra is one of the core areas of universal algebra. It is formally introduced in Chapter 4. Until then, don't obsess about it. You can use an equals sign if you want.

Definition 1.4. A *group* is an algebra $\langle G, \cdot, ^{-1}, e\rangle$ of type $\langle 2, 1, 0\rangle$ such that $\langle G, \cdot, e\rangle$ is a monoid and satisfying

$$x \cdot x^{-1} \approx x^{-1} \cdot x \approx e.$$

A *ring* is an algebra $\langle R, +, \cdot, -, 0\rangle$ such that $\langle R, +, -, 0\rangle$ is a group, $\langle R, \cdot\rangle$ is a semigroup and such that the identities

$$\begin{aligned} x + y &\approx y + x, \\ x \cdot (y + z) &\approx (x \cdot y) + (x \cdot z), \\ (y + z) \cdot x &\approx (y \cdot x) + (z \cdot x) \end{aligned}$$

hold.

Definition 1.5. Let $\mathbf{R}$ be a fixed ring. A (left) $\mathbf{R}$-*module* is an algebra

$$\big\langle M, +, -, 0, \langle \lambda_r : r \in R\rangle\big\rangle$$

such that $\langle M, +, -, 0\rangle$ is an Abelian group, and satisfying, for each $r, s \in R$ the laws

$$\begin{aligned} \lambda_r(x + y) &\approx \lambda_r(x) + \lambda_r(y), \\ \lambda_{r+s}(x) &\approx \lambda_r(x) + \lambda_s(x), \\ \lambda_r(\lambda_s(x)) &\approx \lambda_{rs}(x). \end{aligned}$$

Notice that the similarity type of a module depends upon the ring. If the ring is infinite, so is the similarity type. And notice also that we have employed yet another device to indicate the basic operations on this algebra.

Definition 1.6. A *quasigroup* is an algebra $\langle Q, \cdot, /, \backslash\rangle$ with three binary operations satisfying the laws

$$\begin{aligned} x\backslash(x \cdot y) &\approx y, \quad (x \cdot y)/y \approx x, \\ x \cdot (x\backslash y) &\approx y, \quad (x/y) \cdot y \approx x. \end{aligned}$$

Every group can be made into a quasigroup by taking $x/y = x \cdot y^{-1}$ and $x\backslash y = x^{-1} \cdot y$. Quasigroups provide a generalization of the notion of a group obtained by dropping the associative law. (Exercise: a quasigroup in which "$\cdot$" is associative is a group.) A *loop* is a quasigroup with an identity element.

The expression $x\backslash y$ ("x under y") can be thought of as left-division by x. Dually, x/y ("x over y") is right-division by y.

Much as with groups, quasigroups are often thought of as algebras of type $\langle 2\rangle$ with a "special" binary operation. Specifically, if $\mathbf{Q} = \langle Q, \cdot, /, \backslash\rangle$

is a quasigroup as defined above, then the Cayley table of $\langle Q, \cdot \rangle$ is a *Latin square,* that is, a table in which each entry appears exactly once in each row and column. (This way of thinking makes the most sense for a finite table.) Conversely, each Latin square $\langle Q, \cdot \rangle$ can be expanded to a quasigroup in a unique way. Motivated by this, we define a *Latin square* to be a binar $\langle Q, \cdot \rangle$ such that for every $a, b \in Q$ the equations

$$a \cdot x = b \text{ and } y \cdot a = b$$

have *unique* solutions. In a quasigroup, the element x in the above pair of equations is equal to $a \backslash b$ and y is equal to b/a.

Examples of quasigroups. As we just observed, every group can be transformed into a quasigroup. The set $\mathbb{Z}$ of integers forms a Latin square under the binary operation of subtraction. This is an example of a quasigroup that is not a group. As a more exotic example, define, for points P and Q in the plane, $P \cdot Q$ to be the midpoint of the line segment from P to Q. Then $\langle \mathbb{R}^2, \cdot \rangle$ forms a Latin square. Examples 6.11 and 7.59 exhibit Cayley tables of finite Latin squares that are not associative, so of course, are not groups.

Definition 1.7. A *lattice* is an algebra $\langle L, \wedge, \vee \rangle$ with two binary operations satisfying the identities

$(\mathrm{a}_\wedge)$	$x \wedge (y \wedge z) \approx (x \wedge y) \wedge z$	$(\mathrm{a}_\vee)$	$x \vee (y \vee z) \approx (x \vee y) \vee z$
$(\mathrm{i}_\wedge)$	$x \wedge x \approx x$	$(\mathrm{i}_\vee)$	$x \vee x \approx x$
$(\mathrm{c}_\wedge)$	$x \wedge y \approx y \wedge x$	$(\mathrm{c}_\vee)$	$x \vee y \approx y \vee x$
$(\mathrm{p}_\wedge)$	$x \wedge (x \vee y) \approx x$	$(\mathrm{p}_\vee)$	$x \vee (x \wedge y) \approx x.$

A binary operation satisfying $\mathrm{i}_\wedge$ is called *idempotent.* The identities $\mathrm{p}_\wedge$ and $\mathrm{p}_\vee$ are called the *absorption* laws.

The expression $x \wedge y$ is pronounced "x meet y," and $x \vee y$ is "x join y." One will sometimes see the symbols "$\cdot$" and "$+$" used in place of "$\wedge$" and "$\vee$."

A *semilattice* is an algebra $\langle S, \vee \rangle$ satisfying identities $\mathrm{a}_\vee$, $\mathrm{i}_\vee$, and $\mathrm{c}_\vee$ from the above set. In other words, a semilattice is a commutative, idempotent semigroup.

We will study lattices in some detail in Chapter 2. Among the familiar examples are $\langle \mathbb{Z}, \min, \max \rangle$, where min and max denote the minimum and maximum of two numbers; $\langle \mathbb{Z}^+, \gcd, \mathrm{lcm} \rangle$, where $\mathbb{Z}^+$ denotes the set of positive integers and gcd, lcm denote greatest common divisor and least common multiple respectively; and $\langle \mathrm{Sb}(X), \cap, \cup \rangle$ for any set X. For any vector space $\mathbf{V}$, the set of subspaces of $\mathbf{V}$ forms a lattice under the operations of intersection and sum.

A note on idempotence. The word "idempotent" is used in several related, but distinct, ways in algebra. (1) An element, a, of an algebra is called idempotent if, for every basic operation f, $f(a, a, \ldots, a) = a$. (2) For $n > 1$, an n-ary operation, f, is called idempotent if, for all x, $f(x,x,\ldots,x) = x$. Finally, (3) a unary operation f is idempotent if $f \circ f = f$.

Notice that (3) is really a special case of (1). For a unary operation can be considered an element in the semigroup of self-maps. The condition $f \circ f = f$ is precisely that f is an idempotent element of that semigroup. By contrast, condition (2) applied to a unary operation would assert that for every x, $f(x) = x$, i.e., f is the identity map.

1.3 More about subalgebras, homomorphisms and direct products

99.9% of the time, when we refer to several algebras in the same breath they will all be of the same similarity type. In fact, we frequently neglect to state that hypothesis explicitly. From this perspective it is logical to fix a similarity type and then consider algebras of that type.

Thus, instead of I, let us take for an index set a family $\mathcal{F}$ of *operation symbols,* and let $\rho\colon \mathcal{F} \to \omega$. Then by an algebra $\mathbf{A}$ of type ρ we mean a set A together with, for each operation symbol $f \in \mathcal{F}$, an operation $f^{\mathbf{A}}$ of rank $\rho(f)$. Thus

$$\mathbf{A} = \langle\, A, \mathcal{F}^{\mathbf{A}} \,\rangle \text{ where } \mathcal{F}^{\mathbf{A}} = \langle\, f^{\mathbf{A}} : f \in \mathcal{F} \,\rangle .$$

For example, if we wanted to discuss groups, we might start with operations symbols m, j and e of ranks 2, 1, and 0 respectively. Then, if $\mathbf{A}$ denotes a particular group, the product of a and b would be denoted $m^{\mathbf{A}}(a,b)$ and the inverse of a as $j^{\mathbf{A}}(a)$. (Of course in real life, we prefer to use the "infix" symbol '$\cdot$' instead of a prefix symbol like m. So we might be writing $a \cdot^{\mathbf{A}} b$ instead of $m^{\mathbf{A}}(a,b)$. Obviously, we leave off the superscript whenever we can get away with it.)

Sometimes it is useful to throw away operations. For example, we may want to discuss the additive structure of a ring, or view a group as being merely a monoid. If $\mathbf{A} = \langle\, A, F \,\rangle$ is an algebra and G is a subsequence of F, then the algebra $\mathbf{B} = \langle\, A, G \,\rangle$ is called a *reduct* of $\mathbf{A}$, and $\mathbf{A}$ is called an *expansion* of $\mathbf{B}$.

It is frequently useful to take intersections of subalgebras of an algebra. But there are two technical problems with this. First, strictly speaking, subalgebras are not sets, so we must intersect their respective universes. Second, it is possible for two subalgebras to have disjoint universes, but we are not allowing empty algebras. To cope with these two technicalities, we introduce the following definition.

Definition 1.8. Let $\mathbf{A}$ be an algebra. A subset U of A is called a *subuniverse of* $\mathbf{A}$ if, for every basic operation f of $\mathbf{A}$, with $n = \text{rank}(f)$,

$$u_1, u_2, \ldots, u_n \in U \implies f(u_1, u_2, \ldots, u_n) \in U. \tag{1–3}$$

In common parlance, subuniverses are the subsets that are "closed under the operations" of $\mathbf{A}$. Note that if $\mathbf{B}$ is a subalgebra of $\mathbf{A}$, then B is a nonempty subuniverse of $\mathbf{A}$, and conversely, every nonempty subuniverse is the universe of a subalgebra of $\mathbf{A}$. However, the empty set may be a subuniverse of $\mathbf{A}$, but it is never the universe of a subalgebra. In fact, the empty set is a subuniverse of $\mathbf{A}$ if and only if $\mathbf{A}$ has no nullary basic operations. (Think about this last statement. It is a good exercise for developing an understanding of the boundary case in an implication like (1–3).) The set of subuniverses of $\mathbf{A}$ is denoted $\text{Sub}(\mathbf{A})$.

An injective (i.e., one-to-one) homomorphism is often called an *embedding.* If $h\colon \mathbf{A} \to \mathbf{B}$ is a surjective homomorphism, then $\mathbf{B}$ is called a *homomorphic image* of $\mathbf{A}$. And if h is bijective, then $\mathbf{A}$ and $\mathbf{B}$ are called *isomorphic.* We write $\mathbf{A} \cong \mathbf{B}$ in this latter case. An algebra is called *trivial* if its universe has cardinality 1. Notice that, up to isomorphism, there is exactly one trivial algebra of each similarity type.

As you can easily check, if $\mathbf{A} \xrightarrow{g} \mathbf{B} \xrightarrow{h} \mathbf{C}$ are homomorphisms, then $h \circ g$ is a homomorphism as well. It follows that the set of all *endomorphisms* of $\mathbf{A}$, (that is, homomorphisms from $\mathbf{A}$ to itself) forms a monoid under the binary operation of composition. And, together with Exercise 1.11.2, we see that the set $\text{Aut}(\mathbf{A})$ of *automorphisms* of $\mathbf{A}$ can be made into a group. The corresponding algebra $\langle \text{Aut}(\mathbf{A}), \circ, {}^{-1}, \iota_A \rangle$ should be denoted $\mathbf{Aut}(\mathbf{A})$, although we often cheat a bit on the boldfacing. The algebra $\mathbf{A}$ is called *rigid* if its automorphism group is trivial.

Definition 1.9. Let $\mathcal{S} = \langle S_i : i \in I \rangle$ be a sequence of sets. The *direct, or Cartesian product* of $\mathcal{S}$ is the set

$$\prod_{i \in I} S_i = \Big\{ f\colon I \to \bigcup_{i \in I} S_i \;\Big|\; (\forall i \in I)\ f(i) \in S_i \Big\}.$$

The elements of the direct product are often called *choice functions* since each such function chooses an element from each S_i.

Now let $\mathcal{A} = \langle \mathbf{A}_i : i \in I \rangle$ be a sequence of algebras (all of the same similarity type). The direct product of this sequence is the algebra $\mathbf{B}$ with universe $B = \prod_{i \in I} A_i$ and such that for every (n-ary) basic operation symbol g and $f_1, \ldots, f_n \in B$

$$\big(g^{\mathbf{B}}(f_1, \ldots, f_n)\big)(i) = g^{\mathbf{A}_i}\big(f_1(i), \ldots, f_n(i)\big), \quad \text{for all } i \in I.$$

When I is infinite, direct products are quite difficult to comprehend and work with. (But necessary, nonetheless.) In the finite case, things are not so bad. The product $\mathbf{A}_1 \times \mathbf{A}_2$ is the algebra whose universe is the set of ordered pairs (a_1, a_2) (with $a_i \in A_i$, for $i = 1, 2$) and with operations computed "coordinatewise." In other words, for an n-ary operation f,

$$f^{\mathbf{A}_1 \times \mathbf{A}_2}\big((a_{11}, a_{21}), (a_{12}, a_{22}), \ldots, (a_{1n}, a_{2n})\big) = \big(f^{\mathbf{A}_1}(a_{11}, \ldots, a_{1n}), f^{\mathbf{A}_2}(a_{21}, \ldots, a_{2n})\big).$$

Now, you need to extend this to a direct product of k factors and, more importantly, understand how your definition is a special case of the official definition above. You also ought to think about what happens when $I = \varnothing$.

In the special case that all of the algebras $\mathbf{A}_i$ are equal to the same algebra $\mathbf{A}$, we obtain the *direct power* algebra $\mathbf{A}^I$ with universe A^I of all functions from I to A and with operations acting coordinatewise. The direct power construction is familiar from calculus. We treat the collection of all functions from the reals to itself as a ring, with $(f+g)(x) = f(x)+g(x)$ and $(f \cdot g)(x) = f(x) \cdot g(x)$. In our notation, this ring is $\langle \mathbb{R}, +, -, \cdot, 0 \rangle^{\mathbb{R}}$.

Definition 1.10. Let $\mathcal{K}$ be a class of similar algebras. We write

$\mathsf{H}(\mathcal{K})$ for the class of all homomorphic images of members of $\mathcal{K}$

$\mathsf{S}(\mathcal{K})$ for the class of all algebras isomorphic to a subalgebra of a member of $\mathcal{K}$

$\mathsf{P}(\mathcal{K})$ for the class of all algebras isomorphic to a direct product of members of $\mathcal{K}$.

We say that $\mathcal{K}$ is closed under the formation of homomorphic images if $\mathsf{H}(\mathcal{K}) \subseteq \mathcal{K}$, and similarly for subalgebras and products. Notice that all three of these "class operators" are designed so that for any class $\mathcal{K}$, $\mathsf{H}(\mathcal{K})$, $\mathsf{S}(\mathcal{K})$ and $\mathsf{P}(\mathcal{K})$ are closed under isomorphic images. On those rare occasions that we need it, we can write $\mathsf{I}(\mathcal{K})$ for the class of algebras isomorphic to a member of $\mathcal{K}$.

Finally, we call $\mathcal{K}$ a *variety* if it is closed under each of H, S and P. Varieties have been the primary objects of study in universal algebra since the 1970s. They will increasingly become the focus of the text in the later chapters.

Exercise Set 1.11.

1. Let $h\colon \mathbf{A} \to \mathbf{B}$ be a homomorphism. Prove that $h(A)$ is a subuniverse of $\mathbf{B}$. By $h(A)$ we mean $\{\, h(a) : a \in A \,\}$.

2. Let $h\colon \mathbf{A} \to \mathbf{B}$ be an isomorphism. Prove that the inverse function h^{-1} is a homomorphism (consequently an isomorphism) from $\mathbf{B}$ to $\mathbf{A}$.

3. Let $\mathbf{A}$ and $\mathbf{B}$ be similar algebras, and $h\colon A \to B$ be a function. View h as a set of ordered pairs. Prove that h is a subuniverse of $\mathbf{A} \times \mathbf{B}$ if and only if h is a homomorphism.

4. Let $\mathbf{G}$ be a fixed group. Recall from your first-year algebra text, the idea of $\mathbf{G}$ acting on a set S. Let us call such a set S a $\mathbf{G}$-*set*. Describe the class of $\mathbf{G}$-sets as a class of algebraic structures in the sense of Definition 1.1, and describe the identities that must hold.

5. Prove that every group is isomorphic to $\mathbf{Aut}(\mathbf{A})$ for some unary algebra $\mathbf{A}$. (Hint: use the previous exercise.)

6. Let $\langle \mathbf{A}_i : i \in I \rangle$ be a sequence of similar algebras.

   (a) Prove that for each $j \in I$, the mapping $p_j \colon \prod \mathbf{A}_i \to \mathbf{A}_j$ such that $f \mapsto f(j)$ is a surjective homomorphism.

   (b) Let $\mathbf{B}$ be an algebra similar to the $\mathbf{A}_i$. Suppose that for each $i \in I$, we are given a homomorphism $h_i \colon \mathbf{B} \to \mathbf{A}_i$. Prove that there exists a unique homomorphism $\bar{h} \colon \mathbf{B} \to \prod \mathbf{A}_i$ such that for all $i \in I$, $p_i \circ \bar{h} = h_i$.

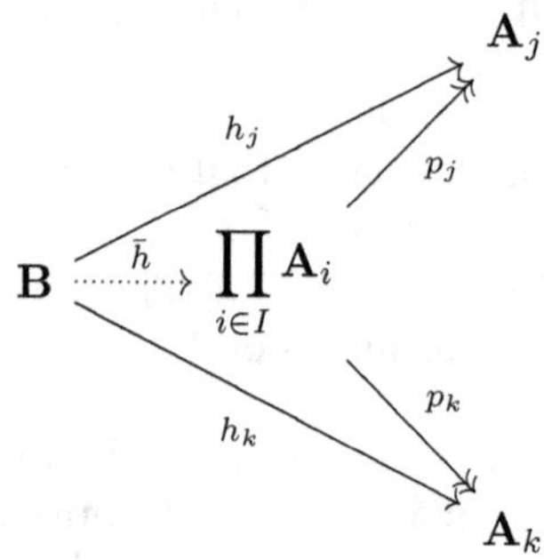

7. A quick game of scattergories, anyone? Fill in the first column of the table in Figure 1.1 with examples of classes of algebras exhibiting closure under the required operators. On "game day," the class will go through everyone's lists. You get one point for each correct example that is different from anybody else's. The instructor will award one point for the example (in each row) that (s)he thinks is the most interesting mathematically.

1.4 Generating subalgebras

Proposition 1.12. *Let* $\mathbf{A}$ *be any algebra, and let* $\mathcal{S}$ *be any nonempty collection of subuniverses of* $\mathbf{A}$. *Then* $\bigcap \mathcal{S}$ *is a subuniverse of* $\mathbf{A}$.

By $\bigcap \mathcal{S}$ we mean the intersection of all the sets in the collection $\mathcal{S}$. One could write $\mathcal{S} = \{ S_i : i \in I \}$ and then $\bigcap_{i \in I} S_i$ for the intersection, but that

	closure under		
class $\mathcal{K}$	**H**	**S**	**P**
	yes	yes	yes
	yes	yes	no
	yes	no	yes
	yes	no	no
	no	yes	yes
	no	yes	no
	no	no	yes
	no	no	no

FIGURE 1.1: Table for Exercise 7

introduces an unnecessary set I and a level of indexing. Note also that $\mathcal{S} \neq \varnothing$ means we are taking the intersection of *something*. But it is entirely possible for the intersection itself to be empty.

Proof. Call the intersection B. Certainly $B \subseteq A$. Let f be a basic n-ary operation of $\mathbf{A}$, and let $b_1, \dots, b_n \in B$. Then for every $S \in \mathcal{S}$, $b_1, \dots, b_n \in S$ so, since S is a subuniverse, $a = f(b_1, \dots, b_n) \in S$. Now $a \in S$ for every $S \in \mathcal{S}$, so $a \in B$. □

Given a set of elements in an algebra we are often interested in the smallest subuniverse containing those elements. The following definition makes that idea precise. Observe that it makes perfect sense even when X (the set of elements) is empty.

Definition 1.13. Let $\mathbf{A}$ be an algebra and $X \subseteq A$. The *subuniverse of* $\mathbf{A}$ *generated by* X is $\mathrm{Sg}^{\mathbf{A}}(X) = \bigcap \{\, U \in \mathrm{Sub}(\mathbf{A}) : X \subseteq U \,\}$.

Notice that in the definition, the collection of subuniverses whose intersection we take is never empty, since it surely contains A itself. Thus Proposition 1.12 ensures that $\mathrm{Sg}(X)$ always exists and must contain X. But it doesn't give us much of a feel about what is going to be in that subuniverse. What we would like is a "bottom-up" description of the subuniverse generated by X. This is supplied by the following construction.

Theorem 1.14. *Let $\mathbf{A} = \langle A, F\rangle$ be an algebra and $X \subseteq A$. Define, by recursion on n, the sets X_n by:*

$$X_0 = X;$$
$$X_{n+1} = X_n \cup \{\, f(a_1, \dots, a_k) : a_1, \dots, a_k \in X_n, f \in F, \text{ and } k = \operatorname{rank} f\,\}.$$

Then $\mathrm{Sg}^{\mathbf{A}}(X) = \bigcup_{n\in\omega} X_n$.

Proof. Let $Y = \bigcup_{n\in\omega} X_n$. Clearly $X_n \subseteq Y \subseteq A$, for every $n \in \omega$. In particular $X = X_0 \subseteq Y$. Let us show that Y is a subuniverse of $\mathbf{A}$. Let f be a basic k-ary operation and $a_1, \dots, a_k \in Y$. From the construction of Y, there is an $n \in \omega$ such that $a_1, \dots, a_k \in X_n$. From its definition, $f(a_1, \dots, a_k) \in X_{n+1} \subseteq Y$. Thus Y is a subuniverse of $\mathbf{A}$ containing X. By Definition 1.13, $\mathrm{Sg}^{\mathbf{A}}(X) \subseteq Y$.

For the opposite inclusion, it is enough to check, by induction on n, that $X_n \subseteq \mathrm{Sg}^{\mathbf{A}}(X)$. Well, $X_0 = X \subseteq \mathrm{Sg}(X)$ from its definition. Assume that $X_n \subseteq \mathrm{Sg}(X)$. If $b \in X_{n+1} - X_n$ then $b = f(a_1, \dots, a_k)$ for a basic k-ary operation f and $a_1, \dots, a_k \in X_n$. But $a_1, \dots, a_k \in \mathrm{Sg}(X)$ and since this latter object is a subuniverse, $b \in \mathrm{Sg}(X)$ as well. □

The argument that appears in Theorem 1.14 is of a type that one encounters frequently throughout algebra. It has two parts. First that Y is a subuniverse containing X. Second that any subuniverse containing X must contain Y. We will explore this phenomenon in more detail in Section 2.4.

Corollary 1.15. *Let $\mathbf{A}$ and X be as in Theorem 1.14, and let $a \in A$. Then $a \in \mathrm{Sg}(X)$ if and only if there is a finite subset Y of X such that $a \in \mathrm{Sg}(Y)$.*

Proof. The right-to-left direction follows from Exercise 1.16.2. For the converse, by Theorem 1.14, there is a natural number n such that $a \in X_n$. Assume that n is minimal. We proceed by induction on n. If $n = 0$ then $a \in X$, so take $Y = \{a\}$. Now suppose that $n = m + 1 > 0$. By definition, there is a basic operation f and $b_1, \dots, b_k \in X_m$ such that $a = f(b_1, \dots, b_k)$. By the induction hypothesis, there are finite sets $Y_1, \dots, Y_k$ such that $b_i \in \mathrm{Sg}(Y_i)$ for $i = 1, 2, \dots, k$. Take $Y = Y_1 \cup Y_2 \cup \cdots \cup Y_k$. □

An algebra $\mathbf{A}$ is called *finitely generated* if there is a finite subset Y such that $A = \mathrm{Sg}^{\mathbf{A}}(Y)$.

Exercise Set 1.16.

1. Let $\mathbf{A}$ be an algebra and $\mathcal{C}$ be a chain of subuniverses of $\mathbf{A}$, i.e., for every $U, V \in \mathcal{C}$, either $U \subseteq V$ or $V \subseteq U$. Prove that $\bigcup \mathcal{C}$ is a subuniverse of $\mathbf{A}$.

2. Let $\mathbf{A}$ be an algebra. For every $X, Y \subseteq A$, verify the following.

   (a) $X \subseteq \mathrm{Sg}(X)$;

   (b) $X \subseteq Y \implies \mathrm{Sg}(X) \subseteq \mathrm{Sg}(Y)$;

(c) $\mathrm{Sg}(\mathrm{Sg}(X)) = \mathrm{Sg}(X)$;
(d) $\mathrm{Sg}(X) = \bigcup\{\,\mathrm{Sg}(Z) : Z \subseteq_\omega X\,\}$.

3. Let $\mathbf{A}$ be a finitely generated algebra. Prove that if $A = \mathrm{Sg}(X)$ for some subset X, then there is a finite subset Y of X such that $A = \mathrm{Sg}(Y)$.

4. Let $\mathbf{A}$ be a semigroup and $X \subseteq A$. Prove the following variation on Theorem 1.14. Define $X_0 = X$ and, for $n \geq 0$, $X_{n+1} = X_n \cup \{\,a \cdot b : a \in X, b \in X_n\,\}$. Then $\mathrm{Sg}^{\mathbf{A}}(X) = \bigcup X_n$. (Remark: while the truth of this exercise should be obvious, a proof is a little slippery. It is a good exercise in formulating proofs by induction.)

5. An element a of an algebra $\mathbf{A}$ is called a *non-generator* of $\mathbf{A}$ if, for every $X \subseteq A$, $A = \mathrm{Sg}(X \cup \{a\})$ implies $A = \mathrm{Sg}(X)$.

   (a) Prove that the set of nongenerators of $\mathbf{A}$ forms a subuniverse, $\mathrm{Frat}(\mathbf{A})$ (called the Frattini subuniverse of $\mathbf{A}$).
   (b) Prove that $\mathrm{Frat}(\mathbf{A})$ is the intersection of all maximal proper subuniverses of $\mathbf{A}$. (If you wish, you can assume that A is finite. To do the infinite case, you will need Zorn's lemma. See page 23.)

6. Let f and g be homomorphisms from $\mathbf{A}$ to $\mathbf{B}$.

   (a) Let $E(f,g) = \{\,a \in A : f(a) = g(a)\,\}$ (the *equalizer* of f and g). Prove that $E(f,g)$ is a subuniverse of $\mathbf{A}$.
   (b) Suppose that $X \subseteq A$ and X generates $\mathbf{A}$. Prove that if $f{\restriction}_X = g{\restriction}_X$ then $f = g$.
   (c) Let $\mathbf{A}$ and $\mathbf{B}$ be algebras and suppose that X generates $\mathbf{A}$. Prove that $|\mathrm{hom}(\mathbf{A},\mathbf{B})| \leq |B|^{|X|}$. (You may assume that both A and B are finite.)

1.5 Congruences and quotient algebras

Let A be a set, and n a natural number. A subset of A^n is called an *n-ary relation* on A. For the moment we will be interested in binary relations, i.e., sets of ordered pairs of elements of A.

Let θ and ψ be binary relations on A. Obviously, both $\theta \cap \psi$ and $\theta \cup \psi$ are binary relations on A as well. We introduce two more ways of producing new binary relations. Let us define

$$\theta \circ \psi = \{\,(x,z) : (\exists y)\ (x,y) \in \theta\ \&\ (y,z) \in \psi\,\} \text{ and}$$
$$\theta^{\smile} = \{\,(y,x) : (x,y) \in \theta\,\}$$

called respectively the *relative product* of θ and ψ and the *converse* of θ. In place of '$(x, y) \in \theta$' it is often convenient to write '$x \,\theta\, y$'.

A binary relation θ is an *equivalence relation* on A if, for all $x, y, z \in A$

$$\begin{aligned} &x \,\theta\, x && \text{(reflexivity)} \\ &x \,\theta\, y \implies y \,\theta\, x && \text{(symmetry)} \\ &x \,\theta\, y \;\&\; y \,\theta\, z \implies x \,\theta\, z && \text{(transitivity).} \end{aligned}$$

The set of all equivalence relations on A is denoted $\mathrm{Eq}(A)$. We distinguish two members of this set. Let $0_A = \{\, (x, x) : x \in A \,\}$ and $1_A = A \times A$. It is easy to see that $0_A, 1_A \in \mathrm{Eq}(A)$, and for every $\theta \in \mathrm{Eq}(A)$ we have $0_A \subseteq \theta \subseteq 1_A$. (The symbols 0 and 1 are used here because, as we will see later, $\mathrm{Eq}(A)$ is a lattice, and it is customary to use those symbols for the smallest and greatest elements of a lattice.) If θ is an equivalence relation, we often use notation such as

$$x \equiv y \pmod{\theta} \qquad \text{or} \qquad x \equiv_\theta y$$

instead of $(x, y) \in \theta$.

One can say quite a bit about a binary relation using just the operations discussed in the previous paragraphs. For example, the three conditions for θ to be an equivalence relation can be expressed as: $0_A \subseteq \theta$, $\theta^\smile \subseteq \theta$ and $\theta \circ \theta \subseteq \theta$. In fact, there is a whole field of "relational algebra." See for example the recent text [Mad06]. The next lemma contains a few useful facts from the calculus of relations. We leave the verifications to the reader.

Lemma 1.17. *Let α, β, γ and θ_i, $i \in I$ be binary relations on a set A.*

(1) $(\alpha \circ \beta) \circ \gamma = \alpha \circ (\beta \circ \gamma)$.

(2) $(\alpha^\smile)^\smile = \alpha$.

(3) $\alpha \circ \bigcup_{i \in I} \theta_i = \bigcup_{i \in I} (\alpha \circ \theta_i)$, *and* $\left(\bigcup_{i \in I} \theta_i\right) \circ \alpha = \bigcup_{i \in I} (\theta_i \circ \alpha)$.

(4) $\left(\bigcup_{i \in I} \theta_i\right)^\smile = \bigcup_{i \in I} \theta_i^\smile$.

(5) $\alpha \subseteq \beta \implies \alpha \circ \gamma \subseteq \beta \circ \gamma \;\&\; \alpha^\smile \subseteq \beta^\smile$.

(6) $(\alpha \circ \beta)^\smile = \beta^\smile \circ \alpha^\smile$.

(7) $\alpha \circ 0_A = \alpha$.

We turn now to another notion that is ubiquitous throughout mathematics. The kernel of a function.

Definition 1.18. Let $f\colon A \to B$ be any function. We define

$$\ker f = \{\, (x, y) \in A^2 : f(x) = f(y) \,\}$$

called the *kernel* of f.

It is easy to see that $\ker f$ is always an equivalence relation on A. One of the central themes of algebra is the converse: every equivalence relation is the kernel of a function. The relevant construction is that of a quotient set.

Let θ be an equivalence relation on A. For $a \in A$ we write

$$a/\theta = \{ x \in A : a \mathrel{\theta} x \},$$

the *equivalence class of a modulo* θ. Notice that $\{ a/\theta : a \in A \}$ is a partition of A. The set of equivalence classes modulo θ is denoted A/θ and is called the *quotient of A by* θ.

There is a natural map $q_\theta \colon A \to A/\theta$ given by $q_\theta(a) = a/\theta$. Notice that q_θ is always surjective. And one easily verifies that $\theta = \ker q_\theta$. Thus every equivalence relation is the kernel of a function.

The cases on the "boundary" are noteworthy. Since $a/0_A = \{a\}$, we have $A/0_A = \{\{a\} : a \in A\}$, with which $\mathbf{A}$ has an obvious bijection. At the opposite extreme, $a/1_A = A$, so that $A/1_A = \{A\}$, a 1-element set.

Now let us return to algebraic structures. Suppose that $h \colon \mathbf{A} \to \mathbf{B}$ is a homomorphism, and let $\theta = \ker h$. Certainly θ is an equivalence relation on A, but does it have any other special properties? Suppose, for the sake of argument, that the similarity type includes a binary operation '$\cdot$'. Let $(x_1, y_1), (x_2, y_2) \in \theta$. This means that $h(x_i) = h(y_i)$ for $i = 1, 2$. Now, since h is a homomorphism, $h(x_1 \cdot x_2) = h(x_1) \cdot h(x_2) = h(y_1) \cdot h(y_2) = h(y_1 \cdot y_2)$. Therefore $(x_1 \cdot x_2) \mathrel{\theta} (y_1 \cdot y_2)$. We formalize this condition in the following definition.

Definition 1.19. Let $\mathbf{A}$ be an algebra and θ a binary relation on A.

(1) We say that θ has the *substitution property* if for every basic operation f, with $n = \operatorname{rank} f$, we have

$$x_1 \mathrel{\theta} y_1 \;\&\; x_2 \mathrel{\theta} y_2 \;\&\; \cdots \;\&\; x_n \mathrel{\theta} y_n \implies f(x_1, \dots, x_n) \mathrel{\theta} f(y_1, \dots, y_n).$$

(2) A *congruence relation on* $\mathbf{A}$ is an equivalence relation on A with the substitution property.

Thus the kernel of every homomorphism is a congruence relation, and as we now show, every congruence relation is the kernel of a homomorphism.

Let θ be a congruence relation on an algebra $\mathbf{A}$. From above, we have a function $q_\theta \colon A \to A/\theta$. We wish to define an algebraic structure on A/θ so that q_θ becomes a homomorphism. There is only one way to do this. Let f be a basic n-ary operation symbol, and let $a_1, \dots, a_n \in A$. We require $q_\theta\big(f^{\mathbf{A}}(a_1, \dots, a_n)\big) = f^{\mathbf{A}/\theta}\big(q_\theta(a_1), \dots, q_\theta(a_n)\big)$. In other words

$$f^{\mathbf{A}/\theta}(a_1/\theta, \dots, a_n/\theta) = f^{\mathbf{A}}(a_1, \dots, a_n)/\theta. \tag{1–4}$$

As is typical when working with quotients, we must verify that equation (1–4) is "well-defined." In other words, if $a_i/\theta = b_i/\theta$ for $i \leq n$, is

$f(\mathbf{a})/\theta = f(\mathbf{b})/\theta$? But this is precisely the content of the substitution property!

Definition 1.20. Let $\mathbf{A}$ be an algebra and θ a congruence relation on $\mathbf{A}$. The *quotient algebra* $\mathbf{A}/\theta$ is the algebra similar to $\mathbf{A}$, with universe A/θ and with basic operations defined by equation (1–4).

From our discussion above, we obtain that q_θ is a homomorphism from $\mathbf{A}$ to $\mathbf{A}/\theta$ with kernel θ. Thus every congruence relation is the kernel of a homomorphism.

For any algebra $\mathbf{A}$, the equivalence relations 0_A and 1_A are always congruences. We note that $\mathbf{A}/0_A \cong \mathbf{A}$ and $\mathbf{A}/1_A$ is a trivial algebra (see page 8). There is, up to isomorphism, one trivial algebra of each similarity type, and it is a homomorphic image of every algebra of its type.

Of course most algebras studied in practice will contain other congruences besides 0 and 1. Those that do not should receive special distinction. A nontrivial algebra possessing exactly two congruences is called *simple*.

All this talk of kernels presumably reminds the reader of earlier courses on groups and rings. What is the connection? Suppose $h\colon \mathbf{G} \to \mathbf{H}$ is a group homomorphism. Recall that the traditional notion of the kernel of h is the set of elements of G that map to the identity of $\mathbf{H}$. Observe that

$$h(x) = h(y) \iff h(y)^{-1} \cdot h(x) = e_H \iff h(y^{-1}x) = e_H.$$

Thus the pair (x, y) lies in our notion of the kernel just in case $y^{-1}x$ is a member of the group-theoretic kernel. As we recall from our group theory classes, there is a strong relationship between kernels and normal subgroups. Here is a precise statement.

Theorem 1.21. *Let* $\mathbf{G}$ *be a group.*

(1) *For every normal subgroup* N*, the relation*

$$\theta_N = \{\, (x, y) \in G \times G : y^{-1} \cdot x \in N \,\}$$

is a congruence on $\mathbf{G}$*. For any* $x \in G$*, the right coset* Nx *coincides with* x/θ_N*.*

(2) *For every congruence* θ*, the equivalence class* e/θ *is a normal subgroup of* $\mathbf{G}$*.*

(3) *The mapping* $N \mapsto \theta_N$ *is a bijection from the set of normal subgroups to the set of congruences, with inverse map* $\theta \mapsto e/\theta$*. For normal subgroups* N *and* M*,* $N \subseteq M$ *if and only if* $\theta_N \subseteq \theta_M$*.*

The verification of Theorem 1.21 is left to the reader. It should be clear from the theorem that the quotient structure $\mathbf{G}/\theta_N$ is the same as the familiar quotient group G/N. In Exercise 1.26.4 you are asked to formulate a similar correspondence between ring congruences and two-sided ideals.

We return to the relationship between homomorphisms and their kernels. The essence of that relationship is captured in the fundamental homomorphism theorem.

Theorem 1.22 (The Fundamental Homomorphism Theorem).
Let $\mathbf{A}$ *and* $\mathbf{B}$ *be similar algebras, and let* $h\colon \mathbf{A} \to \mathbf{B}$ *be a homomorphism with kernel* θ. *There is a unique injective homomorphism* $\bar{h}\colon \mathbf{A}/\theta \to \mathbf{B}$ *such that* $\bar{h} \circ q_\theta = h$. *If* h *is surjective, then* $\bar{h}$ *is an isomorphism.*

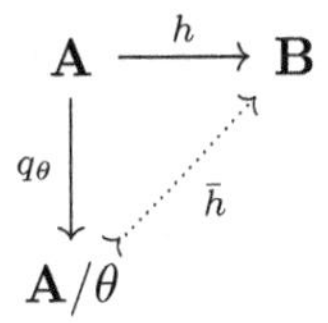

Proof. A typical element of $\mathbf{A}/\theta$ is of the form a/θ, for some $a \in A$. Since $a/\theta = q_\theta(a)$ the condition $\bar{h} \circ q_\theta = h$ requires that $\bar{h}(a/\theta) = h(a)$. That covers the 'uniqueness' clause in the theorem. We must verify that $\bar{h}$ is an injective homomorphism. Consider the chain of equivalences

$$a/\theta = b/\theta \iff (a,b) \in \theta = \ker h \iff h(a) = h(b) \iff \bar{h}(a/\theta) = \bar{h}(b/\theta).$$

The left-to-right direction is exactly the assertion that $\bar{h}$ is well-defined, while the right-to-left directions says that $\bar{h}$ is injective. Finally to check that $\bar{h}$ is a homomorphism, let f be an n-ary operation symbol. Then for all $a_1, \dots, a_n$ in A,

$$\bar{h}(f^{\mathbf{A}/\theta}(\mathbf{a}/\theta)) = \bar{h}(f^{\mathbf{A}}(\mathbf{a})/\theta) = h(f^{\mathbf{A}}(\mathbf{a})) = f^{\mathbf{B}}(h(\mathbf{a})) = f^{\mathbf{B}}(\bar{h}(\mathbf{a}/\theta)).$$

□

We now proceed much as we did with subuniverses. The set of congruences on an algebra $\mathbf{A}$ is denoted $\operatorname{Con}\mathbf{A}$. Our goal is to understand the smallest congruence containing a set of ordered pairs.

Proposition 1.23. *Let* $\mathbf{A}$ *be an algebra and let* Θ *be a nonempty collection of congruences on* $\mathbf{A}$. *Then* $\bigcap\Theta$ *is a congruence on* $\mathbf{A}$.

The proof of this proposition is similar to that of Proposition 1.12. One can use the relationships in Lemma 1.17 to show that $\bigcap\Theta$ is an equivalence relation.

Definition 1.24. Let $\mathbf{A}$ be an algebra and let $\nu \subseteq A \times A$. The *congruence on* $\mathbf{A}$ *generated by* ν is $\operatorname{Cg}^{\mathbf{A}}(\nu) = \bigcap\{\theta \in \operatorname{Con}\mathbf{A} : \nu \subseteq \theta\}$.

Once again, we would like to know how to build $\operatorname{Cg}(\nu)$ from ν. It is helpful to introduce a shorthand. Suppose $\mathbf{a} = \langle a_1, \dots, a_n\rangle$ and $\mathbf{b} = \langle b_1, \dots, b_n\rangle$ are

n-tuples of elements of A. We write $\mathbf{a}\,\theta\,\mathbf{b}$ when $a_i\,\theta\,b_i$ for $i = 1, 2, \ldots, n$. So we can write the substitution property as

$$\mathbf{a}\,\theta\,\mathbf{b} \implies f(\mathbf{a})\,\theta\,f(\mathbf{b})$$

for each basic operation f and all sequences $\mathbf{a}$ and $\mathbf{b}$ from A. The construction is a bit more elaborate than the analogous Theorem 1.14 because we have to ensure that the result is transitive.

Theorem 1.25. *Let* $\mathbf{A} = \langle A, F\rangle$ *be an algebra and* $\nu \subseteq A \times A$. *Define* ν_n *recursively by*

$$\nu_0 = \nu \cup \nu^{\smile} \cup 0_A;$$
$$\nu_{n+1} = (\nu_n \circ \nu_n) \cup \Big\{ (f(\mathbf{a}), f(\mathbf{b})) : f \in F \ \ \& \ \ \mathbf{a}\,\nu_n\,\mathbf{b} \Big\}.$$

Then $\mathrm{Cg}^{\mathbf{A}}(\nu) = \bigcup_{n\in\omega} \nu_n$.

Proof. Let $\psi = \bigcup_{n\in\omega} \nu_n$. We must check three things:

(1) $\nu \subseteq \psi$,

(2) ψ is a congruence,

(3) $\theta \in \mathrm{Con}\,\mathbf{A}$ and $\nu \subseteq \theta$ implies $\psi \subseteq \theta$.

Since $\nu \subseteq \nu_0 \subseteq \psi$ the first of these is true. For the second, let us define $\sigma_n = \big\{ (f(\mathbf{a}), f(\mathbf{b})) : f \in F \ \ \& \ \ \mathbf{a}\,\nu_n\,\mathbf{b} \big\}$. Thus $\nu_{n+1} = (\nu_n \circ \nu_n) \cup \sigma_n$. In the following paragraphs we use the properties in Lemma 1.17 repeatedly.

In order to show that ψ is an equivalence, we first claim that $0_A \subseteq \nu_0 \subseteq \nu_1 \subseteq \cdots \subseteq \psi$. The first inclusion holds by definition. Arguing by induction, $\nu_{n+1} \supseteq \nu_n \circ \nu_n \supseteq \nu_n \circ 0_A = \nu_n$. Similarly, $\nu_{n+1}^{\smile} = ((\nu_n \circ \nu_n) \cup \sigma_n)^{\smile} = (\nu_n^{\smile} \circ \nu_n^{\smile}) \cup \sigma_n^{\smile} = (\nu_n \circ \nu_n) \cup \sigma_n = \nu_{n+1}$.

From the previous paragraph ψ is reflexive and $\psi^{\smile} = \left(\bigcup \nu_n\right)^{\smile} = \bigcup \nu_n^{\smile} = \bigcup \nu_n = \psi$, in other words ψ is symmetric. For transitivity,

$$\psi \circ \psi = \bigcup_{j,k\in\omega} (\nu_j \circ \nu_k) \subseteq \bigcup_{n\in\omega} (\nu_n \circ \nu_n) \subseteq \bigcup_{n\in\omega} \nu_{n+1} = \psi.$$

Thus ψ is transitive, hence an equivalence relation. The argument that ψ has the substitution property is similar to the proof of Theorem 1.14 using the fact that $\sigma_n \subseteq \nu_{n+1}$.

Finally, for the third requirement, assume that $\nu \subseteq \theta \in \mathrm{Con}\,\mathbf{A}$. Since θ is reflexive and symmetric, $\nu_0 \subseteq \theta$. Arguing by induction and using the transitivity of θ and the substitution property, $\nu_{n+1} \subseteq \theta$ for every n. Therefore $\psi \subseteq \theta$ as desired. □

A congruence of the form $\mathrm{Cg}(\nu)$ where ν is a finite set (of pairs) is called *finitely generated*. Of special importance are congruences $\mathrm{Cg}\big(\{(x, y)\}\big)$. These are called *principal congruences*. It is customary to shorten the notation of this congruence to $\mathrm{Cg}(x, y)$.

Exercise Set 1.26.

1. Let **A** be an algebra and let θ and ψ be binary relations on A.
   (a) Prove that θ has the substitution property (with respect to **A**) if and only if θ is a subalgebra of $\mathbf{A} \times \mathbf{A}$.
   (b) Prove that if θ and ψ are subalgebras of $\mathbf{A} \times \mathbf{A}$ then so are $\theta \circ \psi$ and $\theta^{\smile}$.

2. Let θ and ψ be equivalence relations on a set A.
   (a) Show that neither $\theta \cup \psi$ nor $\theta \circ \psi$ is, in general, an equivalence relation.
   (b) Let $\nu = (\theta \circ \psi) \cup (\theta \circ \psi \circ \theta) \cup (\theta \circ \psi \circ \theta \circ \psi) \cup \cdots$. Prove that $\nu \in \mathrm{Eq}(A)$, and that, for all $\beta \in \mathrm{Eq}(A)$, $\theta \cup \psi \subseteq \beta$ iff $\nu \subseteq \beta$.
   (c) Prove that $\theta \circ \psi$ is an equivalence relation iff $\theta \circ \psi = \psi \circ \theta$.

3. Let $h\colon \mathbf{A} \to \mathbf{B}$ be a homomorphism. Prove that h is injective iff $\ker h = 0_A$.

4. Let **R** be a ring. Prove that there is a one-to-one correspondence $I \mapsto \theta_I$ between the (two-sided) ideals of **R** and the congruences of **R**. Prove that $I \subseteq J$ iff $\theta_I \subseteq \theta_J$.

5. Let **S** be a semigroup. A nonempty subset I of S is called an *ideal* if for every $s \in S$ and $a \in I$, both sa and as are elements of I. Prove that, for any ideal I, the binary relation $I^2 \cup 0_S$ is a congruence on **S**, called the *Rees congruence* induced by I.

6. Let **Q** be a quasigroup and θ a congruence on **Q**. Prove that for any $a, b \in Q$ the function $f\colon a/\theta \to b/\theta$ given by $f(x) = a\backslash(x \cdot b)$ is a bijection. Conclude that any quasigroup of prime order is simple.

7. Let **A** be an algebra and $\alpha, \beta \in \mathrm{Con}\,\mathbf{A}$.
   (a) Prove that there is a homomorphism $f\colon \mathbf{A} \to \mathbf{A}/\alpha \times \mathbf{A}/\beta$ given by $f(a) = (a/\alpha, a/\beta)$.
   (b) Prove that $\ker f = \alpha \cap \beta$. Hence f is injective if and only if $\alpha \cap \beta = 0_A$.
   (c) Prove that f is surjective if and only if $\alpha \circ \beta = 1_A$.

8. Let $f\colon \mathbf{A} \to \mathbf{B}$ and $g\colon \mathbf{A} \to \mathbf{C}$ be homomorphisms, with g surjective. Prove that if $\ker g \subseteq \ker f$ then there is a homomorphism $h\colon \mathbf{C} \to \mathbf{B}$ such that $f = h \circ g$.

9. State and prove analogues of Corollary 1.15 and Exercises 1.16.1 and 1.16.2 with "Sg" replaced by "Cg."

Chapter 2

Lattices

Lattices play a curious double role in universal algebra. On the one hand the lattice concept provides a framework for studying many derived objects that apply to any algebraic structure: the collections of all subalgebras, congruences, varieties, and clones all form lattices. On the other hand, lattices are algebras in their own right. As algebras they have some strong properties, and, more important, properties that are rather different from the ones we are familiar with from the study of groups and rings.

Thus we devote a whole chapter to lattices. While we won't delve deeply into the study of lattices-as-algebras, we develop enough of the theory to allow us to use it later in our examples.

2.1 Ordered sets

Let P be a nonempty set, and let σ be a binary relation on P. σ is called a *(partial) ordering* of P if it is reflexive, transitive and *antisymmetric:*

$$(\forall x, y \in P)\ x \mathrel{\sigma} y \ \&\ y \mathrel{\sigma} x \implies x = y$$

(*cf.* the definition of an equivalence relation on page 14). We usually write $x \le y$ instead of $x \mathrel{\sigma} y$. The pair $\langle P, \le \rangle$ is frequently called a *poset.* Naturally, we sometimes write $x \ge y$ instead of $y \le x$. A poset is called *linear,* or sometimes, a *chain* if it satisfies

$$(\forall x, y \in P)\ x \le y \text{ or } y \le x.$$

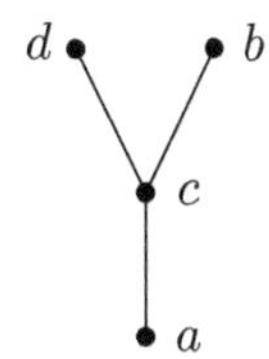

Posets can be depicted as graphs with vertices representing the elements and edges extending upwards to indicate the ordering. These graphs are called *Hasse diagrams.* For example, the poset on $\{a, b, c, d\}$ with $a < c < d$ and $c < b$ has the Hasse diagram to the right.

The *dual* of a poset $\mathbf{P} = \langle P, \sigma \rangle$ is the poset $\langle P, \sigma^{\smile} \rangle$, which we will denote by $\mathbf{P}^{\partial}$. You can think of this as the poset obtained by turning $\mathbf{P}$ upside down. The Hasse diagram for $\mathbf{P}^{\partial}$ is indeed obtained from that of $\mathbf{P}$ in this way.

Let $\langle P, \leq \rangle$ be a poset, $X \subseteq P$ and $a \in P$. Let us write $a \leqslant X$ to indicate that $a \leq x$, for all $x \in X$. When $a \leqslant X$, we call a a *lower bound* of X. The element a is called the *greatest lower bound* or *infimum* of X if

$$a \leqslant X \text{ and } (\forall p \in P)\ p \leqslant X \implies p \leq a.$$

There are analogous definitions of *upper bound* and *least upper bound* (a.k.a. *supremum.*)

Definition 2.1. A poset $\langle P, \leq \rangle$ is called a *lattice* if every pair of elements of P has a supremum and an infimum.

We now have two definitions of a lattice: one just above and the other in Definition 1.7. The two definitions are equivalent in the following sense. Starting with a poset satisfying Definition 2.1 there is a transformation

$$\langle P, \leq \rangle \longmapsto \langle P, \wedge, \vee \rangle; \quad x \wedge y = \inf\{x, y\} \quad \text{and} \quad x \vee y = \sup\{x, y\}$$

yielding an algebra satisfying the conditions in Definition 1.7. Conversely, let $\langle P, \wedge, \vee \rangle$ be a lattice (as an algebra). For any x, y in P, we have $x = x \wedge y$ if and only if $y = x \vee y$ since

$$x = x \wedge y \implies x \vee y = (x \wedge y) \vee y = y$$

with the last equality coming from $(c_\vee)$ and $(p_\vee)$. Now one can easily check that

$$\langle P, \wedge, \vee \rangle \longmapsto \langle P, \leq \rangle; \quad x \leq y \iff x = x \wedge y$$

transforms a lattice in the sense of 1.7 to one as in 2.1. Moreover, these two transformations are inverse to each other.

The smallest element of a lattice $\mathbf{L}$, if it exists, is generally denoted 0_L. In other words, $0_L = \inf L$. Similarly, $1_L = \sup L$ is the largest element of $\mathbf{L}$. If both 0_L and 1_L exist, we call the lattice *bounded.*

We see that every finite lattice is bounded. But there are many familiar unbounded lattices. For example, the lattice of integers (with the standard ordering) and the open interval $\{x \in \mathbb{R} : 0 < x < 1\}$ are not bounded. The poset $\langle \mathrm{Sb}_\omega(\omega), \subseteq \rangle$ has a lower bound but no upper bound.

In a similar manner, every semilattice has an associated ordering. The details are sketched in Exercise 2.4.1.

Definition 2.2. Let $\langle P, \leq^P \rangle$ and $\langle Q, \leq^Q \rangle$ be posets. A function $f \colon P \to Q$ is called *order-preserving* or *isotone* if, for every $x, y \in P$,

$$x \leq^P y \implies f(x) \leq^Q f(y).$$

Now it is immediate that if $\mathbf{P}$ and $\mathbf{Q}$ are lattices and if $f \colon P \to Q$ is a homomorphism (in the algebraic sense), then f is isotone. However, the converse is false. In fact, it is possible for P to be a subposet of Q without being a sublattice. See for example Figure 2.1. However, we do have the following.

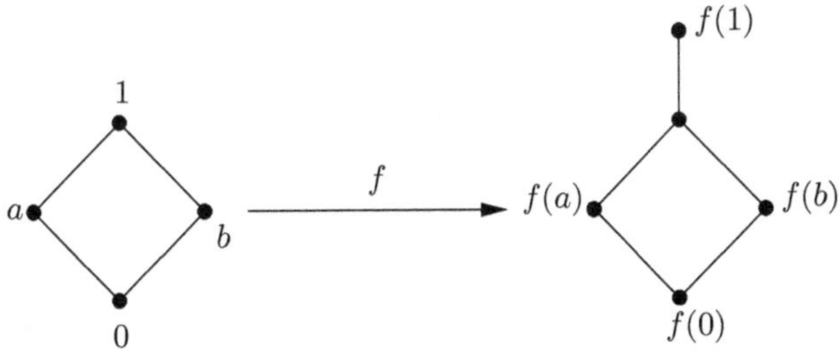

FIGURE 2.1: An isotone map that is not a lattice homomorphism

Proposition 2.3. *Let $f\colon \mathbf{P} \to \mathbf{Q}$ be a bijective, isotone map from the lattice $\mathbf{P}$ to the lattice $\mathbf{Q}$. If f^{-1} is isotone, then f (and consequently f^{-1}) is a lattice isomorphism.*

Proof. Let $a, b \in P$ and $c = a \wedge^P b$. We wish to show that $f(c) = f(a) \wedge^Q f(b)$. Since $c \lll^P \{a, b\}$ and f is isotone, $f(c) \lll^Q \{f(a), f(b)\}$. On the other hand, if $x \lll^Q \{f(a), f(b)\}$ then $f^{-1}(x) \lll^P \{a, b\}$. Since c is the *greatest* lower bound, $f^{-1}(x) \leq^P c$, so $x \leq^Q f(c)$. Thus $f(c)$ is the greatest lower bound of $\{f(a), f(b)\}$. The argument for join is similar to the one just given for meet. □

Notice that the dual of a lattice is another lattice. Thinking of a lattice as a poset, we obtain the dual by turning the lattice upside-down. As an algebra, the dual is obtained by interchanging the operations. As posets: $\langle L, \geq \rangle = \langle L, \leq \rangle^\partial$, as algebras: $\langle L, \wedge, \vee \rangle = \langle L, \vee, \wedge \rangle^\partial$.

If a and b are elements of a poset $\mathbf{P}$, and if $a \leq b$, we write $\mathbf{I}[a, b]$ to denote the subset of $\mathbf{P}$ with universe $\{\, x \in P : a \leq x \leq b \,\}$, called the *interval from a to b*. Observe that if $\mathbf{P}$ is a lattice, then $\mathbf{I}[a, b]$ is a sublattice of $\mathbf{P}$. If $a < b$ and $\mathbf{I}[a, b] = \{a, b\}$, then we say that *a is covered by b* and write $a \prec b$ in this event.

We conclude this section with a statement of Zorn's lemma. Zorn's lemma is known to be equivalent to the axiom of choice, one of our unspoken assumptions about set theory, hence all of mathematics. We will treat Zorn's lemma as an axiom.

Zorn's Lemma. *Let $\mathbf{P}$ be a nonempty poset. Suppose that every chain in P has an upper bound. Then $\mathbf{P}$ has a maximal element.*

Exercise Set 2.4.

1. Let $\langle P, \leq \rangle$ be a poset in which every pair of elements has a greatest lower bound (resp. least upper bound). Show that $\langle P, \wedge \rangle$ (respectively $\langle P, \vee \rangle$) is a semilattice as defined on page 6. Conversely, given a semilattice

$\langle S, \cdot \rangle$, define two binary relations on S by

$$\alpha = \{ (x, y) : x = x \cdot y \}$$
$$\beta = \{ (x, y) : y = x \cdot y \}.$$

Show that α and β are orderings of S and that $\beta = \alpha^{\smile}$.

2. Let $\mathbf{P}$ be a poset. A subset U of P is called an *upset* if

$$x \in U \ \& \ x \leq y \implies y \in U.$$

Similarly, a *downset* is a subset D such that $x \in D \ \& \ y \leq x \implies y \in D$. Let $\mathrm{Dn}(\mathbf{P})$ denote the set of downsets of $\mathbf{P}$ (including the empty set). Obviously, $\mathrm{Dn}(\mathbf{P})$ can be ordered by inclusion.

   (a) Show that $\langle \mathrm{Dn}(\mathbf{P}), \subseteq \rangle$ is a lattice.
   (b) Show that there is an injective, isotone map from $\mathbf{P}$ to $\mathbf{Dn}(\mathbf{P})$.
   (c) Show that $\mathrm{Dn}(\mathbf{P})$ is a topology on P.
   (d) Suppose that $\mathbf{P}$ and $\mathbf{Q}$ are posets. Show that a function $f \colon P \to Q$ is isotone iff it is continuous with respect to the above topologies.
   (e) Obviously, all of the above claims can be "dualized" for the lattice $\mathbf{Up}(\mathbf{P})$ of upsets. Prove or disprove: $\mathbf{Dn}(\mathbf{P}) \cong \mathbf{Up}(\mathbf{P})^{\partial}$.

3. (a) Let $\mathbf{P}$ be a poset. Show that there is no surjective, isotone map from $\mathbf{P}$ to $\langle \mathrm{Dn}(\mathbf{P}), \subseteq \rangle$. (Hint: suppose f were such a map. Consider the set $B = \{ x \in P : (\exists y) \ x \leq y \ \& \ y \notin f(y) \}$.)
   (b) Use part (a) to prove Cantor's theorem: for any set X, there is no surjective function from X to $\mathrm{Sb}(X)$.

4. Let σ be a partial order on a set P. Prove that there is a linear ordering λ of P such that $\sigma \subseteq \lambda$. (Hint: Zorn's lemma.)

5. A subset V of a poset is *convex* if $a, b \in V$ and $a \leq c \leq b$ implies $c \in V$. Prove that every congruence class on a lattice, $\mathbf{L}$, is convex. Prove also that if $\theta \in \mathrm{Con}(\mathbf{L})$ and $\theta \neq 0_L$ then there is a pair $(a, b) \in \theta$ such that $a < b$.

2.2 Distributive and modular lattices

This section is concerned with two of the most important properties of lattices-as-algebras, the distributive and modular laws.

Definition 2.5. Let **L** be a lattice.

(1) **L** is called *distributive* if it satisfies $x \wedge (y \vee z) \approx (x \wedge y) \vee (x \wedge z)$.

(2) **L** is called *modular* if it satisfies $z \leq x \implies x \wedge (y \vee z) \approx (x \wedge y) \vee z$.

Notice that any lattice satisfies

$$x \wedge (y \vee z) \geq (x \wedge y) \vee (x \wedge z) \text{ and}$$
$$z \leq x \implies x \wedge (y \vee z) \geq (x \wedge y) \vee z.$$

So when verifying either modularity or distributivity, we can restrict our attention to the "interesting" inequality.

Proposition 2.6. *Every distributive lattice is modular.*

Proof. Let **L** be distributive and $z \leq x$ in L. Then $x \wedge (y \vee z) = (x \wedge y) \vee (x \wedge z) = (x \wedge y) \vee z$, the first equality coming from distributivity, the second from $z \leq x$. □

Notice that the definition of distributivity asserts that meet distributes over join. One naturally wonders about the dual condition. It is one of the satisfying principles of lattice theory that the two conditions are equivalent.

Proposition 2.7. *Let* **L** *be a lattice.* **L** *is distributive iff it satisfies* $x \vee (y \wedge z) \approx (x \vee y) \wedge (x \vee z)$.

Proof. Assume **L** satisfies the identity in Definition 2.5(1). Consider one side of the dual identity: $(x \vee y) \wedge (x \vee z)$. Set $a = x \vee y$, $b = x$ and $c = y$. Then

$$\begin{aligned}(x \vee y) \wedge (x \vee z) &= a \wedge (b \vee c) \overset{*}{=} (a \wedge b) \vee (a \wedge c)\\ &= ((x \vee y) \wedge x) \vee ((x \vee y) \wedge z)\\ &\overset{*}{=} x \vee (x \wedge z) \vee (y \wedge z) = x \vee (y \wedge z).\end{aligned}$$

The indicated equalities invoke the original definition of distributivity. □

It is not hard to see that any lattice of order less than 5 is distributive. (There are only 5 of them. Why don't you draw them all in the margin.) There are two nondistributive lattices with 5 elements, called $\mathbf{N}_5$ and $\mathbf{M}_3$. They play a central role in the theory of lattices. Notice that $\mathbf{M}_3$ is modular and that $\mathbf{N}_5$ is not. The following theorem tells us that $\mathbf{N}_5$ is the smallest nonmodular lattice in a very strong sense.

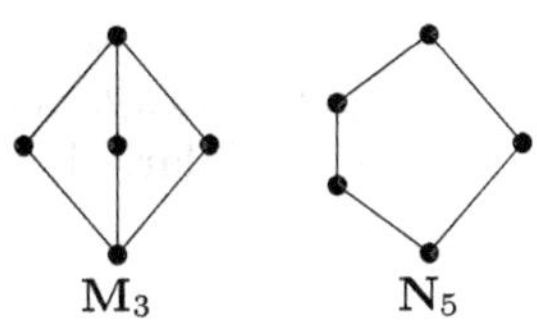

Theorem 2.8 (Dedekind, 1900). *Let* **L** *be a lattice. The following are equivalent.*

(a) **L** *is modular.*

(b) **L** *satisfies* $((x \wedge z) \vee y) \wedge z \approx (x \wedge z) \vee (y \wedge z)$.

(c) **L** *has no sublattice isomorphic to* $\mathbf{N}_5$.

Proof. First assume that **L** is modular and we shall prove (b). Let $x, y, z \in L$ and set $a = z$, $b = y$ and $c = x \wedge z$. Then $c \leq a$ so by modularity $a \wedge (b \vee c) = (a \wedge b) \vee c$. Equivalently, $(x \wedge z) \vee y) \wedge z = (y \wedge z) \vee (x \wedge z)$.

By labeling the elements of N_5 so that $0 < x < z < 1$ and $0 < y < 1$ one sees that $\mathbf{N}_5$ fails to satisfy the identity in (b). Thus (b) implies (c).

Finally assume that **L** is not modular. We shall find a sublattice of **L** isomorphic to $\mathbf{N}_5$. There are elements $a, b, c \in L$ with $a \geq c$ and $a \wedge (b \vee c) > (a \wedge b) \vee c$. We claim that **L** contains the sublattice of Figure 2.2.

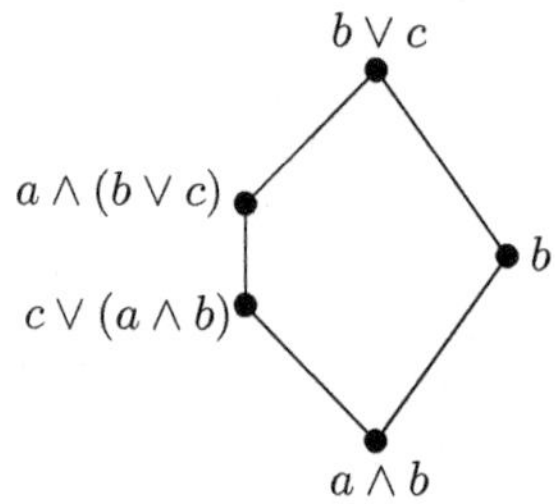

FIGURE 2.2: A sublattice isomorphic to $\mathbf{N}_5$

Notice that we still have some work to do. We must show that the five elements are all distinct and that the meets and joins are as Figure 2.2 indicates. Certainly $a \wedge b \leq b \leq b \vee c$ and $a \wedge b \leq c \vee (a \wedge b) < a \wedge (b \vee c) \leq (b \vee c)$, with the strict inequality coming from our assumption. Also

$$b \vee (c \vee (a \wedge b)) = c \vee b \vee (a \wedge b) = b \vee c$$
$$b \wedge (a \wedge (b \vee c)) = a \wedge b \wedge (b \vee c) = a \wedge b.$$

All of the other meets and joins now follow trivially. Finally, all of the inclusions must be strict or else $c \vee (a \wedge b) = a \wedge (b \vee c)$, which we know to be false. □

The condition defining modularity in Definition 2.5(2) is not an identity, but, rather, an implication. An added bonus of Theorem 2.8 is that it provides an identity equivalent to modularity. As a result, a homomorphic image of a modular lattice is always modular. Furthermore, the third equivalent condition is clearly self-dual. It follows that the dual of a modular lattice is modular.

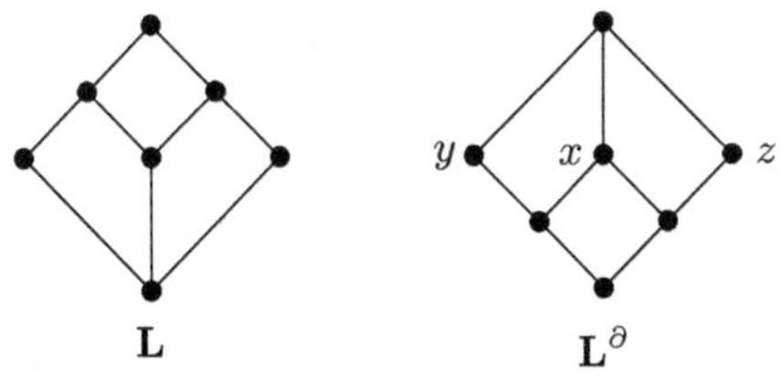

FIGURE 2.3: **L** satisfies $y \wedge \big(z \vee (x \wedge y)\big) \leq x \vee (y \wedge z)$

By the way, it is not generally true that if an identity holds in a lattice, then it holds in the dual lattice. For example consider the lattice **L** of Figure 2.3. The lattice **L** satisfies the identity $y \wedge \big(z \vee (x \wedge y)\big) \leq x \vee (y \wedge z)$, while $\mathbf{L}^{\partial}$ fails this identity with x, y, z chosen as shown.

There is a theorem analogous to 2.8 for distributivity. First let us make an important observation.

Lemma 2.9. $\mathbf{M}_3$ *is a simple lattice.*

Proof. Label the atoms of $\mathbf{M}_3$ by a, b, c, and let $\theta \neq 0$ be a congruence. Suppose first that $a \;\theta\; b$. Then $0 = (a \wedge b) \;\theta\; (b \wedge b) = b$ and similarly, $0 \;\theta\; a$. Therefore $0 = (0 \vee 0) \;\theta\; (a \vee b) = 1$. From this it is easy to see that $\theta = 1$.

On the other hand, suppose that $a \;\theta\; 0$. Then $1 = (a \vee c) \;\theta\; (0 \vee c) = c$ hence $b = (1 \wedge b) \;\theta\; (c \wedge b) = 0$. Thus $a \;\theta\; b$ and we are reduced to the previous paragraph. A similar argument applies if $a \;\theta\; 1$. □

Theorem 2.10 (Birkhoff). *Let* **L** *be a lattice. The following are equivalent.*

(a) **L** *is distributive.*

(b) **L** *satisfies* $(x \wedge y) \vee (x \wedge z) \vee (y \wedge z) \approx (x \vee y) \wedge (x \vee z) \wedge (y \vee z)$.

(c) **L** *has no sublattice isomorphic to either* $\mathbf{N}_5$ *or* $\mathbf{M}_3$.

Proof. First, (a) implies (c) since neither $\mathbf{N}_5$ nor $\mathbf{M}_3$ are distributive. Let us prove (c) implies (b). If **L** contains no copy of $\mathbf{N}_5$, then **L** is modular by Theorem 2.8. Suppose that **L** fails to satisfy the identity in (b). Then there are elements $a, b, c \in L$ such that

$$w = (a \wedge b) \vee (a \wedge c) \vee (b \wedge c) < (a \vee b) \wedge (a \vee c) \wedge (b \vee c) = u.$$

Making the following definitions and applying modularity

$$\begin{aligned} a' &= (w \vee a) \wedge u = w \vee (a \wedge u) \\ b' &= (w \vee b) \wedge u = w \vee (b \wedge u) \\ c' &= (w \vee c) \wedge u = w \vee (c \wedge u) \end{aligned} \tag{2–1}$$

we have the following copy of $\mathbf{M}_3$ as a sublattice of **L**.

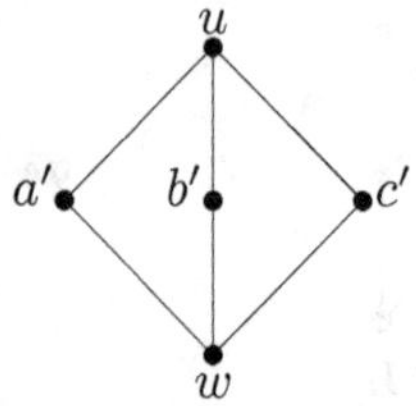

To complete this portion of the proof we must show that all five elements are distinct and that the meets and joins are as in the diagram. Because of the symmetry expressed in equations (2–1), it is enough to show that $a' \wedge b' = w$. For this, observe first that

$$a \vee w = a \vee (a \wedge b) \vee (a \wedge c) \vee (b \wedge c) = a \vee (b \wedge c)$$

hence

$$\begin{aligned} a' = (a \vee w) \wedge u &= \big(a \vee (b \wedge c)\big) \wedge (a \vee b) \wedge (a \vee c) \wedge (b \vee c) \\ &= \big(a \vee (b \wedge c)\big) \wedge (b \vee c) = (b \wedge c) \vee \big(a \wedge (b \vee c)\big). \end{aligned}$$

The last equality is obtained by applying modularity. Similarly $b' = \big(b \vee (a \wedge c)\big) \wedge (a \vee c)$. Therefore

$$\begin{aligned} a' \wedge b' &= \big(a \vee (b \wedge c)\big) \wedge (b \vee c) \wedge \big(b \vee (a \wedge c)\big) \wedge (a \vee c) \\ &\stackrel{2}{=} \big(a \vee (b \wedge c)\big) \wedge \big(b \vee (a \wedge c)\big) \stackrel{3}{=} \Big(a \wedge \big(b \vee (a \wedge c)\big)\Big) \vee (b \wedge c) \\ &\stackrel{4}{=} \big((a \wedge b) \vee (a \wedge c)\big) \vee (b \wedge c) = w. \end{aligned}$$

To see equality "2," observe that $\big(a \vee (b \wedge c)\big) \leq (a \vee c)$ and $\big(b \vee (a \wedge c)\big) \leq (b \vee c)$. Equality "3" follows from $b \vee (a \wedge c) \geq b \wedge c$ and modularity, equality "4" from $a \geq a \wedge c$ and modularity. The proof that the five elements are distinct is the same as for Lemma 2.9.

Finally, assume that **L** satisfies the identity in (b). We first show **L** is modular. Let $x \geq z$. then $x \wedge (y \vee z) = ((x \vee y) \wedge (x \vee z)) \wedge (y \vee z) = (x \wedge y) \vee ((x \wedge z) \vee (y \wedge z)) = (x \wedge y) \vee z$. Now to show distributivity

$$x \wedge (y \vee z) = x \wedge \big((x \vee y) \wedge (x \vee z) \wedge (y \vee z)\big)$$

using (b)

$$= x \wedge \big(((x \wedge y) \vee (x \wedge z)) \vee (y \wedge z)\big)$$

since $x \geq (x \wedge y) \vee (x \wedge z)$ and using modularity

$$\begin{aligned} &= ((x \wedge y) \vee (x \wedge z)) \vee (x \wedge (y \wedge z)) \\ &= (x \wedge y) \vee (x \wedge z). \end{aligned}$$

□

The two expressions in part (b) of the above theorem are of interest in their own right. Let us define

$$\begin{aligned} m_1(x,y,z) &= (x \wedge y) \vee (x \wedge z) \vee (y \wedge z) \\ m_2(x,y,z) &= (x \vee y) \wedge (x \vee z) \wedge (y \vee z). \end{aligned} \tag{2–2}$$

It is easy to see that any lattice satisfies the identities

$$m_i(x,x,y) \approx m_i(x,y,x) \approx m_i(y,x,x) \approx x, \quad \text{for } i = 1,2. \tag{2–3}$$

For this reason, m_1 and m_2 are called *majority terms.* Birkhoff's theorem asserts that a lattice is distributive if and only if m_1 and m_2 agree. Majority terms play a key role in Theorem 2.21 and in Part II.

The notion of complementation in the sense of set theory can be generalized to an arbitrary bounded lattice.

Definition 2.11. Let $\mathbf{L}$ be a bounded lattice.

(1) Elements x and y in L are called *complements* if $x \wedge y = 0$ and $x \vee y = 1$.

(2) $\mathbf{L}$ is called *complemented* if every element has a complement.

(3) $\mathbf{L}$ is a *Boolean lattice* if it is bounded, distributive and complemented.

(4) A *Boolean algebra* is an algebra $\langle B, \wedge, \vee, ', 0, 1\rangle$ in which $\langle B, \wedge, \vee\rangle$ is a distributive lattice with bounds 0 and 1 and for every $x \in B$, x' is a complement of x.

There is no expectation that the complement of an element be unique. The lattices $\mathbf{M}_3$ and $\mathbf{N}_5$ illustrate two ways an element can have multiple complements. (And see Exercise 2.12.4 below.)

Exercise Set 2.12.

1. Let $\mathbf{2} = \langle 0, 1, \wedge, \vee\rangle$ denote the 2-element lattice. Prove that for every set J, $\langle \mathrm{Sb}(J), \cap, \cup\rangle \cong \mathbf{2}^J$.

2. Let $\mathbf{L}$ be a distributive lattice and X a subset of L. Define
$$\begin{aligned} X_1 &= \left\{\bigwedge Y : Y \subseteq_\omega X\right\} \\ X_2 &= \left\{\bigvee Y : Y \subseteq_\omega X_1\right\}. \end{aligned}$$
Show that $X_2 = \mathrm{Sg}^{\mathbf{L}}(X)$.

3. Let $\mathbf{L}$ be a copy of $\mathbf{N}_5$ with elements $\{u, a, b, c, v\}$ and $u < a < b < v$. Let θ be a congruence on $\mathbf{L}$. Prove that $\theta \neq 0_L \implies (a, b) \in \theta$.

4. Prove that a lattice $\mathbf{L}$ is distributive iff for every $a < b$, the sublattice $\mathbf{I}[a, b]$ has the property that every element has at most one complement. (Note that $\mathbf{L}$ is not required to be bounded.)

5. Let $\mathbf{L}$ be a modular lattice, $a, b \in L$. Prove that the mapping $h(x) = x \vee b$ is a lattice isomorphism from $\mathbf{I}[a \wedge b, a]$ to $\mathbf{I}[b, a \vee b]$, with inverse $k(x) = x \wedge a$. We write $\mathbf{I}[c, a] \nearrow \mathbf{I}[b, d]$, and call the intervals *transposes,* if $c = a \wedge b$ and $d = a \vee b$.

2.3 Complete lattices

Definition 2.13. A lattice $\mathbf{L}$ is called *complete* if for every subset X of L, both $\sup X$ and $\inf X$ exist.

Instead of $\sup X$, we often write $\bigvee X$, or, if $X = \{x_i : i \in I\}$, we might write $\bigvee_{i \in I} x_i$. Similarly, $\inf X = \bigwedge X = \bigwedge_{i \in I} x_i$.

It is worth thinking about $\bigvee \varnothing$ and $\bigwedge \varnothing$, which must exist in any complete lattice. If you carefully apply the definition, you will see that $\bigvee \varnothing = 0_L$ and $\bigwedge \varnothing = 1_L$. In other words, every complete lattice must have a smallest and a largest element. Here are some examples for you to contemplate.

(1) Both $\mathbb{Z} \cup \{\pm\infty\}$ and $\mathbb{R} \cup \{\pm\infty\}$ are complete lattices (under the standard ordering);

(2) $\mathbb{Q} \cup \{\pm\infty\}$ is incomplete;

(3) $\langle \mathrm{Sb}(\omega), \cap, \cup \rangle$ is complete;

(4) $\langle \mathrm{Sb}_\omega(\omega), \cap, \cup \rangle$ is incomplete, but $\mathrm{Sb}_\omega(\omega) \cup \{\omega\}$ is complete.

Proposition 2.14. *Let $\mathbf{P}$ be a poset in which $\inf X$ exists for every $X \subseteq P$. Then $\mathbf{P}$ is a complete lattice.*

Proof. By assumption, $\mathbf{P}$ has arbitrary inf's. We just have to show the existence of sup's. Let X be a subset of P. Since $\sup X$ is to be the *least* upper bound of X, let us consider $Y = \{y \in P : y \gg X\}$ and set $a = \inf Y$ which exists by assumption. Let x be any element of X. then for all $y \in Y$, $x \leq y$ by definition of Y. Thus x is a lower bound of Y. Since a is the greatest of all lower bounds, $x \leq a$. Since x was arbitrary, we conclude $a \in Y$. Thus a is the least of all upper bounds of X. □

This is quite a useful proposition. In algebra, we encounter many posets composed of sets ordered by inclusion, and which are closed under arbitrary intersection. It follows that the meet operation coincides with intersection and by Proposition 2.14, our poset is a complete lattice. The next corollary lists the most important applications of this principle. Some others that you have encountered are the lattices of normal subgroups of a group; of left-, right-, and two-sided ideals of a ring; and the closed subsets of a topological space.

Corollary 2.15. *Let* **A** *be an algebra. Then* **Sub(A)** *and* **Con(A)** *are complete lattices. If A is any set, then* **Eq**(A) *is a complete lattice.*

The proof of the corollary for, say, **Sub(A)** follows almost immediately by combining Propositions 1.12 and 2.14. There is one small point to check. Proposition 1.12 asserts that the intersection of a *nonempty* collection of subuniverses is a subuniverse. What of the empty collection? We can't really talk about intersection. But as we noted above, $\inf \varnothing$ must be the largest element in a poset, in this case A itself. Fortunately, every algebra is a subuniverse of itself, so 2.14 applies.

The corollary tells us that, for example, **Sub(A)** is a lattice. But it does not tell us how to compute joins in this lattice. This is a typical problem in algebra. We will give one answer in the next section, and then a better answer later. On the other hand, take another look at Exercise 1.26.2. Part (b) provides a concrete description of $\theta \vee \psi$ in the lattice Eq(A). The next proposition generalizes this to complete joins.

Proposition 2.16. *Let A be a set and $\Theta \subseteq \mathrm{Eq}(A)$. Then in the lattice* **Eq**(A)

$$\bigvee \Theta = 0_A \cup \bigcup \{ \theta_1 \circ \theta_2 \circ \cdots \circ \theta_k : k \in \omega \;\;\&\;\; \theta_i \in \Theta, \textit{for } i \leq k \}.$$

Proof. Let ψ denote the left-hand side and β the right-hand side of the equation in the statement of the proposition. The proof that β is an equivalence relation is similar to the proof of Theorem 1.25. From this it follows that $\psi \subseteq \beta$. On the other hand, for any $\theta_1, \ldots, \theta_k \in \Theta$, $\theta_1 \circ \cdots \circ \theta_k \subseteq \psi \circ \psi \circ \cdots \circ \psi = \psi$. Therefore $\beta \subseteq \psi$. □

One more comment on joins of equivalence relations. It follows from Exercise 1.26.2 that if $\theta \circ \psi = \psi \circ \theta$ then $\theta \vee \psi = \theta \circ \psi$. This is the happiest situation. We say that θ and ψ *permute* if $\theta \circ \psi = \psi \circ \theta$. An algebra has *permuting congruences* if all congruences permute.

Definition 2.17. Let **L** be a complete lattice. A sublattice **M** of **L** is a *complete sublattice* if, for every $X \subseteq M$, $\bigvee X$ and $\bigwedge X$ (as computed in **L**) are elements of M.

Of course, all sorts of combinations are possible. A complete lattice can have incomplete sublattices, and an incomplete lattice can have complete sublattices. It is even possible for a complete lattice to have a sublattice that is a complete lattice, but is not a complete sublattice. Consider for example, $\mathrm{Sb}_\omega(\omega) \cup \{\omega\}$ as a sublattice of $\mathrm{Sb}(\omega)$.

Theorem 2.18. *Let* **A** *be an algebra.* **Con A** *is a complete sublattice of* **Eq** A. *More generally, if* **B** *is a reduct of* **A**, *then* **Con A** *is a complete sublattice of* **Con B**.

Proof. Let $\Theta \subseteq \operatorname{Con} \mathbf{A}$. We must show that $\bigwedge \Theta$ and $\bigvee \Theta$, when computed as members of $\mathbf{Eq}\, A$, lie in $\operatorname{Con} \mathbf{A}$. For the meet, this is easy. $\bigwedge \Theta = \bigcap \Theta$ is a congruence by Proposition 1.23. For the join, let us reuse the notation from the proof of Proposition 2.16. Since we already know that β is an equivalence relation, all that remains is to show that it has the substitution property. So let f be an n-ary operation and $(a_i, b_i) \in \beta$ for $i = 1, 2, \dots, n$. From the definition of β, for every $i \leq n$ there are $\theta_{i,1}, \dots, \theta_{i,k} \in \Theta$ such that $(a_i, b_i) \in \theta_{i,1} \circ \cdots \circ \theta_{i,k} = \alpha_i$. Let $\alpha = \alpha_1 \circ \cdots \circ \alpha_n$. Since every α_j is reflexive, for every $i \leq n$, $(a_i, b_i) \in \alpha$. By Exercise 1.26.1 every α_i, hence α, has the substitution property. So $(f(\mathbf{a}), f(\mathbf{b})) \in \alpha \subseteq \beta$. □

Corollary 2.19. *Let θ and ψ be congruences on an algebra $\mathbf{A}$ and a, b elements of A. Then $(a, b) \in \theta \vee \psi$ if and only if there is an integer k and elements $c_0, c_1, \dots, c_k \in A$ such that $a = c_0 \mathrel{\theta} c_1 \mathrel{\psi} c_2 \mathrel{\theta} c_3 \mathrel{\psi} \cdots \mathrel{\psi} c_k = b$. In practice it is usually immaterial whether the sequence begins or ends with θ or ψ.*

Proof. Take $\Theta = \{\theta, \psi\}$ in Proposition 2.16. Then $(a, b) \in \theta_1 \circ \theta_2 \circ \cdots \circ \theta_k$ where each θ_i is either θ or ψ. Since both θ and ψ are transitive, we can assume that $\theta_1 \circ \theta_2 \circ \cdots \circ \theta_k$ alternates θ and ψ. Since both θ and ψ are reflexive, we can, by repeating an element if necessary, assume that $\theta_1 = \theta$ and $\theta_k = \psi$. □

The structure of the congruence lattice of an algebra often has a profound impact on the structure of the algebra itself. This is the major theme of Part II. We conclude this section with two classical examples. First, we have a theorem that underscores the importance of the modular law (and also explains its name). The second provides an example of a strong property of lattices that is not exhibited by groups.

Theorem 2.20 (Dedekind, 1900). *The congruence lattice of a group is modular.*

Proof. Let us first observe that the congruences on a group permute. For if α and β are congruences and $(x, y) \in \alpha \circ \beta$ then there is some z such that $x \mathrel{\alpha} z \mathrel{\beta} y$. Therefore $x = (xz^{-1}z) \mathrel{\beta} (xz^{-1}y) \mathrel{\alpha} (zz^{-1}y) = y$, so $(x, y) \in \beta \circ \alpha$. It follows from our remarks above that in the congruence lattice $\alpha \vee \beta = \alpha \circ \beta$.

Now let α, β, γ be congruences with $\gamma \subseteq \alpha$. We wish to show that

$$\alpha \wedge (\beta \vee \gamma) \subseteq (\alpha \wedge \beta) \vee \gamma.$$

So let $(x, y) \in \alpha \wedge (\beta \vee \gamma)$. Then $x \mathrel{\alpha} y$ and from the previous paragraph, there is a z such that $x \mathrel{\beta} z \mathrel{\gamma} y$. Since $\gamma \subseteq \alpha$, $z \mathrel{\alpha} y \mathrel{\alpha} x$ thus $z \mathrel{\alpha} x$. Therefore $x \mathrel{(\alpha \wedge \beta)} z \mathrel{\gamma} y$ which is enough to prove the theorem. □

With the notion of a lattice firmly established, we see that the content of Theorem 1.21(3) can be expressed more succinctly as the assertion that the mapping $N \mapsto \theta_N$ is a lattice isomorphism between the lattices $\mathbf{Nml}(\mathbf{G})$

of normal subgroups and **Con(G)**. Thus, a different proof of the previous theorem can be had by showing that **Nml(G)** is modular. You might find it informative to write out that proof directly. But first ask yourself: what is the join of two normal subgroups in the lattice **Nml(G)**?

Theorem 2.21 (Funayama and Nakayama, 1942). *The congruence lattice of any lattice is distributive.*

Proof. Let α, β, γ be congruences on a lattice. We must show that $\alpha \wedge (\beta \vee \gamma) \subseteq (\alpha \wedge \beta) \vee (\alpha \wedge \gamma)$. Let $m = m_1$ be a majority term as discussed in equation (2–2).

Let $(a, b) \in \alpha \wedge (\beta \vee \gamma)$. Unlike the group case, the congruences on a lattice do not generally permute. The best we can say (Corollary 2.19) is that there is an integer n and elements $c_0, c_1, \ldots, c_n$ such that

$$a \; \alpha \; b \text{ and } a = c_0 \; \beta \; c_1 \; \gamma \; c_2 \; \beta \; c_3 \cdots c_{n-1} \; \beta \; c_n = b.$$

For every $i \leq n$, $a = m(a, a, c_i) \; \alpha \; m(a, b, c_i)$. Therefore, for every i, j, $m(a, b, c_i) \; \alpha \; m(a, b, c_j)$. Also using the identities in (2–3)

$$a = m(a, b, c_0) \; \beta \; m(a, b, c_1) \; \gamma \; m(a, b, c_2) \cdots m(a, b, c_{n-1}) \; \beta \; m(a, b, c_n) = b.$$

Therefore

$$\begin{aligned} a =& m(a, b, c_0) \; (\alpha \wedge \beta) \; m(a, b, c_1) \; (\alpha \wedge \gamma) \; m(a, b, c_2) \cdots \\ & m(a, b, c_{n-1}) \; (\alpha \wedge \beta) \; m(a, b, c_n) = b \end{aligned}$$

i.e., $(a, b) \in (\alpha \wedge \beta) \vee (\alpha \wedge \gamma)$. □

There are easy examples to show that the congruence lattice of a group is not necessarily distributive. For instance, $\mathrm{Con}(\mathbb{Z}_2 \times \mathbb{Z}_2) \cong \mathbf{M}_3$. It follows from Theorem 2.21 that the congruence lattice of a lattice is modular. However, unlike for groups, it is not true that congruences on lattices always permute. Consider, for example, the three-element chain.

Exercise Set 2.22.

1. Let **R** be a ring, and let Idl(**R**) denote the set of two-sided ideals of **R**.
   (a) Prove that Idl(**R**) forms a complete lattice.
   (b) Prove that the join in Idl(**R**) of the ideals I and J is $I + J = \{x + y : x \in I, \; y \in J\}$.
   (c) Explain why the mapping given in Exercise 1.26.4 is a lattice isomorphism of Idl(**R**) with Con(**R**).
2. Draw the Hasse diagram for the lattice Eq$\{1, 2, 3\}$. Show that the lattice Eq$\{1, 2, 3, 4\}$ is non-modular.
3. Give an example of a group **G** such that Sub(**G**) is not modular.

4. Let $\mathbf{R}$ be a ring. Suppose that there is a natural number $n > 1$ such that $\mathbf{R}$ satisfies the identity $x^n = x$. Prove that $\text{Con}(\mathbf{R})$ is distributive. (Hint: Exercise 1 will be helpful.)

5. For a group $\mathbf{G}$, let $\text{Nml}(\mathbf{G})$ denote the lattice of normal subgroups of $\mathbf{G}$. Prove or disprove: $\text{Nml}(\mathbf{G})$ is a complete sublattice of $\text{Sub}(\mathbf{G})$.

6. Define a complete homomorphism between complete lattices. Give an example of a homomorphism between complete lattices that is not a complete homomorphism.

7. Prove or disprove: If $\mathbf{L}$ and $\mathbf{M}$ are complete lattices and $h: \mathbf{L} \to \mathbf{M}$ is a lattice isomorphism, then h is a complete isomorphism.

8. (The Tarski fixed-point theorem.) Let $\mathbf{L}$ be a complete lattice, and $f: \mathbf{L} \to \mathbf{L}$ an isotone map. Prove that there is some $a \in L$ such that $f(a) = a$.

2.4 Closure operators and algebraic lattices

Definition 2.23. Let A be a set. A *closure operator* on A is a function $C\colon \text{Sb}(A) \to \text{Sb}(A)$ such that for all $X, Y \subseteq A$

(1) $X \subseteq C(X)$,

(2) $C(C(X)) = C(X)$,

(3) $X \subseteq Y \implies C(X) \subseteq C(Y)$.

From Exercises 1.16.2 and 1.26.9 we see that for any algebra $\mathbf{A}$, $\text{Sg}^{\mathbf{A}}$ and $\text{Cg}^{\mathbf{A}}$ are closure operators on A and A^2 respectively. Of course, the word "closure" comes from topology: if we define $C(X) = \overline{X}$ (the smallest closed set containing X), then C is a closure operator on the topological space.

Carrying this analogy a little further, the closed subsets of a topological space are those subsets X such that $X = C(X)$. This inspires the following definition.

Definition 2.24. Let C be a closure operator on a set A. We define L_C to be the set of all $X \subseteq A$ such that $X = C(X)$. These sets are called the *C-closed* subsets of A.

Thus the $\text{Sg}^{\mathbf{A}}$-closed subsets of A are exactly the subuniverses of $\mathbf{A}$, and the $\text{Cg}^{\mathbf{A}}$-closed subsets of A^2 are the congruences of $\mathbf{A}$. Of course if the operator C is clear from context, we usually just talk about "closed sets."

Theorem 2.25. *Let C be a closure operator on A.*

(1) $L_C = \{C(Y) : Y \subseteq A\}$.

(2) *The intersection of a nonempty family of closed sets is always closed.*

(3) $\langle L_C, \subseteq \rangle$ *is a complete lattice in which the meet operation (on a nonempty family) coincides with intersection, and, for any $\mathcal{X} \subseteq L_C$,*

$$\bigvee \mathcal{X} = C\left(\bigcup \mathcal{X}\right).$$

Proof. The first assertion is easy to see. For the second, let $\mathcal{X}$ be a nonempty collection of closed subsets of A and let $Y = \bigcap \mathcal{X}$. For every $X \in \mathcal{X}$, $Y \subseteq X$ so $C(Y) \subseteq C(X) = X$. From this it follows that $C(Y) \subseteq \bigcap \mathcal{X} = Y \subseteq C(Y)$, so Y is closed.

Now consider (3). A itself is always closed, so by (2) and Proposition 2.14 L_C is a complete lattice. Let $Y = \bigcup \mathcal{X}$. $C(Y)$ is clearly a closed set containing every member of $\mathcal{X}$, hence $\bigvee \mathcal{X} \subseteq Y$. On the other hand, for every $X \in \mathcal{X}$, $X \subseteq \bigvee \mathcal{X}$, so $Y = \bigcup \mathcal{X} \subseteq \bigvee \mathcal{X}$ and therefore $C(Y) \subseteq \bigvee \mathcal{X}$ since the latter set is closed by definition. □

Theorem 2.25 has a converse.

Theorem 2.26. *Let $\mathbf{M}$ be a complete lattice. There is a closure operator C on a set A such that $\mathbf{M} \cong \mathbf{L}_C$.*

Proof. For $x \in M$ define $(x] = \{y \in M : y \leq x\}$. Take $A = M$ and for every subset X of A define $C(X) = \left(\bigvee X\right]$. It is straightforward to check that C is a closure operation on A and that the mapping $f \colon \mathbf{M} \to \mathbf{L}_C$, given by $f(x) = (x]$, is an isomorphism. □

Theorems 2.25 and 2.26 tell us that closure operators and complete lattices are two sides of the same coin. Of course we can not expect every complete lattice $\mathbf{M}$ to be *equal* to a lattice of closed sets since the members of M may not even be sets. But algebraically the two concepts are interchangeable. Studying one is equivalent to studying the other.

Looking again at Exercise 1.16.2, we see that the closure operator $\mathrm{Sg}^{\mathbf{A}}$ has an additional property which somehow expresses the "finitaryness" of the subalgebra construction. We formalize this in the following definition.

Definition 2.27. Let C be a closure operator on a set A. C is called *algebraic* if for every $X \subseteq A$

$$C(X) = \bigcup \{C(Z) : Z \subseteq_\omega X\}.$$

Theorem 2.28. *Let $\mathbf{A}$ be an algebra. Both $\mathrm{Sg}^{\mathbf{A}}$ and $\mathrm{Cg}^{\mathbf{A}}$ are algebraic closure operators.*

Proof. Exercises 1.16.2 and 1.26.9. □

Notice that the closure operator from topology fails to be algebraic. It is natural to wonder what additional properties the lattice of closed subsets will have for an algebraic closure operator.

Definition 2.29. Let $\mathbf{L}$ be a complete lattice.

(1) An element $a \in L$ is called *compact* if for every $X \subseteq L$,

$$a \leq \bigvee X \implies (\exists Z \subseteq_\omega X)\, a \leq \bigvee Z.$$

(2) $\mathbf{L}$ is called an *algebraic lattice* if every element is the join of compact elements.

Theorem 2.30. *Let C be an algebraic closure operator on A. Then $\mathbf{L}_C$ is an algebraic lattice. The compact elements of $\mathbf{L}_C$ are those sets of the form $C(Y)$ for Y a finite subset of A.*

Proof. As an aid to readability, we shall write $\overline{X}$ in place of $C(X)$. First let $Y = \{y_1, \ldots, y_n\}$ be a finite subset. We show that $\overline{Y}$ is compact in $\mathbf{L}_C$. Suppose that $\overline{Y} \subseteq \bigvee \mathcal{X}$ for some family $\mathcal{X}$ of closed subsets. Then

$$\overline{Y} \subseteq \bigvee \mathcal{X} = \overline{\bigcup \mathcal{X}} = \bigcup \left\{ \overline{Z} : Z \subseteq_\omega \bigcup \mathcal{X} \right\}.$$

Therefore, for each $j \leq n$ there is $Z_j \subseteq_\omega \bigcup \mathcal{X}$ such that $y_j \in \overline{Z_j}$.

Let $Z = Z_1 \cup Z_2 \cup \cdots \cup Z_n$. Certainly $Y \subseteq \overline{Z}$, so $\overline{Y} \subseteq \overline{Z}$. Since Z is finite, there is a finite subfamily $\mathcal{Z}$ of $\mathcal{X}$ such that $Z \subseteq \bigcup \mathcal{Z}$. Therefore

$$\overline{Y} \subseteq \overline{Z} \subseteq \overline{\bigcup \mathcal{Z}} = \bigvee \mathcal{Z}$$

so $\overline{Y}$ is compact.

Conversely, let X be any compact element of $\mathbf{L}_C$. Using the facts that X is closed, C is algebraic, and Theorem 2.25(3),

$$X = \overline{X} = \bigcup \{ \overline{Z} : Z \subseteq_\omega X \} = \bigvee \{ \overline{Z} : Z \subseteq_\omega X \}.$$

Since X is compact, the join on the right-hand side can be replaced by a finite subjoin. That is, there are $Z_1, Z_2, \ldots, Z_n \subseteq_\omega X$ such that $X = \bigvee_{i=1}^n \overline{Z_i} = \overline{Z}$ where $Z = Z_1 \cup \cdots \cup Z_n$ is finite.

Finally, to see that $\mathbf{L}_C$ is algebraic, let $X \in L_C$. Since C is algebraic, $X = \overline{X} = \bigvee \{ \overline{Z} : Z \subseteq_\omega X \}$ is a join of compact elements. □

Corollary 2.31. *Let $\mathbf{A}$ be an algebra. Then both $\mathbf{Sub}(\mathbf{A})$ and $\mathbf{Con}(\mathbf{A})$ are algebraic lattices. The compact elements are precisely the finitely generated subalgebras and congruences respectively.*

An important point often overlooked by students is that in an algebraic lattice, every element is the join of the set of all compact elements it dominates. Suppose z lies in the lattice and X is the set of all compact elements below z. Since the lattice is algebraic, there is a set Y of compact elements such that $z = \bigvee Y$. But $Y \subseteq X$ so $z = \bigvee Y \leq \bigvee X \leq z$.

Theorem 2.30 also has a converse. The construction is a bit more involved than the one for Theorem 2.26.

Theorem 2.32. *Let* $\mathbf{M}$ *be an algebraic lattice. Then there is an algebraic closure operator* C *on a set* A *such that* $\mathbf{M} \cong \mathbf{L}_C$.

Proof. Let A be the set of compact elements of $\mathbf{M}$. For $X \subseteq A$ define $C(X) = (\bigvee X] \cap A$. We first check that C is a closure operator on A. Conditions (1) and (3) of the definition are easy to verify. For (2), let $z = \bigvee X$. Observe that $C(X)$ is by definition the set of compact elements dominated by z. By our comments above, $z = \bigvee C(X)$. Therefore

$$CC(X) = (\bigvee C(X)] \cap A = (z] \cap A = (\bigvee X] \cap A = C(X).$$

Now we show that C is algebraic. Let $X \subseteq A$ and let $a \in C(X) = (\bigvee X] \cap A$. Then a is compact and $a \leq \bigvee X$ so there is a finite subset Y of X such that $a \leq \bigvee Y$. Then $a \in (\bigvee Y] \cap A$. Thus C is algebraic.

Finally the map $f\colon \mathbf{M} \to \mathbf{L}_C$ given by $f(x) = (x] \cap A$ is an isomorphism with inverse $X \mapsto \bigvee X$. □

In light of Corollary 2.31 one might wonder whether there is anything stronger that we can say about the lattices $\mathbf{Sub}(\mathbf{A})$ and $\mathbf{Con}(\mathbf{A})$ in general. In other words, are they somehow "more special" than just being algebraic? The answer is no.

Theorem 2.33. *Let* $\mathbf{L}$ *be an algebraic lattice.*

(1) (*Birkhoff-Frink [BF48], 1948*) *There is an algebra* $\mathbf{A}$ *with* $\mathbf{L} \cong \mathbf{Sub}(\mathbf{A})$.

(2) (*Grätzer-Schmidt, 1963*) *There is an algebra* $\mathbf{B}$ *with* $\mathbf{L} \cong \mathbf{Con}(\mathbf{B})$.

Proof. The proof of (2) is long and difficult, so we shall not include it here. See [GS63]. (1) is quite accessible, so let us consider that one. By Theorem 2.32 there is a set A and an algebraic closure operator C on A such that $\mathbf{L} \cong \mathbf{L}_C$. Our goal is to turn A into an algebra in which $\mathbf{L}_C = \mathrm{Sub}(\mathbf{A})$.

For each finite subset S of A (with $n = |S|$) and each $a \in C(S)$ define an n-ary operation f on A by

$$f_{S,a}(x_1, \ldots, x_n) = \begin{cases} a, & \text{if } \{x_1, \ldots, x_n\} = S \\ x_1, & \text{otherwise.} \end{cases}$$

Let $F = \langle f_{S,a} : S \subseteq_\omega A, a \in C(S) \rangle$ and set $\mathbf{A} = \langle A, F \rangle$.

To show $L_C = \mathrm{Sub}(\mathbf{A})$ first take $B \in \mathrm{Sub}(\mathbf{A})$. We wish to show B

is closed. Suppose $a \in C(B) = \bigcup\{C(S) : S \subseteq_\omega B\}$ since C is algebraic. Then there is $S \subseteq_\omega B$ such that $a \in C(S)$. Let $S = \{s_1, \dots, s_n\}$. Then $a = f_{S,a}(s_1, \dots, s_n) \in B$ since B is closed under every member of F. Thus $C(B) \subseteq B$, i.e., $B \in L_C$.

Conversely, suppose that $B = C(B)$. We must show that B is closed under every $f_{S,a}$. Assume $f_{S,a}$ is n-ary and $b_1, \dots, b_n \in B$. There are two possibilities. If $f_{S,a}(b_1, \dots, b_n) = a$, then by definition we must have $S = \{b_1, \dots, b_n\} \subseteq B$, so $a \in C(S) \subseteq C(B) = B$. If $f_{S,a}(b_1, \dots, b_n) \neq a$ then $f_{S,a}(b_1, \dots, b_n) = b_1 \in B$ by assumption. In either case, B is closed under the operation, so $B \in \mathrm{Sub}(\mathbf{A})$. □

Exercise Set 2.34.

1. Let $\mathbf{L}$ be a lattice.
   (a) Prove that the set $\mathrm{Cvx}(\mathbf{L})$ of all convex subsets of $\mathbf{L}$ (see Exercise 2.4.5) forms a complete lattice under inclusion.
   (b) Let Cvg denote the corresponding closure operator ("the convex subset generated by"). Give an explicit description of the members of $\mathrm{Cvg}(X)$ for any $X \subseteq L$.
   (c) Prove that $\mathrm{Cvx}(\mathbf{L})$ is an algebraic lattice.
2. Let C be a closure operator on a set A. Prove that $\mathbf{L}_C$ is closed under finite unions iff $C(X \cup Y) = C(X) \cup C(Y)$.
3. Let $\mathbf{L}$ be a distributive, algebraic lattice. Prove that for every $a \in L$ and $X \subseteq L$, $a \wedge \bigvee X = \bigvee_{x \in X}(a \wedge x)$.
4. A poset is called *up-directed* (or inductive) if every pair of elements has an upper bound. Let C be a closure operator on a set A. Prove that C is algebraic if and only if the union of every up-directed system of closed subsets is closed.
5. Let a and b be compact elements of an algebraic lattice. Prove that $a \vee b$ is compact. Must $a \wedge b$ be compact?

2.5 Galois connections

There is a simple and very general method for producing closure operators. It turns up throughout mathematics. If you keep your eyes open, you will discover it hidden away in almost every corner of our subject.

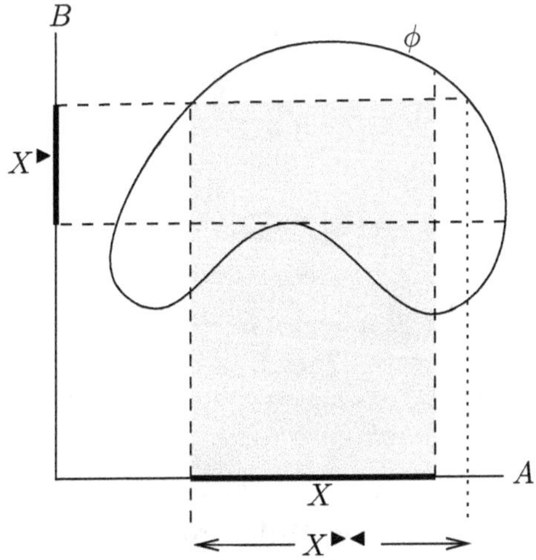

FIGURE 2.4: A Galois connection with X, $X^{\blacktriangleright}$ and $X^{\blacktriangleright\blacktriangleleft}$ illustrated

Let A and B be nonempty sets, and let $\phi \subseteq A \times B$. We define

$$\begin{aligned} X^{\blacktriangleright} &= \{\, b \in B : (\forall x \in X)\ (x, b) \in \phi \,\} \qquad \text{for } X \subseteq A; \\ Y^{\blacktriangleleft} &= \{\, a \in A : (\forall y \in Y)\ (a, y) \in \phi \,\} \qquad \text{for } Y \subseteq B. \end{aligned} \tag{2–4}$$

The functions $X \mapsto X^{\blacktriangleright}$ and $Y \mapsto Y^{\blacktriangleleft}$ are called the *polarities* of ϕ. Figure 2.4 illustrates the action of the polarities on a set X and its polar $X^{\blacktriangleright}$.

The following relationships are easily verified.

1. $(\forall X \subseteq A)\ X \subseteq X^{\blacktriangleright\blacktriangleleft}$;

1′. $(\forall Y \subseteq B)\ Y \subseteq Y^{\blacktriangleleft\blacktriangleright}$;

2. $X_1 \subseteq X_2 \subseteq A \implies X_1^{\blacktriangleright} \supseteq X_2^{\blacktriangleright}$;

2′. $Y_1 \subseteq Y_2 \subseteq B \implies Y_1^{\blacktriangleleft} \supseteq Y_2^{\blacktriangleleft}$;

3. $(\forall X \subseteq A)\ X^{\blacktriangleright} = X^{\blacktriangleright\blacktriangleleft\blacktriangleright}$;

3′. $(\forall Y \subseteq B)\ Y^{\blacktriangleleft} = Y^{\blacktriangleleft\blacktriangleright\blacktriangleleft}$.

In Universal Algebraic jargon, these properties mean that the pair of mappings $\mathrm{Sb}(A) \rightleftarrows \mathrm{Sb}(B)$ form a *Galois connection.* Notice that the last two properties actually follow from the first four. I hope you also get the idea that statements involving a Galois connection always come in pairs, so henceforth we will only state one of them.

For $X \subseteq A$ define $C(X) = X^{\blacktriangleright\blacktriangleleft}$. From properties 1–3 above we see that C is always a closure operator on A. Similarly, $D(Y) = Y^{\blacktriangleleft\blacktriangleright}$ defines a closure operator on B. Thus we obtain two complete lattices of closed subsets. For the closure operator C

$$L_C = \{ X \subseteq A : X = C(X) \} = \{ Y^{\blacktriangleleft} : Y \subseteq B \}.$$

Finally, the mapping $X \mapsto X^{\blacktriangleright}$ is an order-reversing bijection on the closed subsets. Therefore the lattices $\mathbf{L}_C$ and $\mathbf{L}_D^{\partial}$ are isomorphic. We summarize all of this in the following Theorem.

Theorem 2.35. *Let ϕ be a binary relation from A to B, and define the polarities of ϕ as in (2–4) above. Then $C(X) = X^{\blacktriangleright\blacktriangleleft}$ defines a closure operator on A and $D(Y) = Y^{\blacktriangleleft\blacktriangleright}$ defines a closure operator on B. The mappings $X \mapsto X^{\blacktriangleright}$ provides a lattice isomorphism between $\mathbf{L}_C$ and $\mathbf{L}_D^{\partial}$ with inverse $Y \mapsto Y^{\blacktriangleleft}$.*

Example 2.36. (1) Let $p(x) \in \mathbb{Q}[x]$ be a polynomial and let A be its splitting field. Define $B = \mathrm{Aut}_{\mathbb{Q}}(A)$ the group of all automorphisms of A. Take $\phi = \{ (a, f) \in A \times B : f(a) = a \}$. In this context, the polar of a set of automorphisms is the fixed field of that set and the polar of a subset of A is a subgroup of B. This is, of course, the foundation of Galois Theory. The isomorphism guaranteed by Theorem 2.35 is between the subgroups of B and the Galois subextensions of A.

(2) Let $A = \mathbb{C}^n$, $B = \mathbb{Q}[x_1, \ldots, x_n]$ and $\phi = \{ (\mathbf{a}, f) : f(\mathbf{a}) = 0 \}$. Hilbert's Nullstellensatz asserts that the closed subsets of B are precisely the radical ideals of $\mathbb{Q}[x_1, \ldots, x_n]$.

(3) Let V be an inner product space and $A = B = V$. Define $\phi = \{ (v, w) : \langle v, w \rangle = 0 \}$. For any subset X of V, $X^{\blacktriangleright} = X^{\blacktriangleleft}$ is the orthogonal complement of X, usually written $X^{\perp}$. The closed subsets are always subspaces of V. If V is finite-dimensional, every subspace is closed.

(4) Let $\mathbf{M}$ be a lattice, take $A = B = M$ and define $\phi = \{ (x, y) : x \leq y \}$. Then $\{a\}^{\blacktriangleright} = \{ b : a \leq b \} = [a)$ and $\{b\}^{\blacktriangleleft} = (b]$. The lattice of closed subsets of A is complete. There is an embedding of M into this lattice given by $x \mapsto \{x\}^{\blacktriangleleft}$. This is the *Dedekind-MacNeille completion of* $\mathbf{M}$.

(5) Let S be a nonempty set, $A = \mathrm{Sb}(S)$ and B the set of all operations on S. Define $\phi = \{ (X, f) : X \text{ closed under } f \}$. For any $F \subseteq B$, $F^{\blacktriangleleft}$ is the set of subuniverses of the algebra $\langle S, F \rangle$.

2.6 Ideals in lattices

Definition 2.37. Let $\mathbf{L}$ be a lattice. A nonempty subset I of L is called an *ideal* of $\mathbf{L}$ if I is a downset and is closed under finite joins. Equivalently, for all $x, y \in L$

$$(x \in I \implies x \wedge y \in I) \text{ and } (x, y \in I \implies x \vee y \in I).$$

Note the strong similarity to the notion of an ideal in ring theory. Dually, a *filter* of **L** is an upset that is closed under finite meets.

Example 2.38. (1) In the following lattice, $\{b, c, d, 0\}$ is an ideal. Neither $\{a, b, c, 1\}$ nor $\{0, c, d\}$ are ideals.

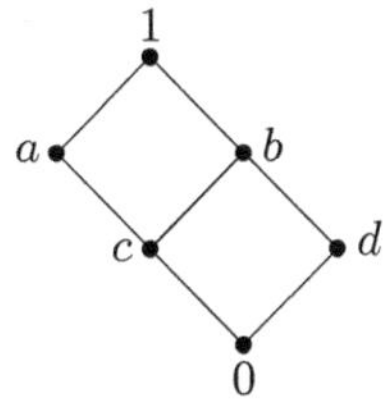

(2) $\mathrm{Sb}_\omega(\omega)$ is an ideal of $\mathrm{Sb}(\omega)$.

(3) Let T be a topology on a set X and fix $x \in X$. Then

$$N_x = \{\, Y \subseteq X : (\exists U \in T)\ x \in U \subseteq Y \,\}$$

is a filter of $\mathrm{Sb}(X)$, called the *neighborhood system* of x.

(4) Let **L** be a lattice with 0 and θ a congruence relation on **L**. Then $0/\theta$ is an ideal of **L**.

As usual, every statement we make about ideals has a dual that is a statement about filters. So we primarily restrict our attention to ideals. If **L** is a lattice and $a \in L$, then the principal ideal generated by a is $(a] = \{\, x \in L : x \leq a \,\}$. Note that every ideal is a sublattice.

We have a dilemma similar to the one we faced with subalgebras. We do not consider the empty set to be an ideal, but we would like to take arbitrary intersections of ideals (which might turn out empty). Thus we define Idl(**L**) to be the set of ideals of **L** if **L** has a lower bound. If **L** has no lower bound, then we define Idl(**L**) to be the set of ideals together with the empty set.

Proposition 2.39. *For any lattice* **L**, **Idl(L)** *is a complete lattice under set-theoretic inclusion. There is a lattice embedding of* **L** *into* **Idl(L)** *given by* $x \mapsto (x]$.

Proof. That Idl(**L**) is closed under intersections is easy to see. Then by Proposition 2.14, we obtain a complete lattice. The last assertion is left as an exercise. □

In fact, **Idl(L)** is an algebraic lattice. See Exercise 2.47.2.

Definition 2.40. A proper ideal I of a lattice **L** is called *prime* if for every $x, y \in L$, $x \wedge y \in I$ implies that $x \in I$ or $y \in I$.

Let **2** denote the two-element chain, viewed as a lattice. We usually write $2 = \{0, 1\}$ in this context. Suppose $h\colon \mathbf{L} \to \mathbf{2}$ is a surjective lattice homomorphism, and let $I = h^{-1}(0)$. Then I is a prime ideal of $\mathbf{L}$. Conversely, for every prime ideal I of a lattice $\mathbf{L}$, there is a surjective homomorphism $h\colon \mathbf{L} \to \mathbf{2}$ such that $I = h^{-1}(0)$. Thus there is a one-to-one correspondence between prime ideals and homomorphisms onto **2**.

Theorem 2.41 (The Prime Ideal Theorem). *Let $\mathbf{L}$ be a distributive lattice, I an ideal and F a filter of $\mathbf{L}$, and suppose that $I \cap F = \varnothing$. Then there is a prime ideal P such that $I \subseteq P$ and $P \cap F = \varnothing$.*

Proof. Let $\mathcal{S} = \{\, J : J \text{ is an ideal, } I \subseteq J,\ J \cap F = \varnothing \,\}$. Surely $I \in \mathcal{S}$, so $\mathcal{S}$ is nonempty. We wish to apply Zorn's lemma to obtain a maximal element of $\mathcal{S}$. Let $\mathcal{C}$ be a chain in $\mathcal{S}$ and take $J = \bigcup \mathcal{C}$. It is straightforward to check that $J \in \mathcal{S}$.

Therefore, $\mathcal{S}$ has a maximal element, P. We must show that P is prime. Suppose that there are elements $a, b \notin P$ such that $a \wedge b \in P$. Define

$$J = \{\, p \vee x : p \in P, x \leq a \,\} \text{ and } K = \{\, p \vee y : p \in P, y \leq b \,\}.$$

Using the distributivity of $\mathbf{L}$ and the fact that P is an ideal it is easy to show that J is an ideal. Also, $P \subseteq J$ since if $p \in P$ then $p = p \vee (p \wedge a) \in J$. Finally since $a = (a \wedge p) \vee a \in J - P$ we see that J is a proper extension of P. By the maximality of P in $\mathcal{S}$ it must be the case that J intersects F. Similarly, $K \cap F \neq \varnothing$.

Thus there are $x \leq a$, $y \leq b$, and $p, q \in P$ such that $p \vee x$ and $q \vee y$ are members of F. Since F is a filter, it must contain $z = (p \vee x) \wedge (q \vee y)$. But

$$z = (p \wedge q) \vee (p \wedge y) \vee (q \wedge x) \vee (x \wedge y) \in P$$

since each of the first three summands lie below either p or q (hence lie in P) and the fourth is dominated by $a \wedge b$ which is in P by assumption. Thus $P \cap F \neq \varnothing$. This contradiction proves the theorem. □

Corollary 2.42. *In a distributive lattice, every proper ideal is an intersection of prime ideals.*

Proof. Let I be a proper ideal of a distributive lattice $\mathbf{L}$. For each $x \in L - I$ let $F = [x)$ and apply the Prime Ideal Theorem. □

For a distributive lattice $\mathbf{L}$, let $\mathrm{Prm}(\mathbf{L})$ denote the set of prime ideals of $\mathbf{L}$. For $a \in L$, let

$$\mathcal{P}_a = \{\, P \in \mathrm{Prm}\,\mathbf{L} : a \notin P \,\}.$$

The following is a weak form of the representation theorem for distributive lattices, proved by Birkhoff (1933) and strengthened by Stone (1936).

Theorem 2.43. *Let $\mathbf{L}$ be a distributive lattice. There is a lattice embedding of $\mathbf{L}$ into $\langle \mathrm{Sb}(\mathrm{Prm}\,\mathbf{L}), \cap, \cup \rangle$.*

Proof. The desired mapping is given by $a \mapsto \mathcal{P}_a$. To see that this map is injective, suppose that $a \neq b$. We can assume that $a \not\leq b$. Apply Theorem 2.41 to $I = (b]$ and $F = [a)$ to obtain a prime ideal $P \in \mathcal{P}_a - \mathcal{P}_b$.

The condition

$$\mathcal{P}_{a \wedge b} = \{ P : a \wedge b \notin P \} = \{ P : a \notin P \} \cap \{ P : b \notin P \} = \mathcal{P}_a \cap \mathcal{P}_b$$

follows from the fact that P is both a downset and is prime while

$$\mathcal{P}_{a \vee b} = \{ P : a \vee b \notin P \} = \{ P : a \notin P \} \cup \{ P : b \notin P \} = \mathcal{P}_a \cup \mathcal{P}_b$$

is a consequence of the fact that P is an ideal. □

Frequently the above theorem is expressed in the form "every distributive lattice is isomorphic to a ring of sets." This reflects the superficial similarity in the definitions of rings and lattices. An analogous statement holds for Boolean algebras.

Corollary 2.44. *Let* $\mathbf{B}$ *be a Boolean algebra. There is a Boolean embedding of* $\mathbf{B}$ *into* $\langle \mathrm{Sb}(\mathrm{Prm}\,\mathbf{B}), \cap, \cup, ', \varnothing, \mathrm{Prm}\,\mathbf{B} \rangle$.

Proof. Using the same embedding as above, $\mathcal{P}_0 = \varnothing$, since every prime ideal must contain 0, while $\mathcal{P}_1 = \mathrm{Prm}\,\mathbf{B}$, since every prime ideal must be proper. From these two equalities it follows that the mapping must preserve complementation as well. □

Using the relationship between prime ideals and homomorphisms onto $\mathbf{2}$, we can reformulate Theorem 2.43 in a way that will generalize to other algebraic structures.

Corollary 2.45. *Let* $\mathbf{L}$ *be a distributive lattice (or Boolean algebra), and let* $\mathbf{2}$ *denote the two-element distributive lattice (resp. Boolean algebra). There is a set* J *and a lattice embedding* $h\colon \mathbf{L} \to \mathbf{2}^J$ *such that for every* $j \in J$, $p_j \circ h$ *is surjective.*

Proof. Take $J = \mathrm{Prm}(\mathbf{L})$ and use Exercise 2.12.1. □

Neither the theorem nor the corollaries are "sharp" in the sense that we have very little idea of what the image of the embedding will look like. For finite distributive lattices we can do better. The generalization of the following theorem to the infinite case is provided by the Stone (for Boolean algebras) and Priestley (distributive lattices) dualities. See the Davey-Priestley book [DP02] for details.

Observe that $\mathrm{Prm}\,\mathbf{L}$ is not just a set, but is partially ordered by set-theoretic inclusion. For any $a \in L$, the set $\mathcal{P}_a$ constructed above is a downset of $\langle \mathrm{Prm}\,\mathbf{L}, \subseteq \rangle$. Therefore the image of the embedding defined in Theorem 2.43 is contained in the sublattice $\mathbf{Dn}(\mathbf{Prm}\,\mathbf{L})$ of $\mathbf{Sb}(\mathrm{Prm}(\mathbf{L}))$. (See Exercise 2.4.2.) As the next theorem shows, in the finite case, the mapping is an isomorphism.

Theorem 2.46. *Let* $\mathbf{L}$ *be a finite distributive lattice.* $\mathbf{L} \cong \mathbf{Dn}(\mathbf{Prm}\,\mathbf{L})$.

Proof. All that remains from Theorem 2.43 is to prove that the assignment $a \mapsto \mathcal{P}_a$ is surjective. So let $\mathcal{D}$ be a downset of $\mathbf{Prm}(\mathbf{L})$. Since the lattice is finite, it has only finitely many prime ideals. So we can write $\mathcal{D} = \{P_1, P_2, \dots, P_n\}$. Let $X = L - \bigcup \mathcal{D} = \{x_1, \dots, x_m\}$ and set $a = x_1 \wedge \cdots \wedge x_m$.

We claim that $\mathcal{P}_a = \mathcal{D}$. First, if $a \in P_i$ for some $i \leq n$, then since P_i is prime, there is some j such that $x_j \in P_i$, which is false. Thus a is excluded from every P_i, which means that $\mathcal{D} \subseteq \mathcal{P}_a$.

On the other hand, let $P \in \mathcal{P}_a$ and set $u = \bigvee P$. Note that $u \in P$ since P is a finite set and an ideal. If $u \in X$ then $a \leq u$ which implies that $a \in P_a$ which is false. Thus $u \notin X$ which means that for some $i \leq n$, $u \in P_i$. Since u is the largest element of P, we get $P \subseteq P_i$. Finally since $\mathcal{D}$ is a downset, we conclude $P \in \mathcal{D}$. □

Exercise Set 2.47.

1. A *Boolean ring* is a ring $\mathbf{R} = \langle R, +, -, \cdot, 0, 1\rangle$ satisfying: $x^2 = x = x \cdot 1$.

   (a) Show that every Boolean ring satisfies: $x + x = 0$ and $x \cdot y = y \cdot x$.

   (b) Let $\mathbf{R}$ be a Boolean ring. Define
   $$x \wedge y = x \cdot y, \quad x \vee y = x + y + x \cdot y, \quad x' = 1 - x.$$
   Then $B(\mathbf{R}) = \langle R, \wedge, \vee, ', 0, 1\rangle$ is a Boolean algebra.

   (c) Let $\mathbf{B} = \langle B, \wedge, \vee, ', 0, 1\rangle$ be a Boolean algebra. Define
   $$x \cdot y = x \wedge y, \quad x + y = (x \wedge y') \vee (x' \wedge y), \quad -x = x.$$
   Then $R(\mathbf{B}) = \langle B, +, -, \cdot, 0, 1\rangle$ is a Boolean ring.

   (d) $R\big(B(\mathbf{R})\big) = \mathbf{R}$ and $B\big(R(\mathbf{B})\big) = \mathbf{B}$.

2. (a) Let $\mathbf{L}$ be a lattice. Prove that for any subset X of L, the ideal generated by X is
   $$\mathrm{Ig}^{\mathbf{L}}(X) = \big\{a \in L : a \leq \bigvee Y, \text{ some } Y \subseteq_\omega X\big\}.$$
   Conclude that $\mathrm{Idl}\,\mathbf{L}$ is an algebraic lattice (see Proposition 2.39.)

   (b) Let $\mathbf{L}$ be distributive, and let I and J be ideals of $\mathbf{L}$. Prove that the join of I and J in the lattice of ideals of $\mathbf{L}$ is
   $$\{x \vee y : x \in I, y \in J\}.$$

   (c) Give an example to show that (b) is false even for modular lattices.

3. In a finite lattice, every ideal is principal. In a distributive lattice, $(a]$ is a prime ideal iff $x \wedge y = a \implies x = a$ or $y = a$.

4. Let **L** be a lattice. Show that Idl(**L**) is distributive iff **L** is distributive.

5. Prove that in a distributive lattice, every maximal ideal is prime. Prove that in a Boolean algebra, every prime ideal is maximal. Give examples of a prime ideal in a distributive lattice that is not maximal, and a maximal ideal in a (nondistributive) lattice that is not prime.

6. (a) Let **L** be a bounded distributive lattice, $a, b \in L$ and suppose that $a \wedge b = 0$, $a \vee b = 1$. Prove that $\mathbf{L} \cong (a] \times (b]$. (Here we are thinking of $(a]$ as a sublattice of **L**.)

   (b) Prove that every finite Boolean algebra is isomorphic to $\mathbf{2}^n$ for some natural number n.

7. Let **L** be a distributive lattice, generated by a subset X.

   (a) Let P and Q be prime ideals of **L**. Show that $P \cap X = Q \cap X \implies P = Q$. (Hint: Exercise 2.12.2.)

   (b) Prove that every finitely generated distributive lattice is finite.

Chapter 3

The Nuts and Bolts of Universal Algebra

Nuts and bolts are among the tools we use to assemble the objects that dominate our lives. In this chapter we develop tools that allow us to understand how algebras and varieties are put together. Often, it is not assembly that is important, but disassembly—that is, recognizing the component parts of an object under study.

An understanding of the complex interplay of subalgebras, homomorphisms and products is essential to the mastery of universal algebra. In Section 3.4 we illustrate many of these ideas with several concrete examples.

3.1 The isomorphism theorems

Let A and B be sets, and $f\colon A \to B$ a function. Then f induces two maps on the subsets of A and B as follows.

$$\begin{aligned} \vec{f}\colon \mathrm{Sb}(A) \to \mathrm{Sb}(B) \quad &\text{given by} \quad X \mapsto \{\, f(x) : x \in X \,\}\,; \\ \overleftarrow{f}\colon \mathrm{Sb}(B) \to \mathrm{Sb}(A) \quad & \qquad\qquad\;\; Y \mapsto \{\, a \in A : f(a) \in Y \,\}\,. \end{aligned}$$

These maps are usually written as $f(X)$ and $f^{-1}(Y)$ instead of $\vec{f}(X)$ and $\overleftarrow{f}(Y)$. But our notation is more precise, and hopefully will help reduce confusion. If we want to use English, we can call these maps the *direct* and *inverse image* under f.

The inverse image behaves very nicely. (The direct image, not so nicely.) The reader might like to verify that for any subsets X and Y of B, and Z and W of A

$$\begin{aligned} \overleftarrow{f}(X \cap Y) &= \overleftarrow{f}(X) \cap \overleftarrow{f}(Y), & \overleftarrow{f}(X \cup Y) &= \overleftarrow{f}(X) \cup \overleftarrow{f}(Y), \\ \overleftarrow{f}(\varnothing) &= \varnothing, & \overleftarrow{f}(B) &= A, \\ \vec{f}(W \cup Z) &= \vec{f}(W) \cup \vec{f}(Z), & & \\ \vec{f}\,\overleftarrow{f}(X) &\subseteq X, & \overleftarrow{f}\,\vec{f}(Z) &\supseteq Z. \end{aligned} \tag{3–1}$$

We remark that the two inclusions at the bottom become equalities if f is surjective (for the first) or injective (for the second).

We now investigate this construction in the case that $\mathbf{A}$ and $\mathbf{B}$ are algebras and f a homomorphism.

Lemma 3.1. *Let $f\colon \mathbf{A} \to \mathbf{B}$ be a homomorphism. If U is a subuniverse of $\mathbf{A}$, then $\vec{f}(U)$ is a subuniverse of $\mathbf{B}$. If V is a subuniverse of $\mathbf{B}$, then $\overleftarrow{f}(V)$ is a subuniverse of $\mathbf{A}$.*

Proof. Exercise. □

Theorem 3.2. *Let $f\colon \mathbf{A} \to \mathbf{B}$ be a homomorphism. Then for any subset, X, of A, $\vec{f}(\mathrm{Sg}^{\mathbf{A}}(X)) = \mathrm{Sg}^{\mathbf{B}}(\vec{f}(X))$.*

Proof. It is easy to check that $\vec{f}$ is order-preserving. Using that observation and the lemma

$$X \subseteq \mathrm{Sg}^{\mathbf{A}}(X) \implies \vec{f}(X) \subseteq \vec{f}(\mathrm{Sg}^{\mathbf{A}}(X)) \implies \mathrm{Sb}^{\mathbf{B}}(\vec{f}(X)) \subseteq \vec{f}(\mathrm{Sg}^{\mathbf{A}}(X)).$$

Conversely, using the lemma and equations (3–1)

$$X \subseteq \overleftarrow{f}\vec{f}(X) \subseteq \overleftarrow{f}(\mathrm{Sg}^{\mathbf{B}}(\vec{f}(X))) \implies \mathrm{Sg}^{\mathbf{A}}(X) \subseteq \overleftarrow{f}(\mathrm{Sg}^{\mathbf{B}}(\vec{f}(X))) \implies$$
$$\vec{f}(\mathrm{Sg}^{\mathbf{A}}(X)) \subseteq \vec{f}\,\overleftarrow{f}(\mathrm{Sg}^{\mathbf{B}}(\vec{f}(X))) \subseteq \mathrm{Sg}^{\mathbf{B}}(\vec{f}(X)).$$

□

Corollary 3.3. *Let $f\colon \mathbf{A} \to \mathbf{B}$ be a homomorphism. The map $\vec{f}\colon \mathrm{Sub}\,\mathbf{A} \to \mathrm{Sub}\,\mathbf{B}$ is join-preserving and $\overleftarrow{f}\colon \mathrm{Sub}\,\mathbf{B} \to \mathrm{Sub}\,\mathbf{A}$ is meet-preserving. Thus, both maps are order-preserving.*

Proof. Follows from equations (3–1) and Theorem 3.2. □

Here are a pair of examples that demonstrate that the above corollary cannot be strengthened. Let $\mathbb{R}$ denote the additive group of real numbers, and define $f\colon \mathbb{R}^2 \to \mathbb{R}$ by $f(x,y) = x - y$. f is easily seen to be a homomorphism. Let $X = \mathbb{R}\times\{0\}$ and $Y = \{0\}\times\mathbb{R}$ be subalgebras of $\mathbb{R}^2$. Then $X \cap Y = \{(0,0)\}$, so $\vec{f}(X \cap Y) = \{0\}$. But $\vec{f}(X) \cap \vec{f}(Y) = \mathbb{R} \cap \mathbb{R} = \mathbb{R}$. Thus $\vec{f}$ does not in general preserve meets.

For the second example, let $f\colon \mathbf{L} \to \mathbf{M}$ be the lattice homomorphism given in Figure 3.1. The map f identifies c and 1 and nothing else. Let $X = \{\bar{0}, \bar{a}\}$ and $Y = \{\bar{0}, \bar{b}\}$ be sublattices of $\mathbf{M}$. Then $X \vee Y = M$ so $\overleftarrow{f}(X \vee Y) = L$. But $\overleftarrow{f}(X) \vee \overleftarrow{f}(Y) = \{0,a\} \vee \{0,b\} = \{0,a,b,c\}$. Thus $\overleftarrow{f}$ is not join-preserving.

We can extend this notation to congruences, or to arbitrary n-ary relations. If $\theta \subseteq A^2$, and $\psi \subseteq B^2$, then we define

$$\vec{f}(\theta) = \{\, (f(x), f(y)) : (x,y) \in \theta \,\} \quad \text{and}$$
$$\overleftarrow{f}(\psi) = \{\, (x,y) \in A^2 : (f(x), f(y)) \in \psi \,\}.$$

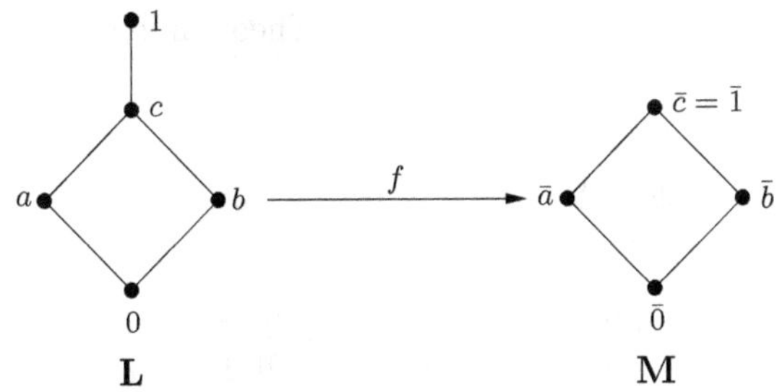

FIGURE 3.1: $\overleftarrow{f}$ not join-preserving

Lemma 3.4. *Let $f\colon \mathbf{A} \to \mathbf{B}$ be a homomorphism. Then for every $\psi \in \operatorname{Con} \mathbf{B}$, $\overleftarrow{f}(\psi)$ is a congruence on* **A**.

The proof of this lemma is a straightforward verification. Since $\ker f = \overleftarrow{f}(\{0_B\})$ we see that for any congruence ψ on **B**, $\overleftarrow{f}(\psi) \supseteq \ker f$.

The direct image of a congruence is not, in general, a congruence. Consider the lattice homomorphism in Figure 3.2. Here, $g(a) = 0$, $g(b) = g(c) = 1$ and $g(d) = 2$. Let $\theta = \operatorname{Cg}^{\mathbf{L}}\{(a,b),\,(c,d)\} = 0_L \cup \{(a,b),(b,a),(c,d),(d,c)\}$. Then $\vec{g}(\theta) = 0_M \cup \{(0,1),(1,0),(1,2),(2,1)\}$, which is not transitive.

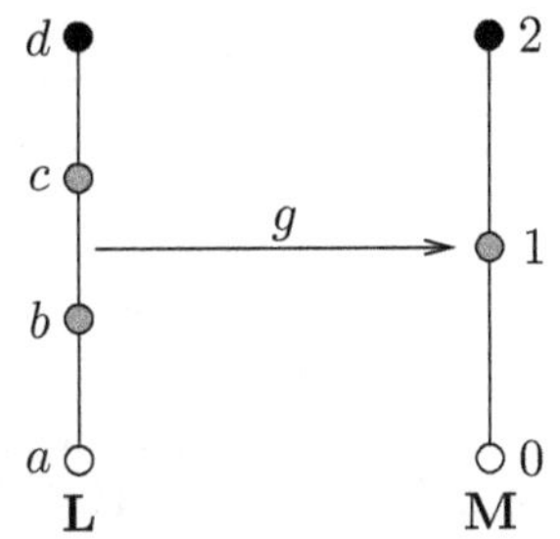

FIGURE 3.2: Counterexample to Lemma 3.4

Now we turn to the isomorphism theorems. You might want to have a look back at the corresponding statements for groups in your first-year algebra text. We proved the Fundamental Homomorphism Theorem in Chapter 1. From this it follows that the homomorphic images of an algebra **A** are (up to isomorphism) precisely the quotients of **A**. The Fundamental Homomorphism Theorem is frequently called the First Isomorphism Theorem. The Second Isomorphism Theorem and the Correspondence Theorem provide more details on the relationship between the congruences of **A** and its homomorphic images.

Let θ and ψ be congruences of **A**, and assume that $\theta \subseteq \psi$. We define a binary relation ψ/θ on A/θ by

$$\psi/\theta = \{\, (x/\theta, y/\theta) : (x,y) \in \psi \,\}.$$

One form of the Second Isomorphism Theorem is given below. Another was given in Exercise 1.26.8.

Theorem 3.5 (The Second Isomorphism Theorem). *Let $\theta \subseteq \psi$ be congruences on an algebra $\mathbf{A}$. Then ψ/θ is a congruence on $\mathbf{A}/\theta$. The algebras $(\mathbf{A}/\theta)/(\psi/\theta)$ and $\mathbf{A}/\psi$ are isomorphic.*

Proof. Define $f\colon \mathbf{A}/\theta \to \mathbf{A}/\psi$ by $f(a/\theta) = a/\psi$. The condition $\theta \subseteq \psi$ is equivalent to the assertion that f is well-defined. Clearly, f is surjective. A straightforward verification shows that f is a homomorphism and that $\ker f = \psi/\theta$. Thus by the fundamental homomorphism theorem, $\mathbf{A}/\psi \cong (\mathbf{A}/\theta)/\ker f = (\mathbf{A}/\theta)/(\psi/\theta)$. □

Theorem 3.6 (The Correspondence Theorem). *Let $\mathbf{A}$ be an algebra and let θ be a congruence on $\mathbf{A}$. Let $q\colon \mathbf{A} \twoheadrightarrow \mathbf{A}/\theta$ be the canonical homomorphism. Then $\vec{q}$ is a lattice isomorphism of the interval $\mathbf{I}[\theta, 1_A]$ of $\mathrm{Con}(\mathbf{A})$ with $\mathrm{Con}(\mathbf{A}/\theta)$ mapping ψ to ψ/θ.*

Proof. Suppose that ψ is a congruence on $\mathbf{A}$ and that $\psi \supseteq \theta$. It follows directly from the definition that $\vec{q}(\psi) = \psi/\theta$. We wish to show that $\vec{q}$ is a lattice isomorphism with inverse $\overleftarrow{q}$.

We already know that both $\vec{q}$ and $\overleftarrow{q}$ are order-preserving. And according to equations (3–1), $\vec{q} \circ \overleftarrow{q}$ is the identity map. For the other direction

$$(a,b) \in \overleftarrow{q}\vec{q}(\psi) \iff (q(a), q(b)) \in \vec{q}(\psi) \iff (a/\theta, b/\theta) \in \psi/\theta \iff (a,b) \in \psi.$$

□

The Third Isomorphism Theorem for groups is usually stated as follows: $\mathbf{H}$ and $\mathbf{N}$ are subgroups of a group $\mathbf{G}$, with $\mathbf{N}$ normal in $\mathbf{G}$. Then $\mathbf{HN}$ is a subgroup of $\mathbf{G}$, $\mathbf{N}$ is normal in $\mathbf{HN}$ and $\mathbf{HN}/\mathbf{N} \cong \mathbf{H}/(\mathbf{H} \cap \mathbf{N})$.

Now, how do we translate this into a statement of universal algebra? Obviously we want to replace N with a congruence of $\mathbf{G}$. But what do we do with HN? Well, note that $HN = \bigcup_{h \in H} hN$ and hN is nothing but the congruence class of h modulo N. This motivates the following definition.

Definition 3.7. Suppose that B is a subset of an algebra $\mathbf{A}$, and θ is a congruence on $\mathbf{A}$. Then $B^\theta = \bigcup\limits_{b \in B} b/\theta$ and $\theta{\restriction}_B = \theta \cap B^2$.

Now it is an easy verification that if $\mathbf{B}$ is a subalgebra of $\mathbf{A}$, then B^θ is a subuniverse of $\mathbf{A}$ containing B and $\theta{\restriction}_B$ is a congruence on $\mathbf{B}$. We denote by $\mathbf{B}^\theta$ the subalgebra of $\mathbf{A}$ with universe B^θ

Theorem 3.8 (The Third Isomorphism Theorem). *Let $\mathbf{B}$ be a subalgebra of $\mathbf{A}$, and θ a congruence on $\mathbf{A}$. Then $\mathbf{B}^\theta/(\theta{\restriction}_{B^\theta}) \cong \mathbf{B}/(\theta{\restriction}_B)$.*

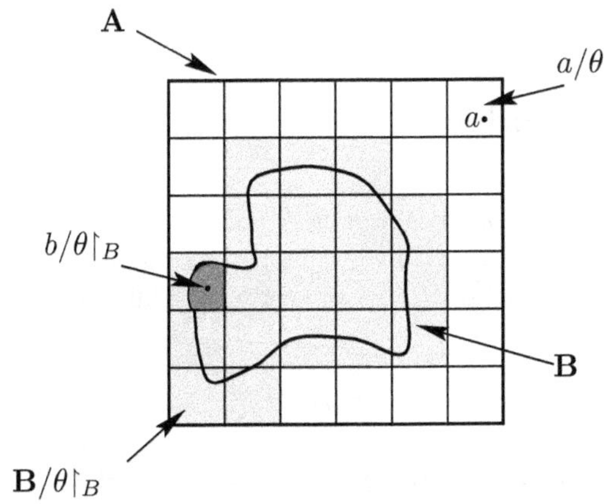

FIGURE 3.3: The Third Isomorphism Theorem

Proof. Define f to be the composition of the homomorphisms $\mathbf{B} \hookrightarrow \mathbf{B}^\theta \twoheadrightarrow \mathbf{B}^\theta/(\theta{\restriction}_{B^\theta})$. To see that f is surjective, choose a typical element of $\mathbf{B}^\theta/(\theta{\restriction}_{B^\theta})$, say $a/(\theta{\restriction}_{B^\theta})$. Since $a \in B^\theta$, there is $b \in B$ such that $a \mathrel{\theta} b$. But then $a/(\theta{\restriction}_{B^\theta}) = b/(\theta{\restriction}_{B^\theta}) = f(b)$. Clearly $\ker f = \theta$. The isomorphism now follows from the Fundamental Homomorphism Theorem. □

Exercise Set 3.9.

1. Let $h\colon \mathbf{A} \to \mathbf{B}$ be a homomorphism, and let $\psi \in \operatorname{Con}\mathbf{B}$. Prove that there is a canonical embedding $h/\psi\colon \mathbf{A}/\overleftarrow{h}(\psi) \to \mathbf{B}/\psi$. Draw an appropriate commuting diagram.

2. Let $g\colon \mathbf{A} \to \mathbf{B}$ be a surjective homomorphism and $\theta \subseteq A^2$. Prove that $\vec{g}(\mathrm{Cg}^{\mathbf{A}}(\theta) \vee \ker(g)) = \mathrm{Cg}^{\mathbf{B}}(\vec{g}(\theta))$.

3. Let $g\colon \mathbf{A} \to \mathbf{B}$ be a surjective homomorphism with $\ker g = \psi$ and θ a congruence on $\mathbf{A}$. Prove that $\vec{g}(\theta)$ is a congruence on $\mathbf{B}$ if and only if $\theta \circ \psi \circ \theta \subseteq \psi \circ \theta \circ \psi$.

4. Let $g\colon A \to B$ be a function. For $X \subseteq A$, define $C(X) = \overleftarrow{g}\vec{g}(X)$. Prove that C is a closure operator on A, that $C(\varnothing) = \varnothing$ and that for any subsets X and Y of A, $C(X \cup Y) = C(X) \cup C(Y)$ and that $C(X \cap C(Y)) = C(X) \cap C(Y)$.

5. Let $f\colon \mathbf{A} \to \mathbf{C}$ be a homomorphism, $B \subseteq \mathbf{A}$ and $\theta = \ker f$. Prove that $B^\theta = \overleftarrow{f}\vec{f}(B)$.

6. Let $\mathcal{V}$ and $\mathcal{W}$ be varieties of groups. Define $\mathcal{V} \cdot \mathcal{W}$ to be the class of all groups $\mathbf{A}$ containing a normal subgroup $\mathbf{B}$ such that $\mathbf{B} \in \mathcal{V}$ and $\mathbf{A}/\mathbf{B} \in \mathcal{W}$. Prove that $\mathcal{V} \cdot \mathcal{W}$ is a variety.

3.2 Direct products

Let $\mathbf{A}_1$ and $\mathbf{A}_2$ be similar algebras, and form their product $\mathbf{A}_1 \times \mathbf{A}_2$. There is a pair of surjective homomorphisms, $p_i\colon \mathbf{A}_1 \times \mathbf{A}_2 \twoheadrightarrow \mathbf{A}_i$ called the *coordinate projections* mapping $(a_1, a_2) \mapsto a_i$ for $i = 1, 2$. We denote by η_i the kernel of p_i (called a projection kernel). Thus for two *pairs* (a_1, a_2) and (b_1, b_2) we have

$$(a_1, a_2) \equiv_{\eta_i} (b_1, b_2) \iff a_i = b_i, \qquad \text{for } i = 1, 2.$$

Definition 3.10. A pair $\{\alpha, \beta\} \subseteq \mathrm{Con}(\mathbf{A})$ is called a pair of *complementary factor congruences* on $\mathbf{A}$ if $\alpha \cap \beta = 0_A$ and $\alpha \circ \beta = 1_A$.

It is easy to see that the projection kernels on a direct product of two algebras always form a pair of complementary factor congruences. The converse of this was the content of Exercise 1.26.7.

Theorem 3.11. *If α and β form a pair of complementary factor congruences on an algebra $\mathbf{A}$, then $\mathbf{A} \cong \mathbf{A}/\alpha \times \mathbf{A}/\beta$ under the map $a \mapsto \langle a/\alpha, a/\beta\rangle$. Conversely, every direct product decomposition arises in this way.*

Definition 3.12. A nontrivial algebra is called *directly indecomposable* if it is not isomorphic to a direct product of two nontrivial algebras.

Equivalently, $\mathbf{A}$ is directly indecomposable if and only if the only pair of complementary factor congruences on $\mathbf{A}$ is $\{0_A, 1_A\}$. It is not hard to see, on cardinality grounds, that every finite algebra is a direct product of directly indecomposable algebras. However, this is not true for infinite algebras.

For example, let $\mathbb{F}_2$ denote the field of order 2 and consider the variety of $\mathbb{F}_2$-vector spaces. The only directly indecomposable vector space is the 1-dimensional space. (In a vector space, every partition of a basis into two pieces gives rise to a direct decomposition.) Every space of dimension n can be decomposed into a product of n one-dimensional spaces.

But consider a space $\mathbf{V}$ with basis $\{\, v_i : i \in \omega \,\}$. This space is not directly indecomposable, since we can split the basis into two nonempty pieces. It is not a product of finitely many 1-dimensional spaces since such a product would be finite, and V is infinite. But $\mathbf{V}$ is also not the product of infinitely many 1-dimensional spaces. The reason is that V has cardinality $\aleph_0$ while, based on a standard set-theoretic argument, the product can not have cardinality $\aleph_0$. Thus $\mathbf{V}$ has no decomposition into directly indecomposable components.

The question of *uniqueness* of direct decompositions is a complex and fascinating subject. We won't even give a precise definition in this text. It turns out that finite groups, rings, and lattices all have a unique decomposition into directly indecomposable algebras. But see Exercise 3.15.7 for an example of a finite algebra with nonunique decomposition.

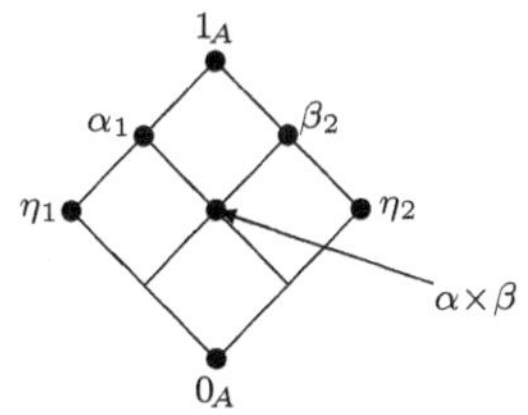

FIGURE 3.4: A product congruence

Suppose that $\mathbf{A} \cong \mathbf{A}_1 \times \mathbf{A}_2$, and let η_1 and η_2 be the projection kernels. Then $\mathbf{A}/\eta_i \cong \mathbf{A}_i$ for $i = 1, 2$. From the Correspondence Theorem it follows that the lattice $\operatorname{Con}(\mathbf{A}_i)$ is isomorphic to the interval $\mathbf{I}[\eta_i, 1_A]$ of $\operatorname{Con}\mathbf{A}$. Let α be a congruence on $\mathbf{A}_1$. Then, under this correspondence, α is mapped to the congruence $\alpha_1 = \overleftarrow{p}_1(\alpha)$. (A better notation might be $\overleftarrow{\alpha}_1$.) Thus

$$\langle a_1, a_2\rangle \equiv_{\alpha_1} \langle b_1, b_2\rangle \iff a_1 \equiv_\alpha b_1.$$

Similarly, a congruence β on $\mathbf{A}_2$ corresponds to a congruence β_2 on $\mathbf{A}$. The congruence $\alpha_1 \cap \beta_2$ is often denoted $\alpha \times \beta$. In other words,

$$\langle a_1, a_2\rangle \equiv_{\alpha\times\beta} \langle b_1, b_2\rangle \iff$$
$$a_1 \equiv_\alpha b_1 \;\;\&\;\; a_2 \equiv_\beta b_2.$$

See Figure 3.4 for a picture. Congruences on $\mathbf{A}_1 \times \mathbf{A}_2$ not of the form $\alpha \times \beta$ are called *skew congruences.*

Theorem 3.11 can be extended to products with finitely many factors, see Exercise 3.15.5. But there is no particularly useful characterization of direct products with infinitely many factors. We salvage as much of the theory as we can.

Let $\mathbf{B} = \prod_{i\in I} \mathbf{A}_i$. Then for each $i \in I$ we have a surjective homomorphism $p_i \colon \mathbf{B} \twoheadrightarrow \mathbf{A}_i$ sending $f \mapsto f(i)$, see Exercise 1.11.6. Let $\eta_i = \ker p_i$. Thus for $f, g \in B$

$$f \equiv_{\eta_i} g \iff f(i) = g(i).$$

It follows that $\bigcap\limits_{i\in I} \eta_i = 0_B$. This inspires the next definition.

Definition 3.13. Let $\mathbf{B}$ and $\mathbf{A}_i$ be algebras, for all $i \in I$, and let $h_i \colon \mathbf{B} \to \mathbf{A}_i$ be a homomorphism for each $i \in I$. We say that the family $\langle h_i : i \in I\rangle$ *separates points* if for every pair x, y of distinct members of B, there is some $i \in I$ such that $h_i(x) \neq h_i(y)$.

Given a family of maps $h_i \colon \mathbf{B} \to \mathbf{A}_i$ for $i \in I$, there is a natural map $h\colon \mathbf{B} \to \prod_{i\in I} \mathbf{A}_i$ given by $h(b) = f$ where the function (that is, the member of the product) f is defined by $f(i) = h_i(b)$. It seems reasonable to write $h = \prod_{i\in I} h_i$. The following easy proposition summarizes the situation.

Proposition 3.14. *Let h_i be a homomorphism from* $\mathbf{B}$ *to* $\mathbf{A}_i$*, for each* $i \in I$*, and let* $h = \prod_{i\in I} h_i$*. Then* $\ker(h) = \bigcap_{i\in I} \ker(h_i)$*. Furthermore, the following are equivalent:*

(a) *The family* $\langle h_i : i \in I \rangle$ *separates points.*

(b) h *is injective.*

(c) $\bigcap_{i\in I} \ker h_i = 0_B$.

Exercise Set 3.15.

1. Let $\mathbf{A}$ and $\mathbf{B}$ be similar algebras.

   (a) Show that the mapping $\langle \alpha, \beta \rangle \mapsto \alpha \times \beta$ gives an embedding of $\mathrm{Con}(\mathbf{A}) \times \mathrm{Con}(\mathbf{B})$ into $\mathrm{Con}(\mathbf{A} \times \mathbf{B})$.

   (b) Prove that if $\mathrm{Con}(\mathbf{A} \times \mathbf{B})$ is distributive, then the embedding is an isomorphism.

   (c) Show by example that the embedding is not necessarily an isomorphism if $\mathrm{Con}(\mathbf{A} \times \mathbf{B})$ is modular.

2. Prove that the only directly indecomposable Boolean algebra is the two-element algebra. (Hint: Let a be any element of a Boolean algebra. Define $x \mathrel{\alpha} y \iff x \vee a = y \vee a$ and $x \mathrel{\beta} y \iff x \vee a' = y \vee a'$.)

3. Prove that every chain is a directly indecomposable lattice.

4. Find a pair of similar algebras, neither of which can be embedded into their product.

5. Let $\mathbf{A}$ be an algebra, n a positive integer, and $\theta_1, \dots, \theta_n \in \mathrm{Con}\,\mathbf{A}$. Suppose that $\bigcap_{i=1}^{n} \theta_i = 0_A$ and, for every $i \leq n$, $\theta_i \circ \bigcap_{j\neq i} \theta_j = 1_A$. Prove that $\mathbf{A} \cong \prod_{i=1}^{n} \mathbf{A}/\theta_i$.

6. Let $\mathbf{A} = \langle A, \cdot \rangle$ be an algebra with a single binary operation. $\mathbf{A}$ is called a *left-zero semigroup* if it satisfies the law $x \cdot y \approx x$. Similarly, $\mathbf{A}$ is called a right-zero semigroup if it satisfies $x \cdot y \approx y$. Finally, $\mathbf{A}$ is called a *rectangular band* if it satisfies the laws

$$(x \cdot y) \cdot z \approx x \cdot (y \cdot z), \qquad x \cdot x \approx x, \qquad (x \cdot y) \cdot z \approx x \cdot z.$$

   (a) Let $\mathbf{A}$ be a rectangular band. Define

$$\theta = \{ (x, y) \in A^2 : (\forall z \in A)\; x \cdot z = y \cdot z \}.$$

   Prove that θ is a congruence on $\mathbf{A}$ and that $\mathbf{A}/\theta$ is a left-zero semigroup.

(b) Prove that $\mathbf{A}$ is a rectangular band if and only if $\mathbf{A} \cong \mathbf{L} \times \mathbf{R}$ for some left-zero semigroup $\mathbf{L}$, and right-zero semigroup $\mathbf{R}$.

7. Let $2 = \{0, 1\}$, '+' denote addition modulo 2, $f(x) = x$ and $g(x) = x+1$. Consider the following two algebras of type $\langle 2, 1 \rangle$. $\mathbf{A} = \langle 2, +, f \rangle$ and $\mathbf{B} = \langle 2, +, g \rangle$.

   (a) Let $d\colon \mathbf{B}^2 \to \mathbf{A}$ by $d(x, y) = x + y$. Show that d is a surjective homomorphism. Consequently $\mathbf{B}^2 / \ker(d) \cong \mathbf{A}$.

   (b) Let $\delta = \ker(d)$. Show that $\{\eta_1, \eta_2\}$, $\{\eta_1, \delta\}$ and $\{\eta_2, \delta\}$ each form a pair of complementary factor congruences on $\mathbf{B} \times \mathbf{B}$.

   (c) Prove that $\mathbf{B} \times \mathbf{B} \cong \mathbf{B} \times \mathbf{A} \ncong \mathbf{A} \times \mathbf{A}$.

3.3 Subdirect products

Direct decompositions, when they exist, are an extremely powerful tool for the analysis of an algebra. Unfortunately, more often than not, a direct decomposition is not available to us. Furthermore, precisely because direct decomposition is so strong, the property of being directly indecomposable is rather weak.

We seek a more generous decomposition that can be applied without restrictions. Birkhoff (1944) proposed the following notion which has turned out to be a useful tool throughout universal algebra.

Definition 3.16. An algebra $\mathbf{B}$ is a *subdirect product* of $\langle \mathbf{A}_i : i \in I \rangle$ if $\mathbf{B}$ is a subalgebra of $\prod_{i \in I} \mathbf{A}_i$ and for every $i \in I$, $p_i{\restriction}_B\colon \mathbf{B} \to \mathbf{A}_i$ is surjective.

An embedding $g\colon \mathbf{B} \to \prod_{i \in I} \mathbf{A}_i$ is called *subdirect* if $\vec{g}(\mathbf{B})$ is a subdirect product of $\langle \mathbf{A}_i : i \in I \rangle$. g is also called a *subdirect representation* of $\mathbf{B}$.

As an example, Corollary 2.45 says precisely that every distributive lattice has a subdirect representation by copies of $\mathbf{2}$.

Proposition 3.17. *Let $\mathbf{A}$ be an algebra and let θ_i be a congruence on $\mathbf{A}$ for every $i \in I$. If $\bigcap_{i \in I} \theta_i = 0_A$ then the natural map $\mathbf{A} \to \prod_{i \in I} \mathbf{A}/\theta_i$ is a subdirect embedding. Conversely, if $g\colon \mathbf{A} \to \prod \mathbf{B}_i$ is a subdirect embedding, then with $\theta_i = \ker(p_i \circ g)$, we have $\bigcap \theta_i = 0_A$ and $\mathbf{A}/\theta_i \cong \mathbf{B}_i$*

Proof. Let $h_i\colon \mathbf{A} \to \mathbf{A}/\theta_i$ be the natural map, for each $i \in I$, and $h = \prod h_i$. Note that each h_i is surjective. From Proposition 3.14,

$$\ker h = \bigcap \ker h_i = \bigcap \theta_i = 0$$

by assumption, so h is an embedding. The projection $p_i\colon \prod A/\theta_j \to \mathbf{A}/\theta_i$ has the property that $p_i \circ h = h_i$. Since h_i is surjective, so is p_i.

For the converse, since g is a subdirect embedding, each $p_i \circ g$ is surjective so $\mathbf{A}/\theta_i \cong \mathbf{B}_i$. Since $\ker g = \bigcap \theta_i$, the intersection must be 0. □

For example, let P denote the set of prime integers, $\mathbb{Z}$ the additive group of integers and, for $p \in P$, θ_p denote "congruence modulo p." Since $\bigcap_{p\in P} \theta_p = 0$, we obtain a subdirect embedding $\mathbb{Z} \rightarrowtail \prod \mathbb{Z}/\theta_p \cong \prod \mathbb{Z}_p$ the product over all cyclic groups of prime order.

In general, we hope to start with an algebra, find a subdirect representation, and then, perhaps find subdirect representations of each of the factors, repeating the process until we can do no further reductions. We need to define precisely what we mean.

Definition 3.18. A nontrivial algebra $\mathbf{A}$ is called *subdirectly irreducible* if for every subdirect embedding $h\colon \mathbf{A} \to \prod_{i\in I} \mathbf{A}_i$, there is a $j \in I$ such that $p_j \circ h\colon \mathbf{A} \to \mathbf{A}_j$ is an isomorphism

Example 3.19. The only subdirectly irreducible distributive lattice is the two-element chain. This is a consequence of Corollary 2.45.

Example 3.20. A finite Abelian group is subdirectly irreducible if and only if it is cyclic of prime-power order. The "only if" direction follows from the structure theory for finite Abelian groups. The converse is a consequence of Theorem 3.23.

As we now show, subdirect irreducibility is really a property of the congruence lattice of an algebra. Once that observation is in hand, the existence of subdirect representations by subdirectly irreducibles is quite easy.

Definition 3.21. Let $\mathbf{L}$ be a complete lattice. An element a is called *meet-irreducible* if $a = b \wedge c$ implies $a = b$ or $a = c$. The element a is *completely meet-irreducible* if $a \neq 1_L$ and whenever $a = \bigwedge_{i\in I} b_i$, there is a $j \in I$ such that $a = b_j$.

Lemma 3.22. *Suppose that a is an element of a complete lattice $\mathbf{L}$. The following are equivalent.*

(a) *a is completely meet-irreducible.*

(b) *There is an element $c \in L$ such that $a < c$ and for every $x \in L$, $a < x$ implies $c \leq x$.*

Proof. Let $a \in L$, $X = \{x \in L : x > a\}$ and $c = \bigwedge X$. If a is completely meet-irreducible, then we must have $a < c$. For any $x \in L$, $a < x$ implies $x \in X$ so $c \leq x$ since c is the infimum of X.

Conversely, suppose the element c exists as in (b). Assume, by way of contradiction, that $a = \bigwedge X$ with $a \notin X$. Then by the assumption on c, $c \leqslant\!\!\leqslant X$ so $a < c \leq \bigwedge X$ which is false. □

When a is completely meet-irreducible, the element c guaranteed by the above lemma is called the *upper cover* of a.

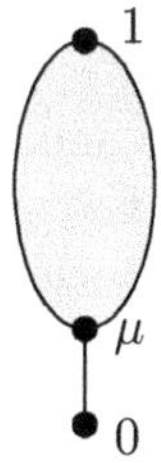

FIGURE 3.5: Con **A** for a subdirectly irreducible algebra **A**

Theorem 3.23. *An algebra* **A** *is subdirectly irreducible if and only if* 0_A *is completely meet-irreducible in* Con **A**. *More generally, if* θ *is a congruence on an algebra* **A**, *then* $\mathbf{A}/\theta$ *is subdirectly irreducible if and only if* θ *is completely meet-irreducible in* Con **A**.

Proof. Suppose first that **A** is subdirectly irreducible. To show that 0_A is completely meet-irreducible, assume that $0_A = \bigwedge_i \theta_i$ where $\theta_i \in \operatorname{Con}\mathbf{A}$, for $i \in I$. We must show that some $\theta_i = 0$. By Proposition 3.17 the natural map $\mathbf{A} \to \prod_i \mathbf{A}/\theta_i$ is a subdirect representation. Therefore, for some $i \in I$, the projection $p_i\colon \mathbf{A} \to \mathbf{A}/\theta_i$ is an isomorphism. Thus $0_A = \ker p_i = \theta_i$, showing that 0_A is completely meet-irreducible.

Conversely, assume that 0_A is completely meet-irreducible and suppose that $g\colon \mathbf{A} \to \prod_{i\in I} \mathbf{B}_i$ is a subdirect representation. By Proposition 3.17 again we obtain $0_A = \bigcap_{i\in I} \theta_i$ for some family of congruences. By complete meet-irreducibility, there is an $i \in I$ such that $\theta_i = 0$. Hence the map $p_i \circ g\colon \mathbf{A} \to \mathbf{B}_i$ is an isomorphism.

For the last statement, let θ be an arbitrary congruence. By the correspondence theorem (Theorem 3.6) the lattices $\mathbf{I}[\theta, 1_A]$ and $\operatorname{Con}(\mathbf{A}/\theta)$ are isomorphic. Thus $\mathbf{A}/\theta$ is subdirectly irreducible iff $0_{A/\theta}$ is completely meet-irreducible in $\operatorname{Con}(\mathbf{A}/\theta)$ iff θ is completely meet-irreducible in Con **A**. □

If **A** is subdirectly irreducible, then according to Lemma 3.22 and Theorem 3.23, there is a least nonzero congruence, μ, on **A**. This congruence is called the *monolith* of **A**. Pick any pair (a, b) in μ with $a \neq b$. Then by the minimality of μ, $\mu = \operatorname{Cg}^{\mathbf{A}}(a, b)$. We say that **A** is (a, b)-*irreducible* when we want to specify a pair of elements that make this so. Figure 3.5 is an illustration of the general form of the congruence lattice of a subdirectly irreducible algebra.

Notice that for a nontrivial algebra, we have the following sequence of implications:

simple $\Longrightarrow$ subdirectly irreducible $\Longrightarrow$ directly indecomposable.

Some examples of subdirectly irreducible algebras.

(1) For $n \geq 5$ the alternating and symmetric groups on n letters are subdirectly irreducible. $\mathbf{A}_n$ because it is simple, $\mathbf{S}_n$ because its normal subgroup lattice is a 3-element chain—with A_n as the monolith.

(2) The group (also the ring) $\mathbb{Z}_{p^k}$ is subdirectly irreducible (p a prime, $k > 0$) since its congruence lattice is a chain of length $k + 1$.

(3) According to Lemma 2.9, the lattice $\mathbf{M}_3$ is simple, hence subdirectly irreducible.

(4) Exercise 2.12.3 tells us that the lattice $\mathbf{N}_5$ is (a, b)-irreducible (in the notation of that exercise).

This brings us to the Subdirect Representation Theorem.

Theorem 3.24 (Birkhoff, 1944). *Every nontrivial algebra is isomorphic to a subdirect product of subdirectly irreducible algebras.*

Proof. We give one proof here. Another is outlined in Exercise 3.26.8. Let $\mathbf{A}$ be a nontrivial algebra and set $I = \{\{a, b\} : a, b \in A, a \neq b\}$. For each $\{a, b\} \in I$ use Zorn's lemma to find a congruence $\theta_{a,b}$ maximal with respect to the exclusion of (a, b). This congruence is completely meet-irreducible. For if $\theta_{a,b} = \bigcap \Psi$ then, since $(a, b) \notin \theta_{a,b}$, there is a $\psi \in \Psi$ with $(a, b) \notin \psi$. Since $\theta_{a,b} \leq \psi$, the maximality of $\theta_{a,b}$ ensures that $\theta_{a,b} = \psi \in \Psi$.

Since $\theta_{a,b}$ is completely meet-irreducible, $\mathbf{A}/\theta_{a,b}$ is subdirectly irreducible. From the definition of these congruences we obtain

$$\bigcap_{\{a,b\} \in I} \theta_{a,b} = 0_A$$

yielding, by Proposition 3.17, a subdirect representation of $\mathbf{A}$ by subdirectly irreducible algebras. □

Notice that the subdirect factors obtained in the Representation Theorem are quotients of the represented algebra. If $\mathbf{A}$ is finite, then all of its quotients are finite, as is the index set I. Thus we have the following corollary.

Corollary 3.25. *Every finite algebra is isomorphic to a subdirect product of a finite number of finite subdirectly irreducible algebras.*

Exercise Set 3.26.

1. Represent the three-element chain as a subdirect product of subdirectly irreducible lattices.

2. Review Exercise 1.26.5 on the Rees congruence. Recall that a semilattice is a semigroup that is commutative and idempotent.

(a) Let $\mathbf{S}$ be a semilattice and a an element of S. Prove that $aS = \{\, a \cdot s : s \in S \,\}$ is an ideal of $\mathbf{S}$. Describe aS in terms of the ordering on $\mathbf{S}$.

(b) Show that the only subdirectly irreducible semilattice is the two-element chain.

3. Let p be a prime. Prove that the Abelian group $\mathbb{Z}$ has a subdirect representation by the groups $\mathbb{Z}_{p^k}$ as k ranges over the natural numbers.

4. Let $\mathbf{A}$ be an algebra and $a, b \in A$. Prove that $\mathbf{A}$ is (a, b)-irreducible if and only if: for every homomorphism f with domain $\mathbf{A}$, f is injective iff $f(a) \neq f(b)$.

5. Let $\mathbf{R}$ be a commutative ring with identity, and assume that $\mathbf{R}$ has no nilpotent elements.

   (a) Let a be a nonzero element of $\mathbf{R}$. Prove that $\mathbf{R}$ has a prime ideal P_a with $a \notin P_a$. (Hint: Find an ideal maximal with respect to the exclusion of a, a^2, a^3,...)

   (b) Prove that $\mathbf{R}$ is a subdirect product of integral domains.

6. Let p be a prime. Prove that a nonabelian group of order p^3 is subdirectly irreducible.

7. Let $\mathbb{Q}$ denote the additive group of rational numbers, and let p be a prime integer.

   (a) Let $\mathbb{Q}_p = \{\, a/p^k : a \in \mathbb{Z} \ \& \ k \geq 0 \,\}$. Prove that $\mathbb{Q}_p$ is a subgroup of $\mathbb{Q}$, and $\mathbb{Z}$ is a subgroup of $\mathbb{Q}_p$.

   (b) Let $\mathbb{Z}(p^\infty) = \mathbb{Q}_p/\mathbb{Z}$. Prove that every element of $\mathbb{Z}(p^\infty)$ has finite order. Denote the element $x + \mathbb{Z}$ of $\mathbb{Z}(p^\infty)$ by $\overline{x}$. Groups of the form $\mathbb{Z}(p^\infty)$ are called *quasi-cyclic.*

   (c) Let H be a subgroup of $\mathbb{Z}(p^\infty)$. Prove that if there is an upper bound on the orders of the elements of H, then H is the cyclic subgroup generated by $\overline{1/p^k}$ for some $k \geq 0$.

   (d) Prove that if a subgroup H has no upper bound on the orders of its elements, then $H = \mathbb{Z}(p^\infty)$.

   (e) Prove that $\mathbb{Z}(p^\infty)$ is subdirectly irreducible. Show that the lattice of subgroups forms an infinite chain: $(0) = H_0 < H_1 < H_2 < \cdots < \mathbb{Z}(p^\infty)$, and that $\mathbb{Z}(p^\infty)/H_k \cong \mathbb{Z}(p^\infty)$ for every $k \geq 0$.

8. Let $\mathbf{L}$ be an algebraic lattice.

   (a) Prove that for every $a < b$ in L, there is a completely meet-irreducible element $m \in L$ such that $a \leq m$ and $b \nleq m$.

(b) Prove that every element of **L** is equal to a meet of completely meet-irreducible elements.

(c) Use part (b) to give another proof of Theorem 3.24.

9. An embedding $h: \mathbf{A} \to \mathbf{B}$ is called *essential* if for every congruence θ on $\mathbf{B}$, $\theta \neq 0_B \implies \overleftarrow{h}(\theta) \neq 0_A$. We often call $\mathbf{B}$ an *essential extension* of $\mathbf{A}$ in this case.

(a) Let $h: \mathbf{A} \to \mathbf{B}$ be an arbitrary embedding. Prove that there is a congruence ψ on $\mathbf{B}$ such that $q_\psi \circ h$ is an essential embedding of $\mathbf{A}$ into $\mathbf{B}/\psi$.

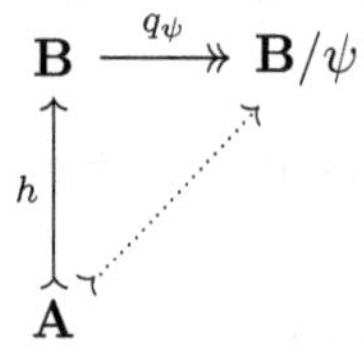

(b) Prove that an essential extension of a subdirectly irreducible algebra is subdirectly irreducible.

3.4 Case studies

Now that we have some of the building blocks of the subject under our belt, we pause to develop several examples that we can use to illustrate the constructions that we have considered so far, as well as those that are to come.

In this section, our interest is in subdirectly irreducible algebras. We already know about distributive lattices and Boolean algebras (Example 3.19). We shall characterize the subdirectly irreducible algebras in several varieties: certain varieties of commutative rings, Abelian groups, and pseudocomplemented distributive lattices.

Definition 3.27. For an integer $n > 1$, $\mathcal{C}r_n$ denotes the class of commutative rings that satisfy the identity $x^n \approx x$.

The members of $\mathcal{C}r_n$ play a key role in the structure theory of rings. These classes also provide important examples in universal algebra.

Theorem 3.28. *Let n be an integer, $n > 1$. The subdirectly irreducible members of $\mathcal{C}r_n$ are those finite fields of order d, where $(d-1) \mid (n-1)$.*

Proof. Since its multiplicative group is cyclic, it is easy to see that a finite field of order d satisfies $x^n \approx x$ if and only if $d-1$ divides $n-1$. And since every field is simple (as a ring), it is certainly subdirectly irreducible.

Conversely, let $\mathbf{R}$ be a subdirectly irreducible member of $\mathcal{C}r_n$. We first show that $\mathbf{R}$ is simple.

Let x be any element of R, and let $e = x^{n-1}$. Then $e^2 = e$ since (if $n > 2$)

$$e^2 = (x^{n-1})^2 = x^{2n-2} = x^n \cdot x^{n-2} = x \cdot x^{n-2} = x^{n-1} = e.$$

Since $\mathbf{R}$ is subdirectly irreducible, it has a least nontrivial ideal M (Exercise 2.22.1). Pick $a \in M - \{0\}$, and let $e = a^{n-1}$. Since $ea = a \neq 0$, it follows that $e \in M - \{0\}$.

Let $I = \{ x \in R : xe = 0 \}$. Since $\mathbf{R}$ is commutative, I is an ideal of $\mathbf{R}$. Since $e^2 = e \neq 0$, we have $e \notin I$, hence $M \nsubseteq I$. But M is contained in every nonzero ideal of $\mathbf{R}$, so it follows that $I = \{0\}$. Now, for all $x \in R$,

$$(xe - x)e = xe - xe = 0 \implies xe - x = 0 \implies xe = x.$$

In other words, e is an identity element of $\mathbf{R}$. But e is an element of M, so $M = R$. Thus the only ideals of $\mathbf{R}$ are $\{0\}$ and R itself. Therefore $\mathbf{R}$ is simple.

But a simple commutative ring is a field. And since $\mathbf{R}$ satisfies the identity $x^n \approx x$, we deduce that every element of R is a root of the polynomial $x^n - x$. Therefore, $\mathbf{R}$ is a finite field of order, say, d, and as we remarked at the beginning of the proof, we must have $(d-1) \mid (n-1)$.

□

In fact, Theorem 3.28 is true without the assumption of commutativity. This is a famous theorem due to Jacobson (about 1945). It involves more ring theory than we care to get into right now. There is a proof in [MMT87, Corollary 4, page 177].

A characterization of all subdirectly irreducible rings, even the commutative ones, seems to be out of reach at the present time. The situation is similar for groups. However, we can provide a complete description of all subdirectly irreducible Abelian groups.

Theorem 3.29. *An Abelian group is subdirectly irreducible if and only if it is cyclic of prime-power order or quasi-cyclic.*

The *quasi-cyclic* groups are the groups $\mathbb{Z}(p^\infty)$ defined in Exercise 3.26.7.

Proof. Recall from 1.21 that the congruence lattice of a group is isomorphic to its lattice of normal subgroups. Furthermore, in an Abelian group, every subgroup is normal. Thus in our quest for subdirectly irreducible Abelian groups, we can focus on the lattice of subgroups instead of the lattice of congruences. Applying this observation together with Theorem 3.23, an Abelian group is subdirectly irreducible if and only if it has a least nontrivial subgroup, which we are justified in calling the monolith.

For any prime p and positive integer k, the cyclic group $\mathbb{Z}_{p^k}$ is subdirectly irreducible since its lattice of subgroups is a chain of length $k+1$, so it has a monolith, namely the unique subgroup of order p. According to Exercise 3.26.7, the group $\mathbb{Z}(p^\infty)$ is subdirectly irreducible (since its subgroup lattice forms a chain isomorphic to ω).

Now we consider the converse. From the classical structure theory, every finite Abelian group is a direct (hence subdirect) product of Abelian groups of prime-power order. Thus every finite subdirectly irreducible Abelian group is of the form $\mathbb{Z}_{p^k}$. So let $\mathbf{A}$ be an infinite subdirectly irreducible Abelian group. We must show that $\mathbf{A} \cong \mathbb{Z}(p^\infty)$ for some prime p. Let M denote the monolith of $\mathbf{A}$. We structure the proof as a series of observations.

Claim 1. For some prime p, $\mathbf{M} \cong \mathbb{Z}_p$.
Proof. As the monolith of $\mathbf{A}$, $\mathbf{M}$ can have no proper nontrivial subgroups. From elementary group theory we know that $\mathbf{M} \cong \mathbb{Z}_p$ for some prime p.

Claim 2. Every nontrivial subgroup of $\mathbf{A}$ is subdirectly irreducible with monolith M.
Proof. Let B be a nontrivial subgroup of $\mathbf{A}$, and let C be a nontrivial subgroup of B. Since M is the monolith of A, $M \subseteq C \subseteq B$. Thus M is also the monolith of B.

Claim 3. Every element of A has finite order.
Proof. Let $a \in A$, $a \neq e$ and $B = \mathrm{Sg}^{\mathbf{A}}(a)$. By Claim 2, $\mathbf{B}$ is subdirectly irreducible. Since $\mathbb{Z}$ is not subdirectly irreducible, (see the example following Theorem 3.17) B must be finite.

Claim 4. Every finite subgroup of $\mathbf{A}$ is isomorphic to p^k for some integer k.
Proof. let B be a finite, nontrivial subgroup. By Claim 2 $\mathbf{B}$ is subdirectly irreducible. So by our earlier observation, $\mathbf{B} \cong \mathbb{Z}_{q^k}$ for some prime q. Since $M \subseteq B$, we must have, by Lagrange's theorem, $p = q$.

Claim 5. For any subgroups $\mathbf{B}$ and $\mathbf{C}$ of $\mathbf{A}$, either $B \subseteq C$ or $C \subseteq B$.
Proof. Suppose not. Let $b \in B - C$ and $c \in C - B$. Let $D = \mathrm{Sg}^{\mathbf{A}}\{b, c\}$. By Claim 3, b and c have finite order, so D is finite. Therefore $\mathbf{D} \cong \mathbb{Z}_{p^k}$ by Claim 4, so $\mathrm{Sub}(\mathbf{D})$ is a chain. Therefore either $B \cap D \subseteq C \cap D$ or vice versa. This is impossible due to our choice of b and c.

Claim 6. Every proper subgroup of $\mathbf{A}$ is finite.
Proof. Let $\mathbf{B}$ be an infinite subgroup of $\mathbf{A}$. For any $a \in A$, the order of a is finite, so certainly $B \not\subseteq \mathrm{Sg}(a)$. Therefore by Claim 5, $a \in B$. Thus $B = A$.

Claim 7. $\mathrm{Sub}(\mathbf{A})$ forms a chain $B_0 \subset B_1 \subset B_2 \subset B_3 \subset \cdots \subseteq A$, in which every $\mathbf{B}_k \cong \mathbb{Z}_{p^k}$.
Proof. By Claim 5, $\mathrm{Sub}(\mathbf{A})$ forms a chain. By Claims 4 and 6, every proper subgroup must be of the form $\mathbb{Z}_{p^k}$. Suppose there were a largest proper subgroup B_k. Choose $a \in A - B_k$. Since $\mathrm{Sg}(a)$ is finite, we must have $a \in B_k$, which is impossible.

Now we shall define, by recursion, embeddings $f_k\colon \mathbf{B}_k \to \mathbb{Z}(p^\infty)$, such that f_{k+1} extends f_k, for $k \in \omega$. We have $B_0 = \{0\}$, so define $f_0(0) = \bar{0}$. Assume f_k is defined. Since $\mathbf{B}_k$ is cyclic of order p^k and f_k is an embedding there is $b \in B_k$ such that $B_k = \mathrm{Sg}(b)$ and $f_k(b) = \overline{1/p^k}$. From Claim 7

$\mathbf{B}_k < \mathbf{B}_{k+1} \cong \mathbb{Z}_{p^{k+1}}$, so there is a generator c of $\mathbf{B}_{k+1}$ such that $pc = b$. Thus we define, for $0 \leq n < p^{k+1}$, $f_{k+1}(nc) = \overline{n/p^{k+1}}$. This defines the embedding f_{k+1} and, since $f_{k+1}(b) = f_{k+1}(pc) = \overline{p/p^{k+1}} = f_k(b)$, f_{k+1} extends f_k. Finally define $f = \bigcup_k f_k$. This map is easily seen to be an isomorphism of $\mathbf{A}$ with $\mathbb{Z}(p^\infty)$. □

We are now going to develop the theory of an algebraic structure that is not typically part of a first-year algebra course. It forms a nice contrast to the Abelian group and commutative ring examples.

Recall that a *bounded distributive lattice* is an algebra $\langle L, \wedge, \vee, 0, 1\rangle$ satisfying the following identities:

$$\begin{aligned}
&x \wedge (y \wedge z) \approx (x \wedge y) \wedge z, && x \vee (y \vee z) \approx (x \vee y) \vee z,\\
&x \wedge y \approx y \wedge x, && x \vee y \approx y \vee x,\\
&x \wedge x \approx x, && x \vee x \approx x,\\
&x \wedge (x \vee y) \approx x, && x \vee (x \wedge y) \approx x,\\
&x \wedge 0 \approx 0, && x \vee 1 \approx 1,\\
&x \wedge (y \vee z) \approx (x \wedge y) \vee (x \wedge z).
\end{aligned} \tag{3–2}$$

There are, of course, many variations on this definition. For example, the idempotent laws can simply be dropped, and the two laws defining the bounds can be replaced with: $x \vee 0 \approx x \approx x \wedge 1$.

Definition 3.30. Let $\mathbf{L}$ be a bounded distributive lattice, $a, b \in L$. b is called the *pseudocomplement* of a if, for every x in L, $x \wedge a = 0 \iff x \leq b$. (That is, b is the largest element of L disjoint from a.)

Finally, a bounded distributive lattice is called *pseudocomplemented* if every element has a pseudocomplement. In such a lattice, we denote the pseudocomplement of a by a^*.

Example 3.31. (1) Every complemented distributive lattice is pseudocomplemented.

(2) Every finite distributive lattice is pseudocomplemented. Let $\mathbf{L}$ be such a lattice and let $a \in L$. Then

$$a^* = \bigvee_{x \wedge a = 0} x.$$

By distributivity

$$a^* \wedge a = \bigvee_{x \wedge a = 0} (x \wedge a) = 0.$$

(3) Let X be a set, and Ω a topology on X. Then $\mathbf{\Omega} = \langle \Omega, \cap, \cup, \varnothing, X\rangle$ forms the bounded distributive lattice of open subsets. For $A \in \Omega$, $(X - A)^\circ = \bigcup\{B : B \subseteq X - A,\ B \in \Omega\}$ (the interior of the complement of A) is the pseudocomplement of A in $\mathbf{\Omega}$.

(4) Let **C** be a bounded chain, (with bounds 0 and 1, as usual). Then **C** is pseudocomplemented. The pseudocomplement of 0 is 1, and the pseudocomplement of any other element is 0.

(5) Let **A** be any distributive lattice with smallest element 0, and let $\mathbf{L} = \mathbf{Idl}(\mathbf{A})$ (the lattice of ideals of **A**). By Exercise 2.47.4 **L** is distributive and bounded. For $I \in L$, the pseudocomplement of I is

$$I^* = \{\, x \in A : (\forall y \in I)\ x \wedge y = 0 \,\}.$$

(6) For any algebra **A**, if $\mathbf{Con}(\mathbf{A})$ is a distributive lattice, then it is pseudocomplemented. (Combine Example 2 with Exercise 2.34.3.)

(7) Let $L = \mathrm{Sb}_\omega(\omega) \cup \{\omega\}$. Then L is a subuniverse of the pseudocomplemented distributive lattice $\mathbf{Sb}(\omega)$. However, no element of L besides $\varnothing$ and ω has a pseudocomplement in L. Thus **L** is not pseudocomplemented.

As the last example shows, the class of bounded distributive lattices that are pseudocomplemented is not closed under the formation of sublattices. Consider also the lattice **L**, whose Hasse diagram is

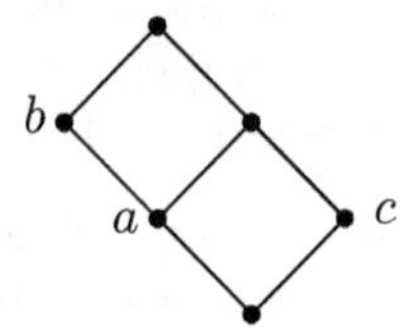

The pseudocomplement of c is b. But in the sublattice $L - \{b\}$ (which is also pseudocomplemented), $c^* = a$.

In universal algebra, the standard way to avoid these pathologies is to add additional operations to the type. In this case, we add a single unary operation which returns the pseudocomplement of an element.

Definition 3.32. A *p-algebra* is an algebra $\langle L, \wedge, \vee, {}^*, 0, 1\rangle$ of similarity type $\langle 2, 2, 1, 0, 0\rangle$ satisfying all of the identities (3–2), and further, the condition:

$$(\forall a)\,(\forall x) \quad x \wedge a = 0 \iff x \leq a^*. \tag{3–3}$$

$\mathcal{Pa}$ denotes the class of all p-algebras.

In the literature, the algebraic structure we have defined is generally called a distributive p-algebra (since the lattice is distributive), or frequently, a pseudocomplemented distributive lattice. We will just call them p-algebras since we won't be considering anything more general than that.

Note that every Boolean algebra is a p-algebra (by Example 3.31.1). As the other examples show, the class $\mathcal{Pa}$ is strictly larger than the class $\mathcal{Ba}$ of Boolean algebras.

We begin with some properties that hold in any p-algebra.

Lemma 3.33. *Let* $\mathbf{L}$ *be a p-algebra,* $x, y \in L$.

(1) $0^* = 1,\ 1^* = 0$;

(2) $x \leq x^{**}$;

(3) $x \leq y \implies x^* \geq y^*$;

(4) $x^* = x^{***}$;

(5) $x \wedge y = 0 \implies x^{**} \wedge y = 0$;

(6) $(x \vee y)^* = x^* \wedge y^*$;

(7) $(x \wedge y)^{**} = x^{**} \wedge y^{**}$.

Proof. (1–4) are left as exercises. For (5), $x \wedge y = 0 \implies x \leq y^* \implies$ (by (3, 4)) $x^{**} \leq y^* \implies x^{**} \wedge y = 0$.

For (6), let $z = x^* \wedge y^*$. Then

$z \leq x^*,\ z \leq y^*$;	
$z \wedge x = z \wedge y = 0$;	definition of pseudocomplement;
$z \wedge (x \vee y) = 0$	by distributivity;
$z \leq (x \vee y)^*$	definition of pseudocomplement.

Conversely,

$$x \vee y \geq x, y \implies (x \vee y)^* \leq x^*, y^* \implies (x \vee y)^* \leq x^* \wedge y^*.$$

Finally, for (7), $x \wedge y \leq x, y \implies (x \wedge y)^{**} \leq x^{**} \wedge y^{**}$, while $x \wedge y \wedge (x \wedge y)^* = 0$ implies (by (5)) $x^{**} \wedge y \wedge (x \wedge y)^* = 0$, thus by (5) again $x^{**} \wedge y^{**} \wedge (x \wedge y)^* = 0$, so $x^{**} \wedge y^{**} \leq (x \wedge y)^{**}$. □

Recall that a variety of algebras is a class of algebras closed under the formation of homomorphic images, subalgebras and products. Unlike most of the classes we have studied, it is not obvious whether $\mathcal{Pa}$ is a variety, since it is not defined by a set of equations. Nevertheless, it is in fact a variety, as the next theorem demonstrates.

Theorem 3.34. $\mathcal{Pa}$ *is a variety of algebras.*

Proof. As all of the conditions in (3–2) are equational, they are preserved by H, S and P. It remains to prove that condition (3–3) is preserved by these operators. Let $\mathbf{L} \in \mathcal{Pa}$ and let $\mathbf{A}$ be a subalgebra of $\mathbf{L}$. Suppose x and a are elements of A. Then $x \wedge a = 0$ (in $\mathbf{A}$) iff $x \wedge a = 0$ (in $\mathbf{L}$) iff $x \leq_{\mathbf{L}} a^*$ (since $\mathbf{L}$ is a p-algebra) iff $x \leq_{\mathbf{A}} a^*$, (since $\mathbf{A}$ is a subalgebra of $\mathbf{L}$, a^* is the same element in both $\mathbf{A}$ and $\mathbf{L}$). Therefore, $\mathsf{S}(\mathcal{Pa}) = \mathcal{Pa}$.

Secondly, suppose $\mathbf{L}_i \in \mathcal{Pa}$, for $i \in I$. Let $\mathbf{A} = \prod_{i \in I} \mathbf{L}_i$. Then for $x, a \in A$, $x \wedge a = 0 \iff (\forall i \in I)\ x_i \wedge a_i = 0 \iff (\forall i \in I)\ x_i \leq a_i^* \iff x \leq a^*$.

Finally, we check that $\mathcal{Pa}$ is closed under homomorphic images. Let $\mathbf{L} \in \mathcal{Pa}$ and let $h\colon \mathbf{L} \to \mathbf{A}$ be a surjective homomorphism. (This implies that h preserves $\wedge, \vee, {}^*, 0$ and 1.) Then $h(0_L) = 0_A$, and $1_A = h(1_L) = h(0_L^*) = h(0_L)^* = 0_A^*$. Suppose $x \wedge a = 0$ in $\mathbf{A}$. Then $(x \wedge a)^* = 0_{\mathbf{A}}^* = 1_{\mathbf{A}} = h(1_{\mathbf{L}})$. Since h is surjective, there are $\bar{x}, \bar{a} \in L$ such that $h(\bar{x}) = x$ and $h(\bar{a}) = a$. We have $(\bar{x} \wedge \bar{a}) \wedge (\bar{x} \wedge \bar{a})^* = 0$ since $\mathbf{L} \in \mathcal{Pa}$. By associativity, $\big(\bar{x} \wedge (\bar{x} \wedge \bar{a})^*\big) \wedge \bar{a} = 0$, hence $\bar{x} \wedge (\bar{x} \wedge \bar{a})^* \leq \bar{a}^*$. Now, applying h,

$$h(\bar{x}) \wedge \big(h(\bar{x}) \wedge h(\bar{a})\big)^* \leq h(\bar{a})^*,$$

in other words, $x \wedge (x \wedge a)^* \leq a^*$. But $(x \wedge a)^* = 1$, so $x \leq a^*$.

Conversely, if $x \leq a^*$ (in $\mathbf{A}$), then $x \wedge a \leq a^* \wedge a = h(\bar{a})^* \wedge h(\bar{a}) = h(0_{\mathbf{L}}) = 0_{\mathbf{A}}$. □

In [Lak71], H. Lakser characterized all of the subdirectly irreducible members of $\mathcal{Pa}$. We present a modified version of his proof.

Definition 3.35. Let $\mathbf{L} \in \mathcal{Pa}$. Then $D(\mathbf{L}) = \{\, x \in L : x^* = 0 \,\}$. The elements of $D(\mathbf{L})$ are called *dense* elements of $\mathbf{L}$.

Observe that in the case that $\mathbf{L}$ is the p-algebra of open subsets of a topological space (see Example 3.31(3)), the dense elements are precisely the (topologically) dense open subsets.

Lemma 3.36. *For any $\mathbf{L} \in \mathcal{Pa}$, $D(\mathbf{L})$ is a filter of the distributive lattice $\mathbf{L}$, and $D(\mathbf{L}) \cup \{0\}$ is a sub-p-algebra of $\mathbf{L}$. Furthermore, for any $x \in L$, $x \vee x^*$ is dense.*

In the theory of p-algebras, there is a deep connection between the (p-algebra) congruences of $\mathbf{L}$, and (distributive lattice) congruences of $D(\mathbf{L})$. We do not need to develop the whole theory here. We prove one lemma that suffices for our purposes. It follows from Lemma 3.36 that $D(\mathbf{L})$ is a distributive lattice with 1.

Lemma 3.37. *Let $\mathbf{L}$ be a p-algebra and let θ be a distributive lattice congruence on $D(\mathbf{L})$. Define*

$$\tilde{\theta} = \{\, (x, y) \in L^2 : x^* = y^* \;\; \& \;\; (\forall u \in D(\mathbf{L}))\; (x \vee u)\; \theta\; (y \vee u) \,\}.$$

Then $\tilde{\theta}$ is a congruence on $\mathbf{L}$, and $\tilde{\theta} \cap D(\mathbf{L})^2 = \theta$.

Proof. $\tilde{\theta}$ is clearly an equivalence relation on L. To verify that it is a congruence, we need to check that it is preserved by all three operations of $\mathbf{L}$. I.e., assume that $x_1 \,\tilde{\theta}\, y_1$ and $x_2 \,\tilde{\theta}\, y_2$. We need to show that $(x_1 \vee x_2) \,\tilde{\theta}\, (y_1 \vee y_2)$, $(x_1 \wedge x_2) \,\tilde{\theta}\, (y_1 \wedge y_2)$ and $x_1^* \,\tilde{\theta}\, y_1^*$.

For the first of these,

$$(x_1 \vee x_2)^* = x_1^* \wedge x_2^* = y_1^* \wedge y_2^* = (y_1 \vee y_2)^*$$

by 3.33(6). Let $u \in D(\mathbf{L})$. Then

$$(x_1 \vee x_2) \vee u = (x_1 \vee u) \vee (x_2 \vee u) \mathrel{\theta} (y_1 \vee u) \vee (y_2 \vee u) = (y_1 \vee y_2) \vee u.$$

Observe that since $x_i \mathrel{\tilde{\theta}} y_i$, it follows that $x_i^* = y_i^*$, and therefore $x_1^* \mathrel{\tilde{\theta}} y_1^*$ since $\tilde{\theta}$ is reflexive.

Finally for meet,

$$(x_1 \wedge x_2)^* = (x_1 \wedge x_2)^{***} = (x_1^{**} \wedge x_2^{**})^* = (y_1^{**} \wedge y_2^{**})^* = (y_1 \wedge y_2)^*$$

using 3.33(7), and for $u \in D(\mathbf{L})$,

$$(x_1 \wedge x_2) \vee u = (x_1 \vee u) \wedge (x_2 \vee u) \mathrel{\theta} (y_1 \vee u) \wedge (y_2 \vee u) = (y_1 \wedge y_2) \vee u$$

as desired.

The claim that $\tilde{\theta} \cap D(\mathbf{L})^2 = \theta$ is easily verified. □

We are now in a position to describe the subdirectly irreducible p-algebras. Let $\mathbf{B} = \langle B, \wedge, \vee, ', 0, e\rangle$ be a Boolean algebra. (Note that the largest element of $\mathbf{B}$ is called e here.) Define a p-algebra $\overline{\mathbf{B}}$ with universe $B \cup \{1\}$ (where $1 \notin B$) and extend the ordering of $\mathbf{B}$ so that $1 > x$ for all $x \in B$. It is easy to verify that with this ordering $\overline{\mathbf{B}}$ will be a p-algebra in which

$$x^* = \begin{cases} 1 & \text{if } x = 0, \\ 0 & \text{if } x = 1, \\ x' & \text{otherwise.} \end{cases}$$

Also, for any natural number n, let $\mathbb{B}_n$ denote the (unique) Boolean algebra with n atoms. In other words, $\mathbb{B}_n$ will be isomorphic to the power set of an n-element set. Pictures of $\mathbb{B}_n$ and $\overline{\mathbb{B}}_n$ for $n < 4$ are included for your perusal (Figure 3.6). Notice that $\mathbb{B}_1 \cong \overline{\mathbb{B}}_0$.

Theorem 3.38. *Let* $\mathbf{L}$ *be a p-algebra. Then* $\mathbf{L}$ *is subdirectly irreducible iff* $\mathbf{L}$ *is isomorphic to the algebra* $\overline{\mathbf{B}}$ *for some Boolean algebra* $\mathbf{B}$.

Proof. Let $\mathbf{B}$ be a Boolean algebra. We first show that $\overline{\mathbf{B}}$ is subdirectly irreducible, in fact, $(e, 1)$-irreducible. Let $\theta \in \operatorname{Con}(\overline{\mathbf{B}})$, $\theta \neq 0$. We need to verify that $e \mathrel{\theta} 1$.

Claim. There is $c > 0$ such that $0 \mathrel{\theta} c$.
Proof. By assumption, $\theta \neq 0$, so there are $a < b$ in $\overline{\mathbf{B}}$, such that $a \mathrel{\theta} b$. We must have $a \neq 1$. If $a = 0$ then we can take $c = b$. So suppose $a \neq 0$. Then a^* is the complement of a in $\mathbf{B}$, so $0 = (a \wedge a^*) \mathrel{\theta} (b \wedge a^*) \neq 0$, so we take $c = b \wedge a^*$.

Applying the claim, $0 \mathrel{\theta} c \implies 1 = (0^* \vee e) \mathrel{\theta} (c^* \vee e) = e$.

Now for the converse. Let $\mathbf{L}$ be a subdirectly irreducible p-algebra, with monolith μ. Let $\nu = \mu \cap D(\mathbf{L})^2$. Then ν is a (lattice) congruence of $D(\mathbf{L})$. If $|L| = 2$, then $\mathbf{L} = \overline{\mathbb{B}}_0$. So we may assume that $L \supset \{0, 1\}$.

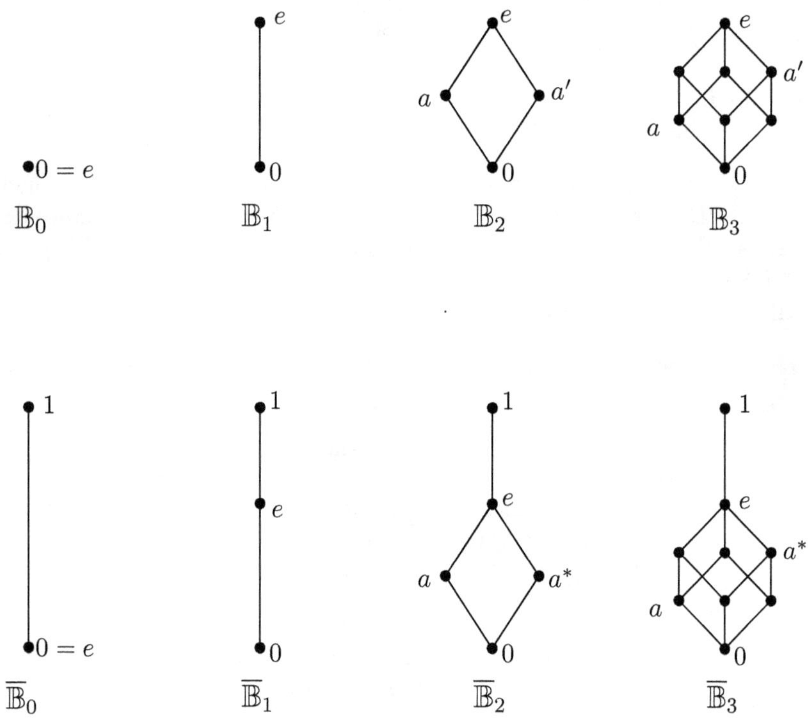

FIGURE 3.6: The first few subdirectly irreducible p-algebras

We first claim that if $x \in L - \{0,1\}$, then $x \vee x^* < 1$. For if $x \vee x^* = 1$, then x and x^* are complements. In that case, it follows as in Exercise 3.15.2 that

$$\begin{aligned} \alpha &= \{ (y,z) \in L^2 : x \vee y = x \vee z \} \\ \beta &= \{ (y,z) \in L^2 : x^* \vee y = x^* \vee z \} \end{aligned} \tag{3–4}$$

form a pair of nontrivial congruences of $\mathbf{L}$ such that $\alpha \cap \beta = 0$, contradicting the subdirect irreducibility of $\mathbf{L}$.

We now wish to show that $\nu \neq 0$ on $D(\mathbf{L})$. Since $\mu \neq 0$, there is $(a,b) \in \mu$ with $a < b$. There are two cases.

Case 1. $a = 0$.
Then $b \neq 0$ and we can assume that $b \neq 1$. (If $b = 1$, replace b with any third element of $\mathbf{L}$. We still have $(a,b) \in \mu$.) Since μ is a p-algebra congruence, $(a \vee a^*)\ \mu\ (b \vee b^*)$. But $a = 0$, so $a \vee a^* = 1$. From the previous claim, $1 \neq (b \vee b^*) \in D(\mathbf{L})$. Thus $\nu \neq 0$.

Case 2. $a \neq 0$.
Then $(a \vee a^*)\ \mu\ (b \vee a^*)$, and $a \vee a^* \in D(\mathbf{L})$. Since $a < b$, it follows that $b \vee a^*$ is dense as well. Therefore, $(a \vee a^*)\ \nu\ (b \vee a^*)$, so if these two elements are distinct, we have $\nu \neq 0$. On the other hand, if $a \vee a^* = b \vee a^*$, then $b \neq 1$ (since $a \vee a^* \neq 1$), so, since $\{0, a, b, a^*, a \vee a^*\}$ can not form an $\mathbf{N}_5$, $0 = a \wedge a^* < b \wedge a^*$. Now Case 1 applies since $0 = (a \wedge a^*)\ \mu\ (b \wedge a^*)$.

Let θ be any nontrivial congruence on $D(\mathbf{L})$. Then by Lemma 3.37, $\tilde{\theta}$ is a nontrivial congruence on $\mathbf{L}$, so $\tilde{\theta} \supseteq \mu$. Therefore, $\theta = \tilde{\theta} \cap D(\mathbf{L})^2 \supseteq \nu$. In other words, ν is the monolith of $D(\mathbf{L})$ (and $D(\mathbf{L})$ is a subdirectly irreducible distributive lattice). It follows from Example 3.19 that $D(\mathbf{L}) = \{e, 1\}$, for some element $e \neq 1$.

Let $B = L - \{1\}$. For any $x \in B - \{0\}$, $x \vee x^*$ is dense and can not equal 1, so we have $x \vee x^* = e$. If we now define $x' = x^*$ (for $x \neq 0$), and $0' = e$, then $\langle B, \wedge, \vee, ', 0, e\rangle$ is a Boolean algebra, and $\mathbf{L} = \overline{\mathbf{B}}$, as desired. □

So, in contrast to the situation for distributive lattices, there are infinitely many subdirectly irreducible p-algebras. However, as we will see later, they are pretty well organized. The four smallest subdirectly irreducible p-algebras are pictured in the bottom row of Figure 3.6

By the Subdirect Representation Theorem (Theorem 3.24), every p-algebra is a subdirect product of algebras of the form $\overline{\mathbf{B}}$. Let's look at a few very small examples. The two- and three-element chains are subdirectly irreducible. There are two 4-element distributive lattices. Each has a unique expansion to a p-algebra. One is the 4-element chain. This is isomorphic to a subdirect product of $\overline{\mathbb{B}}_1 \times \overline{\mathbb{B}}_1$ with universe

$$\{(0,0),\ (e,e),\ (e,1),\ (1,1)\}.$$

The other 4-element p-algebra is 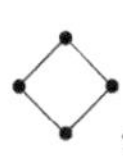, which is isomorphic to $\overline{\mathbb{B}}_0 \times \overline{\mathbb{B}}_0$.

As another example, consider the distributive lattice **L**:

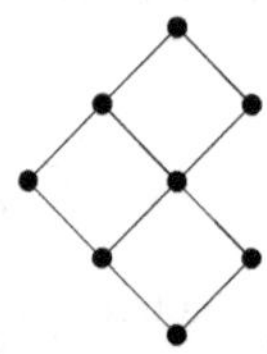

(Remember, every finite distributive lattice is pseudocomplemented, so it has a unique expansion to a p-algebra. Why not take a moment and label the points and their pseudocomplements in **L**.) This is clearly not of the form $\overline{\mathbf{B}}$ for a Boolean algebra **B**, so it must have a subdirect representation. Instead of explicitly describing a subdirect representation, we can work with congruences. In the diagram below, the classes of two congruences, α and β on **L** are illustrated. Convince yourself that these are indeed p-algebra congruences, that $\alpha \cap \beta = 0$ and that $\mathbf{L}/\alpha \cong \overline{\mathbb{B}}_2$ and $\mathbf{L}/\beta \cong \overline{\mathbb{B}}_1$. Describe the subdirect representation directly.

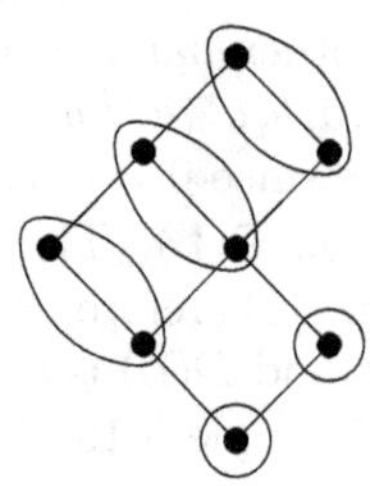

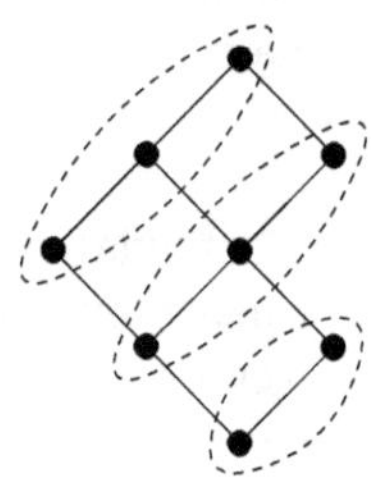

Exercise Set 3.39.

1. Let $\mathcal{V}$ be the variety of commutative rings satisfying the additional identity $x^2y \approx xy$. Prove that every subdirectly irreducible member of $\mathcal{V}$ is either a field of order 2 or a *zero-ring* (which means it satisfies the identity $xy \approx 0$).
   Hint: let **R** be subdirectly irreducible with minimal ideal M. Let $a \in M - \{0\}$.

   (a) Suppose first that the annihilator of a, ann(a), is (0). Then arguing as in Theorem 3.28, a is an identity element of **R**, and $\mathbf{R} \cong \mathbb{Z}_2$.

   (b) On the other hand, if ann$(a) \neq (0)$, then $a^2 = 0$, so $aR = (0)$. Now let $b \in R$. If $a = b^2c$ then $0 = ab = b^2c = a$, a contradiction. Therefore $(0) = bR = b^2R$.)

2. Verify the claim in Example 3.31.5 that the lattice of ideals of a lower-bounded distributive lattice is pseudocomplemented. Extend this result by dropping the assumption that the lattice **A** has a lower bound.

3. Prove Lemma 3.36. Also prove that $\mathbf{L}$ is a Boolean algebra if and only if $D(\mathbf{L}) = \{1\}$.

4. Although the algebras $\overline{\mathbf{B}}$ are subdirectly irreducible, their duals are not (except for $\overline{\mathbb{B}}_0$ and $\overline{\mathbb{B}}_1$)! Find a subdirect representation for $\overline{\mathbb{B}}_2^{\partial}$ and $\overline{\mathbb{B}}_3^{\partial}$.

5. "Wait a minute," I hear a student exclaim. "Theorem 3.38 shows that the three element chain is subdirectly irreducible. But we argued earlier in the course (cf. Exercise 3.26.1) that $\mathbf{C}_3$ was a subdirect product of two copies of $\mathbf{C}_2$, so it is not subdirectly irreducible (see the picture)." Can you explain the discrepancy?

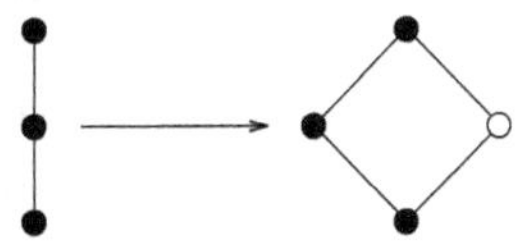

3.5 Varieties and other classes of algebras

Recall from Definition 1.10 the operators H, S and P that can be applied to a class $\mathcal{K}$ of similar algebras. That is, $\mathsf{H}(\mathcal{K})$, $\mathsf{S}(\mathcal{K})$ and $\mathsf{P}(\mathcal{K})$ represent the closure of the class $\mathcal{K}$ under the formation of homomorphic images, subalgebras and products, respectively. (In practice, $\mathcal{K}$ is usually not a set but rather a proper class. There are some set-theoretic difficulties inherent in any discussion of a class of classes. But putting those aside, we observe that each of these three are closure operators on the class of all classes of algebras of any particular similarity type.)

To be more precise, we require each of the classes $\mathsf{H}(\mathcal{K})$, $\mathsf{S}(\mathcal{K})$ and $\mathsf{P}(\mathcal{K})$ to be closed under isomorphisms. Thus $\mathsf{S}(\mathcal{K})$ consists of all algebras that can be embedded in a member of $\mathcal{K}$, and $\mathsf{P}(\mathcal{K})$ consists of all algebras that have a direct representation by members of $\mathcal{K}$. Furthermore, we define the "empty product" to be a trivial algebra, so that $\mathsf{P}(\mathcal{K})$ contains all trivial algebras. Let us add one more operator to our list.

Definition 3.40. For any class $\mathcal{K}$ of algebras, let $\mathsf{P_s}(\mathcal{K})$ denote the class of all algebras isomorphic to a subdirect product of members of $\mathcal{K}$.

These operators can be composed, forming yet more closure operators. If O_1 and O_2 are closure operators on classes, we can say that $\mathsf{O}_1 \leq \mathsf{O}_2$ if, for every class $\mathcal{K}$, we have $\mathsf{O}_1(\mathcal{K}) \subseteq \mathsf{O}_2(\mathcal{K})$. The following lemma, whose proof is left as an exercise, gives a few important relationships.

Lemma 3.41. $\mathsf{SH} \leq \mathsf{HS}$, $\mathsf{PS} \leq \mathsf{SP}$ *and* $\mathsf{PH} \leq \mathsf{HP}$. *Trivially,* $\mathsf{P_s} \leq \mathsf{SP}$.

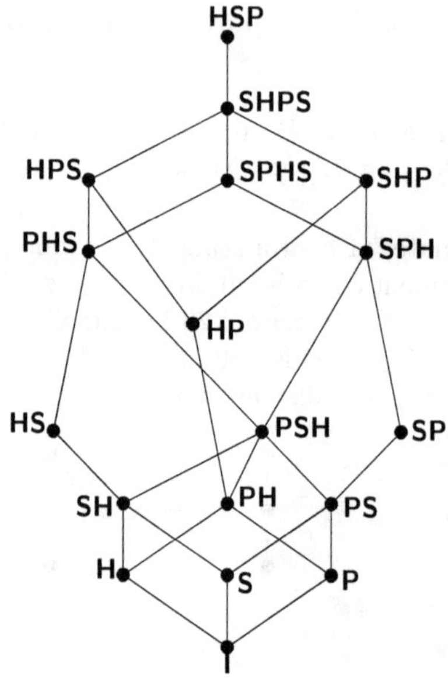

FIGURE 3.7: Operators generated by $\{\mathsf{H}, \mathsf{S}, \mathsf{P}\}$

In 1972, Pigozzi [Pig72] determined the entire poset of operators generated by $\{\mathsf{H}, \mathsf{S}, \mathsf{P}\}$, see Figure 3.7. Interestingly, it just barely fails to be a lattice ordering.

Now, a variety is a class closed under each of H, S and P. It is obvious, that for any fixed similarity type ρ, the class $(\mathbf{L}_{\mathsf{V}})_\rho$ of all varieties of type ρ is ordered by inclusion, and closed under arbitrary intersection. (Again, there are some set-theoretic difficulties here. Ignore them for the moment. Later, we will show how to circumvent them.) It follows that $(\mathbf{L}_{\mathsf{V}})_\rho$ is a complete lattice. There is a corresponding closure operator.

Definition 3.42. Let $\mathcal{K}$ be a class of similar algebras. The *variety generated by* $\mathcal{K}$, denoted $\mathsf{V}(\mathcal{K})$, is the smallest variety containing $\mathcal{K}$.

If $\mathcal{K} = \{\mathbf{A}_1, \mathbf{A}_2, \dots, \mathbf{A}_n\}$ is a finite set of algebras, then we often write $\mathsf{V}(\mathbf{A}_1, \dots, \mathbf{A}_n)$ instead of $\mathsf{V}(\mathcal{K})$. As usual, we would like to know how to construct $\mathsf{V}(\mathcal{K})$ directly from $\mathcal{K}$. Offhand it would seem as if one would have to iteratively apply the operators H, S and P infinitely many times to generate a variety. The following theorem of Birkhoff's [Bir35] shows that in fact, it is enough to apply each operator once, as long as one does it in the correct order.

Theorem 3.43. $\mathsf{V} = \mathsf{HSP}$.

Proof. Let $\mathcal{K}$ be some class of algebras. To see that $\mathsf{HSP}(\mathcal{K})$ is a variety, we use Lemma 3.41 to compute

$$\mathsf{H}(\mathsf{HSP}) = \mathsf{HSP}, \quad \mathsf{S}(\mathsf{HSP}) \leq \mathsf{HS}^2\mathsf{P} = \mathsf{HSP}, \quad \mathsf{P}(\mathsf{HSP}) \leq \mathsf{HSP}^2 = \mathsf{HSP}.$$

Thus $\mathsf{HSP} \geq \mathsf{V}$. On the other hand, with $\mathcal{V} = \mathsf{V}(\mathcal{K})$,

$$\mathsf{HSP}(\mathcal{K}) \subseteq \mathsf{HSP}(\mathcal{V}) = \mathcal{V} = \mathsf{V}(\mathcal{K})$$

so $\mathsf{HSP} \leq \mathsf{V}$. □

If $\mathcal{K}$ is any class of algebras, we let $\mathcal{K}_{\text{si}}$ denote the collection of subdirectly irreducible members of $\mathcal{K}$. The importance of subdirect irreducibility is captured in the following theorem and corollary.

Theorem 3.44. *Let $\mathcal{V}$ be a variety. Then $\mathcal{V}$ is exactly the class of all algebras isomorphic to subdirect products of subdirectly irreducible members of $\mathcal{V}$. In symbols: $\mathcal{V} = \mathsf{P_s}(\mathcal{V}_{\text{si}})$.*

Proof. Let $\mathbf{A}$ be an algebra in $\mathcal{V}$. By Theorem 3.24 $\mathbf{A}$ has a subdirect representation $\prod_{i\in I} \mathbf{A}_i$ by subdirect irreducibles. By assumption, for each $i \in I$ the projection map $p_i\colon \mathbf{A} \to \mathbf{A}_i$ is a surjective homomorphism. Therefore $\mathbf{A}_i \in \mathsf{H}(\mathbf{A}) \subseteq \mathcal{V}$. Thus $\mathbf{A} \in \mathsf{P_s}(\mathcal{V}_{\text{si}})$. This shows $\mathcal{V} \subseteq \mathsf{P_s}(\mathcal{V}_{\text{si}})$. For the converse, $\mathsf{P_s}(\mathcal{V}_{\text{si}}) \subseteq \mathsf{SP}(\mathcal{V}_{\text{si}}) \subseteq \mathsf{SP}(\mathcal{V}) \subseteq \mathcal{V}$. □

Corollary 3.45. *A variety is determined by its subdirectly irreducible members. More generally, for any two varieties $\mathcal{V}$ and $\mathcal{W}$ of the same similarity type, $\mathcal{V} \subseteq \mathcal{W}$ iff $\mathcal{V}_{\text{si}} \subseteq \mathcal{W}_{\text{si}}$.*

Example 3.46. The variety of distributive lattices is equal to $\mathsf{P_s}(\mathbf{2})$. In fact, if $\mathbf{L}$ is any nontrivial lattice, then $\mathsf{V}(\mathbf{L})$ contains the variety of all distributive lattices.

Proof. Let $\mathcal{D}$ denote the variety of distributive lattices. Since $\mathcal{D}_{\text{si}} = \{\mathbf{2}\}$ (up to isomorphism), the first sentence follows from Corollary 3.45. For the second, any nontrivial lattice contains $\mathbf{2}$ as a sublattice. Thus $\mathcal{D} = \mathsf{P_s}(\mathbf{2}) \subseteq \mathsf{P_s}\mathsf{S}(\mathbf{L}) \subseteq \mathsf{V}(\mathbf{L})$. □

Example 3.47. By definition, a p-algebra is a Boolean algebra if and only if it satisfies the identity $x \vee x^* \approx 1$. According to Lemma 3.33, every p-algebra satisfies one of the "DeMorgan laws", namely $(x \vee y)^* \approx x^* \wedge y^*$. What about the other one? Is it satisfied in every p-algebra?

Let $\mathcal{S}$ denote the class of p-algebras that satisfy the DeMorgan identity

$$(x \wedge y)^* \approx x^* \vee y^*. \qquad (\epsilon)$$

Since $\mathcal{P}a$ is a variety (Theorem 3.34) and identities are always preserved under HSP, it follows that $\mathcal{S}$ is a subvariety. By the corollary just above, varieties are determined by their subdirect irreducibles. So to determine whether $\mathcal{S} = \mathcal{P}a$

it is enough to check whether $\mathcal{S}_{si} = \mathcal{P}a_{si}$. By Theorem 3.38 this boils down to the question of whether $\overline{\mathbf{B}} \in \mathcal{S}$ for every Boolean algebra $\mathbf{B}$.

As you can easily check, if $\mathbf{B}$ is a Boolean algebra of cardinality greater than 2 then $\overline{\mathbf{B}}$ fails to satisfy (ϵ). Thus $\mathcal{S}_{si} = \{\overline{\mathbb{B}}_0, \overline{\mathbb{B}}_1\}$. So, far from being valid in every p-algebra, (ϵ) holds in only a small subset. On the other hand, since $\overline{\mathbb{B}}_1$ is not Boolean, (ϵ) does define a class slightly larger than the class of Boolean algebras. The class $\mathcal{S}$ is called the variety of *Stone algebras,* and has been extensively studied.

As another example, consider the identity $x \approx x^{**}$. Which p-algebras satisfy this identity? Well, which subdirectly irreducible p-algebras satisfy it? Notice that if $|B| > 1$, then $e^{**} = 1 \neq e$ in $\overline{\mathbf{B}}$. Thus the only subdirectly irreducible algebra that satisfies $x^{**} \approx x$ is $\overline{\mathbb{B}}_0$. Therefore, an arbitrary p-algebra satisfies $x \approx x^{**}$ iff it is a subdirect product of copies of $\overline{\mathbb{B}}_0$, which is to say, it is Boolean.

We shall consider these ideas in detail once we have completed a rigorous development of identities and their satisfaction. For now, we conclude with a pair of important properties a variety may possess.

Definition 3.48. (1) An algebra is called *locally finite* if every finitely generated subalgebra is finite. A variety is locally finite if every member is locally finite.

(2) A variety is called *finitely generated* if it is of the form $\mathbf{V}(\mathcal{K})$ in which $\mathcal{K}$ is a finite set of finite algebras.

For example, the Abelian group $\mathbb{Z}$ is not locally finite since for any $n \neq 0$, $\mathrm{Sg}(n)$ is infinite. On the other hand, the groups $\mathbb{Z}(p^\infty)$ are locally finite. This follows from Exercise 3.26.7. As another example, in Exercise 2.47.7 you proved that the variety of distributive lattice is locally finite.

Part (2) of the definition may seem odd. One would expect "finitely generated variety" to be a variety generated by a finite number of algebras. However, we shall show in Chapter 4 that every variety is generated by a single algebra. So the custom has been to use the phrase "finitely generated variety" as we have above.

Theorem 3.49. *Every finitely generated variety is locally finite.*

Proof. Let $\mathcal{V}$ be a finitely generated variety. Thus $\mathcal{V} = \mathbf{V}(\mathcal{K})$ where $\mathcal{K} = \{\mathbf{A}_1, \mathbf{A}_2, \ldots, \mathbf{A}_n\}$, n is a natural number and each A_k is finite. Let $\mathbf{B}$ be a finitely generated member of $\mathcal{V}$, say $\mathbf{B}$ is generated by the finite set Y. We must show that B is finite.

Since $\mathbf{B} \in \mathcal{V} = \mathbf{HSP}(\mathcal{K})$ (Theorem 3.43), there is an algebra $\mathbf{C}$, a surjective homomorphism $h\colon \mathbf{C} \to \mathbf{B}$ and algebras $\mathbf{C}_i \in \mathcal{K}$, for $i \in I$, such that $\mathbf{C}$ is isomorphic to a subalgebra of the product $\prod_i \mathbf{C}_i$. See Figure 3.8. Since h is surjective, we can find a subset X of C such that h is a bijection of X with Y. Thus X is finite. By replacing $\mathbf{C}$ with $\mathrm{Sg}^{\mathbf{C}}(X)$, we can assume that $\mathbf{C}$ is finitely generated. It suffices to show that C is finite.

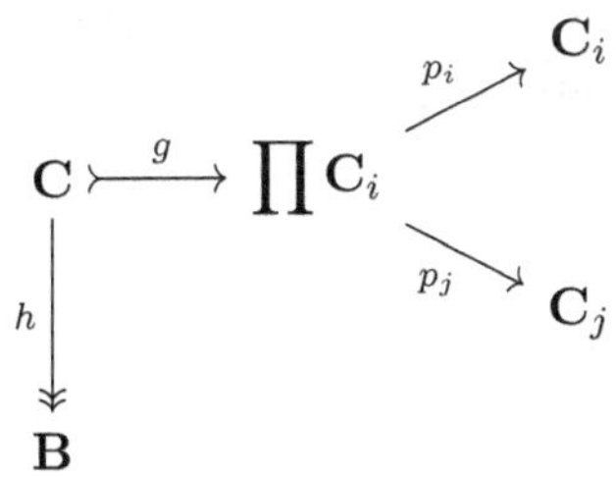

FIGURE 3.8: $\mathbf{B} \in \mathsf{HSP}(\mathcal{K})$

Let g denote the embedding of $\mathbf{C}$ into the product and, for each $i \in I$, p_i denote the projection of the product onto the ith factor. For each $k \leq n$, the algebra $\mathbf{A}_k$ is finite. Since $\mathbf{C}$ is finitely generated, by Exercise 1.16.6, $\mathrm{Hom}(\mathbf{C}, \mathbf{A}_k)$ is a finite set. Since $\{\, p_i \circ g : i \in I \,\} \subseteq \bigcup_{k=1}^{n} \mathrm{Hom}(\mathbf{C}, \mathbf{A}_k)$, there are actually only finitely many maps $p_i \circ g$. Thus there is a finite subset J of I such that $\{\, p_i \circ g : i \in I \,\} = \{\, p_j \circ g : j \in J \,\}$. We have $0_C = \bigcap \{\, \ker(p_i \circ g) : i \in I \,\} = \bigcap \{\, \ker(p_j \circ g) : j \in J \,\}$. Thus $\mathbf{C}$ embeds into the product $\prod_{j \in J} \mathbf{C}_j$. Since this is a finite product of finite algebras, C, hence B, is finite. □

The converse of Theorem 3.49 is false. As we will show in Corollary 4.55, the variety of p-algebras is locally finite but not finitely generated.

Recall that if an algebra $\mathbf{A}$ is an expansion of an algebra $\mathbf{B}$, then $\mathbf{Con\,A}$ will be a sublattice of $\mathbf{Con\,B}$. Now, we know that every lattice has a distributive congruence lattice. Since every p-algebra is an expansion of a lattice, it follows that every p-algebra is congruence-distributive. Varieties in which every algebra has a distributive congruence lattice ("congruence-distributive varieties") have powerful properties. The following theorem is a typical example.

Theorem 3.50. *Let $\mathcal{K} = \{\mathbf{A}_1, \mathbf{A}_2, \dots, \mathbf{A}_n\}$ in which n is a natural number and every $\mathbf{A}_j$ is a finite algebra. Suppose that $\mathcal{V} = \mathbf{V}(\mathcal{K})$ is a congruence-distributive variety. Then every finite subdirectly irreducible member of $\mathcal{V}$ is in $\mathsf{HS}(\mathcal{K})$.*

Proof. The proof begins in much the same way as for 3.49. One can again refer to Figure 3.8. Let $\mathbf{B}$ be a finite subdirectly irreducible member of $\mathcal{V}$. Since $\mathcal{V} = \mathsf{HSP}(\mathcal{K})$, there are a set I, algebras $\mathbf{C}$ and $\mathbf{C}_i$ for $i \in I$ and homomorphisms g, h such that

$$\mathbf{C}_i \in \mathcal{K} \quad \text{for all } i \in I,$$
$$g\colon \mathbf{C} \rightarrowtail \prod_{i \in I} \mathbf{C}_i,$$
$$h\colon \mathbf{C} \twoheadrightarrow \mathbf{B}.$$

Let $\theta = \ker h$. Then $\mathbf{B} \cong \mathbf{C}/\theta$ by the Fundamental Homomorphism Theorem.

Since $\mathcal{K}$ is a finite set of finite algebras, $\mathcal{V}$ is locally finite (Theorem 3.49). Therefore since B is finite, C can be taken to be finite as well. Let $p_i \circ g\colon \mathbf{C} \to \mathbf{C}_i$ be the canonical projection, and let $\eta_i = \ker(p_i \circ g)$. We have $\bigcap_{i \in I} \eta_i = 0$ on $\mathbf{C}$. Now certainly $\operatorname{Con} \mathbf{C}$ is a finite set, so $\{\, \eta_i : i \in I \,\}$ must be finite as well. Therefore, there is a finite $J \subseteq I$ such that $\bigcap_{j \in J} \eta_j = 0$.

By the distributivity of $\operatorname{Con} \mathbf{C}$

$$\theta = \theta \vee 0 = \theta \vee \bigcap_{j \in J} \eta_j = \bigcap_{j \in J} (\theta \vee \eta_j).$$

Since $\mathbf{B}$ is subdirectly irreducible, θ is meet-irreducible in $\operatorname{Con} \mathbf{C}$. It follows that there is some $j \in J$ such that $\theta = \theta \vee \eta_j$, in other words, $\theta \supseteq \eta_j$.

Let $\mathbf{D}$ be the image of $p_j \circ g$. Note that $\mathbf{D} \leq \mathbf{C}_i \in \mathcal{K}$. By the Fundamental Homomorphism Theorem, $\mathbf{C}/\eta_j \cong \mathbf{D} \leq \mathbf{C}_j$. By the Correspondence Theorem, there is a congruence σ on $\mathbf{D}$ such that $\mathbf{B} \cong \mathbf{C}/\theta \cong \mathbf{D}/\sigma$. Therefore $\mathbf{B} \in \mathsf{HS}(\mathcal{K})$ as desired. □

As an application of Theorem 3.50 consider p-algebras. As expansions of lattices, they are congruence-distributive. Note that for every n, $\overline{\mathbf{B}}_n$ is a subalgebra of $\overline{\mathbf{B}}_{n+1}$. Thus $\mathsf{V}(\overline{\mathbf{B}}_n) \subseteq \mathsf{V}(\overline{\mathbf{B}}_{n+1})$. On the other hand, if $\overline{\mathbf{B}}_{n+1} \in \mathsf{V}(\overline{\mathbf{B}}_n)$, then by the theorem, $\overline{\mathbf{B}}_{n+1} \in \mathsf{HS}(\overline{\mathbf{B}}_n)$ which is impossible on cardinality grounds. Thus

$$\mathsf{V}(\overline{\mathbf{B}}_0) \subset \mathsf{V}(\overline{\mathbf{B}}_1) \subset \mathsf{V}(\overline{\mathbf{B}}_2) \subset \cdots$$

A similar argument can be applied to commutative rings to obtain, for any prime p

$$\mathsf{V}(\mathbb{F}_p) \subset \mathsf{V}(\mathbb{F}_{p^2}) \subset \mathsf{V}(\mathbb{F}_{p^4}) \subset \cdots$$

but $\mathbb{F}_{p^2} \notin \mathsf{V}(\mathbb{F}_{p^3})$.

Exercise Set 3.51.

1. Prove Lemma 3.41. Give examples to show that each inclusion is proper.

2. Prove that for any sequence of algebras $\mathbf{A}_1, \ldots, \mathbf{A}_n$, $\mathsf{V}(\mathbf{A}_1, \ldots, \mathbf{A}_n) = \mathsf{V}(\mathbf{A}_1 \times \cdots \times \mathbf{A}_n)$.

3. Let $\mathcal{K}$ be a class of Abelian groups. Prove that $\mathsf{HS}(\mathcal{K}) = \mathsf{SH}(\mathcal{K})$. Now step back and try to formulate a universal algebraic condition called CEP such that if $\mathcal{V}$ is a variety with CEP and $\mathcal{K} \subseteq \mathcal{V}$ then $\mathsf{HS}(\mathcal{K}) = \mathsf{SH}(\mathcal{K})$. Figure out what CEP stands for.

4. Let $\mathbf{L}$ and $\mathbf{M}$ be bounded lattices. Define $\mathbf{L} \oplus \mathbf{M}$ to be the lattice with universe $L \cup M$ and ordered so that every member of L lies below every member of M (the "ordinal sum"). Prove that $\mathsf{V}(\mathbf{L}, \mathbf{M}) = \mathsf{V}(\mathbf{L} \oplus \mathbf{M})$.

5. For a natural number n, recall that $\mathcal{C}r_n$ denotes the variety of commutative rings satisfying $x^n \approx x$. Prove that $\mathcal{C}r_{95} = \mathcal{C}r_3$.

6. Which of the following identities, together with the axioms of p-algebras, define the varieties of Boolean algebras? Stone algebras?

    (a) $x^* \vee x^{**} \approx 1$

    (b) $(x \wedge y)^* \vee (x^* \wedge y)^* \vee (x \wedge y^*)^* \approx 1$

    (c) $(x^{**} \vee y^{**}) \approx (x \vee y)^{**}$

7. Let n be a prime power and $\mathbb{F}_n$ the field of order n. Find all values of n for which $\mathbf{V}(\mathbb{F}_n) = \mathcal{C}r_n$.

Chapter 4

Clones, Terms, and Equational Classes

In Chapter 3 we developed the structural side of our subject: subalgebras, homomorphism, quotients and products. These are notions familiar from your first-year course in abstract algebra. But what distinguishes universal algebra from the traditional study of groups and rings is its attention to semantics. We often characterize algebras by the operations they possess and the identities that they satisfy. That is the focus of the present chapter. The most beautiful and striking results in the subject (at least in the opinion of this author) provide bridges between the structural and the semantic. We shall meet several of these theorems in this chapter and many more in Part II.

4.1 Clones

Let A be a set. For every natural number n let $\mathrm{Op}_n(A)$ denote the set of all n-ary operations on A. Put another way, $\mathrm{Op}_n(A) = A^{(A^n)}$. Let $\mathrm{Op}(A) = \bigcup_{n\in\omega} \mathrm{Op}_n(A)$ be the set of all operations on A. For any $k \leq n$ there is an n-ary operation $p^n_k(x_1, \ldots, x_n) = x_k$, called the k-th *projection operation.*

Let n and k be natural numbers, and suppose that $f \in \mathrm{Op}_n(A)$ and $g_1, \ldots g_n \in \mathrm{Op}_k(A)$. Then we define a new k-ary operation $f[g_1, \ldots, g_n]$ by

$$(x_1, x_2, \ldots, x_k) \mapsto f\big(g_1(x_1, \ldots, x_k), \ldots, g_n(x_1, \ldots, x_k)\big)$$

called the *generalized composite* of f with $g_1, \ldots, g_n$. Note that, unlike the ordinary composition of unary operation, the generalized composite only exists when all of the ranks match up correctly.

There is one peculiarity that should be pointed out. Suppose that f is a nullary operation with value $c \in A$. Then $f[\ ]$, the result of composing f with 0-many n-ary operations is the n-ary operation $(x_1, \ldots, x_n) \mapsto c$. Thus, starting from a nullary operation with constant value c, we can construct an n-ary operation with constant value c for all n.

Just as the set of unary operations forms a monoid under the operation of composition, we can form a kind of algebraic structure whose elements are members of $\mathrm{Op}(A)$ with the operation of generalized composition.

Definition 4.1. Let A be a nonempty set. A *clone* on A is a subset $\mathcal{C}$ of

Op(A) that contains all projection operations and is closed under generalized composition.

Example 4.2. Here are some examples of clones to get us started.

(1) Op(A) and $\mathrm{Proj}(A) = \{\, p_k^n : 1 \le k \le n \in \omega \,\}$ are clones on any set A.

(2) The set of $\mathbb{Z}$-*linear functions* on $\mathbb{Q}$, i.e., the set of all operations of the form $f(x_1, \ldots, x_n) = a_1x_1 + \cdots + a_nx_n$ with $a_i \in \mathbb{Z}$ forms a clone on $\mathbb{Q}$. Similarly, the set of $\mathbb{Z}$-affine functions on $\mathbb{Q}$ forms a clone. $f(x_1, \ldots, x_n)$ is *affine* if it is of the form $a_1x_1 + \cdots + a_nx_n + b$ for some $a_1, \ldots, a_n \in \mathbb{Z}$ and $b \in \mathbb{Q}$.

(3) The "polynomial ring" $\mathbb{Z}[x_1, x_2, \ldots]$ is a clone over $\mathbb{Q}$ (or any commutative ring with identity, for that matter).

(4) The set $\mathcal{E}(A)$ of all idempotent operations on A is a clone. Recall that an operation f is *idempotent* if $f(x, x, \ldots, x) = x$ for all x.

(5) The set $\mathcal{U}(A)$ of all essentially unary operations on A is a clone. An n-ary operation f is *essentially unary* if for some $k \le n$ and some $g \in \mathrm{Op}_1(A)$, $f(x_1, \ldots, x_n) = g(x_k)$.

(6) Let $\langle P, \le \rangle$ be a poset. The set of all isotone operations is a clone on P. An operation f is *isotone* if $x_i \le y_i$ for $i \le n$ implies $f(x_1, x_2, \ldots, x_n) \le f(y_1, \ldots, y_n)$.

As a rule, a clone is a very complicated object, containing operations of every rank. Unlike the above examples, most clones can not be described easily as "the set of all f such that blah-blah-blah." It is a major challenge just to find techniques for characterizing clones. We shall provide a couple of illustrations in this and the next section.

First, we see a now-familiar pattern. Let A be a fixed set. The collection of clones on A is ordered by inclusion, and is closed under arbitrary intersection. Therefore, by Proposition 2.14, it forms a complete lattice. The smallest and largest elements of this lattice are Proj(A) and Op(A) respectively. There is an associated closure operator, Clo^A. In other words, for any set $F \subseteq \mathrm{Op}(A)$, $\mathrm{Clo}^A(F)$ is the smallest clone on A containing F. We denote this lattice $\mathbf{L}_{\mathrm{Clo}^A}$. In fact, it turns out that $\mathbf{L}_{\mathrm{Clo}^A}$ is not just complete, but is both algebraic and co-algebraic. (Naturally, we leave off the superscript "A" when we think that no one will yell at us.)

Post [Pos41] proved in 1941 that the lattice of clones on a two-element set is countable. By contrast, if $|A| > 2$ then A has uncountably many clones (Janov and Mucnik [JM59], 1959; and Hulanicki and Świerczkowski [HŚ60], 1960). For infinite sets A, the lattice of clones seems to be so complicated that almost nothing is known about it.

For any natural number n, the set of n-ary members of a clone $\mathcal{C}$ will be denoted $\mathcal{C}_{(n)}$. Similarly the set of n-ary members of Clo(F) is denoted Clo$_n(F)$.

As always, we seek a "bottom-up" description of the members of Clo(F). By thinking of a clone as a kind of algebra it is clear that a description analogous to Theorem 1.14 ought to be possible. But recall that function composition, even generalized composition, is associative. This makes it possible to provide a slightly slicker formulation, in the manner of Exercise 1.16.4.

Theorem 4.3. *Let A be a set and F a set of operations on A. Define*

$$\begin{aligned} F_0 &= \operatorname{Proj}(A);\\ F_{n+1} &= F_n \cup \{\, f[g_1, \dots, g_k] : f \in F,\ k = \operatorname{rank} f \\ &\qquad \text{and } g_1, \dots, g_k \in F_n \cap \operatorname{Op}_m(A), \text{ some } m \in \omega \,\}, \quad \textit{for } n \in \omega. \end{aligned}$$

Then $\operatorname{Clo}^A(F) = \bigcup_n F_n$.

Proof. Let $\overline{F} = \bigcup_n F_n$. It is easy to argue by induction that every $F_n \subseteq$ Clo(F). Thus $\overline{F} \subseteq$ Clo(F). For the converse, we must show that $\overline{F}$ is a clone containing F. Since $F_0 \subseteq \overline{F}$, we see that $\overline{F}$ contains the projection operations. Also, for any k-ary operation $f \in F$ we have $f = f[p_1^k, p_2^k, \dots, p_k^k] \in F_1 \subseteq \overline{F}$. We are reduced to showing that $\overline{F}$ is closed under generalized composition. This follows from the following claim.

Claim. If $f \in F_n$ is k-ary and if $g_1, \dots, g_k \in F_m$ are all of the same rank, then $f[g_1, \dots, g_k] \in F_{n+m}$.

Proof. We prove the claim by induction on n. If $n = 0$ then f is a projection, so $f[g_1, \dots, g_k] = g_i \in F_{0+m}$ for some $i \leq k$. So assume the claim holds for n and that $f \in F_{n+1} - F_n$. From the definition, there are operations $f_1 \in F$ of rank t and $h_1, \dots, h_t \in F_n$ such that $f = f_1[h_1, \dots, h_t]$. Note that the rank of each h_i must be equal to the rank of f, namely k. By the induction hypothesis, for each $i \leq k$, $h_i' = h_i[g_1, \dots, g_k] \in F_{n+m}$. Applying the definition, $f_1[h_1', \dots, h_t'] \in F_{n+m+1} = F_{(n+1)+m}$. Since

$$f_1[h_1', \dots, h_t'] = f_1[h_1[\mathbf{g}], \dots, h_t[\mathbf{g}]] = f[\mathbf{g}]$$

the claim is proved. □

Corollary 4.4. *For any set A, $\mathbf{L}_{\mathrm{Clo}^A}$ is an algebraic lattice.* □

If we take another look at the definition of F_{n+1} in Theorem 4.3, we see that the rank of the newly constructed operation $f[g_1, \dots, g_k]$ does not depend on f, but only on the rank of $g_1, \dots, g_k$, which already appear in F_n. Thus, if F_0 were to contain only m-ary operations (for a fixed m), then we would never obtain anything but m-ary operations from this construction. Using this observation we can directly build the set Clo$_m$ A.

Corollary 4.5. *Let A be a set and F a set of operations on A. For any natural number m, define*

$$G_0 = \{p_1^m, p_2^m, \ldots, p_m^m\};$$
$$G_{n+1} = G_n \cup \{\, f[g_1, \ldots, g_k] : f \in F, g_1, \ldots, g_k \in G_n \,\}, \quad \textit{for } n \in \omega.$$

Then $\mathrm{Clo}_m A = \bigcup_n G_n$.

Proof. A simple inductive argument shows that for every natural number n, $G_n = F_n \cap \mathrm{Op}_m(A)$, where F_n comes from Theorem 4.3. □

As an example, let F consist of the operations of addition, negation, multiplication and 1 (a nullary operation) on the set $\mathbb{Q}$ of rational numbers. What are the members of $\mathrm{Clo}_2^{\mathbb{Q}}(F)$? The set G_0 consists of the two projection operations, which we can abbreviate to $\{x, y\}$. At level one, we apply each of the members of F to G_0 obtaining

$$\begin{aligned} G_1 &= \{x + x, x + y, y + y, -x, -y, x \cdot x, x \cdot y, y \cdot y, 1\} \\ &= \{2x, x + y, 2y, -x, -y, x^2, xy, y^2, 1\}. \end{aligned}$$

The 1 in the above set is the binary constant operation with value 1. G_2 is constructed from G_1 in the same way. There are already too many functions to list, but among them are $3x$, x^3, $2x + y$, $x^2 + 2y$, $y + 1$, etc. Note that we will obtain the binary constant operations 0 (as $x + (-x)$) and 2. Continuing in this way, we see that we can construct integer powers and integer multiples of x and y, together with their sums and products. In other words

$$\mathrm{Clo}_2^{\mathbb{Q}}(F) = \mathbb{Z}[x, y].$$

Of course, a set A together with a family F of operations is nothing but an algebra. This motivates the next definition.

Definition 4.6. Let $\mathbf{A} = \langle A, F \rangle$ be an algebra. The clone of *term operations* on $\mathbf{A}$ is $\mathrm{Clo}^A(F)$, which we denote $\mathrm{Clo}(\mathbf{A})$. The clone of *polynomial operations* on $\mathbf{A}$, $\mathrm{Pol}(\mathbf{A})$ is the clone on A generated by $F \cup \mathrm{Op}_0(A)$.

One of the underlying principles of our approach to universal algebra is that there is nothing particularly special about the basic operations of an algebra. They are merely the generators of the clone of term operations. In other words, in most contexts one can work interchangeably with the algebras $\langle A, F \rangle$ and $\langle A, \mathrm{Clo}\, F \rangle$.[1]

Theorem 4.3 gives us a recipe for constructing the term operations from

[1]Actually, we have bent our own rules here. Definition 1.1 requires that the basic operations of an algebra be given as a sequence. But $\mathrm{Clo}\, F$ is just a set. That is easily remedied by choosing any convenient indexing for $\mathrm{Clo}\, F$. Most constructions (subalgebra, congruence, etc.) work perfectly well in this setting. We shall have more to say about this construction in Chapter 8.

the basic operations. The same theorem could obviously be used to construct the polynomial operations. Here is an alternate description of the polynomial clone obtained directly from the terms.

Proposition 4.7. *Let* **A** *be an algebra and let* n *be a natural number. An operation* g *is in* $\mathrm{Pol}_n(\mathbf{A})$ *if and only if for some* $m \in \omega$, $f \in \mathrm{Clo}_{n+m}(\mathbf{A})$ *and* $a_1 \ldots, a_m \in A$,

$$g(x_1, \ldots, x_n) = f(x_1, \ldots, x_n, a_1, \ldots, a_m).$$

Proof. Let $\mathcal{C}$ be the set of all $f(x_1, \ldots, x_n, a_1, \ldots, a_m)$ as in the theorem. Since every member of $\mathcal{C}$ is built from basic operations and constants, $\mathcal{C}$ is a subset of $\mathrm{Pol}(\mathbf{A})$. On the other hand, $\mathcal{C}$ is clearly closed under generalized composition, it contains the basic operations and projections (take $m = 0$) and all constant nullary operations (since for $a \in A$, $p_1^1(a)$ is the nullary operation with value a). Thus $\mathcal{C} \supseteq \mathrm{Pol}(\mathbf{A})$. □

Example 4.8. In general it is quite difficult to describe the clone of term operations of an algebra. Here are a few examples in which a characterization is possible.

(1) Let $\mathbb{Q}$ denote the ring of rational numbers, with identity. Our discussion following Corollary 4.5 can be interpreted as asserting that $\mathrm{Clo}_2(\mathbb{Q}) = \mathbb{Z}[x_1, x_2]$. More generally, the clone of all term operations is the set $\mathbb{Z}[x_1, x_2, x_3, \ldots]$. To argue this more carefully, first we observe that every member of $\mathbb{Z}[x_1, x_2, \ldots]$ is a term operation on $\mathbb{Q}$. Thus $\mathbb{Z}[x_1, x_2, \ldots] \subseteq \mathrm{Clo}(\mathbb{Q})$. Second, $\mathbb{Z}[x_1, x_2, \ldots]$ is a clone and it contains the basic operations of the ring. This gives the reverse inclusion $\mathrm{Clo}(\mathbb{Q}) \subseteq \mathbb{Z}[x_1, x_2, \ldots]$.

Suppose instead we considered $\mathbb{Q}$ as a ring without identity. Our earlier arguments still apply, except that now we can not build any constant operations, except for 0. Thus the clone of term operations is $\{\, f \in \mathbb{Z}[x_1, x_2, \ldots] : f(0, 0, \ldots, 0) = 0 \,\}$.

What about $\mathrm{Pol}(\mathbb{Q})$? (Notice that it doesn't matter whether we consider $\mathbb{Q}$ as a ring with or without identity.) According to Proposition 4.7, we simply replace some of the variables in the term operations with rational numbers. In this case, that has the effect of producing polynomials with rational coefficients. Thus $\mathrm{Pol}(\mathbb{Q}) = \mathbb{Q}[x_1, x_2, \ldots]$.

(2) Consider now the group $\mathbb{Q}$ of rational numbers under addition. It is clear that every $\mathbb{Z}$-linear function on $\mathbb{Q}$ (see Example 4.2.2) is a term operation of $\mathbb{Q}$. Conversely, the clone of $\mathbb{Z}$-linear functions contains the basic operations of the group. Therefore, $\mathrm{Clo}(\mathbb{Q})$ is the clone of $\mathbb{Z}$-linear functions on $\mathbb{Q}$. Then by Proposition 4.7, $\mathrm{Pol}(\mathbb{Q})$ is equal to the clone of $\mathbb{Z}$-affine operations on $\mathbb{Q}$.

(3) Taking **2** to be the two-element Boolean algebra, Clo(**2**) = Op(2). This is an algebraic version of the well-known fact of propositional logic that every truth table can be realized by a formula in conjunctive normal form. A finite algebra **A** with the property that Clo(**A**) = Op(A) is called *primal*. Primal algebras are considered in more detail in Section 6.1.

(4) Let $\mathbf{2} = \langle 2, \wedge, \vee \rangle$. Then Pol(**2**) is the clone of isotone operations on $\langle 2, \leq \rangle$, while Clo(**2**) consists of the operations that are both isotone and idempotent (see Examples 4.2.4 and 6).

The verification of this assertion requires a bit more work. Certainly the operations "$\wedge$," "$\vee$," "0," and "1" are all isotone and the first two are idempotent. One direction of each claim follows from this. Now suppose that $f(x_1, x_2, \dots, x_n)$ is both idempotent and isotone. In this context idempotence simply says that $f(0, 0, \dots, 0) = 0$ and $f(1, 1, \dots, 1) = 1$. Now for any $\mathbf{a} = (a_1, a_2, \dots, a_n) \in 2^n$, define $t_{\mathbf{a}}(x_1, \dots, x_n) = x_{j_1} \wedge x_{j_2} \wedge \cdots \wedge x_{j_k}$ where $\{j_1, j_2, \dots, j_k\} = \{\, i : a_i = 1 \,\}$. Let $\mathbf{a}^1, \mathbf{a}^2, \dots, \mathbf{a}^p$ be the minimal members of $\overleftarrow{f}(1)$. Note that p is nonzero since f is idempotent. Then using the fact that f is isotone we easily check that

$$f(x_1, \dots, x_n) = t_{\mathbf{a}^1} \vee t_{\mathbf{a}^2} \vee \cdots \vee t_{\mathbf{a}^p}$$

showing that $f \in$ Clo(**2**). Finally, if f is isotone but not idempotent, then either $f(0, 0, \dots, 0) = 1$ or $f(1, 1, \dots, 1) = 0$. But the only isotone map satisfying the first equality is the constant function 1, the second, the constant function 0. Thus $f \in$ Pol(**2**).

For any algebra **A**, and $n \in \omega$, $\mathrm{Clo}_n(\mathbf{A}) \subseteq \mathrm{Op}_n(A) = A^{(A^n)}$. While this relationship is obvious, it can be difficult to picture $\mathrm{Clo}_n(\mathbf{A})$ as a bunch of elements in a huge direct power. But thinking of it this way has some important consequences. The set $A^{(A^n)}$ is the universe of the algebra $\mathbf{A}^{(A^n)}$. Corollary 4.5 tells us that $\mathrm{Clo}_n(\mathbf{A})$ is a subuniverse of this algebra.

Theorem 4.9. *For any algebra* **A** *and natural number* n, $\mathrm{Clo}_n(\mathbf{A})$ *is the subuniverse of* $\mathbf{A}^{(A^n)}$ *generated by the projection functions* $p_1^n, \dots, p_n^n$. $\mathrm{Pol}_n(\mathbf{A})$ *is the subuniverse of* $\mathbf{A}^{(A^n)}$ *generated by the projection functions and all* n*-ary constant operations.*

Proof. Compare the construction in Corollary 4.5 to that of Theorem 1.14. We see that they are exactly the same once we realize that for a basic operation symbol f

$$f^{\mathbf{A}}[g_1, \dots, g_k] = f^{\mathbf{A}^{(A^n)}}(g_1, \dots, g_k) \tag{4–1}$$

where, on the left-hand side, the g_i's are thought of as operations on A and on the right as elements of A^{A^n}. □

If the similarity type of $\mathbf{A}$ contains no nullary operation symbols, then $\mathrm{Clo}_0(\mathbf{A})$ is empty. But except for that, Theorem 4.9 yields an algebra that we can denote $\mathbf{Clo}_n(\mathbf{A})$ or $\mathbf{Pol}_n(\mathbf{A})$ as appropriate. Actually equation (4–1) tells us a bit more. If $\mathcal{C}$ is any clone on A containing F then $\mathcal{C}_{(n)}$ is a subalgebra of $\mathbf{A}^{(A^n)}$.

We will have more to say about these algebras as we proceed. For now we provide a few simple examples just to illustrate the idea. Of necessity, the examples are exceedingly small since the size of $\mathrm{Clo}_n(\mathbf{A})$ usually grows exponentially or even doubly-exponentially in the size of A and n.

Consider an algebra $\mathbf{A} = \langle A, f, g\rangle$ in which $A = \{0,1,2\}$ and f and g are unary operations. Algebras in this similarity type can be conveniently depicted as directed graphs with a solid edge denoting the action of f and a dotted one for g. Figure 4.1(a) illustrates an example of such an algebra $\mathbf{A}$.

From above, $\mathbf{Clo}_1(\mathbf{A})$ is the subalgebra of $\mathbf{A}^{(A^1)} \cong \mathbf{A}^3$ generated by p^1_1. Any unary operation, h, on A can be described as an ordered triple, namely $\langle h(0), h(1), h(2)\rangle$. Thus $p^1_1 = \langle 0,1,2\rangle$, $f = \langle 0,0,2\rangle$ and $g[f] = \langle 1,1,1\rangle$. Note that we traverse edges by applying the basic operations coordinatewise to our triples. Figure 4.1(b) shows the entire algebra $\mathbf{Clo}_1(\mathbf{A})$. It contains six of the 9 elements in $\mathbf{A}^{(A^1)}$.

Extending this to $\mathrm{Clo}_2(\mathbf{A})$ requires us to describe a binary operation h as an element of $A^{(A^2)}$. We could write h as a 9-tuple. But it seems more natural to write h as a 3×3 matrix $\left\langle \begin{smallmatrix} h(00) & h(01) & h(02) \\ h(10) & h(11) & h(12) \\ h(20) & h(21) & h(22) \end{smallmatrix} \right\rangle$. The algebra $\mathbf{Clo}_2(\mathbf{A})$ is generated by $p^2_1 = \left\langle \begin{smallmatrix} 0&0&0\\1&1&1\\2&2&2 \end{smallmatrix}\right\rangle$ and $p^2_2 = \left\langle \begin{smallmatrix} 0&1&2\\0&1&2\\0&1&2 \end{smallmatrix}\right\rangle$. Strictly speaking the unary operation g is not a member of $\mathrm{Clo}_2(\mathbf{A})$, but $g[p^2_1] = \left\langle \begin{smallmatrix} 1&1&1\\0&0&0\\1&1&1 \end{smallmatrix}\right\rangle$ and $g[p^2_2] = \left\langle \begin{smallmatrix} 1&0&1\\1&0&1\\1&0&1 \end{smallmatrix}\right\rangle$ are members. Figure 4.1(c) shows all 10 points in $\mathrm{Clo}_2(\mathbf{A})$, but only some of the edges are drawn to keep the picture readable. The high degree of symmetry in this example is deceptive. It is a consequence of the fact that all basic operations are unary.

As another example, consider the p-algebra $\overline{\mathbb{B}}_1$ (Figure 3.6, page 68). Since this is also a 3-element algebra, unary operations can be denoted by an ordered triple as in the previous example. The projection operation $p^1_1 = \langle 0, e, 1\rangle$ is the generator of $\mathrm{Clo}_1(\overline{\mathbb{B}}_1)$. The reader can verify that the six solid dots in Figure 4.2 form the clone of term operations. For example, $\langle 0,e,1\rangle^* = \langle 1,0,0\rangle$ and then $\langle 0,e,1\rangle \vee \langle 1,0,0\rangle = \langle 1,e,1\rangle$. Since the nullary operations 0 and 1 are part of the similarity type, the unary operations $\langle 0,0,0\rangle$ and $\langle 1,1,1\rangle$ are members of Clo_1.

For the clone of polynomial operations we add the remaining constant operation $\langle e,e,e\rangle$. $\mathrm{Pol}_1(\overline{\mathbb{B}}_1)$ is the entire 12-element algebra shown in the figure.

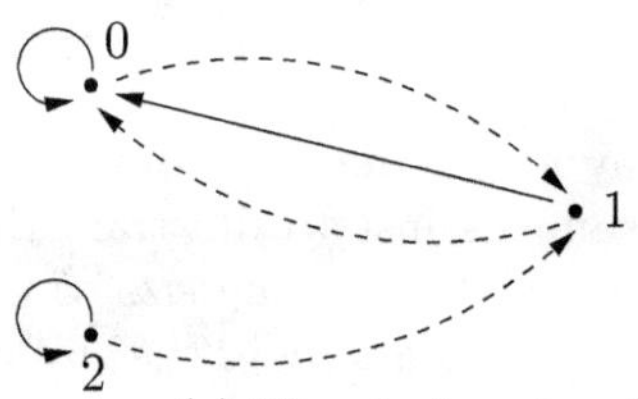

	f	g
0	0	1
1	0	0
2	2	1

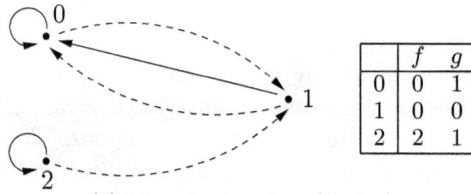

(a) The algebra $\mathbf{A} = \langle A, f, g \rangle$

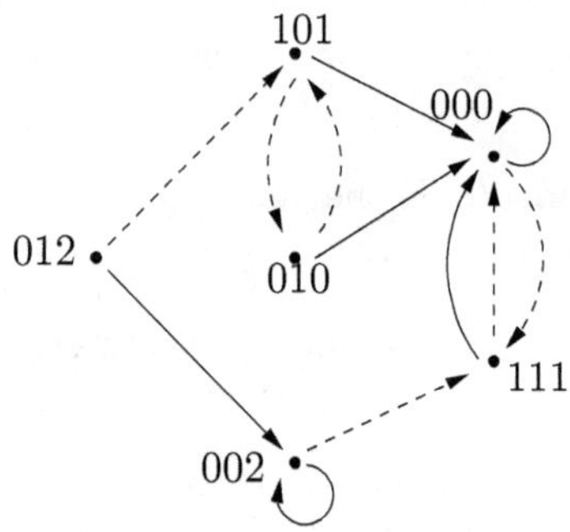

(b) $\mathrm{Clo}_1(\mathbf{A})$

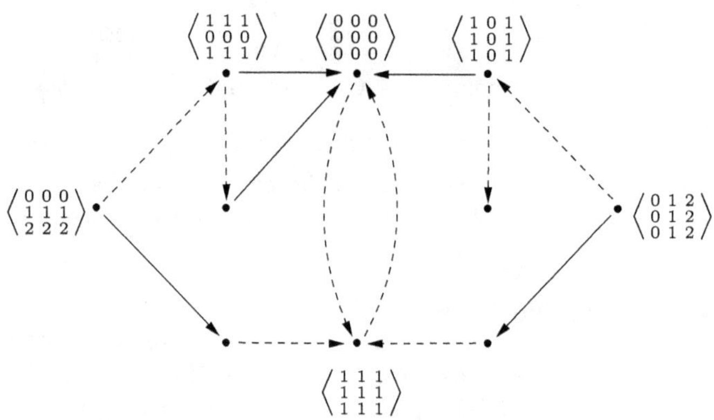

(c) $\mathrm{Clo}_2(\mathbf{A})$ (partial)

FIGURE 4.1: The clone of a unary algebra

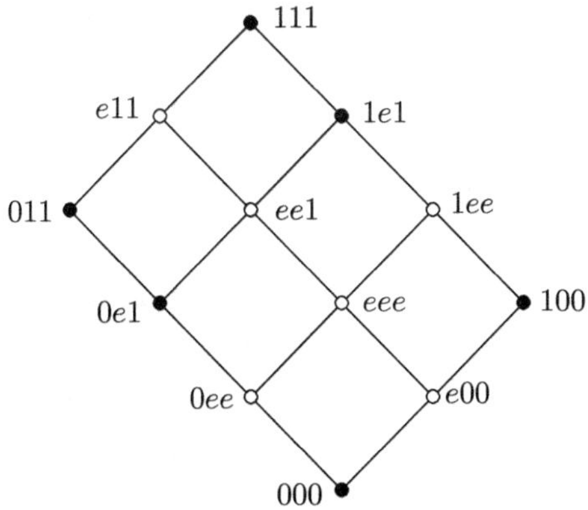

FIGURE 4.2: The algebras $\mathrm{Clo}_1(\overline{\mathbb{B}}_1)$ (filled dots) and $\mathrm{Pol}_1(\overline{\mathbb{B}}_1)$ (all dots)

Exercise Set 4.10.

1. (McKenzie, McNulty, Taylor) Construct an algebra $\mathbf{A}$ such that for $n < 9$, $\mathrm{Clo}_n(\mathbf{A}) \subseteq \mathrm{Proj}(A)$, but $\mathrm{Clo}_9(\mathbf{A})$ contains a non-projection.

2. Consider again Example 4.8.4. Let $\mathbf{3}$ denote the 3-element chain (as a lattice). Is it true that $\mathrm{Pol}(\mathbf{3})$ coincides with the clone of all isotone maps on 3?

3. For every $n > 0$ and $S \subseteq \{1, 2, \ldots, n\}$, let f_S^n denote the n-ary operation on $\{0, 1\}$ given by

$$f_S^n(x_1, \ldots, x_n) = \begin{cases} 1 & \text{if } x_i = 1 \text{ for all } i \in S \\ 0 & \text{otherwise.} \end{cases}$$

Let $\mathcal{C} = \{f_S^n : n > 0, \varnothing \neq S \subseteq \{1, 2, \ldots, n\}\}$.

   (a) Prove that $\mathcal{C}$ is a clone on 2.

   (b) Prove that $\mathcal{C} = \mathrm{Clo}(\langle 2, \wedge \rangle)$.

   (c) Prove that $\mathcal{C}$ is a minimal nontrivial element in the lattice of clones on 2.

4. Let '$\overline{\wedge}$' be the binary operation on $2 = \{0, 1\}$ shown in Figure 4.3. Show that $\overline{\wedge} \notin \mathrm{Clo}(\langle 2, \wedge \rangle)$. Show that $\{', \wedge, \vee\} \subseteq \mathrm{Clo}(\langle 2, \overline{\wedge} \rangle)$. Consequently, $\mathrm{Clo}(\langle 2, \overline{\wedge} \rangle)$ is the clone of all operations on 2. (Note the contrast to the previous exercise.)

5. Consider the binary operation '$\rightarrow$', shown in Figure 4.3.

   (a) Show that $\vee \in \mathrm{Clo}(\langle 2, \rightarrow \rangle)$.

$\wedge$	0	1
0	0	0
1	0	1

$\bar{\wedge}$	0	1
0	1	1
1	1	0

$\to$	0	1
0	1	1
1	0	1

FIGURE 4.3: Three binary operations on $\{0,1\}$

(b) Write out the Cayley tables for $\mathrm{Clo}_n(\langle 2, \to\rangle)$ for $n = 1, 2$. Also draw them as join-semilattices.

(c) Show that an n-ary operation f is a member of $\mathrm{Clo}(\langle 2, \to\rangle)$ if and only if there is $i \le n$ such that $f(x_1, \ldots, x_n) \ge x_i$.

6. Let $\mathbf{A} = \langle\{0,1,2\}, \cdot\rangle$ be the binar with Cayley table

$\cdot$	0	1	2
0	0	0	0
1	0	0	0
2	0	0	2

(a) Determine $\mathrm{Clo}_n(\mathbf{A})$ for $n \le 3$.

(b) Show that $|\mathrm{Clo}_n(\mathbf{A})| = n + 2^n - 1$. (Berman)

7. Let f be an n-ary operation on a set A, and let $1 \le i \le n$. We say that f *depends on* x_i if there are $a_1, \ldots, a_n, b \in A$ such that

$$f(a_1, \ldots, a_i, \ldots, a_n) \ne f(a_1, \ldots, b, \ldots, a_n)$$

(with b in the ith position). Prove that f is essentially unary (cf. 4.2.5) if and only if there is at most one $i \le n$ such that f depends on x_i.

8. Prove that on a finite set A, $\mathrm{Clo}^A(\mathrm{Op}_2(A)) = \bigcup_{n>0} \mathrm{Op}_n(A)$. (Hint: consider the ring operations on $\mathbb{Z}_k$ together with the unary operations $t_i(x)$ in which $t_i(i) = 1$ and $t_i(x) = 0$ otherwise.) (Sierpiński, [Sie45])

4.2 Invariant relations

There is another approach to describing clones. For each $n > 0$, let $\mathrm{Rel}_n(A) = \mathrm{Sb}(A^n)$ be the set of all n-ary relations on A. Set $\mathrm{Rel}(A) = \bigcup_{n>0} \mathrm{Rel}_n(A)$. It is common to identify a unary relation with the corresponding subset of A. Thus we write $\{a, b, c\}$ instead of $\{(a), (b), (c)\} \in \mathrm{Rel}_1(A)$.

Definition 4.11. Let A be a set, $f \in \mathrm{Op}_n(A)$ and $\theta \in \mathrm{Rel}_k(A)$. We say that

f *preserves* θ and write $f \mid\colon \theta$ if

$$(x_{11}, x_{12}, \dots, x_{1k}), (x_{21}, \dots, x_{2k}), \dots (x_{n1}, \dots, x_{nk}) \in \theta \implies \\ \big(f(x_{11}, x_{21}, \dots, x_{n1}), \dots, f(x_{1k}, \dots, x_{nk})\big) \in \theta.$$

In place of "f preserves θ" we also say "θ is invariant under f."

Although this definition seems imposing, it has a simple structure. One way to visualize it is to put the nk variables in a matrix:

$$\begin{array}{ccccccc} x_{11} & x_{12} & \dots & x_{1k} & \in & \theta \\ x_{21} & x_{22} & \dots & x_{2k} & \in & \theta \\ \vdots & \vdots & & \vdots & & \vdots \\ x_{n1} & x_{n2} & \dots & x_{nk} & \in & \theta \\ \downarrow f & \downarrow f & & \downarrow f & & \\ \star & \star & \dots & \star & \in & \theta \end{array} \tag{4–2}$$

The antecedent says that each of the rows of the matrix is an element of θ. The consequent asserts that if we apply f to the columns of the matrix, the resulting row is again an element of θ. Another way to demystify the preservation relationship is via the following lemma.

Lemma 4.12. *Let θ be a k-ary relation and f an operation on A. Then f preserves θ if and only if θ is a subuniverse of $\langle A, f\rangle^k$.*

Proof. Let $\mathbf{x}_i = (x_{i1}, x_{i2}, \dots, x_{ik})$, for $i = 1, \dots, n$, represent elements of A^k. To say that θ is a subalgebra of $\langle A, f\rangle^k$ is to assert that

$$\mathbf{x}_1, \dots, \mathbf{x}_n \in \theta \implies f^{\mathbf{A}^k}(\mathbf{x}_1, \dots, \mathbf{x}_n) \in \theta.$$

But $f^{\mathbf{A}^k}(\mathbf{x}_1, \dots, \mathbf{x}_n) = \big(f^{\mathbf{A}}(x_{11}, \dots, x_{n1}), \dots, f^{\mathbf{A}}(x_{1k}, \dots, x_{nk})\big)$ which is precisely the condition that f preserve θ. □

One can, of course, extend this notion to sets of operations and relations. For $F \subseteq \mathrm{Op}(A)$ and $\Theta \subseteq \mathrm{Rel}(A)$, write $F \mid\colon \Theta$ iff for every $f \in F$ and $\theta \in \Theta$, $f \mid\colon \theta$.

The preservation relation gives rise to a Galois connection between the sets $\mathrm{Op}(A)$ and $\mathrm{Rel}(A)$. We name the polarities in the next definition.

Definition 4.13. Let A be a set, $F \subseteq \mathrm{Op}(A)$ and $\Theta \subseteq \mathrm{Rel}(A)$. We define

$$\mathcal{R}(F) = \big\{\sigma \in \mathrm{Rel}(A) \colon F \mid\colon \sigma\big\} \text{ and} \\ \mathcal{F}(\Theta) = \big\{g \in \mathrm{Op}(A) \ : g \mid\colon \Theta\big\}.$$

$\mathcal{R}(F)$ is called the set of relations *invariant under F*. $\mathcal{F}(\Theta)$ is sometimes called the set of *polymorphisms* of Θ.

From the theory of Galois connections discussed in Section 2.5, we obtain a closure operation on Op(A). The closed subsets of Op(A) are those of the form $\mathcal{F}(\Theta)$ for some set Θ of relations, and F is closed if and only if $F = \mathcal{F}(\mathcal{R}(F))$. Similar statements hold for the closed subsets of Rel(A). Furthermore, there is an anti-isomorphism between the lattices of closed subsets of Op(A) and Rel(A) (see Theorem 2.35).

Theorem 4.14. *For any set A and $\Theta \subseteq \mathrm{Rel}(A)$, $\mathcal{F}(\Theta)$ is a clone on A. If A is finite, every clone is of the form $\mathcal{F}(\Theta)$ for some set Θ of relations on A.*

In other words, for a finite set A, the closed subsets of Op(A) are precisely the clones. This approach to the study of clones has many applications. For example, the clone of isotone operations of a poset $\langle P, \leq \rangle$ (see Example 4.2.6) is nothing but $\mathcal{F}(\leq)$. The exercises discuss several of the other examples from this perspective.

Proof. The proof that on a finite set every clone is closed is left as Exercise 4.18.7. We shall show that every set of the form $\mathcal{F}(\Theta)$ is a clone. It is easy to see that the projection operations preserve every relation. Thus we are reduced to showing that for any relation θ and operations f, $g_1, \ldots, g_m$

$$f, g_1, g_2, \ldots, g_m \mathrel{|\!:} \theta \implies f[g_1, g_2, \ldots, g_m] \mathrel{|\!:} \theta.$$

So let f be an m-ary and $g_1, \ldots, g_m$ be n-ary operations preserving the k-ary relation θ. Further let $\mathbf{a}_i = (a_{i1}, \ldots, a_{ik}) \in \theta$, for $i = 1, \ldots, n$. For $j = 1, 2, \ldots, k$, let $\hat{\mathbf{a}}_j = (a_{1j}, \ldots, a_{mj})$. That is, $\hat{\mathbf{a}}_j$ is the jth column of the matrix in (4–2). We wish to show that $\big(f[g_1, \ldots, g_m](\hat{\mathbf{a}}_1), \ldots, f[g_1, \ldots, g_m](\hat{\mathbf{a}}_k)\big) \in \theta$. But by assumption, every g_i preserves θ, so

$$\begin{array}{cccccc} g_1(\hat{\mathbf{a}}_1) & g_1(\hat{\mathbf{a}}_2) & \cdots & g_1(\hat{\mathbf{a}}_k) & \in & \theta \\ g_2(\hat{\mathbf{a}}_1) & g_2(\hat{\mathbf{a}}_2) & \cdots & g_2(\hat{\mathbf{a}}_k) & \in & \theta \\ \vdots & \vdots & & \vdots & & \\ g_m(\hat{\mathbf{a}}_1) & g_m(\hat{\mathbf{a}}_2) & \cdots & g_m(\hat{\mathbf{a}}_k) & \in & \theta \end{array}$$

Applying f to the columns of this matrix (and using the fact that f preserves θ) we obtain $\big(f(g_1(\hat{\mathbf{a}}_1), \ldots, g_m(\hat{\mathbf{a}}_1)), \ldots, f(g_1(\hat{\mathbf{a}}_k), \ldots, g_m(\hat{\mathbf{a}}_k))\big) \in \theta$. But $f\big(g_1(\hat{\mathbf{a}}_j), \ldots, g_m(\hat{\mathbf{a}}_j)\big) = f[g_1, \ldots, g_m](\hat{\mathbf{a}}_j)$ from which the result follows. □

Suppose once again that f is an n-ary, and g a k-ary, operation on a set A. The condition that f be a homomorphism $\langle A, g\rangle^n \to \langle A, g\rangle$ is precisely

$$\begin{aligned} f\big(g(a_{11}, \ldots, a_{1k}), g(a_{21}, \ldots, a_{2k}), \ldots, g(a_{n1}, \ldots, a_{nk})\big) = \\ g\big(f(a_{11}, \ldots, a_{n1}), f(a_{12}, \ldots, a_{n2}), \ldots, f(a_{1k}, \ldots, a_{nk})\big). \quad (4\text{–}3) \end{aligned}$$

But this equality is also equivalent to the assertion that $g\colon \langle A, f\rangle^k \to \langle A, f\rangle$ is a homomorphism. When the condition in (4–3) holds, we say that f and g *commute*.

If we write $g^\square$ for the "graph" of g, i.e.,

$$g^\square = \{ (a_1, \ldots, a_k, b) : b = g(a_1, \ldots, a_k) \}$$

then f commutes with g if and only if f preserves the relation $g^\square$. It follows that the set of operations that commute with g forms a clone, called the *centralizer of* g. More generally, for a subset G of $\mathrm{Op}(A)$, the centralizer of G is the clone $\mathcal{F}(\{ g^\square : g \in G \})$.

It is possible, but not necessary, for an operation to commute with itself. A unary operation always commutes with itself. For a binary operation '$\cdot$', the condition that it commute with itself is

$$(x \cdot y) \cdot (z \cdot w) \approx (x \cdot z) \cdot (y \cdot w).$$

An operation with this property has been variously called *medial* or *entropic*. For this reason, an algebra $\langle A, F \rangle$ is called entropic if F is contained in its own centralizer. It is easy to see that a monoid or group is entropic if and only if it is commutative.

Many important concepts of Universal Algebra, such as subuniverses, automorphisms, congruences, etc., can be described as invariant relations with some special property. For example a subuniverse is nothing but an invariant unary relation. An automorphism is an invariant binary relation that is the graph of a permutation. Here is an application of this principle to congruences.

Definition 4.15. Let $\mathbf{A} = \langle A, F \rangle$ be an algebra. An *elementary translation* of $\mathbf{A}$ is a unary operation of the form $f(a_1, a_2, \ldots, a_{i-1}, x, a_{i+1}, \ldots, a_n)$ where $f \in F$ and $a_1, \ldots, a_n \in A$. The set of elementary translations of $\langle A, F \rangle$ is denoted $F_{(A)}$.

Thus, the elementary translations are the unary polynomials built from a single basic operation.

Theorem 4.16. *Let $\mathbf{A} = \langle A, F \rangle$ be an algebra, and θ an equivalence relation on A. The following are equivalent.*

(a) *θ is a congruence on $\mathbf{A}$;*

(b) $\mathrm{Clo}(\mathbf{A}) \mathrel{|:} \theta$;

(c) $\mathrm{Pol}(\mathbf{A}) \mathrel{|:} \theta$;

(d) $\mathrm{Pol}_1(\mathbf{A}) \mathrel{|:} \theta$;

(e) $F_{(A)} \mathrel{|:} \theta$.

Proof. By assumption, θ is an equivalence relation. To be a congruence relation on $\mathbf{A}$, θ must have the substitution property, which is precisely the condition $F \mathrel{|:} \theta$. Since $F \subseteq \mathrm{Clo}(\mathbf{A}) \subseteq \mathrm{Pol}(\mathbf{A})$ and since the $\mathcal{R}$ operator is order-reversing, we obtain $\mathcal{R}(\mathrm{Pol}(\mathbf{A})) \subseteq \mathcal{R}(\mathrm{Clo}(\mathbf{A})) \subseteq \mathcal{R}(F)$. This

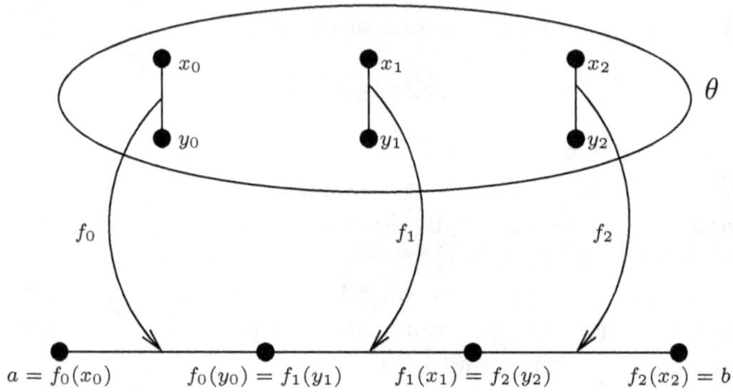

FIGURE 4.4: Illustration of Theorem 4.17. $(a, b) \in \mathrm{Cg}(\theta)$.

shows (c) $\Rightarrow$ (b) $\Rightarrow$ (a). Similarly since $F_{(A)} \subseteq \mathrm{Pol}_1(\mathbf{A}) \subseteq \mathrm{Pol}(\mathbf{A})$ we get (c) $\Rightarrow$ (d) $\Rightarrow$ (e).

Now we show (a) $\Rightarrow$ (c). From Theorem 4.14, $\mathcal{F}(\theta)$ is a clone. Since θ is reflexive, $\mathrm{Op}_0(A) \subseteq \mathcal{F}(\theta)$ and (a) implies that $F \subseteq \mathcal{F}(\theta)$. Therefore $\mathrm{Pol}(\mathbf{A}) = \mathrm{Clo}^A(\mathrm{Op}_0(A) \cup F) \subseteq \mathcal{F}(\theta)$ which is equivalent to (c).

Finally assume (e). We show that $F \mid: \theta$. Let $f \in F$ and $a_i \equiv_\theta b_i$, for $i = 1, \ldots, n$. We need $f(\mathbf{a}) \equiv_\theta f(\mathbf{b})$. For $i \leq n$ define

$$g_i(x) = f(b_1, b_2, \ldots, b_{i-1}, x, a_{i+1}, \ldots, a_n).$$

Each g_i is an elementary translation, so by assumption $g_i(a_i) \equiv_\theta g_i(b_i)$. Then using the transitivity of θ

$$f(\mathbf{a}) = g_1(a_1) \; \theta \; g_1(b_1) = g_2(a_2) \; \theta \; g_2(b_2) \; \theta \cdots \theta \; g_n(b_n) = f(\mathbf{b}). \quad \square$$

Theorem 1.25 gave a method of constructing the congruence generated by a set of pairs. Using the equivalent conditions in the previous theorem we can give another construction that is often useful in practice. In the statement of the theorem, $F^*_{(A)}$ denotes the monoid of self-maps of A generated by the members of $F_{(A)}$ under composition.

Theorem 4.17 (Maltsev). *Let* $\mathbf{A}$ *be an algebra and* $\theta \subseteq A^2$. *The congruence generated by* θ *is equal to the set of all pairs* (a, b) *such that*

$$\exists z_0, \ldots, z_n \in A, \; \exists (x_0, y_0), \ldots, (x_{n-1}, y_{n-1}) \in \theta, \; \exists f_0, \ldots, f_{n-1} \in F^*_{(A)}$$

such that

$$a = z_0, \quad b = z_n \text{ and}$$
$$\{z_i, z_{i+1}\} = \{f_i(x_i), f_i(y_i)\}, \quad \text{for } i < n.$$

Proof. Let ψ be the binary relation described in the theorem. Figure 4.4 attempts to illustrate the construction. Since $\theta \subseteq \mathrm{Cg}(\theta)$ and the latter relation is preserved by $F_{(A)}$, we obtain $\psi \subseteq \mathrm{Cg}(\theta)$. Conversely, ψ is clearly designed to be an equivalence relation. Since the identity mapping is a member of $F^*_{(A)}$ by definition, $\theta \subseteq \psi$. Once we verify that ψ is invariant under $F_{(A)}$, we can conclude that ψ is a congruence so $\mathrm{Cg}(\theta) \subseteq \psi$.

To see this, let $(a, b) \in \psi$ via the intermediate points $z_0, z_1, \dots, z_n$ and polynomials $f_0, \dots, f_{n-1} \in F^*_{(A)}$. Then for any $g \in F_{(A)}$ we get $g(a) \equiv_\psi g(b)$ via the intermediate points $g(z_0), \dots, g(z_n)$ and mappings $gf_0, \dots, gf_{n-1}$. Note that each $gf_i \in F^*_{(A)}$ since this set of polynomials is closed under composition. □

It is easy to see that we can replace, in Theorem 4.17, the monoid $F^*_{(A)}$ with the larger monoid $\mathrm{Pol}_1(\mathbf{A})$ and obtain the same result. This is the way the theorem is usually stated in the literature.

Theorem 4.17 will see a good deal of use in our study of finitely based algebras (Section 5.5). For now we content ourselves with a simple example to illustrate the theorem.

Let $\mathbf{R}$ be a commutative ring with identity and let c, d be elements of R. We wish to apply Theorem 4.17 to determine the congruence generated by (c, d). The elementary translations are those operations of the form $f_r(x) = rx$ and $g_s(x) = x + s$, for each $r, s \in R$. It follows that the set $F^*_{(R)}$ from the theorem will be equal to $\{rx+s : r, s \in R\}$, which, in this case, coincides with $\mathrm{Pol}_1(\mathbf{R})$.

The theorem yields a sequence $z_0, z_1, \dots, z_n$ in which each pair $\{z_i, z_{i+1}\}$ is of the form $\{r_i c+s_i, r_i d+s_i\}$ and allows us to conclude that $(z_0, z_n) \in \mathrm{Cg}(c, d)$. In the case of rings, the sequences all collapse. For example, suppose that $n = 2$. We have relationships of the form

$$\begin{aligned} z_0 &= r_0 c + s_0 \\ z_1 &= r_0 d + s_0 = r_1 c + s_1 \\ z_2 &= r_1 d + s_1. \end{aligned}$$

Then $s_0 = r_1 c + s_1 - r_0 d$ so that

$$z_0 = (r_0 + r_1)c + (s_1 - r_0 d) \text{ while } z_2 = (r_0 + r_1)d + (s_1 - r_0 d).$$

Proceeding by induction on n, one can conclude that

$$\mathrm{Cg}^{\mathbf{R}}(c, d) = \{\, (rc + s,\ rd + s) : r, s \in R \,\}.$$

Exercise Set 4.18.

1. Let '$<$' be the strict less-than relation on $\{0, 1\}$. Show that $\mathcal{F}(<) = \mathcal{F}(\{0\}, \{1\})$. Is this clone the same as $\mathcal{F}(\{0, 1\})$?

2. An n-ary operation f on a set A is called *conservative* if, for all elements $a_1, \ldots, a_n$ in A, we have $f(a_1, \ldots, a_n) \in \{a_1, \ldots, a_n\}$. Prove that the set of all conservative operations forms a clone on A. Prove, in fact, that it is the clone $\mathcal{F}(\mathrm{Rel}_1(A))$.

3. Let n be a positive integer and ϵ an equivalence relation on $\{1, \ldots, n\}$. For any set A, we define the relation

$$\theta_\epsilon = \{\, (x_1, \ldots, x_n) : (i,j) \in \epsilon \implies x_i = x_j \,\}.$$

θ_ϵ is called a *diagonal relation.* Let $\mathcal{C}$ be a clone on A. Prove that $\mathcal{C} = \mathrm{Op}(A)$ if and only if $\mathcal{R}(\mathcal{C})$ is the set of all diagonal relations on A.

4. Let $\mathbf{A} = \langle A, F \rangle$ be an algebra and $\theta \in \mathcal{R}(F)$ be binary. Prove or disprove: The equivalence relation on A generated by θ is invariant under F.

5. Let A be a set and define $\nu = \{\, (x,y,z) : x = y \text{ or } y = z \,\}$. Prove that the clone of essentially unary operations, $\mathcal{U}(A)$, is equal to $\mathcal{F}(\nu)$.

6. Let A be a set. Find a set Σ of relations on A so that the clone $\mathcal{E}(A)$ of idempotent operations is equal to $\mathcal{F}(\Sigma)$.

7. Finish the proof of Theorem 4.14. Here is an outline. Let A be a finite set, and let $\mathcal{C}$ be a clone on A. Let n be a natural number. Enumerate the elements of A^n: $\mathbf{a}_1, \ldots, \mathbf{a}_t$. For each n-ary member, g, of $\mathcal{C}$, let $\hat{g} = \langle g(\mathbf{a}_1), \ldots, g(\mathbf{a}_t) \rangle \in A^t$. Now define $\theta_n = \{\, \hat{g} : g \in \mathcal{C}_{(n)} \,\}$. Thus $\theta_n \in \mathrm{Rel}_t(A)$. Prove that $\mathcal{C} = \mathcal{F}\{\, \theta_n : n \in \omega \,\}$.

8. Let α and β be partial orders of a set A. Prove that if $\mathcal{F}(\alpha) = \mathcal{F}(\beta)$ then either $\alpha = \beta$ or $\alpha = \beta^{\smile}$. (Hint: it suffices to work with unary operations whose range has cardinality at most 2.) (Berman)

4.3 Terms and free algebras

Suppose that $\mathbf{G} = \langle\, G, \cdot, ^{-1}, e \,\rangle$ and $\mathbf{H} = \langle\, H, \circ, ', d \,\rangle$ are groups. $\mathrm{Clo}(\mathbf{G})$ contains a binary term operation mapping (x,y) to $x \cdot y \cdot x^{-1} \cdot y^{-1}$. Similarly, $\mathrm{Clo}(\mathbf{H})$ contains the binary term operation $(x,y) \mapsto x \circ y \circ x' \circ y'$. Of course, these two operations are completely different. After all, they are honest-to-goodness functions on different sets. Nevertheless, a group theorist would refer to both of them as the commutator operation on the respective group. We emphasize that both of them are built out of the basic operations in the same way.

Recall that a similarity type is a function $\rho\colon \mathcal{F} \to \omega$ where the members of the index set $\mathcal{F}$ are called the *operation symbols* of ρ. If f is an operation symbol and $\mathbf{A}$ is an algebra of type ρ, we usually write $f^{\mathbf{A}}$ to denote the basic operation on A indexed by f. Just as we can build other operations in $\mathrm{Clo}(\mathbf{A})$ by composing the basic operations, we wish to extend the idea of operation symbol to include formal objects that correspond, in any given algebra, to the term operations on the algebra.

Fix a similarity type $\rho\colon \mathcal{F} \to \omega$, and let X be a set disjoint from $\mathcal{F}$. The elements of X are called *variables*. For every $n \in \omega$, let $\mathcal{F}_n = \overleftarrow{\rho}(n)$, in other words, the set of n-ary operation symbols. By a *word* on $X \cup \mathcal{F}$ we mean a nonempty, finite sequence of members of $X \cup \mathcal{F}$. We denote the concatenation of sequences by simple juxtaposition.

Definition 4.19. Let $\rho\colon \mathcal{F} \to \omega$ be a similarity type and X a set disjoint from $\mathcal{F}$. We define, by recursion on n, the sets T_n of words on $X \cup \mathcal{F}$ by

$$T_0 = \{\, \langle w \rangle : w \in X \cup \mathcal{F}_0 \,\};$$
$$T_{n+1} = T_n \cup \{\, f s_1 s_2 \ldots s_k : f \in \mathcal{F},\ k = \rho(f) \text{ and } s_1, \ldots, s_k \in T_n \,\}.$$

Finally, we define $T_\rho(X) = \bigcup_{n\in\omega} T_n$, called the set of *terms of type ρ over X*.

Many (but not all) mathematicians will write $f(s_1, \ldots, s_n)$ instead of $f s_1 \ldots s_n$. If w is an element of $X \cup \mathcal{F}_0$, we usually identify w and $\langle w \rangle$. Moreover, if f is a unary or binary symbol, we would probably write something like s' or $s \cdot t$, or whatever is appropriate to the context.

A complicated term is a sequence consisting of several operation symbols interspersed with variables. We use the notation $t(x_1, \ldots, x_n)$ to denote a term whose variables are among $x_1, \ldots, x_n$. For example, if f is a binary term and g is ternary, then the term

$$f(g(x, y, z), f(x, z)) \text{ which is formally } \langle\, f, g, x, y, z, f, x, z \,\rangle$$

might be denoted $t(x, y, z)$ if we don't care to indicate the exact form of the term.

Referring again to Definition 4.19, if w is a term, we define $|w|$ to be the least n such that $w \in T_n$, called the *height* of w. The height is a useful index for recursion and induction.

Notice that the set $T_\rho(X)$ is nonempty if and only if either X or $\mathcal{F}_0$ is nonempty. As long as $T_\rho(X)$ is nonempty, we can impose an algebraic structure of type ρ.

Definition 4.20. For every basic n-ary operation symbol f of type ρ let $f^{\mathbf{T}_\rho(X)}$ be the n-ary operation on $T_\rho(X)$ that maps an n-tuple $(t_1, \ldots, t_n)$ to the term $f t_1 \ldots t_n$. We define $\mathbf{T}_\rho(X)$ to be the algebra with universe $T_\rho(X)$ and with basic operations $f^{\mathbf{T}_\rho(X)}$ for $f \in \mathcal{F}$.

The construction of $\mathbf{T}_\rho(X)$ may seem to be making something out of nothing. But it plays a crucial role in the theory.

Theorem 4.21. *Let ρ be a similarity type.*

(1) $\mathbf{T}_\rho(X)$ *is generated by* X.

(2) *For every algebra* $\mathbf{A}$ *of type* ρ *and every function* $h\colon X \to A$ *there is a unique homomorphism* $\bar{h}\colon \mathbf{T}_\rho(X) \to \mathbf{A}$ *such that* $\bar{h}\restriction_X = h$.

Proof. The definition of $T_\rho(X)$ exactly parallels the construction in Theorem 1.14. That accounts for (1). For (2), we define $\bar{h}(t)$ by induction on $|t|$. Suppose $|t| = 0$. Then $t \in X \cup \mathcal{F}_0$. If $t \in X$ then define $\bar{h}(t) = h(t)$. For $t \in \mathcal{F}_0$, $\bar{h}(t) = t^{\mathbf{A}}$. Note that since $\mathbf{A}$ is an algebra of type ρ and t is a nullary operation symbol, $t^{\mathbf{A}}$ is defined.

For the inductive step, let $|t| = n + 1$. Then $t = f(s_1, \dots, s_k)$ for some $f \in \mathcal{F}_k$ and $s_1, \dots, s_k$ each of height at most n. We define

$$\bar{h}(t) = f^{\mathbf{A}}\big(\bar{h}(s_1), \dots, \bar{h}(s_k)\big).$$

By its very definition, $\bar{h}$ is a homomorphism.

Finally, the uniqueness of $\bar{h}$ follows from Exercise 1.16.6. □

We will generally leave off the subscript ρ when no confusion will result. The properties attributed to $\mathbf{T}(X)$ in the previous theorem are of fundamental importance. We pause to develop the relevant concepts.

Definition 4.22. Let $\mathcal{K}$ be a class of algebras and $\mathbf{U}$ an algebra of the same type as the members of $\mathcal{K}$. Let X be a subset of U.

(1) We say that $\mathbf{U}$ has *the universal mapping property for* $\mathcal{K}$ *over* X if for every $\mathbf{A} \in \mathcal{K}$ and every function $h\colon X \to A$, there is a homomorphism $\bar{h}\colon \mathbf{U} \to \mathbf{A}$ such that $\bar{h}\restriction_X = h$.

(2) We say that $\mathbf{U}$ is *free for* $\mathcal{K}$ *over* X if $\mathbf{U}$ has the universal mapping property and furthermore, $\mathbf{U}$ is generated by X.

(3) Finally, $\mathbf{U}$ is free *in* $\mathcal{K}$ *over* X if $\mathbf{U}$ is free for $\mathcal{K}$ over X and $\mathbf{U}$ is a member of $\mathcal{K}$.

Let $\mathcal{Alg}_\rho$ denote the class of all algebras of type ρ. Then Theorem 4.21 can be restated as: $\mathbf{T}_\rho(X)$ is free in $\mathcal{Alg}_\rho$ over X. This is sometimes expressed by saying that $\mathbf{T}_\rho(X)$ is *absolutely free* over X.

Proposition 4.23. *Let* $\mathbf{U}$ *be free for* $\mathcal{K}$ *over* X. *Then for every* $\mathbf{A} \in \mathcal{K}$ *and* $h\colon X \to A$, *the extension* $\bar{h}$ *of* h *to* $\mathbf{U}$ *is unique.*

Proof. The universal mapping property asserts the existence of $\bar{h}$. The fact that $\mathbf{U}$ is generated by X accounts for uniqueness (Exercise 1.16.6). □

Proposition 4.24. *Let* $\mathbf{U}_1$ *and* $\mathbf{U}_2$ *be free in* $\mathcal{K}$ *over* X_1 *and* X_2 *respectively. If* $|X_1| = |X_2|$ *then* $\mathbf{U}_1 \cong \mathbf{U}_2$.

Proof. Assuming X_1 and X_2 have the same cardinality, there is a bijection $h\colon X_1 \to X_2$. Since $\mathbf{U}_2 \in \mathcal{K}$ and $\mathbf{U}_1$ is free for $\mathcal{K}$, there is a homomorphism $\bar{h}_1\colon \mathbf{U}_1 \to \mathbf{U}_2$ extending h. Similarly, since $\mathbf{U}_2$ is free for $\mathcal{K}$ and $\mathbf{U}_1 \in \mathcal{K}$, there is a homomorphism $\bar{h}_2\colon \mathbf{U}_2 \to \mathbf{U}_1$ extending h^{-1}.

Now $\bar{h}_2 \circ \bar{h}_1$ is an endomorphism of $\mathbf{U}_1$ extending the identity map on X_1. Of course, the identity map on U_1 is also an extension of the identity on X_1. Therefore by Proposition 4.23, $\bar{h}_2 \circ \bar{h}_1$ is the identity on $\mathbf{U}_1$. A similar argument shows that $\bar{h}_1 \circ \bar{h}_2$ is the identity on $\mathbf{U}_2$. Thus $\mathbf{U}_1 \cong \mathbf{U}_2$. □

Note that Proposition 4.24 applies to algebras that are free *in* $\mathcal{K}$. It tells us that up to isomorphism, a free algebra is determined by the cardinality of a free generating set. According to the next proposition, freeness extends upwards from a class $\mathcal{K}$ to its generated variety.

Proposition 4.25. *Let* $\mathbf{U}$ *be free for* $\mathcal{K}$ *over* X. *Then* $\mathbf{U}$ *is free for* $\mathsf{HSP}(\mathcal{K})$ *over* X.

Proof. Let $\mathbf{U}$ be free for $\mathcal{K}$ over X. It is enough to show that if O is one of H, S or P then $\mathbf{U}$ has the universal mapping property for $\mathsf{O}(\mathcal{K})$ over X.

Consider the case $\mathbf{A} \in \mathsf{H}(\mathcal{K})$. Then there is $\mathbf{B} \in \mathcal{K}$ and a surjective homomorphism $f\colon \mathbf{B} \to \mathbf{A}$. Let $h\colon X \to A$ be a function. We must find an extension of h to a homomorphism from $\mathbf{U}$ to $\mathbf{A}$. For each $x \in X$ choose an element $b_x \in \overleftarrow{f}(h(x))$. Define the function $g\colon X \to B$ by $g(x) = b_x$. Since $\mathbf{U}$ is free for $\mathcal{K}$ and $\mathbf{B} \in \mathcal{K}$, g extends to a homomorphism $\bar{g}\colon \mathbf{U} \to \mathbf{B}$. Then $f \circ \bar{g}$ is the desired extension of h.

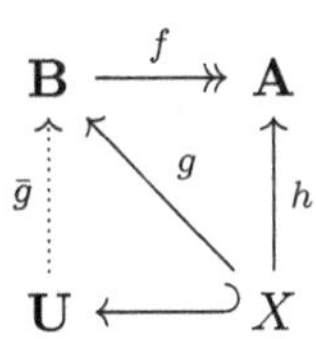

Now assume that $\mathbf{A} \in \mathsf{S}(\mathcal{K})$. Thus $\mathbf{A} \leq \mathbf{B} \in \mathcal{K}$. A mapping h from X to A is automatically a mapping to B. By freeness, h extends to a homomorphism $h^*\colon \mathbf{U} \to \mathbf{B}$. Since X generates $\mathbf{U}$,

$$\overrightarrow{h^*}(U) = \overrightarrow{h^*}(\mathrm{Sg}^{\mathbf{U}}(X)) = \mathrm{Sg}^{\mathbf{B}}(\vec{h}(X)) \subseteq A$$

by Theorem 3.2. Consequently, h^* is actually a map from $\mathbf{U}$ to $\mathbf{A}$.

Finally, assume that $\mathbf{A} = \prod \mathbf{B}_i$ with each $\mathbf{B}_i \in \mathcal{K}$. If $h\colon X \to A$ then for each i, $p_i \circ h$ is a mapping from X to B_i. By freeness we get homomorphisms $\bar{h}_i\colon \mathbf{U} \to \mathbf{B}_i$ which can be reassembled to obtain $\bar{h} = \prod \bar{h}_i\colon \mathbf{U} \to \mathbf{A}$ □

As an example, let $\mathbb{Z}_n$ denote the cyclic group of order n. The group $\mathbb{Z}_{30}$ is free for $\mathcal{K} = \{\mathbb{Z}_2, \mathbb{Z}_3\}$ over $\{1\}$. Therefore, $\mathbb{Z}_{30}$ is free for $\mathcal{A}_6 = \mathsf{HSP}(\mathcal{K})$, which happens to be the variety of Abelian groups satisfying the identity $6x \approx 0$. But note that $\mathbb{Z}_{30}$ is not a member of $\mathcal{A}_6$. In order to construct an algebra free *in* $\mathcal{A}_6$, we must resort to a homomorphic image of $\mathbb{Z}_{30}$ that retains the universal mapping property, in this case $\mathbb{Z}_6$. This approach applies quite generally. Let us study the details.

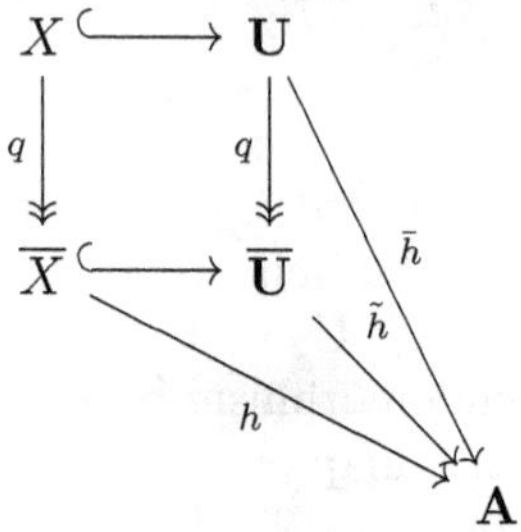

FIGURE 4.5: Illustration of Theorem 4.28

Definition 4.26. Let $\mathcal{K}$ be a class of algebras and $\mathbf{A}$ an algebra of the same type. We define

$$\Lambda_{\mathcal{K}}^{\mathbf{A}} = \{\, \theta \in \operatorname{Con}\mathbf{A} : \mathbf{A}/\theta \in \mathsf{S}(\mathcal{K}) \,\} \text{ and}$$
$$\lambda_{\mathcal{K}}^{\mathbf{A}} = \bigwedge \Lambda_{\mathcal{K}}^{\mathbf{A}}.$$

Of course, we leave out the superscripts and subscripts when we can get away with it. Notice that it is possible for Λ to be empty. In that case $\lambda^{\mathbf{A}} = 1_A$ and the quotient $\mathbf{A}/\lambda$ will be a trivial algebra.

Lemma 4.27. *Let $\mathcal{K}$ be a class of algebras, and $\mathbf{A}$ an algebra of the same type. Then $\mathbf{A}/\lambda_{\mathcal{K}} \in \mathsf{SP}(\mathcal{K})$.*

Proof. $\mathbf{A}/\lambda_{\mathcal{K}} \hookrightarrow \prod \langle\, \mathbf{A}/\theta : \theta \in \Lambda \,\rangle \in \mathsf{PS}(\mathcal{K})$ so $\mathbf{A}/\lambda_{\mathcal{K}} \in \mathsf{SPS}(\mathcal{K}) = \mathsf{SP}(\mathcal{K})$. □

Theorem 4.28. *Let $\mathbf{U}$ be free for $\mathcal{K}$ over X. Then $\mathbf{U}/\lambda_{\mathcal{K}}$ is free in $\mathsf{SP}(\mathcal{K})$ over $X/\lambda_{\mathcal{K}}$.*

Proof. Let $\overline{\mathbf{U}} = \mathbf{U}/\lambda$ and $\overline{X} = X/(\lambda\restriction_X)$. Since X generates $\mathbf{U}$, it follows from Theorem 3.2 that $\overline{X}$ generates $\overline{\mathbf{U}}$. From Lemma 4.27, $\overline{\mathbf{U}} \in \mathsf{SP}(\mathcal{K})$. If we can show that $\overline{\mathbf{U}}$ has the universal mapping property for $\mathcal{K}$, then by Proposition 4.25 $\overline{\mathbf{U}}$ will be free in $\mathsf{SP}(\mathcal{K})$ over $\overline{X}$.

Let $\mathbf{A} \in \mathcal{K}$ and $h\colon \overline{X} \to A$. Let q be the canonical map from $\mathbf{U}$ to $\overline{\mathbf{U}}$. Since $\mathbf{U}$ is free over X, there is a homomorphism $\bar{h}\colon \mathbf{U} \to \mathbf{A}$ extending $h \circ q$ (see Figure 4.5). Let $\psi = \ker \bar{h}$. By the fundamental homomorphism theorem, $\mathbf{U}/\psi \in \mathsf{S}(\mathbf{A}) \subseteq \mathsf{S}(\mathcal{K})$, so $\psi \in \Lambda_{\mathcal{K}}^{\mathbf{U}}$, and therefore, $\lambda_{\mathcal{K}}^{\mathbf{U}} \leq \psi$. Since $\lambda = \ker q$, by Exercise 1.26.8, there is a homomorphism $\tilde{h}\colon \overline{U} \to \mathbf{A}$ such that $\bar{h} = \tilde{h} \circ q$. But now it follows easily that $\tilde{h}\restriction_{\overline{X}} = h$. □

Although the statement of the above theorem suggests that $\lambda_{\mathcal{K}}^{\mathbf{U}}$ might identify two of the free generators of $\mathbf{U}$, in practice that does not occur. Exercise 4.34.2 shows that if $\mathcal{K}$ contains a nontrivial algebra then $\lambda_{\mathcal{K}}\restriction_X = 0_X$.

We now apply these ideas to our algebra of terms. Since $\mathbf{T}(X)$ is free for $\mathcal{A}lg$, it follows that for any subclass $\mathcal{K}$ of $\mathcal{A}lg$, $\mathbf{T}(X)$ is free for $\mathcal{K}$ over X, although it is not necessarily a member of $\mathcal{K}$. Taking the quotient modulo $\lambda_{\mathcal{K}}$ we obtain an algebra free in $\mathsf{SP}(\mathcal{K})$.

Definition 4.29. Let $\mathcal{K}$ be a class of algebras of type $\rho\colon \mathcal{F} \to \omega$ and let X be a set. Assume that either X or $\mathcal{F}_0$ is nonempty. We define $\mathbf{F}_{\mathcal{K}}(X) = \mathbf{T}(X)/\lambda_{\mathcal{K}}$.

Notice that, strictly speaking, X is not a subset of $\mathbf{F}_{\mathcal{K}}(X)$ so it doesn't make sense to talk about X generating $\mathbf{F}_{\mathcal{K}}(X)$. As we remarked above, as long as $\mathcal{K}$ contains a nontrivial algebra, $\lambda{\restriction}_X = 0_X$. In light of Proposition 4.24, we can identify X with $X/\lambda_{\mathcal{K}}$ in $F_{\mathcal{K}}(X)$. All of these considerations are accounted for in the following corollary.

Corollary 4.30. *Let $\mathcal{V}$ be a variety of algebras of type ρ. If $X \cup \mathcal{F}_0 \neq \varnothing$ then $\mathbf{F}_{\mathcal{V}}(X)$ is free in $\mathcal{V}$ over X. If $\mathcal{V} = \mathsf{V}(\mathcal{K})$, then $\mathbf{F}_{\mathcal{V}}(X) = \mathbf{F}_{\mathcal{K}}(X) \in \mathsf{SP}(\mathcal{K})$.*

Terms are built from the operation symbols of a similarity type in much the same way that the term operations of an algebra are built from its basic operations. It seems reasonable to expect each term to represent a term operation in any given algebra. In other words, given a term, an algebra, and a sequence of elements, we want to "evaluate" the term at those elements. But in general, there is no unique way to assign "positions" to the variables in a term. For example, if $X = \{\clubsuit, \heartsuit\}$, how is the term $\clubsuit \cdot (\clubsuit + \heartsuit)$ to be evaluated at $\langle 3, 5\rangle$ in the ring $\mathbb{R}$? As $3 \cdot (3+5)$ or as $5 \cdot (5+3)$? The most natural solution is to fix a set $X_\omega = \{x_1, x_2, x_3, \dots\}$ of variables and work with terms over X_ω. Under this restriction, the term $x_1 \cdot (x_1 + x_2)$ is uniquely evaluated as $3 \cdot (3+5)$ on the pair $\langle 3, 5\rangle$ since 3 is the first element of the pair. It will also be convenient to define, for each natural number n, the set X_n to be $\{x_1, x_2, \dots, x_n\}$.

For any similarity type ρ, we have the following relationships:

- $\mathbf{T}_\rho(X_n)$ is the subalgebra of $\mathbf{T}_\rho(X_\omega)$ generated by X_n;
- $T_\rho(X_0) \subseteq T_\rho(X_1) \subseteq T_\rho(X_2) \subseteq \cdots$;
- $T_\rho(X_\omega) = \bigcup_{n\in\omega} T_\rho(X_n)$.

Thus, for every term t there is a least integer n such that $t \in T(X_n)$, so in that sense, we can think of the term t as being n-ary. However, when it suits us, we can consider t to be a member of $T(X_m)$ for any $m > n$.

Definition 4.31. Let $t(x_1, \dots, x_n) \in T_\rho(X_n)$ and let $\mathbf{A}$ be an algebra of type ρ. We define an n-ary operation $t^{\mathbf{A}}$ on A by recursion on $|t|$, as follows:

(i) if t is the variable x_i then $t^{\mathbf{A}}(a_1, \dots, a_n) = a_i$;

(ii) if $t = f s_1 s_2 \dots s_k$ where $f \in \mathcal{F}_k$ and $s_1, \dots, s_k$ are terms, then

$$t^{\mathbf{A}}(a_1, \dots, a_n) = f^{\mathbf{A}}\big(s_1^{\mathbf{A}}(a_1, \dots, a_n), \dots, s_k^{\mathbf{A}}(a_1, \dots, a_n)\big).$$

The operation $t^{\mathbf{A}}$ is called the *term operation on* $\mathbf{A}$ *induced by* t.

Notice that the definition of $t^{\mathbf{A}}$ above could be written more succinctly as (i) $(x_i)^{\mathbf{A}} = p_i^n$ (projection operation) and (ii) $(fs_1 \dots s_k)^{\mathbf{A}} = f^{\mathbf{A}}[s_1^{\mathbf{A}}, \dots, s_k^{\mathbf{A}}]$. In particular, $t^{\mathbf{A}}$ is always a member of $\mathrm{Clo}(\mathbf{A})$. This is the origin of the name "term operation" for the members of the clone.

If we define a function $h\colon X_n \to A^{(A^n)}$ by $h(x_i) = p_i^n$, for $i = 1, \dots, n$, then by Theorem 4.21 there is a unique homomorphism $\bar{h}\colon \mathbf{T}(X_n) \to \mathbf{A}^{(A^n)}$ extending h. It is easy to see (by inducting on the height) that for any $t \in T(X_n)$, $\bar{h}(t) = t^{\mathbf{A}}$.

If we take $\mathbf{A} = \mathbf{T}(X_n)$ and h to be the identity map then $\bar{h}$ will be the identity on $\mathbf{T}(X_n)$. From the discussion in the previous paragraph we obtain, for any term t, that $t = \bar{h}(t) = t^{\mathbf{T}(X_n)}$.

Theorem 4.32. *Let* $\mathbf{A}$ *and* $\mathbf{B}$ *be algebras of type* ρ.

(1) *For every* n*-ary term* t *and homomorphism* $g\colon \mathbf{A} \to \mathbf{B}$,
$g\big(t^{\mathbf{A}}(a_1, \dots, a_n)\big) = t^{\mathbf{B}}\big(g(a_1), \dots, g(a_n)\big)$.

(2) *For every term* $t \in T_\rho(X_\omega)$ *and every* $\theta \in \mathrm{Con}(\mathbf{A})$,
$\mathbf{a} \equiv_\theta \mathbf{b} \implies t^{\mathbf{A}}(\mathbf{a}) \equiv_\theta t^{\mathbf{A}}(\mathbf{b})$.

(3) *For every subset* Y *of* A,
$\mathrm{Sg}^{\mathbf{A}}(Y) = \big\{\, t^{\mathbf{A}}(a_1, \dots, a_n) : t \in T(X_n),\ n \in \omega, \text{ and } a_i \in Y, \text{ for } i \le n \,\big\}$.

Proof. The first statement is an easy induction on $|t|$. The second statement follows from the first by taking $\mathbf{B} = \mathbf{A}/\theta$ and g the canonical homomorphism. Now consider (3). Arguing again by induction on the height of t, every subalgebra must be closed under the action of $t^{\mathbf{A}}$. Thus the right-hand side is contained in the left. On the other hand, the right-hand side is clearly a subalgebra containing the elements of Y (take $t = x_1$) from which the reverse inclusion follows. □

Corollary 4.33. *Let* $\mathbf{A}$ *be an algebra of type* ρ, *and* n *a natural number. Then* $\mathrm{Clo}_n(\mathbf{A}) = \big\{\, t^{\mathbf{A}} : t \in T_\rho(X_n) \,\big\}$.

Proof. Recall that we have a homomorphism from $\mathbf{T}(X_n)$ to $\mathbf{A}^{(A^n)}$ given by $t \mapsto t^{\mathbf{A}} \in \mathrm{Clo}_n(\mathbf{A})$. Since the image of this homomorphism contains the projection operations, and since these projections generate the algebra $\mathbf{Clo}_n(\mathbf{A})$, the image must be equal to $\mathrm{Clo}_n(\mathbf{A})$. □

It is natural, at this point in the development, to seek examples of free algebras in order to gain an understanding of the constructions in this section. It can also be very useful in practice to have a good understanding of a free algebra for some particular class of interest. Unfortunately, as we shall soon see, in some sense, a free algebra is the most general member of a variety.

There are at least two ways to gain some insight into the structure of a free algebra. One is through Exercise 4.34.3 which says that if $\mathcal{V} = \mathbf{V}(\mathbf{A})$, then $\mathbf{F}_{\mathcal{V}}(X_n) \cong \mathbf{Clo}_n(\mathbf{A})$ for all $n \in \omega$. Another is to recall that $\mathbf{F}_{\mathcal{V}}(X_n) = \mathbf{T}(X_n)/\lambda_{\mathcal{V}}$. Thus each element of the free algebra is an equivalence class of terms. Two terms are equivalent precisely if they induce the same term operation on every member of $\mathcal{V}$. In the following examples, we utilize both approaches.

(1) Let $\mathcal{A}$ be the variety of Abelian groups, with basic operation symbols $\langle +, -, 0 \rangle$. Terms of this similarity type are actually quite complex. The terms $(x_1+(x_2+x_1))+(x_2-x_1)$ and $x_1+(x_2+x_2)$ are distinct. However, they must represent the same element of $\mathbf{F}_{\mathcal{A}}(X_2)$ because they induce the same term operation, namely $x_1 + 2x_2$, on every Abelian group. By contrast, the terms x_1+x_1 and 0 must represent different elements of the free algebra since there are Abelian groups on which the corresponding term operations are different.

Using the properties of Abelian groups, it is quite easy to see that

$$F_{\mathcal{A}}(X_n) = \{\, k_1x_1 + \cdots k_nx_n : k_1, \dots, k_n \in \mathbb{Z} \,\} \cong \mathbb{Z}^n.$$

One might recognize the elements of $\mathbf{F}_{\mathcal{A}}(X_n)$ as resembling the $\mathbb{Z}$-linear operations on $\mathbb{Q}$ that we discussed in Examples 4.2(2) and 4.8(2). This is not surprising as $\mathcal{A} = \mathsf{V}(\mathbb{Q})$, so by Exercise 4.34.3, $\mathbf{F}_{\mathbf{A}}(X_n) \cong \mathbf{Clo}_n(\mathbb{Q})$.

(2) The situation for commutative rings with identity is similar. This time, of course, we can add and multiply variables. It is fairly easy to see that the free commutative ring with identity on $\{x_1, \dots, x_n\}$ is isomorphic to $\mathbb{Z}[x_1, \dots, x_n]$. Suppose, on the other hand, that we drop the requirement that the rings have an identity. Then we lose the nullary operation 1. As a result, our derived operations will have no constant term. Thus

$$\mathbf{F}_{\mathcal{CR}}(x_1, \dots, x_n) = \\ \{f(x_1, \dots, x_n) \in \mathbb{Z}[x_1, \dots, x_n] : f(0, 0, \dots, 0) = 0\}.$$

(3) Consider the variety $\mathcal{D}$ of distributive lattices. Because of the idempotent laws we see immediately that $\mathbf{F}_{\mathcal{D}}(x) = \{x\}$ is a trivial lattice. It is equally easy to see that $\mathbf{F}_{\mathcal{D}}(x, y) = \{x,\, y,\, x \wedge y,\, x \vee y\}$. One can show this by observing that every term is equivalent modulo $\lambda_{\mathcal{D}}$ to one of these four. Or we can use the fact that $\mathcal{D} = \mathsf{V}(\mathbf{2})$ and apply Exercise 4.34.3 and Example 4.8(4).

But already these techniques become difficult to apply to the algebra $\mathbf{F}_{\mathcal{D}}(x, y, z)$, whose Hasse diagram is pictured in Figure 4.6. The three atoms of that lattice are $x \wedge y$, $x \wedge z$ and $y \wedge z$. The join of these three (represented by the square in the diagram) is the term we called $m_1(x, y, z)$ in Equation (2–2) on page 29. Dually, the meet of the three coatoms is $m_2(x, y, z)$. Theorem 2.10 showed that in any distributive lattice, $m_1 = m_2$. (In fact, that equality characterizes distributivity.)

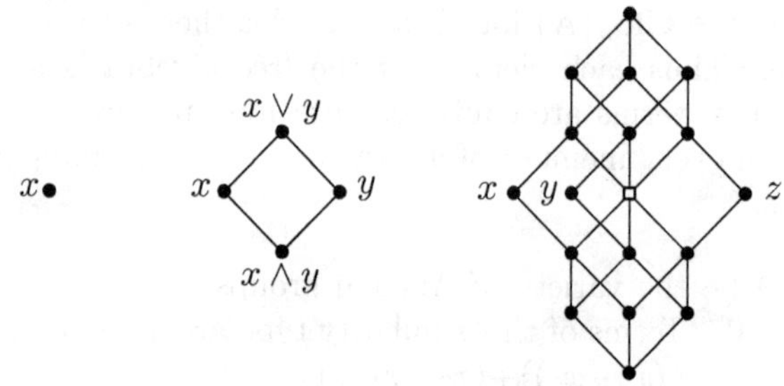

F(1) **F**(2) **F**(3)

FIGURE 4.6: the free distributive lattices on 1, 2 and 3 generators

It is interesting to recall that Example 4.8(4) provides an explicit description of the members of $\mathrm{Clo}_n(\mathbf{2})$, which, as an algebra, is isomorphic to $\mathbf{F}_{\mathcal{D}}(X_n)$. Despite this concrete characterization, the determination of $\mathbf{F}_{\mathcal{D}}(X_n)$ is a difficult combinatorial problem. As of this writing, even the cardinalities are known only for $n \leq 8$.

(4) Let us determine the free p-algebra on one generator. To begin with, we have the constant terms 0 and 1 and the generator x. Applying pseudocomplementation, we can produce x^* and x^{**}. By Lemma 3.33 (page 65), $x \wedge x^* = x^* \wedge x^{**} = 0$, $x \leq x^{**}$ and $x^{***} = x^*$. We can apply joins to produce $x \vee x^{**}$ and $x^* \vee x^{**}$. But that is the end of the line since $(x \vee x^{**})^* = x^* \wedge x^{***} = x^*$ and $(x^* \vee x^{**})^* = x^{**} \wedge x^* = 0$. The resulting lattice is pictured in Figure 4.7. We know all of the elements are distinct because, well, the algebra in the figure is a perfectly good p-algebra generated by x, so the congruence $\lambda_{\mathcal{P}a}$ can not identify any pair of these elements.

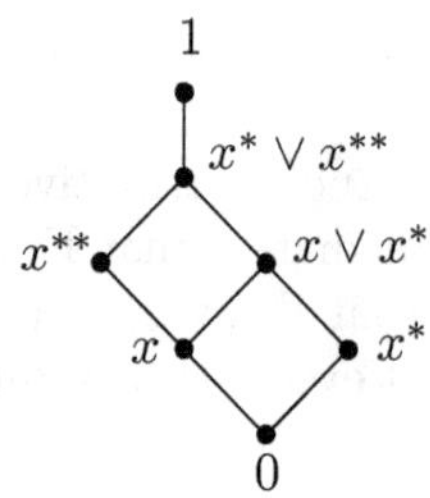

FIGURE 4.7: $\mathbf{F}_{\mathcal{P}a}(x)$

Exercise Set 4.34.

1. Let $\mathbf{U}$ be free for $\mathcal{K}$ over X. Prove that for every $\mathbf{A} \in \mathcal{K}$ there is a bijection between the sets $\hom(\mathbf{U}, \mathbf{A})$ and A^X.

2. Let $\mathbf{U}$ be free for the class $\mathcal{K}$ over a nonempty set X. Show that $\mathcal{K}$ contains a nontrivial algebra if and only if the congruence $\lambda_{\mathcal{K}}^{\mathbf{U}}$ does not identify any two distinct elements of X.

3. Let $\mathbf{A}$ be an algebra and $\mathcal{V} = \mathsf{V}(\mathbf{A})$. Then $\mathbf{F}_{\mathcal{V}}(X_n) \cong \mathbf{Clo}_n(\mathbf{A})$. (Hint: In view of Propositions 4.25 and 4.24 it is enough to show that $\mathbf{Clo}_n(\mathbf{A})$ is free for $\{\mathbf{A}\}$ over $\{p_1^n, \dots, p_n^n\}$. Alternately, one could argue that the kernel of the map in Corollary 4.33 is $\lambda_{\{\mathbf{A}\}}^{\mathbf{T}(X_n)}$.)

4. Let ρ be the similarity type consisting of a single unary operation symbol, f.

   (a) Describe the algebras $\mathbf{T}_\rho(\{x\})$ and $\mathbf{T}_\rho(\{x, y\})$. (Draw a picture similar to that of Figure 4.1.)

   (b) Let $\mathcal{V}$ be the subvariety of $\mathcal{Alg}_\rho$ defined by the single identity $f^6(x) \approx f^2(x)$. Describe the free algebra $\mathbf{F}_{\mathcal{V}}(\{x\})$.

5. Let $\mathcal{V}$ be the variety of binars satisfying the identities

$$x \cdot x \approx x \text{ and } (x \cdot y) \cdot z \approx (z \cdot y) \cdot x.$$

   (a) Show that every member of $\mathcal{V}$ satisfies the following identities.

$$\begin{aligned}
(x \cdot y) \cdot (z \cdot w) &\approx (x \cdot z) \cdot (y \cdot w) \\
x \cdot (y \cdot z) &\approx (x \cdot y) \cdot (x \cdot z) \\
(y \cdot z) \cdot x &\approx (y \cdot x) \cdot (z \cdot x) \\
y \cdot (x \cdot y) &\approx (y \cdot x) \cdot y \\
(y \cdot x) \cdot x &\approx x \cdot y.
\end{aligned}$$

   (b) Let $\mathcal{W}$ be the subvariety of $\mathcal{V}$ defined by the additional identity $y \cdot (x \cdot y) \approx x$. Determine $\mathbf{F}_{\mathcal{W}}(x, y)$. Write out the Cayley table. (Save it for Chapter 7.)

6. Let $\mathcal{V}$ be the variety of commutative semigroups satisfying the identity $x^2 \approx x^3$. Show that $|\mathbf{F}_{\mathcal{V}}(n)| = 3^n - 1$.

7. Let $\mathcal{V}$ be a variety. Prove that the monomorphisms of $\mathcal{V}$ are precisely the injective homomorphisms. (A homomorphism $f \colon \mathbf{A} \to \mathbf{B}$ is a *monomorphism* if for every $\mathbf{C} \in \mathcal{V}$ and every pair of homomorphisms $g_1, g_2 \colon \mathbf{C} \to \mathbf{A}$, $f \circ g_1 = f \circ g_2 \implies g_1 = g_2$.)

8. Let $\mathbf{G}$ be a group and let $\mathcal{A}$ denote the variety of Abelian groups. Then the normal subgroup $e/\lambda_{\mathcal{A}}^{\mathbf{G}}$ is equal to $\mathrm{Sg}^{\mathbf{G}}(\{\, [x, y] : x, y \in G \,\})$.

9. Let $\mathbf{L}$ be a lattice and $\mathcal{D}$ the variety of distributive lattices. Then

$$\begin{aligned}
\lambda_{\mathcal{D}}^{\mathbf{L}} = \mathrm{Cg}^{\mathbf{L}}\{\, \big(x \wedge (y \vee z), (x \wedge y) \vee (x \wedge z)\big) : x, y, z \in L \,\} = \\
\bigcap \{\, \theta \in \mathrm{Con}\,\mathbf{L} : |L/\theta| = 2 \,\}.
\end{aligned}$$

10. (Jónsson-Tarski, [JT61]) Let $\mathcal{JT}$ be the variety of algebras $\langle A, \cdot, \ell, r\rangle$ of similarity type $\langle 2, 1, 1\rangle$ satisfying the identities
$$\ell(x \cdot y) \approx x, \quad r(x \cdot y) \approx y, \quad \ell(x) \cdot r(x) \approx x.$$
Let $\mathbf{A}$ be a nontrivial algebra in $\mathcal{JT}$, $n > 0$, $X = \{a_1, \dots, a_n\} \subseteq A$, and $X' = \{(a_1 \cdot a_2), a_3, a_4, \dots, a_n\}$.
    (a) Show that every nontrivial member of $\mathcal{JT}$ is infinite.
    (b) Prove that $\mathbf{A}$ is generated by X iff $\mathbf{A}$ is generated by X'.
    (c) Prove that $\mathbf{A}$ has the universal mapping property for $\mathcal{JT}$ over X iff $\mathbf{A}$ has the universal mapping property for $\mathcal{JT}$ over X'.

    Conclude that for any positive integers m and n, $\mathbf{F}_{\mathcal{JT}}(n) \cong \mathbf{F}_{\mathcal{JT}}(m)$.

4.4 Identities and Birkhoff's theorem

Now, finally, we can formalize the idea we have been using since the first page of this text.

Definition 4.35. Let ρ be a similarity type. An *identity* or *equation* of type ρ is an ordered pair of terms, written $p \approx q$, from $T_\rho(X_\omega)$. If $\mathbf{A}$ is an algebra of type ρ we say that $\mathbf{A}$ *satisfies* $p \approx q$ if $p^{\mathbf{A}} = q^{\mathbf{A}}$. In this situation, we write
$$\mathbf{A} \models p \approx q.$$
If $\mathcal{K}$ is a class of algebras of type ρ, we write $\mathcal{K} \models p \approx q$ if, for every $\mathbf{A} \in \mathcal{K}$, $\mathbf{A} \models p \approx q$. Finally, if Σ is a set of equations, we write $\mathcal{K} \models \Sigma$ if every member of $\mathcal{K}$ satisfies every member of Σ.

For example, consider a similarity type with a single binary operation symbol. Let p be the term $x \cdot (y \cdot z)$ and q the term $(x \cdot y) \cdot z$. These terms are distinct. It would never be correct to write $p = q$. However, in any semigroup $\mathbf{A}$ the terms should somehow represent the same quantity. That quantity is precisely the term operation $p^{\mathbf{A}}$ and it *is* correct to write $p^{\mathbf{A}} = q^{\mathbf{A}}$. Observe that the equality of those two term operations exactly captures the fact that for every $a, b, c \in A$, $a \cdot (b \cdot c) = (a \cdot b) \cdot c$.

The satisfaction of an identity is preserved by the formation of subalgebras, homomorphic images and products. This principle is formalized in the following lemma whose proof, in light of Theorem 3.43, is a straightforward verification.

Lemma 4.36. *For any class $\mathcal{K}$, each of the classes $\mathsf{S}(\mathcal{K})$, $\mathsf{H}(\mathcal{K})$, $\mathsf{P}(\mathcal{K})$ and $\mathsf{V}(\mathcal{K})$ satisfy exactly the same identities as does $\mathcal{K}$.* □

While Definition 4.35 gives the intuitively right definition of satisfaction, the following characterization is often more useful in practice.

Lemma 4.37. *$\mathcal{K} \models p \approx q$ if and only if for every $\mathbf{A} \in \mathcal{K}$ and every homomorphism $h\colon \mathbf{T}(X_\omega) \to \mathbf{A}$, we have $h(p) = h(q)$.*

Proof. First assume that $\mathcal{K} \models p \approx q$. Pick $\mathbf{A}$ and h as in the theorem. Then

$$\mathbf{A} \models p \approx q \implies p^{\mathbf{A}} = q^{\mathbf{A}} \implies$$
$$p^{\mathbf{A}}(h(x_1), \ldots, h(x_n)) = q^{\mathbf{A}}(h(x_1), \ldots, h(x_n)).$$

Since h is a homomorphism, we get $h(p^{\mathbf{T}}(x_1, \ldots, x_n)) = h(q^{\mathbf{T}}(x_1, \ldots, x_n))$, i.e., $h(p) = h(q)$.

To prove the converse we must take any $\mathbf{A} \in \mathcal{K}$ and $a_1, \ldots, a_n \in A$ and show that $p^{\mathbf{A}}(a_1, \ldots, a_n) = q^{\mathbf{A}}(a_1, \ldots, a_n)$. Let $h_0\colon X_\omega \to A$ be a function with $h_0(x_i) = a_i$, for $i \le n$. By Theorem 4.21, h_0 extends to a homomorphism h from $\mathbf{T}(X_\omega)$ to $\mathbf{A}$. By assumption $h(p) = h(q)$. Since $h(p) = h(p^{\mathbf{T}}(x_1, \ldots, x_n)) = p^{\mathbf{A}}(h(x_1), \ldots, h(x_n)) = p^{\mathbf{A}}(a_1, \ldots, a_n)$ (and similarly for q) the result follows. □

Theorem 4.38. *Let $\mathcal{K}$ be a class of algebras and $p \approx q$ an equation. The following are equivalent.*

(a) *$\mathcal{K} \models p \approx q$.*

(b) *(p, q) belongs to the congruence $\lambda_{\mathcal{K}}$ on $\mathbf{T}(X_\omega)$.*

(c) *$\mathbf{F}_{\mathcal{K}}(X_\omega) \models p \approx q$.*

Proof. We shall show (a) ⇒ (c) ⇒ (b) ⇒ (a). Throughout the proof we write $\mathbf{F}$ for $\mathbf{F}_{\mathcal{K}}(X_\omega)$, $\mathbf{T}$ for $\mathbf{T}(X_\omega)$ and λ for $\lambda_{\mathcal{K}}$. Recall that $\mathbf{F} = \mathbf{T}/\lambda \in \mathsf{SP}(\mathcal{K})$. From (a) and Lemma 4.36 we get $\mathsf{SP}(\mathcal{K}) \models p \approx q$. Thus (c) holds.

From (c), $p^{\mathbf{F}}(\bar{x}_1, \ldots, \bar{x}_n) = q^{\mathbf{F}}(\bar{x}_1, \ldots, \bar{x}_n)$ where $\bar{x}_i = x_i/\lambda$. From the definition of $\mathbf{F}$, $p^{\mathbf{T}}(x_1, \ldots, x_n) \equiv_\lambda q^{\mathbf{T}}(x_1, \ldots, x_n)$ from which (b) follows since $p = p^{\mathbf{T}}(x_1, \ldots, x_n)$ and $q = q^{\mathbf{T}}(x_1, \ldots, x_n)$.

Finally assume (b). We wish to apply Lemma 4.37. Let $\mathbf{A} \in \mathcal{K}$ and $h\colon \mathbf{T} \to \mathbf{A}$ a homomorphism. Then $\mathbf{T}/\ker h \in \mathsf{S}(\mathbf{A}) \subseteq \mathsf{S}(\mathcal{K})$ so $\ker h \supseteq \lambda$. Then (b) implies that $h(p) = h(q)$ hence (a) holds. □

Theorem 4.38 tells us that we can determine whether an identity is true in a variety by consulting a particular algebra, namely $\mathbf{F}(X_\omega)$. Sometimes it is convenient to work with algebras free on other generating sets besides X_ω. The following corollary takes care of that for us.

Corollary 4.39. *Let $\mathcal{K}$ be a class of algebras, p and q n-ary terms, Y a set and $y_1, y_2, \ldots, y_n$ distinct elements of Y. Then $\mathcal{K} \models p \approx q$ if and only if $p^{\mathbf{F}_{\mathcal{K}}(Y)}(y_1, \ldots, y_n) = q^{\mathbf{F}_{\mathcal{K}}(Y)}(y_1, \ldots, y_n)$. In particular, $\mathcal{K} \models p \approx q$ if and only if $\mathbf{F}_{\mathcal{K}}(X_n) \models p \approx q$.*

Proof. Since $\mathbf{F}_{\mathcal{K}}(Y) \in \mathbf{SP}(\mathcal{K})$, the left-to-right direction uses the same argument as in Theorem 4.38. So assume that $p^{\mathbf{F}}(y_1, \ldots, y_n) = q^{\mathbf{F}}(y_1, \ldots, y_n)$, where we have written $\mathbf{F}$ in place of $\mathbf{F}_{\mathcal{K}}(Y)$. To show that $\mathcal{K} \models p \approx q$, let $\mathbf{A} \in \mathcal{K}$ and $a_1, \ldots, a_n \in A$. We must show $p^{\mathbf{A}}(a_1, \ldots, a_n) = q^{\mathbf{A}}(a_1, \ldots, a_n)$. There is a homomorphism $h\colon \mathbf{F} \to \mathbf{A}$ such that $h(y_i) = a_i$ for $i \leq n$. Then

$$p^{\mathbf{A}}(a_1, \ldots, a_n) = p^{\mathbf{A}}(hy_1, \ldots, hy_n) = h(p^{\mathbf{F}}(y_1, \ldots, y_n)) =$$
$$h(q^{\mathbf{F}}(y_1, \ldots, y_n)) = q^{\mathbf{A}}(a_1, \ldots a_n).$$

□

The alert reader will have noticed that '⊨' constitutes a binary relation between the class of all algebras and the set of all equations (of some fixed similarity type). And as a good universal algebraist, said reader will have immediately thought: "Galois Connection!" Indeed it is, and perhaps the most important one in the subject. Let us give names to the polarities.

Definition 4.40. Let $\mathcal{K}$ be a class of algebras and Σ a set of equations, each of similarity type ρ. We define

$$\mathrm{Id}(\mathcal{K}) = \{\, p \approx q : \mathcal{K} \models p \approx q \,\} \text{ and}$$
$$\mathrm{Mod}(\Sigma) = \{\, \mathbf{A} : \mathbf{A} \models \Sigma \,\}.$$

Classes of the form $\mathrm{Mod}(\Sigma)$ are called *equational classes,* and Σ is called an *equational base* or an *axiomatization* of the class. $\mathrm{Mod}(\Sigma)$ is called the class of *models* of Σ. Dually, a set of identities of the form $\mathrm{Id}(\mathcal{K})$ is called an *equational theory.*

From the general theory of Galois connections in Section 2.5, $\mathrm{Mod} \circ \mathrm{Id}$ and $\mathrm{Id} \circ \mathrm{Mod}$ are closure operators and there is a dual-isomorphism between the two lattices of closed sets. The equational classes are precisely the closed sets of algebras; the equational theories are the closed sets of equations. The theory of Galois connections does not tell us more than that. In particular, it does not tell us whether either of these closure operators is algebraic. We shall address that question in the next section.

It follows from Lemma 4.36 that every equational class is a variety, a principle we have been tacitly applying since page 1. In 1935, Birkhoff [Bir35] demonstrated that the converse was true.

Theorem 4.41. *Every variety is an equational class.*

Proof. Let $\mathcal{W}$ be a variety. We must find a set of equations that axiomatizes $\mathcal{W}$. The obvious choice is to use the set of all equations that hold in $\mathcal{W}$. To this end, take $\Sigma = \mathrm{Id}(\mathcal{W})$. Let $\overline{\mathcal{W}} = \mathrm{Mod}(\Sigma)$. From the relations following equations (2–4) $\mathcal{W} \subseteq \overline{\mathcal{W}}$. We shall prove the reverse inclusion.

Let $\mathbf{A} \in \overline{\mathcal{W}}$ and Y a set of cardinality $\max(|A|, |\omega|)$. Then we may choose a surjection $h_0\colon Y \to A$. By Theorem 4.21, h_0 extends to a (surjective) homomorphism $h\colon \mathbf{T}(Y) \to \mathbf{A}$. Furthermore, since $\mathbf{F}_{\mathcal{W}}(Y) = \mathbf{T}(Y)/\lambda_{\mathcal{W}}$, there is a surjective homomorphism $g\colon \mathbf{T}(Y) \to \mathbf{F}_{\mathcal{W}}(Y)$.

We claim that $\ker g \subseteq \ker h$. If the claim is true then by Exercise 1.26.8 there is a map $f\colon \mathbf{F}_{\mathcal{W}}(Y) \to \mathbf{A}$ such that $f \circ g = h$. Since h is surjective, so is f. Hence $\mathbf{A} \in \mathsf{H}(\mathbf{F}_{\mathcal{W}}(Y)) \subseteq \mathcal{W}$ completing the proof.

Now we prove the claim. Let $u, v \in T(Y)$ and assume that $g(u) = g(v)$. Since $\mathbf{T}(Y)$ is generated by Y (Theorem 4.21) there is an integer n, terms $p, q \in T(X_n)$ and $y_1, \ldots, y_n \in Y$ such that $u = p^{\mathbf{T}(Y)}(y_1, \ldots, y_n)$ and $v = q^{\mathbf{T}(Y)}(y_1, \ldots, y_n)$ (Theorem 4.32). Applying the homomorphism g,

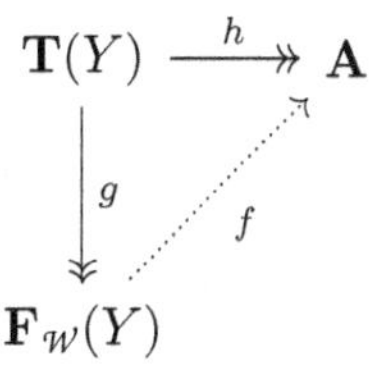

$$p^{\mathbf{F}_{\mathcal{W}}(Y)}(y_1, \ldots, y_n) = g(u) = g(v) = q^{\mathbf{F}_{\mathcal{W}}(Y)}(y_1, \ldots, y_n).$$

Then by Corollary 4.39, $\mathcal{W} \models p \approx q$, hence $(p \approx q) \in \Sigma$. Since $\mathbf{A} \in \overline{\mathcal{W}} = \mathrm{Mod}(\Sigma)$ we get $\mathbf{A} \models p \approx q$. Therefore

$$h(u) = p^{\mathbf{A}}(h_0(y_1), \ldots, h_0(y_n)) = q^{\mathbf{A}}(h_0(y_1), \ldots, h_0(y_n)) = h(v)$$

as desired. □

Theorem 4.41 was the first truly fundamental result of universal algebra. It is probably the reason that the subject exists at all. It says that the classes closed under **H**, **S** and **P** are precisely the ones definable by equations. Birkhoff's theorem was also the first "preservation theorem" in logic. It was followed in later years by the Łoś-Tarski theorem [**Lo**55a, Tar54] (a first-order theory is preserved by the formation of subalgebras if and only if it is equivalent to a universal theory) and Lyndon's theorem [Lyn59] (a first-order theory is preserved by homomorphic images if and only if it is equivalent to a positive theory), among others.

The following corollary is a restatement of the theorem. Think of it as giving another description of the algebras that lie in the variety generated by a class $\mathcal{K}$.

Corollary 4.42. *For any class $\mathcal{K}$ of algebras, $\mathsf{V}(\mathcal{K}) = \mathrm{Mod}(\mathrm{Id}(\mathcal{K}))$.*

Proof. Since the notions of variety and equational class are identical, so are the corresponding closure operators. □

Corollary 4.43. *Let Y be an infinite set and $\mathcal{V}$ a variety. Then $\mathcal{V} = \mathsf{V}(\mathbf{F}_{\mathcal{V}}(Y))$. Thus, every variety is generated by a single algebra.*

Proof. Let $\Sigma = \mathrm{Id}(\mathcal{V})$. By Theorem 4.41, $\mathcal{V} = \mathrm{Mod}(\Sigma)$. By Theorem 4.38, $\mathrm{Id}(\mathbf{F}_{\mathcal{V}}(X_\omega)) = \Sigma$. Therefore by Corollary 4.42, $\mathcal{V} = \mathsf{V}(\mathbf{F}_{\mathcal{V}}(X_\omega))$.

Now, since Y is infinite, there is a surjection from Y to X_ω, hence a surjective homomorphism from $\mathbf{F}_{\mathcal{V}}(Y)$ to $\mathbf{F}_{\mathcal{V}}(X_\omega)$. Therefore $\mathcal{V} = \mathsf{V}(\mathbf{F}_{\mathcal{V}}(X_\omega)) \subseteq \mathsf{V}(\mathbf{F}_{\mathcal{V}}(Y))$. But $\mathbf{F}_{\mathcal{V}}(Y)) \in \mathsf{SP}(\mathcal{V}) = \mathcal{V}$ (4.27 and 4.29) from which the reverse inclusion follows. □

By Proposition 4.24, the isomorphism type of the free algebra $\mathbf{F}_{\mathcal{K}}(X)$ depends only on the cardinality of X. It is customary to write $\mathbf{F}_{\mathcal{K}}(\omega)$ or $\mathbf{F}_{\mathcal{K}}(n)$ instead of $\mathbf{F}_{\mathcal{K}}(X_\omega)$ or $\mathbf{F}_{\mathcal{K}}(X_n)$. We will also occasionally denote an equation by a Greek letter, usually ε if we don't care about the constituent terms.

Because of its centrality to the subject, we should exhibit a few examples to illustrate Birkhoff's theorem. This is complicated by the fact that most of the varieties with which we are familiar (groups, rings, lattices, etc.) are initially defined as equational classes. But we have set the stage by presenting, in the first few chapters of this text, a couple of examples of classes closed under **H**, **S** and **P**, but not obviously defined by equations.

Our first example is the class of p-algebras. We proved in Theorem 3.34 that $\mathcal{Pa}$ is closed under **H**, **S**, and **P**. Therefore by Theorem 4.41, $\mathcal{Pa}$ is axiomatized by some set of equations. (Recall that (3–3) is an implication, not an equation, so it doesn't count.) In fact, it is an easy exercise to verify that an equational base for $\mathcal{Pa}$ consists of the distributive lattice identities, (3–2), together with the identities

$$\begin{gathered} 0^* \approx 1, \quad 1^* \approx 0 \\ x \wedge (x \wedge y)^* \approx x \wedge y^*. \end{gathered} \tag{4–4}$$

A second set of examples comes from group theory. Let $\mathcal{V}$ and $\mathcal{W}$ be varieties of groups. In Exercise 3.9.6 we constructed a class $\mathcal{V} \cdot \mathcal{W}$ and proved that it was a variety. Thus it is defined by a set of equations, but can we get our hands on such a set? There is a method to construct $\mathrm{Id}(\mathcal{V} \cdot \mathcal{W})$ from $\mathrm{Id}(\mathcal{V})$ and $\mathrm{Id}(\mathcal{W})$, but that isn't helpful in practice since we usually can't enumerate all of $\mathrm{Id}(\mathcal{V})$ and $\mathrm{Id}(\mathcal{W})$. (See [Neu67, 21.12] for details.)

For example, taking $\mathcal{A}$ to be the variety of Abelian groups, we shall show that $\mathcal{A} \cdot \mathcal{A}$ is defined by the group laws together with the additional identity $[x, y] \cdot [u, v] \approx [u, v] \cdot [x, y]$, where $[x, y]$ is shorthand for the term $x^{-1}y^{-1}xy$. First, let $\mathbf{G} \in \mathcal{A} \cdot \mathcal{A}$. This means that there is a normal subgroup N of $\mathbf{G}$ such that both $\mathbf{N}$ and $\mathbf{G}/N$ are Abelian. The abelianness of the quotient implies that every commutator $[x, y]$ is a member of N. But then, since the members of N commute, $[x, y] \cdot [u, v] = [u, v] \cdot [x, y]$.

Conversely, assume that $\mathbf{G}$ is a group satisfying the identity in question. Let $N = e/\lambda_{\mathcal{A}}^{\mathbf{G}}$. By Exercise 4.34.8, N is generated by all of the elements $[x, y]$. Also, $\mathbf{G}/N = \mathbf{G}/\lambda_{\mathcal{A}}$ is Abelian. Since the generators of $\mathbf{N}$ commute, $\mathbf{N}$ is Abelian. Thus $\mathbf{G} \in \mathcal{A} \cdot \mathcal{A}$.

Suppose now we define $\mathcal{G}_n$ to be the class of groups satisfying $x^n \approx e$ (groups of exponent n). Consider the variety $\mathcal{G}_3 \cdot \mathcal{G}_5$? It is easy to show that every member of this variety has exponent 15. Is $\mathcal{G}_3 \cdot \mathcal{G}_5 = \mathcal{G}_{15}$? Does this variety have a finite equational base? (The answer to this may well be known, but not to the author of this text.)

Of course one can produce arbitrary examples of varieties of the form $\mathbf{V}(\mathcal{K})$, or even $\mathbf{V}(\mathbf{A})$, for an individual algebra $\mathbf{A}$. Birkhoff's theorem ensures that such a variety has an equational base, but finding one can be challenging, if not impossible. (On the face of it, there is no reason for an equational base to be decidable, let alone finite.) The literature contains numerous results devoted to equational bases that arise in this way.

Consider the variety generated by the eight-element dihedral group (the group of symmetries of the square). MacDonald and Street [MS72] proved in 1972 that this variety is axiomatized by the usual axioms of group theory, together with the identities

$$x^4 \approx e, \quad x^2 y \approx y x^2.$$

Surprisingly, this variety is also generated by the eight-element quaternion group.

As another example, the variety generated by the lattice $\mathbf{M}_3$ has as an equational base the axioms for modular lattices together with the identity

$$x \wedge (y \vee z) \wedge (y \vee w) \wedge (z \vee w) \leq (x \wedge y) \vee (x \wedge z) \vee (x \wedge w)$$

(Jónsson, 1968 [Jón68]). McKenzie [McK72] found in 1972 that the identities

$$x \wedge (y \vee z) \wedge (y \vee w) \leq \big(x \wedge \big(y \vee (z \wedge w)\big)\big) \vee (x \wedge z) \vee (x \wedge w) \tag{4–5}$$

$$x \wedge \big(y \vee (z \wedge (x \vee w))\big) \approx \big(x \wedge \big(y \vee (x \wedge z)\big)\big) \vee \big(x \wedge \big((x \wedge y) \vee (z \wedge w)\big)\big) \tag{4–6}$$

together with the lattice identities form an equational base for $\mathbf{V}(\mathbf{N}_5)$. These results are not easy to see. By contrast, we will present in Theorem 5.28 a 4-element binar $\mathbf{A}$ such that $\mathbf{V}(\mathbf{A})$ has no finite equational base at all.

We close this section with one application of Birkhoff's theorem that will be useful later on.

Theorem 4.44. *Let $\mathcal{V}$ be a variety and $\mathbf{A}$ an algebra. Then $\mathbf{A}$ is a member of $\mathcal{V}$ if and only if every finitely generated subalgebra of $\mathbf{A}$ is a member of $\mathcal{V}$.*

Proof. One direction follows from the fact that $\mathcal{V}$ is closed under subalgebra. For the other direction, suppose that every finitely generated subalgebra of $\mathbf{A}$ lies in $\mathcal{V}$. Let $\Sigma = \mathrm{Id}(\mathcal{V})$. To show $\mathbf{A} \in \mathcal{V}$ it is enough to show that $\mathbf{A} \models \Sigma$. Let $p, q \in T(X_n)$ and suppose that $(p \approx q) \in \Sigma$. To show $p^{\mathbf{A}} = q^{\mathbf{A}}$ let $a_1, \ldots, a_n \in A$ and $B = \mathrm{Sg}^{\mathbf{A}}\{a_1, \ldots, a_n\}$. Then $\mathbf{B}$ is a finitely generated subalgebra of $\mathbf{A}$, so by assumption $\mathbf{B} \in \mathcal{V}$. Then

$$p^{\mathbf{A}}(a_1, \ldots, a_n) = p^{\mathbf{B}}(a_1, \ldots, a_n) = q^{\mathbf{B}}(a_1, \ldots, a_n) = q^{\mathbf{A}}(a_1, \ldots, a_n). \quad \square$$

Exercise Set 4.45.

1. Prove the claim above that the equations in (3–2) together with (4–4) form a base for $\mathcal{Pa}$.

2. Let $\mathbb{F}_9$ denote the field of order 9. Show that $\mathbf{V}(\mathbb{F}_9)$ is defined by the axioms for $\mathcal{Cr}_9$ together with the identity $9x \approx 0$.

3. Let $\mathcal{A}_n$ denote the variety of Abelian groups satisfying $x^n \approx x$. Show that the variety $\mathcal{A}_3 \cdot \mathcal{A}_2$ is defined by the group laws together with the identities
$$x^6 \approx e, \quad [x^2, y^2] \approx e, \quad [x, y]^3 \approx e.$$

4. Show that the variety $\mathcal{A}_2 \cdot \mathcal{A}_2$ (see the previous exercise) is defined by the group laws together with the single identity $(x^2y^2)^2 \approx e$. (H. Neumann [Neu67, pg. 24])

5. Let **A** be the following semigroup.

$\cdot$	0	1	2
0	0	0	0
1	0	0	1
2	0	1	2

Prove that $\mathsf{V}(\mathbf{A})$ is the variety of commutative semigroups satisfying $x^2 \approx x^3$ (cf. Exercise 4.34.6). (Berman)

6. For a Boolean algebra **B**, let $x \to y$ denote the binary term operation $x' \vee y$. Define $\mathbf{B}^{\to}$ to be the algebra $\langle B, \to\rangle$. The variety $\mathit{Imp} = \mathsf{HSP}\{\,\mathbf{B}^{\to} : \mathbf{B}$ a Boolean algebra$\,\}$ is called the variety of *implication algebras*. Our goal is to prove that the following three identities, Σ, form an equational base for *Imp*.
$$(x \to y) \to x \approx x,$$
$$(x \to y) \to y \approx (y \to x) \to x,$$
$$x \to (y \to z) \approx (x \to y) \to (x \to z).$$

(a) Derive, from Σ, the identity $a \to a \approx (a \to b) \to (a \to b)$. [Hint: $a \to a \approx ((a \to b) \to a) \to a \approx$ $(a \to (a \to b)) \to (a \to b) \approx$ $(((a \to b) \to a) \to (a \to b)) \to (a \to b) \approx (a \to b) \to (a \to b)$. Amazingly, you only need the first two axioms to do this.]

(b) Derive the identity $x \to x \approx y \to y$. [Hint: let $z = (y \to x) \to x = (x \to y) \to y$. Then $x \to x \approx (y \to x) \to (y \to x) \approx z \to z$.]

(c) Let 1 denote the constant term $x \to x$. Derive the identities $1 \to x \approx x$ and $x \to 1 \approx 1$.

(d) Define $x \leq y \iff (x \to y = 1)$. Show that for any model $\langle A, \to\rangle$ of Σ, '$\leq$' is an ordering of A with largest element 1. [This is the first time you use the third axiom.]

(e) Let $\mathbf{A} \models \Sigma$ and $a \in A - \{1\}$. Define $\theta_a = \{\,(x, y) : a \to x = a \to y\,\}$. Prove that θ_a is a congruence on **A** and $a/\theta_a = [a)$.

(f) Prove that the only subdirectly irreducible model of Σ is $\mathbb{B}_1^{\rightarrow}$ (i.e., the reduct of the two-element Boolean algebra).
[Hint: If $|A| > 2$, show that $\bigcap_{a \neq 1} \theta_a = 0_A$.]

(g) Conclude that $\mathcal{Imp} = \mathrm{Mod}(\Sigma) = \mathbf{V}(\mathbb{B}_1^{\rightarrow})$.

Note that $\mathbb{B}_1^{\rightarrow}$ is the same algebra considered in Exercise 4.10.5.

4.5 The lattice of subvarieties

For any variety $\mathcal{V}$, the collection of subvarieties of $\mathcal{V}$ forms a complete lattice, $\mathbf{L_V}(\mathcal{V})$ under inclusion. Although this lattice is often hopelessly complicated, any insight into its structure is an important tool in the analysis of $\mathcal{V}$. The two characterizations of varieties — closure under **HSP** and as equational classes — gives us two very different ways of approaching this analysis. In this section, we illustrate the interplay of these techniques with several examples.

First of all, let $\rho\colon \mathcal{F} \to \omega$ be a similarity type, and let $\kappa = \max\big(|\mathcal{F}|, \aleph_0\big)$. Then the set of equations of type ρ has cardinality κ. Consequently, there are at most 2^κ many equational theories. Because of the bijection between varieties and equational theories, we conclude that $|\mathbf{L_V}(\mathcal{Alg}_\rho)| \leq 2^\kappa$. In particular, a variety of finite similarity type has at most $2^{\aleph_0}$ subvarieties. At the very bottom is the variety t of all trivial algebras of type ρ. At the top of $\mathbf{L_V}(\mathcal{V})$ is, of course, $\mathcal{V}$ itself.

The meet operation in $\mathbf{L_V}(\mathcal{V})$ is simply intersection. The join of two subvarieties $\mathcal{W}_1$ and $\mathcal{W}_2$ is more complicated to determine. One way is to use the general theory of closure operators. From this we see that

$$\mathcal{W}_1 \vee \mathcal{W}_2 = \mathbf{HSP}(\mathcal{W}_1 \cup \mathcal{W}_2) = \mathbf{HS}\,\{\, \mathbf{A}_1 \times \mathbf{A}_2 : \mathbf{A}_1 \in \mathcal{W}_1,\ \mathbf{A}_2 \in \mathcal{W}_2 \,\}\,.$$

We can also apply Birkhoff's theorem together with the Galois connection. If $\Sigma_1 = \mathrm{Id}(\mathcal{W}_1)$ and $\Sigma_2 = \mathrm{Id}(\mathcal{W}_2)$ then $\mathcal{W}_1 \vee \mathcal{W}_2 = \mathrm{Mod}(\Sigma_1 \cap \Sigma_2)$.

Let $\mathcal{A}$ denote the variety of all Abelian groups, written additively. For any natural number n, let $\mathcal{A}_n$ denote the class of all Abelian groups satisfying the identity $nx \approx 0$. Note that each $\mathcal{A}_n$ is a subvariety of $\mathcal{A}$, $\mathcal{A} = \mathcal{A}_0$ and $\mathcal{A}_1 = t$ is the trivial variety.

Theorem 4.46. $\mathbf{L_V}(\mathcal{A}) = \{\, \mathcal{A}_n : n \in \omega \,\}$, *ordered by* $\mathcal{A}_n \subseteq \mathcal{A}_m$ *if and only if* n *divides* m.

Proof. The fact that $\mathcal{A}_n \subseteq \mathcal{A}_m$ if and only if $n \mid m$ is elementary group theory. We must show that every subvariety of $\mathcal{A}$ is equal to $\mathcal{A}_n$ for some $n \in \omega$.

Let Γ be a set of equations that define the variety $\mathcal{A}$. Two sets of identities Σ_1 and Σ_2 are *equivalent relative to* $\mathcal{A}$ if $\mathrm{Mod}(\Gamma \cup \Sigma_1) = \mathrm{Mod}(\Gamma \cup \Sigma_2)$.

Throughout the proof, we shall write "equivalent" instead of "equivalent relative to $\mathcal{A}$."

Suppose $\mathcal{W}$ is a subvariety of $\mathcal{A}$. From Birkhoff's theorem there is a set Σ of equations such that $\mathcal{W} = \mathrm{Mod}(\Gamma \cup \Sigma)$. We wish to show that Σ is equivalent to the single identity $nx \approx 0$ for some n. The following four claims accomplish this task.

Claim 1. Every identity $s \approx t$ is equivalent to the identity $s - t \approx 0$ relative to $\mathcal{A}$.

Claim 2. Every identity is equivalent to an identity of the form $m_1x_1 + m_2x_2 + \cdots + m_kx_k \approx 0$ with $m_1, \ldots, m_k \in \mathbb{Z}$.
Proof. By claim 1, every identity is equivalent to one of the form $s - t \approx 0$. But using the axioms of Abelian groups, the term $s - t$ can be rewritten in the form $m_1x_1 + \cdots + m_kx_k$.

Claim 3. The identity $m_1x_1 + \cdots + m_kx_k \approx 0$ is equivalent to the set of identities $\{\, m_ix \approx 0 : i \leq k \,\}$.
Proof. Suppose an Abelian group $\mathbf{A}$ satisfies $m_1x_1 + \cdots + m_kx_k \approx 0$. By setting every x_j equal to 0 for $j \neq i$, we see that $\mathbf{A} \models m_ix \approx 0$. Conversely, if $\mathbf{A}$ satisfies each of the identities in the set, then for any $a_1, \ldots, a_k \in A$, $m_1a_1 + m_2a_2 + \cdots + m_ka_k = 0 + 0 + \cdots + 0 = 0$. Thus $\mathbf{A}$ satisfies the original k-variable identity.

Claim 4. Let M be a set of natural numbers and $n = \gcd(M)$. Then the set $\{\, mx \approx 0 : m \in M \,\}$ is equivalent to the identity $nx \approx 0$.
Proof. Suppose $\mathbf{A} \models nx \approx 0$. For every $m \in M$, n is a divisor of m, so $\mathbf{A} \models mx \approx 0$. Conversely, assume $\mathbf{A}$ satisfies every identity $mx \approx 0$ for $m \in M$. We know from elementary number theory that there is a finite subset $\{m_1, \ldots, m_k\}$ of M and integers $p_1, \ldots p_k$ such that $n = p_1m_1 + \cdots + p_km_k$. Then for any $a \in A$,

$$na = p_1(m_1a) + p_2(m_2a) + \cdots + p_k(m_ka) = 0$$

by assumption. Thus $\mathbf{A} \models nx \approx 0$. □

As an example, consider the variety of Abelian groups defined by the additional two identities $\{8x + 4y \approx 2x + y,\ 20x + 9y \approx 5x\}$. Applying Claim 1 we obtain the equivalent set $\{6x + 3y \approx 0,\ 15x + 9y \approx 0\}$. From Claim 3 we get the equivalent set $\{6x \approx 0, 3x \approx 0, 15x \approx 0, 9x \approx 0\}$ and then from Claim 4 the single identity $3x \approx 0$. Thus our original pair of identities defines the variety $\mathcal{A}_3$.

Joins and meets in the lattice $\mathbf{L_V}(\mathcal{A})$ are easily computed: $\mathcal{A}_n \vee \mathcal{A}_m = \mathcal{A}_{\mathrm{lcm}(n,m)}$ and $\mathcal{A}_n \wedge \mathcal{A}_m = \mathcal{A}_{\gcd(n,m)}$. Another way to think about this is that $\mathbf{L_V}(\mathcal{A})$ is dually isomorphic to the lattice of ideals of the ring $\mathbb{Z}$. See Exercise 4.56.2 for a generalization.

Corollary 4.47. *$\mathbf{L_V}(\mathcal{A})$ is not an algebraic lattice.*

Proof. Observe that $\mathbb{Z} \in \mathbf{V}\{\mathbb{Z}_n : n > 1\}$ but for any positive integer m, $\mathbb{Z} \notin \mathbf{V}\{\mathbb{Z}_2, \mathbb{Z}_3, \ldots, \mathbb{Z}_m\} = \mathcal{A}_p$ where $p = \mathrm{lcm}(2, 3, \ldots, m)$. □

One consequence of Theorem 4.46 is that every subvariety of $\mathcal{A}$ is defined by a finite set of identities, and furthermore, there are only countably many subvarieties. By contrast, the lattice of varieties of all groups is uncountable (Vaughan-Lee, Ol′shanskiĭ [VL70, Ol′70]). Since there are only countably many finite sets of equations, "most" varieties of groups are not finitely based.

Let us move on to a couple of examples of varieties whose lattice of subvarieties is particularly small.

Theorem 4.48. *The variety of distributive lattices has only one proper subvariety, namely, the trivial variety.*

Proof. Let $\mathcal{D}$ denote the variety of distributive lattices. Recall from Example 3.19 that the only subdirectly irreducible distributive lattice is $\mathbf{2}$, the two-element chain. Suppose that $\mathcal{W}$ is a subvariety of $\mathcal{D}$. By Corollary 3.45 $\mathcal{W}_{\mathrm{si}} \subseteq \mathcal{D}_{\mathrm{si}}$. Consequently, either $\mathcal{W}_{\mathrm{si}} = \{\mathbf{2}\}$, in which case $\mathcal{W} = \mathcal{D}$ or $\mathcal{W}$ has no subdirectly irreducible algebras, which is to say, $\mathcal{W}$ is the trivial variety. □

Theorem 4.48 tells us that the variety of distributive lattices is an atom in the lattice of varieties. Applying the Galois connection, the equational theory of distributive lattices is a maximal proper theory. Such a theory is called *equationally complete.* It is customary to call minimal varieties equationally complete as well.[2] Other examples of equationally complete varieties are those of Boolean algebras, semilattices, and the varieties $\mathcal{A}_p$ of Abelian groups, for p a prime. We shall prove in Proposition 7.61 that every nontrivial variety contains an equationally complete subvariety.

Here is another easy example.

Theorem 4.49. *The variety $\mathcal{R}b$ of rectangular bands (Exercise 3.15.6) has four subvarieties:*

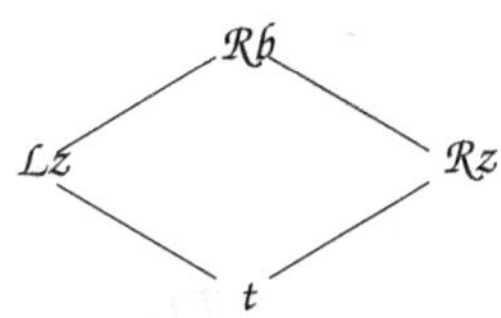

where $\mathcal{L}z$ and $\mathcal{R}z$ are the varieties of left- and right-zero semigroups, respectively.

Proof. It is easy to see that every equivalence relation on a left-zero semigroup is a congruence. Thus the only subdirectly irreducible left-zero semigroup is

[2]The word "complete" comes from logic. In our context it can be taken to mean that adding any one additional identity to the theory would allow us to derive every possible identity, including $x \approx y$.

the 2-element algebra, which we shall denote $\mathbf{L}_2$. Using the same argument as in the previous theorem, we deduce that the variety $\mathcal{Lz}$ is equationally complete. Similarly, the only subdirectly irreducible right-zero semigroup is the two-element algebra $\mathbf{R}_2$, and $\mathcal{Rz}$ is equationally complete.

From Exercise 3.15.6 we know that every rectangular band is the product of a left-zero and a right-zero semigroup. Since every direct product is a subdirect product, a subdirectly irreducible rectangular band must lie in either $\mathcal{Lz}$ or $\mathcal{Rz}$. Thus $\mathcal{Rb}_{\text{si}} = \{\mathbf{L}_2, \mathbf{R}_2\}$. From this it follows that there can be no other varieties of rectangular bands and the lattice of subvarieties is as claimed. □

Notice that the method of proof we used for the varieties of Abelian groups was purely semantic, while for distributive lattices and rectangular bands, we relied on algebraic considerations. Now let us turn to the variety of p-algebras. This example is interesting because our approach will involve a consideration of both the subdirectly irreducible algebras (characterized in Theorem 3.38) and the fact that every variety is an equational class.

We proved in Theorem 3.34 that $\mathcal{Pa}$ is a variety. We have already isolated two subvarieties: Boolean algebras and Stone algebras. First, we need several facts about Boolean algebras. Most of these have come up before, but it is good to have them all in one place. The reader may want to review the discussion of p-algebras in Section 3.4.

Lemma 4.50. (1) *The only directly indecomposable Boolean algebra is* $\mathbb{B}_1$.

(2) *Every finite Boolean algebra is isomorphic to* $(\mathbb{B}_1)^n$, *for some natural number* n. $(\mathbb{B}_1)^n \cong \text{Sb}(\{1, 2, \ldots, n\}) \cong \mathbb{B}_n$.

(3) *Every Boolean algebra is locally finite. In fact, a* k*-generated Boolean algebra has cardinality at most* 2^{2^k}.

(4) *If* $\mathbf{B}$ *is an infinite Boolean algebra, then* $\mathbf{B}$ *contains subalgebras isomorphic to* $\mathbb{B}_n$ *for all positive integers* n.

Proof. (1) was Exercise 3.15.2. (2) follows immediately from (1). According to Corollary 2.45, the only subdirectly irreducible Boolean algebra is $\mathbb{B}_1$. Therefore, the variety $\mathcal{Ba}$ of all Boolean algebras is generated by $\mathbb{B}_1$, so by Theorem 3.49 is locally finite. A k-generated Boolean algebra is a homomorphic image of the free algebra $\mathbf{F}_{\mathcal{Ba}}(k)$. According to Exercise 4.34.3, $\mathbf{F}_{\mathcal{Ba}}(k) \cong \text{Clo}_k(\mathbb{B}_1) \leq \mathbb{B}_1^{(2^k)}$. This bounds the cardinality of a k-generated algebra. (In fact, $|\text{Clo}_k(\mathbb{B}_1)| = 2^{(2^k)}$, but we don't even need equality here.)

We prove the last assertion by induction on n. If $n = 1$, then $\{0, 1\}$ forms a subalgebra of $\mathbf{B}$ isomorphic to $\mathbb{B}_1$. Assume the result holds for an integer n. Since $\mathbf{B}$ is infinite, it is not directly indecomposable. Therefore, $\mathbf{B} \cong \mathbf{A} \times \mathbf{C}$, where both $\mathbf{A}$ and $\mathbf{C}$ are nontrivial, and at least one, say $\mathbf{C}$, is infinite. Then $\mathbf{A}$ has a subalgebra $\mathbf{A}'$ isomorphic to $\mathbb{B}_1$, and by the induction hypothesis, $\mathbf{C}$ has a subalgebra $\mathbf{C}'$ isomorphic to $\mathbb{B}_n$, whence $\mathbf{B}$ has a subalgebra isomorphic to $\mathbf{A}' \times \mathbf{C}' \cong \mathbb{B}_1 \times \mathbb{B}_n \cong \mathbb{B}_1 \times \mathbb{B}_1^n \cong \mathbb{B}_1^{n+1} \cong \mathbb{B}_{n+1}$. □

The proof of the next proposition is left as an exercise.

Proposition 4.51. *Every subdirectly irreducible p-algebra is locally finite. Every subalgebra of a subdirectly irreducible p-algebra is itself subdirectly irreducible.*

Proposition 4.52. (1) *Let* $\mathbf{B}$ *and* $\mathbf{C}$ *be Boolean algebras, and suppose that* $\mathbf{B}$ *can be embedded in* $\mathbf{C}$*. Then* $\overline{\mathbf{B}}$ *can be embedded in* $\overline{\mathbf{C}}$*.*

(2) *If* $n < m$ *then* $\overline{\mathbb{B}}_n$ *can be embedded in* $\overline{\mathbb{B}}_m$*.*

(3) *Let* $\mathbf{L}$ *be an infinite, subdirectly irreducible p-algebra. Then* $\mathbf{L}$ *contains a copy of every finite subdirectly irreducible p-algebra as a subalgebra.*

Proof. (1) follows from the definition of $\overline{\mathbf{B}}$ in Section 3.4. The other two claims follow from Lemma 4.50. □

We are now in a position to describe the lattice of subvarieties of $\mathcal{Pa}$. For each $n \in \omega$, let $\mathcal{Pa}_n = \mathsf{V}(\overline{\mathbb{B}}_n)$. Each $\mathcal{Pa}_n$ is a subvariety of $\mathcal{Pa}$ since $\overline{\mathbb{B}}_n \in \mathcal{Pa}$. These, together with t, the trivial variety, turn out to be the only proper subvarieties.

Theorem 4.53. *The lattice of subvarieties of* $\mathcal{Pa}$ *forms a chain:*

$$t \subset \mathcal{Pa}_0 \subset \mathcal{Pa}_1 \subset \mathcal{Pa}_2 \subset \cdots \mathcal{Pa}$$

in which all inclusions are proper.

We first prove the following lemma.

Lemma 4.54. *For every natural number* n*, the subdirectly irreducible members of* $\mathcal{Pa}_n$ *are* $\overline{\mathbb{B}}_0, \overline{\mathbb{B}}_1, \ldots, \overline{\mathbb{B}}_n$ *(up to isomorphism).*

Proof. Suppose $\overline{\mathbb{B}}_m \in \mathcal{Pa}_n$. Then by Theorem 3.50 $\overline{\mathbb{B}}_m$ lies in $\mathsf{HS}(\overline{\mathbb{B}}_n)$, from which it follows on cardinality grounds that $m \leq n$. On the other hand, if $m \leq n$, then $\overline{\mathbb{B}}_m$ is isomorphic to a subalgebra of $\overline{\mathbb{B}}_n \in \mathcal{Pa}_n$, so we do have $\overline{\mathbb{B}}_m \in \mathcal{Pa}_n$.

Finally, if $\mathbf{L}$ is an infinite subdirectly irreducible algebra in $\mathcal{Pa}_n$, then by Proposition 4.52(3), $\overline{\mathbb{B}}_{n+1} \in \mathcal{Pa}_n$, contradicting the previous paragraph. □

Proof of Theorem 4.53. By definition, each $\mathcal{Pa}_n$ is a subvariety of $\mathcal{Pa}$. By Lemma 4.54 and Corollary 3.45, $\mathcal{Pa}_n \subsetneq \mathcal{Pa}_{n+1}$ for every $n \in \omega$. Therefore each $\mathcal{Pa}_n$ is a proper subvariety of $\mathcal{Pa}$.

Now we wish to show that these are the only proper subvarieties of $\mathcal{Pa}$. So, let $\mathcal{V}$ be any proper nontrivial subvariety of $\mathcal{Pa}$. There must be some identity ϵ such that $\mathcal{V} \models \epsilon$ but $\mathcal{Pa} \not\models \epsilon$. Therefore, there is an algebra $\mathbf{L} \in \mathcal{Pa}$ such that $\mathbf{L} \not\models \epsilon$. $\mathbf{L}$ has a subdirect representation by subdirectly irreducible algebras, say $\langle \mathbf{L}_j : j \in J \rangle$. Thus, there must be some $j \in J$ such that $\mathbf{L}_j \not\models \epsilon$.

Now, ϵ is an expression involving the variables $x_1, x_2, \ldots, x_k$ for some natural number k. To say that $\mathbf{L}_j \not\models \epsilon$ is to claim that there are elements

$a_1, \dots, a_k \in L_j$ such that ϵ does not hold when a_i is substituted for x_i, $i = 1, 2, \dots, k$. Let $\mathbf{C}$ be the subalgebra of $\mathbf{L}_j$ generated by $a_1, \dots, a_k$. By Proposition 4.51, $\mathbf{C}$ is a finite, subdirectly irreducible p-algebra, that is to say, $\mathbf{C} \cong \overline{\mathbb{B}}_n$ for some integer n. It follows that $\overline{\mathbb{B}}_n \not\models \epsilon$, which implies that $\overline{\mathbb{B}}_n \notin \mathcal{V}$. By Proposition 4.52 for every $m > n$, $\overline{\mathbb{B}}_m \notin \mathcal{V}$. On the other hand, since $\mathcal{V}$ is nontrivial, $\overline{\mathbb{B}}_0 \in \mathcal{V}$.

Let u be the *largest* integer m such that $\overline{\mathbb{B}}_m \in \mathcal{V}$. (By the previous paragraph, u is well defined.) Then $\mathcal{Pa}_u \subseteq \mathcal{V}$. On the other hand, the maximality of u implies that $\mathcal{V}_{\mathrm{si}} \subseteq \mathsf{I}\{\overline{\mathbb{B}}_0, \overline{\mathbb{B}}_1, \dots, \overline{\mathbb{B}}_u\} \subseteq \mathcal{Pa}_u$. It follows that $\mathcal{V} = \mathcal{Pa}_u$, completing the proof. □

In the above description, $\mathcal{t}$ is the variety of trivial (i.e., 1-element algebras), $\mathcal{Pa}_0$ is the variety of Boolean algebras and $\mathcal{Pa}_1$ is the variety of Stone algebras. The other subvarieties get progressively more complicated to describe. K.B. Lee [Lee70] found a single identity that axiomatizes $\mathcal{Pa}_n$ relative to $\mathcal{Pa}$ for each natural number n.

Knowing the lattice of subvarieties can be quite helpful. Here is an example. Observe that each of the varieties $\mathcal{Pa}_n$ is finitely generated, hence locally finite. However, $\mathcal{Pa}$ itself is not finitely generated (why not?). Nevertheless, it is locally finite, as the next corollary shows.

Corollary 4.55. *The variety $\mathcal{Pa}$ is locally finite.*

Proof. Let $\mathbf{L}$ be a p-algebra generated by $a_1, a_2, \dots, a_k$. Represent $\mathbf{L}$ as a subdirect product of subdirectly irreducible algebras $\langle \mathbf{L}_i : i \in I \rangle$, and let $p_i \colon \mathbf{L} \twoheadrightarrow \mathbf{L}_i$ be the canonical projection. Then each $\mathbf{L}_i$ is generated by $p_i(a_1), \dots, p_i(a_k)$, so $|L_i| \le 2^{2^k} + 1$. It follows that every $\mathbf{L}_i$ lies in the subvariety $\mathcal{Pa}_{2^k}$ (this is Lemma 4.54 again). But then $\mathbf{L} \in \mathcal{Pa}_{2^k}$ since every variety is closed under subdirect products. Since the variety $\mathcal{Pa}_n$ is locally finite, L is finite. □

It is interesting to compare this last result on p-algebras with the corresponding question for Abelian groups. Every proper subvariety of $\mathcal{A}$ is finitely generated, hence locally finite. But the variety $\mathcal{A}$ itself is not locally finite. (After all, the group $\mathbb{Z}$ is finitely generated.) Can you pinpoint the place where the proof of Corollary 4.55 goes wrong when applied to Abelian groups?

The question of whether a particular variety is locally finite can be difficult. Recall that $\mathcal{G}_n$ denotes the variety of groups of exponent n. The *Burnside problem* asks for a characterization of those n for which $\mathcal{G}_n$ is locally finite. Burnside posed the problem in 1902. It was not until 1968 that any value of n was found for which $\mathcal{G}_n$ is not locally finite. We now know that $\mathcal{G}_n$ is not locally finite for odd values of n greater than 665, and for even values of n greater than 2^{13}. At the time of this writing, it is not known, on the other hand, whether $\mathcal{G}_5$ is locally finite.

Exercise Set 4.56.

1. Determine the lattice of subvarieties of $\mathcal{Cr}_4$. Find an equational base for each subvariety.

2. Let $\mathbf{R}$ be a ring and ${}_{\mathbf{R}}\mathcal{Mod}$ the variety of all unital left $\mathbf{R}$-modules. Generalize Theorem 4.46 to prove that the lattice $\mathbf{L_V}({}_{\mathbf{R}}\mathcal{Mod})$ is isomorphic to the dual of the lattice of (two-sided) ideals of $\mathbf{R}$.

3. Prove Proposition 4.51. Prove also that if a subdirectly irreducible p-algebra is k-generated, then it has cardinality at most $2^{(2^k)} + 1$.

4. Prove that the variety $\mathcal{Pa}$ is not finitely generated.

5. Prove that every variety is generated by its finitely generated subdirectly irreducible members. (Hint: Theorem 4.44.)

6. Let $\mathbf{L}$ be a complete lattice. A *splitting* of $\mathbf{L}$ is a pair (x, y) of elements of L such that L is the disjoint union of $(x]$ and $[y)$.

   (a) Let (x, y) be a splitting of $\mathbf{L}$. Prove that x is completely meet-irreducible and y is completely join-irreducible in $\mathbf{L}$. (See Definition 3.22.)

   (b) Let $\mathcal{V}$ be a variety with equational base Σ and $(\mathcal{W}_1, \mathcal{W}_2)$ a splitting of $\mathbf{L_V}(\mathcal{V})$. Prove that there is a single equation ϵ and a single finitely generated, subdirectly irreducible algebra $\mathbf{A}$ such that $\mathcal{W}_1 = \text{Mod}(\Sigma \cup \epsilon)$ and $\mathcal{W}_2 = \mathbf{V}(\mathbf{A})$.

4.6 Equational theories and fully invariant congruences

By utilizing the Galois correspondence between algebras and identities, we can obtain information on the lattice of varieties by studying instead the lattice of equational theories. Recall that equational theories are sets of equations and that an equation is an ordered pair of elements from $T(X_\omega)$. It turns out that the equational theories can be characterized directly from the algebraic structure of $\mathbf{T}(X_\omega)$. For any algebra $\mathbf{A}$, we denote by $\text{End}(\mathbf{A})$ the set of endomorphisms of $\mathbf{A}$.

Definition 4.57. Let $\mathbf{A}$ be an algebra. A congruence θ on $\mathbf{A}$ is called *fully invariant* if

$$\forall f \in \text{End}(\mathbf{A}) \;\; (a, b) \in \theta \implies (f(a), f(b)) \in \theta.$$

We write $\text{Con}_{\text{fi}}(\mathbf{A})$ for the set of fully invariant congruences on $\mathbf{A}$.

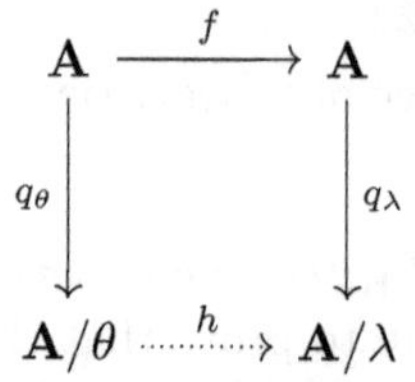

FIGURE 4.8: Diagram for Theorem 4.59

For an algebra $\mathbf{A} = \langle A, F\rangle$, let us write (in this section only) $\mathbf{A}^+ = \langle A, F \cup \mathrm{End}(\mathbf{A})\rangle$. Then from the definition, we see that $\mathrm{Con}_{\mathrm{fi}}(\mathbf{A}) = \mathrm{Con}(\mathbf{A}^+)$.

Proposition 4.58. $\mathrm{Con}_{\mathrm{fi}}(\mathbf{A})$ *is a complete sublattice of* $\mathrm{Con}(\mathbf{A})$ *and is an algebraic lattice.*

Proof. Since the algebra $\mathbf{A}$ is a reduct of the algebra $\mathbf{A}^+$, Theorem 2.18 tells us that $\mathrm{Con}_{\mathrm{fi}}(\mathbf{A}) = \mathrm{Con}(\mathbf{A}^+)$ is a complete sublattice of $\mathrm{Con}(\mathbf{A})$. And as a congruence lattice $\mathrm{Con}_{\mathrm{fi}}(\mathbf{A})$ is algebraic. □

Let $\mathcal{V}$ be a variety, and $\mathbf{A}$ an algebra of the same similarity type as $\mathcal{V}$. Recall the congruence $\lambda_{\mathcal{V}}^{\mathbf{A}}$ of $\mathbf{A}$ from Definition 4.26. This congruence is difficult to grasp since it involves not just the structure of $\mathbf{A}$, but the variety $\mathcal{V}$ as well. The following theorem gives us a necessary condition on such a congruence.

Theorem 4.59. *Let $\mathcal{V}$ be a variety, and* $\mathbf{A}$ *an algebra. Then $\lambda_{\mathcal{V}}$ is a fully invariant congruence of* $\mathbf{A}$.

Proof. Since $\mathcal{V}$ is a variety, we can write $\Lambda_{\mathcal{V}} = \{\theta \in \mathrm{Con}\,\mathbf{A} : \mathbf{A}/\theta \in \mathcal{V}\}$. It follows from Lemma 4.27 that $\mathbf{A}/\lambda_{\mathcal{V}} \in \mathcal{V}$, so $\lambda_{\mathcal{V}}$ is the smallest member of $\Lambda_{\mathcal{V}}$.

Let $f \in \mathrm{End}(\mathbf{A})$ and $(a, b) \in \lambda$. We must show that $(f(a), f(b)) \in \lambda$. Let $q_\lambda \colon \mathbf{A} \twoheadrightarrow \mathbf{A}/\lambda$ be the canonical surjection and $\theta = \ker(q_\lambda \circ f)$. By the fundamental homomorphism theorem there is an embedding $h\colon \mathbf{A}/\theta \to \mathbf{A}/\lambda$ such that $h \circ q_\theta = q_\lambda \circ f$. See Figure 4.8. Since h is an embedding

$$\mathbf{A}/\lambda \in \mathcal{V} \implies \mathbf{A}/\theta \in \mathcal{V} \implies \theta \in \Lambda \implies \lambda \subseteq \theta$$

and therefore

$$(a, b) \in \lambda \implies (a, b) \in \theta \implies q_\lambda \circ f(a) = q_\lambda \circ f(b) \implies (f(a), f(b)) \in \lambda$$

as desired. □

The converse of Theorem 4.59 is false in general, see Exercise 4.63.2. However, it does hold when applied to free algebras.

Theorem 4.60. *Let $\mathcal{V}$ be a variety, $\mathbf{F}$ the free algebra in $\mathcal{V}$ on X and θ a fully invariant congruence on $\mathbf{F}$. Then $\theta = \lambda_{\mathcal{W}}$ for some subvariety $\mathcal{W}$ of $\mathcal{V}$.*

Proof. Let $\mathbf{B} = \mathbf{F}/\theta$ and $\mathcal{W} = \mathsf{V}(\mathbf{B})$. We wish to show that $\theta = \lambda_{\mathcal{W}}$. Since $\mathbf{F}/\theta \in \mathcal{W}$ it follows by definition that $\theta \in \Lambda^{\mathbf{F}}_{\mathcal{W}}$ hence $\lambda_{\mathcal{W}} \subseteq \theta$.

For the converse, let $(a, c) \in \theta$. We must show $(a, c) \in \lambda_{\mathcal{W}}$. Since a and c are members of F, there are terms p and q and $x_1, \ldots, x_n \in X$ such that $a = p^{\mathbf{F}}(x_1, \ldots, x_n)$ and $c = q^{\mathbf{F}}(x_1, \ldots, x_n)$.

We claim that $\mathbf{B} \models p \approx q$. To see this let $b_1, \ldots, b_n \in B$. We will show $p^{\mathbf{B}}(b_1, \ldots, b_n) = q^{\mathbf{B}}(b_1, \ldots, b_n)$. For every $i \leq n$, $b_i = b_i'/\theta$ for some $b_i' \in F$. Since $\mathbf{F}$ is free on X, there is an endomorphism f such that $f(x_i) = b_i'$ for $i = 1, \ldots, n$. Then

$$p^{\mathbf{B}}(b_1, \ldots, b_n) = p^{\mathbf{F}}(b_1', \ldots, b_n')/\theta = f\big(p^{\mathbf{F}}(x_1, \ldots, x_n)\big)/\theta = f(a)/\theta$$

and similarly $q^{\mathbf{B}}(b_1, \ldots, b_n) = f(c)/\theta$. Since $(a, c) \in \theta$ and θ is fully invariant, the claim is proved.

Since $\mathcal{W} = \mathsf{V}(\mathbf{B}) = \mathrm{Mod}(\mathrm{Id}(\mathbf{B}))$, it follows from the claim that $\mathcal{W} \models p \approx q$. The algebra $\mathbf{A} = \mathbf{F}/\lambda_{\mathcal{W}}$ lies in $\mathcal{W}$ by Lemma 4.27 so

$$a/\lambda = p^{\mathbf{A}}(x_1/\lambda, \ldots, x_n/\lambda) = q^{\mathbf{A}}(x_1/\lambda, \ldots, x_n/\lambda) = c/\lambda$$

proving the theorem. □

The above two theorems can be applied to the term algebra $\mathbf{T} = \mathbf{T}_\rho(X_\omega)$. As this algebra is free for the variety of all algebras of similarity type ρ, we obtain

$$\mathrm{Con}_{\mathrm{fi}}(\mathbf{T}) = \{\, \lambda_{\mathcal{V}} : \mathcal{V} \text{ is a variety of algebras of type } \rho \,\}.$$

Finally, we can tie this to the lattice of equational theories. In our formulation, an equation is nothing but an ordered pair of elements of T. An equational theory is a set of equations of the form $\mathrm{Id}(\mathcal{K})$, for some class $\mathcal{K}$ of algebras. But if $\mathcal{V}$ is the variety generated by $\mathcal{K}$, then by Birkhoff's theorem, $\mathcal{V} = \mathrm{Mod}(\mathrm{Id}(\mathcal{K}))$, hence

$$\mathrm{Id}(\mathcal{K}) = \mathrm{Id}(\mathrm{Mod}(\mathrm{Id}(\mathcal{K}))) = \mathrm{Id}(\mathcal{V}) = \lambda_{\mathcal{V}}$$

the first equality coming from the properties of a Galois connection, the last from Theorem 4.38. Thus we obtain the following description of the lattice of equational theories.

Theorem 4.61. *Let ρ be a similarity type. The lattice of equational theories of type ρ is isomorphic to $\mathrm{Con}_{\mathrm{fi}}(\mathbf{T}_\rho(X_\omega))$. The lattice of equational theories is algebraic.*

It follows from this theorem that the closure operator on sets of equations is algebraic. From Theorem 2.30 we know that the compact elements of an algebraic lattice are the closures of finite sets. Combining these two observations, the compact equational theories are precisely the finitely based theories and every equational theory is a join of finitely based theories. In light of the fact that the closure operator in question is $\mathrm{Id} \circ \mathrm{Mod}$ we can rephrase this observation about compactness in the following way.

Corollary 4.62. *Let Σ be a set of equations and ϵ an equation of the same similarity type. If $\mathrm{Mod}(\Sigma) \models \epsilon$ then there is a finite subset Σ_0 of Σ such that $\mathrm{Mod}(\Sigma_0) \models \epsilon$.*

These observations should be seen in contrast to Corollary 4.47, in which we showed that, in general, the lattice of varieties is not algebraic. From the Galois connection we know that there is a dual-isomorphism between the lattices of varieties and equational theories. Thus, although $\mathbf{L_V}(\mathcal{V})$ is not algebraic, it is co-algebraic and the "co-compact" elements are precisely the finitely based varieties.

Theorem 4.61 also gives us insight into the process of generating an equational theory from a set of equations. Suppose Σ is a set of equations. According to the theorem, the equational theory generated by Σ is precisely the fully invariant congruence on $\mathbf{T}$ generated by Σ, and this in turn can be described as $\mathrm{Cg}^{\mathbf{T}^+}(\Sigma)$. We can now apply Theorem 4.17 to actually construct the members of the equational theory.

Let us consider an example. We know from group theory that the identity $x_1 \cdot x_2 \approx x_2 \cdot x_1$ "follows from" the identity $x_1^2 \approx e$, which is to say, the former identity lies in the equational theory generated by the latter (together with the identities defining groups). Let Σ be the following set of identities

$$\begin{aligned} e \cdot x_1 &\approx x_1, & x_1 \cdot e &\approx x_1, \\ x_1^{-1} \cdot x_1 &\approx e, & x_1 \cdot x_1^{-1} &\approx e, \\ x_1 \cdot (x_2 \cdot x_3) &\approx (x_1 \cdot x_2) \cdot x_3, & x_1 \cdot x_1 &\approx e. \end{aligned}$$

Here is one way to show that $x_1 \cdot x_2 \equiv x_2 \cdot x_1 \pmod{\mathrm{Cg}^{\mathbf{T}^+}(\Sigma)}$ using the machinery of Theorem 4.17.

$$\begin{aligned} x_1 \cdot x_2 &\overset{1}{\equiv} (x_1 \cdot x_2) \cdot e \overset{2}{\equiv} (x_1 \cdot x_2) \cdot (x_1 \cdot x_1) \\ &\overset{3}{\equiv} ((x_1 \cdot x_2) \cdot x_1) \cdot x_1 \equiv (((x_1 \cdot x_2) \cdot x_1) \cdot e) \cdot x_1 \\ &\overset{5}{\equiv} (((x_1 \cdot x_2) \cdot x_1) \cdot (x_2 \cdot x_2)) \cdot x_1 \\ &\equiv ((((x_1 \cdot x_2) \cdot x_1) \cdot x_2) \cdot x_2) \cdot x_1 \\ &\equiv (((x_1 \cdot x_2) \cdot (x_1 \cdot x_2)) \cdot x_2) \cdot x_1 \equiv (e \cdot x_2) \cdot x_1 \equiv x_2 \cdot x_1. \end{aligned} \tag{4–7}$$

Keep in mind that the basic operations of $\mathbf{T}^+$ are the members of $\{\cdot, {}^{-1}, e\}$ together with every endomorphism of $\mathbf{T}$. Let f_1 be an endomorphism that maps x_1 to $x_1 \cdot x_2$. The first step in the above sequence is obtained by applying f_1 to the element $x_1 \approx x_1 \cdot e$ from Σ. (Remember that 4.17 allows us to reverse the two sides of an identity.) We get the second step by applying the mapping $f_2(z) = (x_1 \cdot x_2) \cdot z$ to the identity $x_1 \cdot x_1 \approx e$. (Note that f_2 is an elementary translation, specifically $f_2(z) = t(x_1 \cdot x_2, z)$ where $t(y, z) = y \cdot z$ and $x_1 \cdot x_2$ is an element of the algebra $\mathbf{T}$.)

Most of the other steps are similar, but the fifth step is worth singling out since it is not simply an elementary translation. Let

$$g_5(z) = (((x_1 \cdot x_2) \cdot x_1) \cdot z) \cdot x_1$$

and let h_5 be an endomorphism mapping x_1 to $x_2 \cdot x_2$. Then $f_5 = g_5 \circ h_5$ applied to $x_1 \cdot x_1 \approx e$ yields the fifth step.

Notice that the sequence of equivalences in (4–7) amount to a derivation of the identity $x_1 \cdot x_2 \approx x_2 \cdot x_1$ from the set Σ. As the above discussion indicates, such a derivation is always possible. Although we won't make it precise here, this is the content of the *equational completeness* theorem. See [BS81, II.14] for a thorough treatment.

Exercise Set 4.63.

1. Prove that the monolith of a finite subdirectly irreducible algebra is always a fully invariant congruence.

2. Let p be a prime. Prove that every congruence of the group $\mathbb{Z}(p^\infty)$ is fully invariant, but that no congruence, except for 0 and 1, is of the form $\lambda_{\mathcal{V}}$ for a variety $\mathcal{V}$.

3. Let $\mathbf{A}$ be an algebra, $\mathcal{V}$ and $\mathcal{W}$ varieties. Prove that $\lambda^{\mathbf{A}}_{\mathcal{V}\wedge\mathcal{W}} = \lambda^{\mathbf{A}}_{\mathcal{V}} \vee \lambda^{\mathbf{A}}_{\mathcal{W}}$ and $\lambda^{\mathbf{A}}_{\mathcal{V}\vee\mathcal{W}} \subseteq \lambda^{\mathbf{A}}_{\mathcal{V}} \wedge \lambda^{\mathbf{A}}_{\mathcal{W}}$.

4. Let $\mathcal{V}$ be a variety, $\mathbf{A}$ an algebra and $\gamma \in \mathrm{Con}(\mathbf{A})$. Prove that $\lambda^{\mathbf{A}/\gamma}_{\mathcal{V}} = (\lambda^{\mathbf{A}}_{\mathcal{V}} \vee \gamma)/\gamma$.

5. Recall the variety *Imp* of implication algebras from Exercise 4.45.6.

   (a) Show that *Imp* $\models x \to (y \to z) \approx y \to (x \to z)$. (Hint: consider $\mathbb{B}_1^{\to}$.)

   (b) Therefore it must be possible to derive this identity from the three identities that define *Imp*. See if you can do so.

4.7 Maltsev conditions

Recall that two congruences α and β are said to *permute* if $\alpha \circ \beta = \beta \circ \alpha$. Permuting congruences play a crucial role in the direct decomposition of an algebra (Section 3.2). If α and β permute, then $\alpha \circ \beta$ is the join of α and β in the congruence lattice. An algebra is called *congruence-permutable* if every pair of its congruences permute. A variety is congruence-permutable if every member is a congruence-permutable algebra.

The following theorem initiated the study of congruence-permutable varieties. Besides providing one of the most important tools in universal algebra, it is also one of its most beautiful results, linking the algebraic and logical sides of the subject.

Theorem 4.64 (Maltsev [Mal54]). *Let $\mathcal{V}$ be a variety of algebras. The following are equivalent.*

(a) *$\mathcal{V}$ is congruence-permutable.*

(b) *$\mathbf{F}_{\mathcal{V}}(3)$ is congruence-permutable.*

(c) *There is a ternary term $q(x, y, z)$ such that*

$$\mathcal{V} \models q(x, y, y) \approx x \approx q(y, y, x).$$

Proof. Condition (a) implies (b) *a fortiori*. Assume q is a term satisfying condition (c) and let $\mathbf{A}$ be a member of $\mathcal{V}$, α and β congruences on $\mathbf{A}$, and assume that $(a, b) \in \alpha \circ \beta$. We shall show $(a, b) \in \beta \circ \alpha$ which is enough to prove (a).

Since $(a, b) \in \alpha \circ \beta$, there is an element $c \in A$ such that $a \; \alpha \; c \; \beta \; b$. Then using condition (c)

$$a = q^{\mathbf{A}}(a, b, b) \; \beta \; q^{\mathbf{A}}(a, c, b) \; \alpha \; q^{\mathbf{A}}(a, a, b) = b$$

thus $(a, b) \in \beta \circ \alpha$.

Finally we must prove (b) implies (c). Let $\mathbf{F} = \mathbf{F}_{\mathcal{V}}(x, y, z)$ and define $\alpha = \mathrm{Cg}^{\mathbf{F}}(x, y)$ and $\beta = \mathrm{Cg}^{\mathbf{F}}(y, z)$. Then $(x, z) \in \alpha \circ \beta$. Applying (b) we obtain $(x, z) \in \beta \circ \alpha$ which means that there is $u \in F$ such that $x \; \beta \; u \; \alpha \; z$. Since $\mathbf{F}$ is generated by $\{x, y, z\}$, there is a term q such that $u = q^{\mathbf{F}}(x, y, z)$.

Now, to prove (c), let $\mathbf{A}$ be an arbitrary member of $\mathcal{V}$ and $a, b \in A$. By the freeness of $\mathbf{F}$, there is a homomorphism $g \colon \mathbf{F} \to \mathbf{A}$ such that $g(x) = g(y) = a$ and $g(z) = b$. Then $(x, y) \in \ker g$ from which it follows that $\alpha \subseteq \ker g$. From above we have $(u, z) \in \alpha$. Hence

$$b = g(z) = g(u) = g(q^{\mathbf{F}}(x, y, z)) = q^{\mathbf{A}}(g(x), g(y), g(z)) = q^{\mathbf{A}}(a, a, b).$$

This demonstrates the satisfaction of one of the two identities in (c). The other is proved analogously. □

Theorem 4.64 has had a considerable impact on universal algebra. Congruence permutable varieties are now often called Maltsev varieties, and a term satisfying condition (c) is generally called a Maltsev term. Furthermore, characterizations of a flavor similar to 4.64 are called *Maltsev conditions.* We discuss several more Maltsev conditions below.

We know several examples of Maltsev varieties. The motivating example is the variety of groups. For a Maltsev term we can take $q(x, y, z) = x \cdot y^{-1} \cdot z$. Similarly, for rings we can use $q(x, y, z) = x - y + z$. A more exotic example is the variety of quasigroups, see Definition 1.6. An appropriate Maltsev term is $q(x, y, z) = (x/(y\backslash y)) \cdot (y\backslash z)$ (see Exercise 4.75.1). On the other hand, the three-element chain is not congruence-permutable, so no nontrivial variety of lattices has a Maltsev term.

It is sometimes useful to visualize Theorem 4.64 geometrically. Imagine the elements of an algebra as points in a space, the classes of the congruence β as horizontal lines and the classes of α as diagonal lines. In the figure below, we have three points with $a \mathrel{\alpha} b \mathrel{\beta} c$. The Maltsev term "completes the parallelogram" by producing a fourth vertex $q(a, b, c)$.

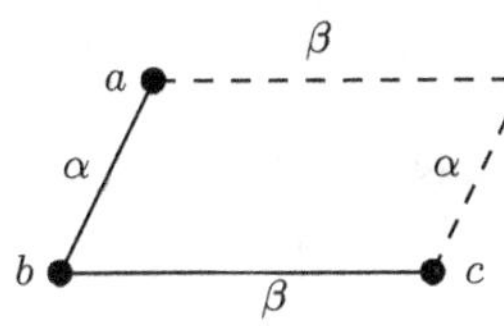

Here is a useful result whose proof is quite easy when a Maltsev term is employed. To improve readability, we write $(\mathbf{u}, \mathbf{v}) \in \theta$, or even $\mathbf{u} \mathrel{\theta} \mathbf{v}$ to indicate that $(u_i, v_i) \in \theta$ for $i = 1, \ldots, n$, where $\mathbf{u}$ is shorthand for $\langle u_1, \ldots, u_n \rangle$.

Theorem 4.65. *Let $\mathcal{V}$ be a Maltsev variety and $\mathbf{A} \in \mathcal{V}$.*

(1) *Every reflexive subalgebra of $\mathbf{A}^2$ is a congruence on $\mathbf{A}$.*

(2) *Let $\theta \subseteq A^2$ and $a, b \in A$. The following are equivalent.*

   (a) $(a, b) \in \mathrm{Cg}^{\mathbf{A}}(\theta)$;

   (b) *There are $(\mathbf{u}, \mathbf{v}) \in \theta$, $\mathbf{w} \in A^m$ and $t \in \mathrm{Clo}_{n+m}(\mathbf{A})$ such that $a = t(\mathbf{u}, \mathbf{w})$ and $b = t(\mathbf{v}, \mathbf{w})$.*

   (c) *There are $(\mathbf{u}, \mathbf{v}) \in \theta$ and $p \in \mathrm{Pol}_n(\mathbf{A})$ such that $a = p(\mathbf{u})$ and $b = p(\mathbf{v})$.*

Proof. For the purposes of this proof, we shall write the elements of A^2 as vertical pairs $\binom{x}{y}$ for $x, y \in A$. Since $\mathcal{V}$ is Maltsev, there is a ternary term q satisfying the identities of Theorem 4.64(c). Let θ be a reflexive subalgebra of $\mathbf{A}^2$. To prove the first claim we must show that θ is symmetric and transitive. For symmetry, let $\binom{a}{b} \in \theta$. Since θ is reflexive, both $\binom{a}{a}$ and $\binom{b}{b}$ are in θ as well. Since θ is a subalgebra of $\mathbf{A}^2$

$$\begin{pmatrix} b \\ a \end{pmatrix} = \begin{pmatrix} q^{\mathbf{A}}(a, a, b) \\ q^{\mathbf{A}}(a, b, b) \end{pmatrix} = q^{\mathbf{A}^2}\left(\begin{pmatrix} a \\ a \end{pmatrix}, \begin{pmatrix} a \\ b \end{pmatrix}, \begin{pmatrix} b \\ b \end{pmatrix}\right) \in \theta.$$

For transitivity, assume that $\binom{a}{b}, \binom{b}{c} \in \theta$. Then

$$\binom{a}{c} = \binom{q^{\mathbf{A}}(a,b,b)}{q^{\mathbf{A}}(b,b,c)} = q^{\mathbf{A}^2}\left(\binom{a}{b}, \binom{b}{b}, \binom{b}{c}\right) \in \theta.$$

Now for the second claim of the theorem. (b) ⇒ (c) ⇒ (a) is trivial. We shall prove (a) ⇒ (b). Let

$$\bar{\theta} = \left\{ \binom{t(\mathbf{u},\mathbf{w})}{t(\mathbf{v},\mathbf{w})} : t \in \mathrm{Clo}(\mathbf{A}),\ \mathbf{u}\,\theta\,\mathbf{v},\ \mathbf{w} \in A^m \right\}.$$

Clearly $\theta \subseteq \bar{\theta} \subseteq \mathrm{Cg}(\theta)$. It is enough to show that $\bar{\theta}$ is a congruence on $\mathbf{A}$. But this follows from part (1) since it is easy to see that $\bar{\theta}$ is a reflexive subalgebra of $\mathbf{A}^2$. □

We now turn to congruence-distributive varieties. Recall that $\mathcal{V}$ is called congruence-distributive if every member has a distributive congruence lattice. We have already seen that such varieties have striking properties (Theorem 3.50) and we will pursue this further in Part II. For now we prove a theorem that is similar in flavor to Maltsev's characterization of congruence-permutable varieties.

Theorem 4.66 (Jónsson [Jón67]). *Let $\mathcal{V}$ be a variety of algebras. The following are equivalent.*

(a) *$\mathcal{V}$ is congruence-distributive.*

(b) *$\mathbf{F}_{\mathcal{V}}(3)$ is congruence-distributive.*

(c) *There is a positive integer n and ternary terms $p_0, p_1, \ldots, p_n$ such that $\mathcal{V}$ satisfies the identities*

(i)	$p_i(x,y,x) \approx x,$	*for* $0 \le i \le n,$
(ii)	$p_0(x,y,z) \approx x,$	
(iii)	$p_n(x,y,z) \approx z,$	
(iv)	$p_i(x,x,y) \approx p_{i+1}(x,x,y),$	*for i even,*
(v)	$p_i(x,y,y) \approx p_{i+1}(x,y,y),$	*for i odd.*

Proof. (a) ⇒ (b) is immediate. Let us prove (b) ⇒ (c). Let $\mathbf{F}$ denote the free algebra on $\{x, y, z\}$ and define

$$\alpha = \mathrm{Cg}^{\mathbf{F}}(x,y), \quad \beta = \mathrm{Cg}^{\mathbf{F}}(y,z), \quad \gamma = \mathrm{Cg}^{\mathbf{F}}(x,z).$$

By congruence-distributivity $(\alpha \vee \beta) \wedge \gamma = (\alpha \wedge \gamma) \vee (\beta \wedge \gamma)$. Since (x,z) clearly lies in the left-hand side of this equation, it lies in the right-hand side as well. Therefore, by Corollary 2.19 there are elements $u_0, u_1, \ldots, u_n \in F$ such that

$$x = u_0\ (\alpha \wedge \gamma)\ u_1\ (\beta \wedge \gamma)\ u_2\ (\alpha \wedge \gamma)\ u_3 \cdots (\beta \wedge \gamma)\ u_n = z. \tag{4–8}$$

Since $\mathbf{F}$ is generated by $\{x, y, z\}$, there are terms p_i for $i = 0, \ldots, n$ such that $u_i = p_i^{\mathbf{F}}(x, y, z)$.

Now to verify that the identities of condition (c) hold in $\mathcal{V}$, let $\mathbf{A} \in \mathcal{V}$ and $a, b, c \in A$. By the freeness of $\mathbf{F}$, there are homomorphisms $f, g, h\colon \mathbf{F} \to \mathbf{A}$ such that

$$\begin{aligned} f(x) = f(y) = a, &\quad f(z) = b \\ g(y) = g(z) = b, &\quad g(x) = a \\ h(x) = h(z) = a, &\quad h(y) = b. \end{aligned}$$

Note that

$$\begin{aligned} \ker f &\supseteq \alpha \supseteq \alpha \wedge \gamma \\ \ker g &\supseteq \beta \supseteq \beta \wedge \gamma \\ \ker h &\supseteq \gamma \supseteq \alpha \wedge \gamma,\ \beta \wedge \gamma. \end{aligned} \tag{4–9}$$

For every $i \leq n$, $h(u_i) = h(p_i^{\mathbf{F}}(x, y, z)) = p_i^{\mathbf{A}}(a, b, a)$. Applying h to the expressions in (4–8) and using (4–9)

$$h(x) = h(u_0) = h(u_1) = h(u_2) = \cdots = h(u_n) = h(z)$$

i.e., $a = p_0^{\mathbf{A}}(a, b, a) = p_1^{\mathbf{A}}(a, b, a) = \cdots = p_n^{\mathbf{A}}(a, b, a)$, which is (i). The fact that $x = u_0$ and $z = u_n$ gives us (ii) and (iii). For (iv), let i be an even index. Then from (4–8), $u_i\ (\alpha \wedge \gamma)\ u_{i+1}$. Applying f and using (4–9) $p_i^{\mathbf{A}}(fx, fy, fz) = p_{i+1}^{\mathbf{A}}(fx, fy, fz)$, i.e., $p_i^{\mathbf{A}}(a, a, b) = p_{i+1}^{\mathbf{A}}(a, a, b)$. A similar argument works for the odd indices.

Finally, assume that the identities in condition (c) hold in $\mathcal{V}$, and let us prove that $\mathcal{V}$ is congruence-distributive. Let $\mathbf{A} \in \mathcal{V}$ and $\alpha, \beta, \gamma \in \mathrm{Con}(\mathbf{A})$. We wish to show that $(\alpha \vee \beta) \wedge \gamma \subseteq (\alpha \wedge \gamma) \vee (\beta \wedge \gamma)$. Call the right-hand side ν and let $(a, b) \in (\alpha \vee \beta) \wedge \gamma$.

Then $(a, b) \in \gamma$ and there are elements $c_0, c_1, \ldots, c_k$ in A such that $a = c_0\ \alpha\ c_1\ \beta\ c_2\ \alpha \cdots \beta\ c_k = b$. Now let $i \leq n$ and $j \leq k$. We have $p_i(a, c_j, b) \equiv_\gamma p_i(a, c_j, a) = a$, using (i) and $b \equiv_\gamma a$, therefore $p_i(a, c_j, b) \equiv_\gamma p_i(a, c_{j+1}, b)$ for all $j < k$. Also $p_i(a, c_0, b) \equiv_\alpha p_i(a, c_1, b) \equiv_\beta p_i(a, c_2, b) \cdots$. Combining these two observations

$$p_i(a, c_j, b) \equiv_\nu p_i(a, c_{j+1}, b), \quad \text{for } j < k. \tag{4–10}$$

We claim that for all $i < n$

$$\begin{aligned} p_i(a, a, b) &\equiv_\nu p_i(a, b, b) \\ p_i(a, b, b) &\equiv_\nu p_{i+1}(a, b, b). \end{aligned} \tag{4–11}$$

To see the first of these observe that from (4–10) $p_i(a, a, b) = p_i(a, c_0, b) \equiv_\nu p_i(a, c_1, b) \equiv_\nu \cdots \equiv_\nu p_i(a, c_k, b) = p_i(a, b, b)$. For the second

$$\begin{aligned} &p_i(a, b, b) = p_{i+1}(a, b, b) && \text{by (v) for } i \text{ odd} \\ &p_i(a, b, b) \equiv_\nu p_i(a, a, b) = p_{i+1}(a, a, b) \equiv_\nu p_{i+1}(a, b, b) && \text{by (iv) for } i \text{ even.} \end{aligned}$$

Using the two relationships in (4–11) we obtain

$$a = p_0(a,b,b) \equiv_\nu p_1(a,b,b) \equiv_\nu p_2(a,b,b) \cdots \equiv_\nu p_n(a,b,b) = b$$

as desired. □

The terms occurring in Theorem 4.66 are usually called *Jónsson terms.* Theorem 4.66 differs qualitatively from 4.64 in that the number of terms required to satisfy condition 4.66(c) depends on the particular variety, while Maltsev's theorem requires exactly one. As n gets larger (in 4.66), the condition gets weaker.

Notice that if $n = 1$ in Theorem 4.66 then in $\mathcal{V}$, $x \approx p_0(x,x,y) \approx p_1(x,x,y) \approx y$, thus $\mathcal{V}$ is a trivial variety. The most important case is $n = 2$. This is equivalent to the identities

$$p_1(x,x,y) \approx p_1(x,y,x) \approx p_1(y,x,x) \approx x.$$

Such a term p_1 is called a *majority term.* In the variety of lattices, the term $m(x,y,z) = (x \wedge y) \vee (x \wedge z) \vee (y \wedge z)$ and its dual both behave as majority terms. If you look carefully at the proof of Theorem 2.21 you will see that it is just a specialization of the proof of the above theorem to the case $n = 2$. Thus the following corollary is immediate.

Corollary 4.67. *If a variety admits a majority term then it is congruence-distributive.*

It is possible to construct, for each $n > 1$, a variety possessing a set of n, but not $n - 1$ Jónsson terms. Exercise 4.75.7 provides an example of this phenomenon for the case $n = 3$.

There is a theorem analogous to 4.66 for congruence modularity. We will not use this theorem in the remainder of the text, so we shall not prove it here. Since the characterization was discovered by Alan Day [Day69], the terms are called *Day terms.*

Theorem 4.68. *Let $\mathcal{V}$ be a variety of algebras. The following are equivalent.*

(a) *$\mathcal{V}$ is congruence-modular.*

(b) *$\mathbf{F}_\mathcal{V}(4)$ is congruence-modular.*

(c) *There is a positive integer n and quatenary terms $p_0, p_1, \ldots, p_n$ such that $\mathcal{V}$ satisfies the identities*

$$\begin{aligned}
&p_i(x,y,y,x) \approx x, && \text{for } 0 \le i \le n,\\
&p_0(x,y,z,u) \approx x, &&\\
&p_n(x,y,z,u) \approx u, &&\\
&p_i(x,x,y,y) \approx p_{i+1}(x,x,y,y), && \text{for } i \text{ even},\\
&p_i(x,y,y,z) \approx p_{i+1}(x,y,y,z), && \text{for } i \text{ odd}.
\end{aligned}$$

We know that each of the conditions of congruence-distributivity and congruence-permutability is important in its own right. The combination of the two properties turns out to be particularly potent. For one thing it is characterized by a nice Maltsev condition.

Definition 4.69. An algebra is called *arithmetical* if it is both congruence-permutable and congruence-distributive. A variety is arithmetical if every member is arithmetical.

Theorem 4.70 (Pixley [Pix63]). *Let $\mathcal{V}$ be a variety. The following are equivalent.*

(a) *$\mathcal{V}$ is arithmetical.*

(b) *$\mathbf{F}_{\mathcal{V}}(3)$ is arithmetical.*

(c) *There is a term $p(x, y, z)$ such that $\mathcal{V}$ satisfies the identities*

$$p(x, y, x) \approx p(x, y, y) \approx p(y, y, x) \approx x.$$

The term p described in part (c) of the above theorem is called, not surprisingly, a *Pixley term.* The proof of Theorem 4.70 is left as an exercise.

Since every arithmetical variety is congruence-distributive by definition, it should satisfy the conditions of Theorem 4.66, which is to say, it will have a set of Jónsson terms. It turns out that in this case we can always take $n = 2$.

Corollary 4.71. *Every arithmetical variety has a majority term.*

Proof. It is easy to see that if p is a Pixley term, then the term

$$m(x, y, z) = p(x, p(x, y, z), z)$$

is a majority term. □

Here's a puzzle: given a Maltsev term and a majority term for a variety, construct a Pixley term.

Among the examples of arithmetical varieties are those of Boolean algebras and the varieties $\mathcal{C}r_n$ of rings. These are discussed in the exercises. Later we shall exhibit several finite quasigroups that generate arithmetical varieties.

Example 4.72. We introduce one other arithmetical variety that is interesting in its own right. We call this the variety of r-lattices. This, by the way, is one of the few examples of a naturally occurring operation of rank greater than 2 that is known to the author. A lattice is called *relatively complemented* if every interval forms a complemented lattice. In other words, for every $x \leq y \leq z$ there is an element w such that $y \wedge w = x$ and $y \vee w = z$. Note that the element w need not be unique. Furthermore, a sublattice of a relatively complemented lattice need not be relatively complemented.

There are several ways to turn a relatively complemented lattice into an algebraic structure. Let us define an *r-lattice* to be an algebra $\langle L, \wedge, \vee, r\rangle$ of type $\langle 2, 2, 3\rangle$ such that

$$\begin{aligned}&\langle L, \wedge, \vee\rangle \text{ is a lattice,}\\ &y \wedge r(x, y, z) \approx x \wedge y \wedge z\\ &y \vee r(x, y, z) \approx x \vee y \vee z.\end{aligned}$$

Notice that whenever $x \leq y \leq z$, $r(x, y, z)$ will be a relative complement of y in the interval $\mathbf{I}[x, z]$.

The variety of r-lattices is arithmetical. For a Pixley term, we can take

$$p(x, y, z) = r(x,\ x \vee y,\ x \vee y \vee z) \wedge r(z,\ z \vee y,\ z \vee y \vee x).$$

For then

$$\begin{aligned}p(x, x, z) &\approx r(x, x, x \vee z) \wedge r(z, z \vee x, z \vee x) \approx (x \vee z) \wedge z \approx z\\ p(x, z, z) &\approx r(x, x \vee z, x \vee z) \wedge r(z, z, z \vee x) \approx x \wedge (z \vee x) \approx x\\ p(x, y, x) &\approx r(x, x \vee y, x \vee y) \wedge r(x, x \vee y, x \vee y) \approx x \wedge x \approx x.\end{aligned}$$

Every Boolean algebra can obviously be made into an r-lattice by defining $r(x, y, z) = (y' \vee (x \wedge z)) \wedge (x \vee z)$. Thus Boolean algebras are arithmetical.

As an application of Pixley's theorem, we provide the following result of Michler and Wille [MW70] from 1970. The proof depends on Lemma 4.74 together with Exercise 4.75.6. The theorem is true even without the assumption of commutativity. However, the proof involves some ring theory that would take us off-track.

Theorem 4.73 (Michler-Wille, 1970). *Let $\mathcal{V}$ be a congruence-distributive variety of commutative rings. Then for some positive integer n, $\mathcal{V}$ is contained in the variety $\mathcal{C}r_n$.*

As we discussed in Section 4.3, the free commutative ring on one generator (with no identity assumed) consists of those members of $\mathbb{Z}[x]$ with no constant term. Since these polynomials are all multiples of x, we can conveniently notate that set as $x\mathbb{Z}[x]$.

Lemma 4.74. *Let $\mathcal{V}$ be a congruence-distributive variety of commutative rings. Then there is a polynomial $f(x) \in x\mathbb{Z}[x]$ such that $\mathcal{V} \models x \approx xf(x)$.*

Proof of Lemma. Every variety of rings is Maltsev. Thus if $\mathcal{V}$ is congruence-distributive, then it is arithmetical. Therefore by Corollary 4.71, it has a majority term, $p(x, y, z)$. We can write

$$p(x, y, z) = p_1(x) + p_2(y) + p_3(z) + p_4(x, y) + p_5(x, z) + p_6(y, z) + p_7(x, y, z)$$

where p_4 is a sum of monomials in which both the variables x and y must appear, and similarly for p_5, p_6 and p_7. Thus for example, $p_4(x, 0) = p_5(x, 0) = p_6(y, 0) = p_7(x, y, 0) = 0$.

Since p is a majority term we see that in $\mathcal{V}$, $0 \approx p(x,0,0) = p_1(x)$. Similarly, $p_2(y) \approx p_3(z) \approx 0$. Therefore

$$x \approx p(x,x,0) \approx p_4(x,x) + p_5(x,0) + p_6(x,0) + p_7(x,x,0) \approx p_4(x,x).$$

Finally, $p_4(x,x)$ can be written as $xf(x)$ for some $f(x) \in x\mathbb{Z}[x]$. □

Exercise Set 4.75.

1. Let $\mathcal{Q}$ be the variety of quasigroups.
   (a) Prove that $\mathcal{Q} \models x/(x\backslash x) \approx x$.
   (b) Prove that $p(x,y,z) = (x/(y\backslash y)) \cdot (y\backslash z)$ is a Maltsev term for $\mathcal{Q}$.
2. Prove Theorem 4.70. (It is possible to do this by directly combining Maltsev and Jónsson terms, but it would be more educational for you to do it by an argument similar to that of Theorem 4.64.)
3. Let p be an n-ary term for a variety $\mathcal{V}$. Call p $\mathcal{V}$-surjective (resp. $\mathcal{V}$-bijective) if, for every algebra $\mathbf{A}$ in $\mathcal{V}$, $p^{\mathbf{A}}$ is a surjective (resp. bijective) function from A^n to A.
   (a) Prove that p is $\mathcal{V}$-surjective if and only if there are unary terms $u_1, \dots, u_n$ such that $\mathcal{V} \models p\big(u_1(x), \dots, u_n(x)\big) \approx x$.
   (b) Prove that p is $\mathcal{V}$-bijective if and only if in addition to the above, the identities $u_i(p(x_1, \dots, x_n)) \approx x_i$ hold in $\mathcal{V}$, for $i = 1, \dots, n$.
4. According to Exercise 2.22.4, for any n, the variety $\mathcal{C}r_n$ is congruence-distributive. And being a variety of rings, it is Maltsev. Therefore it is arithmetical, so by 4.70, it has a Pixley term. Find one.
5. We say that an algebra $\mathbf{A}$ has the *Chinese remainder condition* if, for all $\{\theta_1, \dots, \theta_n\} \subseteq_\omega \mathrm{Con}(\mathbf{A})$ and all $a_1, \dots, a_n \in A$, if

   $$a_i \equiv a_j \pmod{\theta_i \vee \theta_j}, \quad \text{for all } 1 \le i, j \le n$$

   then there is $x \in A$ such that

   $$x \equiv a_i \pmod{\theta_i}, \quad \text{for all } 1 \le i \le n.$$

   Prove that $\mathbf{A}$ is an arithmetical algebra if and only if it satisfies the Chinese remainder condition.
6. Prove Theorem 4.73. (Hint: follow the line of argument used in Theorem 3.28. You may use Lemma 4.74.)
7. Recall the variety $\mathcal{I}mp$ of implication algebras from Exercise 4.45.6.

(a) Prove that $p_0(x,y,z) = x$, $p_1(x,y,z) = (y \to (z \to x)) \to x$, $p_2(x,y,z) = (x \to y) \to z$, $p_3(x,y,z) = z$ form a set of Jónsson terms for *Imp*.

(b) Prove that *Imp* has no majority term. (Hint: Let $A = \{0,1\}^3 - \{(0,0,0)\}$. Then $\mathbf{A}$ is a subalgebra of $(\mathbb{B}_1^{\to})^3$. But $\mathbf{A}$ can not have a majority term operation since $m((001), (010), (100)) = (000)$.)

(c) Prove that *Imp* is not a Maltsev variety.

8. Let $\mathcal{V}$ be a congruence-permutable variety, $\mathcal{V} = \mathcal{V}_1 \vee \mathcal{V}_2$ and $\mathcal{V}_1 \wedge \mathcal{V}_2 = t$.

   (a) Prove that there is a binary term $b(x,y)$ so that $\mathcal{V}_i = \{\, \mathbf{A} \in \mathcal{V} : \mathbf{A} \models b(x_1,x_2) \approx x_i \,\}$, for $i = 1,2$.

   (b) Prove that every algebra $\mathbf{A} \in \mathcal{V}$ decomposes as $\mathbf{A}_1 \times \mathbf{A}_2$ with $\mathbf{A}_i \in \mathcal{V}_i$.

   (Hint: Consider $\lambda_i = \lambda^{\mathbf{F}}_{\mathcal{V}_i}$, where $\mathbf{F}$ is a free algebra. Show that $\lambda_1 \circ \lambda_2 = 1_F$ and $\lambda_1 \cap \lambda_2 = 0_F$.)

9. (Csákány [Csá70]) Let $\mathcal{V}$ be a variety and c a constant unary or nullary operation symbol of $\mathcal{V}$. We say that $\mathcal{V}$ *is weakly congruence-regular at* c if for all $\mathbf{A} \in \mathcal{V}$ and $\theta, \psi \in \mathrm{Con}(\mathbf{A})$, $c/\theta = c/\psi \implies \theta = \psi$. For example, groups are weakly congruence-regular at e.

   (a) Prove that $\mathcal{V}$ is weakly congruence-regular at c if and only if there are binary terms $b_1(x,y), \dots, b_n(x,y)$ such that

   $$\mathcal{V} \models \big(b_1(x,y) \approx c \;\&\; \cdots \;\&\; b_n(x,y) \approx c\big) \leftrightarrow x \approx y.$$

   (Hint: Let $\mathbf{F} = \mathbf{F}_{\mathcal{V}}(x,y)$, $\theta = \mathrm{Cg}(x,y)$ and for all $b \in c/\theta$ let $\theta_b = \mathrm{Cg}(b,c)$. Then $\theta = \bigvee_b \theta_b$.)

   (b) Show that for the variety of groups, we can take $n = 1$ in the above definition.

   (c) Show that *Imp* is weakly congruence-regular at 1, by finding terms b_1 and b_2 satisfying the condition of part (a).

4.8 Interpretations

Algebra is replete with transformations[3] that turn one kind of structure into another. Some examples that we have encountered so far in this text include the following.

[3] In this section we use the word "transformation" as an informal term for any mapping from one class of algebras to another.

(1) Turn a ring into an Abelian group by throwing away multiplication.

(2) Every Abelian group can be turned into a ring by defining $x \cdot y = 0$.

(3) Every group can be turned into a quasigroup by defining $x/y = x \cdot y^{-1}$ and $x\backslash y = x^{-1} \cdot y$.

(4) Every Boolean algebra can be turned into a Boolean ring and vise versa, using the definitions in Exercise 2.47.1.

What these constructions all have in common is that the basic operations of the target object are defined as terms of the source object. Before we formalize this idea, let us address one point. We generally consider a Boolean algebra to be of the form $\mathbf{B} = \langle B, \wedge, \vee, ', 0, 1\rangle$. It is not uncommon to leave out the nullary operations: $\mathbf{B}' = \langle B, \wedge, \vee, '\rangle$. We can transform $\mathbf{B}'$ into $\mathbf{B}$ by defining 0 to be $x \wedge x'$. But $x \wedge x'$ is a *unary* operation with constant value 0, not a *nullary* operation. The prevailing opinion in universal algebra is that it is more convenient to include this transformation in our definition than to exclude it. We handle this with a somewhat roundabout definition.

Definition 4.76. Let $\mathcal{V}$ be a variety of similarity type $\rho\colon \mathcal{F} \to \omega$ and $\mathcal{W}$ a variety of similarity type $\sigma\colon \mathcal{G} \to \omega$.

(1) A *strict interpretation of $\mathcal{V}$ in $\mathcal{W}$* is a mapping $D\colon \mathcal{F} \to T_\sigma(X_\omega)$ such that

- For every $f \in \mathcal{F}$, $\rho(f) = \sigma(D(f))$ and
- For every $\mathbf{A} \in \mathcal{W}$, the algebra $\mathbf{A}^D = \langle A, D(f)^{\mathbf{A}}\rangle_{f\in\mathcal{F}}$ is a member of $\mathcal{V}$.

(2) An *insignificant interpretation* on $\mathcal{V}$ replaces nullary operation symbols with their corresponding constant unary terms.

(3) An *interpretation of $\mathcal{V}$ in $\mathcal{W}$* consists of a strict interpretation optionally preceded by an insignificant interpretation.

It may seem strange that "interpretation of $\mathcal{V}$ in $\mathcal{W}$" entails a transformation from $\mathcal{W}$ to $\mathcal{V}$. By thinking of it as a mapping from terms of $\mathcal{V}$ to terms of $\mathcal{W}$, it makes sense.

It should be clear that all of the examples presented at the beginning of the section are interpretations. The fourth example is noteworthy as it involves two interpretations that are inverses of each other. Note, on the other hand, that the first two examples are not inverses.

Definition 4.77. The varieties $\mathcal{V}$ and $\mathcal{W}$ are *term-equivalent* if there are interpretations D of $\mathcal{V}$ into $\mathcal{W}$, and E of $\mathcal{W}$ into $\mathcal{V}$ such that for every $\mathbf{A} \in \mathcal{V}$ and $\mathbf{B} \in \mathcal{W}$, $\mathbf{A}^{ED} = \mathbf{A}$ and $\mathbf{B}^{DE} = \mathbf{B}$.

Term-equivalent varieties are "the same" for most universal-algebraic purposes. Subalgebras correspond to subalgebras, products to products. Even the free algebras of one variety will be transformed into the free algebras of the other under a term-equivalence. (Except perhaps for a free algebra on 0 generators.)

Two important classes of interpretations are the formation of reducts (see page 7) and subvarieties. In fact, just as every function factors into a surjection followed by an injection, every interpretation can be obtained as a reduct plus an inclusion of a subvariety into a variety, with perhaps a term-equivalence thrown in.

Of course not all familiar transformations are interpretations. For example the mapping that assigns to each ring with identity its group of units is not an interpretation, since it does not preserve the underlying set of the ring. The following theorem provides a characterization of those transformations that do arise from interpretations.

Theorem 4.78. *Let $\mathcal{V}$ and $\mathcal{W}$ be varieties with no nullary operation symbols, and let $T\colon \mathcal{W} \to \mathcal{V}$ be a transformation. The following are equivalent.*

(1) *There is an interpretation D of $\mathcal{V}$ in $\mathcal{W}$ such that for every $\mathbf{B} \in \mathcal{W}$, $T(\mathbf{B}) = \mathbf{B}^D$.*

(2) *For every $\mathbf{A}, \mathbf{B} \in \mathcal{W}$ and homomorphism $h\colon \mathbf{A} \to \mathbf{B}$, the underlying universes of $\mathbf{A}$ and $T(\mathbf{A})$ are the same and $h\colon T(\mathbf{A}) \to T(\mathbf{B})$ is a homomorphism.*

The important point in condition (2) is that it is the same function h that is operating on both $\mathbf{A}$ and on $T(\mathbf{A})$. Of course this wouldn't even make sense if the underlying universes were not the same. Condition (2) can be stated succinctly in the language of category theory by saying that T is a functor that commutes with the forgetful functor. Exercise 4.80.1 further shows the importance of the requirement that T preserve underlying sets.

Proof. (1) $\Rightarrow$ (2) is immediate since a $\mathcal{W}$-homomorphism must preserve every $\mathcal{W}$-term. So assume that the transformation T satisfies condition (2). Let f be a basic n-ary operation symbol of $\mathcal{V}$ and $\mathbf{F} = \mathbf{F}_{\mathcal{W}}(X_n)$. By assumption, $\mathbf{F}$ and $T(\mathbf{F})$ share the same universe, so $X_n \subseteq T(F)$. Let $f^{T(\mathbf{F})}(x_1, \ldots, x_n) = a \in T(F)$. Since $\mathbf{F}$ is generated by X_n (as a $\mathcal{W}$-algebra), there is a $\mathcal{W}$-term $t_f(x_1, \ldots, x_n)$ such that $t_f^{\mathbf{F}}(x_1, \ldots, x_n) = a$. We define $D(f) = t_f$.

Now to verify condition (1), let $\mathbf{B} \in \mathcal{W}$. By assumption, $\mathbf{B}$ and $T(\mathbf{B})$ have the same universe, as does $\mathbf{B}^D$. Let f again be an n-ary operation symbol of $\mathcal{V}$, and $b_1, \ldots, b_n \in B$. To show $T(\mathbf{B}) = \mathbf{B}^D$ we must check that $f^{T(\mathbf{B})}(\mathbf{b}) = f^{(\mathbf{B}^D)}(\mathbf{b})$. By definition, $f^{(\mathbf{B}^D)} = t_f^{\mathbf{B}}$. Let $h\colon \mathbf{F} \to \mathbf{B}$ such that $h(x_i) = b_i$, for $i \leq n$. (h is guaranteed to exist because $\mathbf{F}$ is free.) Then

$$t_f^{\mathbf{B}}(\mathbf{b}) = t_f^{\mathbf{B}}(h(\mathbf{x})) = h(t_f^{\mathbf{F}}(\mathbf{x})) = h(a) = \\ h(f^{T(\mathbf{F})}(\mathbf{x})) \overset{*}{=} f^{T(\mathbf{B})}(h(\mathbf{x})) = f^{T(\mathbf{B})}(\mathbf{b}).$$

The indicated equality follows from the assumption that $h\colon T(\mathbf{F}) \to T(\mathbf{B})$ is a homomorphism. □

If D is an interpretation of $\mathcal{V}$ in $\mathcal{W}$, then for every $\mathbf{A} \in \mathcal{W}$ and $n > 0$ we have $\mathrm{Clo}_n(\mathbf{A}^D) \subseteq \mathrm{Clo}_n(\mathbf{A})$. In particular, if D is a term-equivalence, then this inclusion is an equality. This relationship is important enough to deserve its own definition.

Definition 4.79. Let $\mathbf{A}$ and $\mathbf{B}$ be algebras (not necessarily of the same similarity type).

(1) $\mathbf{A}$ and $\mathbf{B}$ are *term-equivalent* if, for every $n > 0$, $\mathrm{Clo}_n(\mathbf{A}) = \mathrm{Clo}_n(\mathbf{B})$. Note that this condition requires that the algebras have the same underlying universe.

(2) $\mathbf{A}$ and $\mathbf{B}$ are *weakly isomorphic* if $\mathbf{A}$ is isomorphic to an algebra term-equivalent to $\mathbf{B}$.

(3) $\mathbf{A}$ and $\mathbf{B}$ are *polynomially equivalent* if $\mathrm{Pol}(\mathbf{A}) = \mathrm{Pol}(\mathbf{B})$.

If $\mathbf{A}$ and $\mathbf{B}$ are term-equivalent (or just weakly isomorphic) algebras, then $\mathsf{V}(\mathbf{A})$ and $\mathsf{V}(\mathbf{B})$ are term-equivalent varieties (Exercise 4.80.5). Now combine this with the result of Exercise 4.80.6. For any finite Latin square $\mathbf{Q}$, $\mathsf{V}(\mathbf{Q})$ is term-equivalent to a variety of quasigroups.

The notion of interpretation provides a nice framework to discuss Maltsev conditions. Consider the variety $\mathcal{CP}$ consisting of one ternary operation symbol q defined by the identities

$$q(x, y, y) \approx q(y, y, x) \approx x.$$

According to Theorem 4.64 a variety $\mathcal{V}$ is congruence-permutable if and only if $\mathcal{CP}$ can be interpreted in $\mathcal{V}$. One could say that $\mathcal{CP}$ is the most general Maltsev variety.

Since interpretations can be composed, it follows that if $\mathcal{V}$ is congruence-permutable and if there is an interpretation of $\mathcal{V}$ in $\mathcal{W}$, then $\mathcal{W}$ must be congruence permutable.

Jónsson's theorem (4.66) on congruence-distributive varieties is a bit more complex. For each positive integer n, Condition (c) of that theorem describes a variety, $\mathcal{CD}_n$, defined using n terms. These varieties get progressively bigger in the sense that $\mathcal{CD}_n$ is interpretable in $\mathcal{CD}_{n+1}$. A variety $\mathcal{V}$ is congruence-distributive if, for some n, $\mathcal{CD}_n$ is interpretable in $\mathcal{V}$. For example, $\mathcal{CD}_2$ is interpretable in the variety of lattices. $\mathcal{CD}_3$ (but not $\mathcal{CD}_2$) is interpretable in the variety of implication algebras.

This perspective has been instrumental in the characterization and analysis of Maltsev conditions, see [Tay73]. Here is an elementary example. A variety is called *semisimple* if every subdirectly irreducible member is simple. Is semisimplicity a Maltsev condition? In other words, can it be described in a manner similar to that of congruence-permutability or distributivity? Well, the variety of sets is semisimple since the only subdirectly irreducible set has two elements. The variety of sets can be interpreted into, say, groups (since it is a reduct). But groups are not semisimple, so the answer to our question is no, semisimplicity is not a Maltsev condition.

Exercise Set 4.80.

1. A *square band* is an algebra $\langle A, \cdot, ' \rangle$ of type $\langle 2, 1 \rangle$ satisfying the identities

$$(x \cdot y) \cdot (u \cdot v) \approx x \cdot v, \quad x \cdot x \approx x,$$
$$(x \cdot y)' \approx y' \cdot x', \quad x'' \approx x.$$

   (a) Show that there is an interpretation of the variety of rectangular bands in that of square bands.

   (b) Show that, up to isomorphism, there is a one-to-one correspondence between sets and square bands. (Hint: For a square band $\mathbf{A}$, consider $S(\mathbf{A}) = \{x : x = x'\} = \{x \cdot x' : x \in A\}$.)

   (c) Show that the variety of sets is not term-equivalent to the variety of square bands.

2. Let $\mathcal{S}$ denote the variety of binars satisfying the identities

$$(x * y) * (u * v) \approx y^2 * u^2, \quad (x^2)^2 \approx x.$$

   (Here, x^2 denotes $x * x$.) Show that $\mathcal{S}$ is term-equivalent to the variety of square bands.

3. (a) Show that the definition $x \vee y = (x \to y) \to y$ constitutes an interpretation of the variety of semilattices into that of implication algebras (cf. Exercise 4.45.6).

   (b) Show that there is no interpretation of implication algebras in semilattices.

4. We defined the variety $\mathcal{C}r_n$ to be rings with no assumption about an identity element. In Theorem 3.28 we showed that, in fact, every member of this variety has an identity. Let $\mathcal{C}r_n^1$ denote the variety of commutative rings with identity satisfying $x^n \approx x$. Are the varieties $\mathcal{C}r_n$ and $\mathcal{C}r_n^1$ term-equivalent? (Note: according to Definition 4.76, it doesn't matter whether the identity element is defined as a nullary or a unary operation.)

5. Let $\mathbf{A}$ and $\mathbf{B}$ be term-equivalent algebras. Prove that $\mathsf{V}(\mathbf{A})$ is term-equivalent to $\mathsf{V}(\mathbf{B})$.

6. Let $\mathbf{Q} = \langle Q, \cdot \rangle$ be a finite Latin square. Prove that $\mathbf{Q}$ is term-equivalent to a quasigroup. (Hint: let $n = |Q|!$. For each $a \in Q$ the self-map $L_a(x) = a \cdot x$ has $(L_a)^n = \iota_Q$.)

Part II

Selected Topics

Introduction to Part II

We have thus far discussed several properties that a variety may or may not possess. For example, a variety $\mathcal{V}$ is called locally finite if every finitely generated member is finite, and $\mathcal{V}$ is finitely generated if it is of the form $\mathbf{V}(\mathcal{K})$ where $\mathcal{K}$ is a finite set of finite algebras. We have also seen some implications that hold among these properties. In Theorem 3.49 we proved that every finitely generated variety is locally finite.

This aspect of universal algebra is the primary focus of Part II. More precisely, the central theme is the structure of congruence lattices. More than any other, this has proved to be a key organizing principle in the study of varieties.

Recall that an algebra is called congruence-distributive if its congruence lattice is distributive. Similarly, it is called congruence-permutable if its congruences permute. Not surprisingly, we call an algebra *congruence-modular* if its congruence lattice is modular. Finally, an algebra that is both congruence-distributive and congruence-permutable is called arithmetical.

We extend these definitions to varieties in a natural way. If X is one of the above properties, we say that a variety is X if and only if every member is X. Congruence-permutable varieties are also called *Maltsev varieties.*

A word of caution is in order here. It is not true that just because an algebra has property X that its generated variety has X. For example the two-element semilattice $\mathbf{2}$ is simple, so it is both congruence-distributive and congruence-permutable. However, $\mathbf{V}(\mathbf{2})$ is the variety of all semilattices, which is neither congruence-distributive nor permutable.

Since every distributive lattice is modular, congruence-distributivity automatically implies congruence-modularity. But the following proposition, which is not strictly speaking a lattice-theoretic fact, is unexpected.

Proposition. *Every congruence-permutable algebra is congruence-modular.*

Proof. Let $\mathbf{A}$ be an algebra with permuting congruences and let α, β, γ be members of $\mathrm{Con}(\mathbf{A})$ with $\alpha \subseteq \beta$. We wish to show that $\beta \wedge (\alpha \vee \gamma) \subseteq \alpha \vee (\beta \wedge \gamma)$. Applying permutability, this is equivalent to

$$\beta \cap (\alpha \circ \gamma) \subseteq \alpha \circ (\beta \cap \gamma).$$

So let $(a, b) \in \beta \cap (\alpha \circ \gamma)$. Thus $a \mathrel{\beta} b$ and there is $c \in A$ such that $a \mathrel{\alpha} c \mathrel{\gamma} b$. Since $\alpha \subseteq \beta$ we get $a \mathrel{\beta} c$ and since $a \mathrel{\beta} b$ we have $b \mathrel{\beta} c$. Therefore $a \mathrel{\alpha} c \mathrel{(\beta \cap \gamma)} b$ as desired. □

We can draw a simple Venn diagram illustrating the relationships among these definitions.

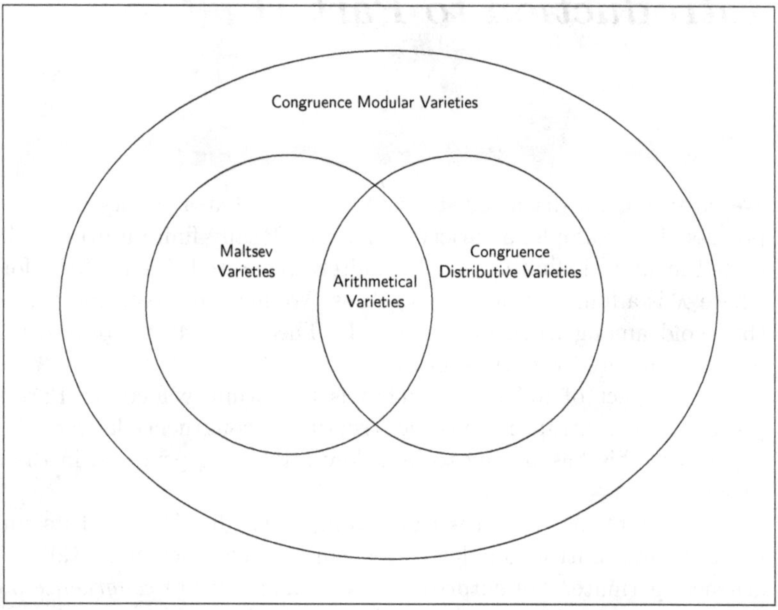

Our plan is to first discuss congruence-distributive varieties, then add permutability and consider arithmetical varieties. Finally, we will turn to Maltsev varieties. Although we shall not do so in this text, much of the work we cover on Maltsev varieties can be extended to congruence-modular varieties.

Chapter 5

Congruence Distributive Varieties

In lattice theory, distributivity is the single strongest property. Distributive lattices have a relatively simple structure and often the key to the structure of an arbitrary lattice lies in isolating distributive sublattices.

It should not be too surprising then to find that algebras with distributive congruences often have nice properties. As we shall see in this chapter, varieties in which every algebra has a distributive congruence lattice lend themselves to deep analysis. The central tool is a discovery of B. Jónsson that imposes considerable structure on the subdirectly irreducible members. This allows an analysis of the subvarieties of a variety, as well as the equations satisfied by its members.

5.1 Ultrafilters and ultraproducts

Let $\mathbf{B}$ be a Boolean algebra. Recall that a proper filter on $\mathbf{B}$ is a nonempty, proper upset that is closed under meets. Equivalently, F is a proper filter on $\mathbf{B}$ if

$$\begin{array}{ll} (i) & 1 \in F \ \& \ 0 \notin F; \\ (ii) & x \in F \ \& \ x \leq y \implies y \in F; \\ (iii) & x, y \in F \implies x \wedge y \in F. \end{array}$$

We can, of course, dualize Definition 2.30 to get the notion of a prime filter. It follows from Exercise 2.47.5 that on a Boolean algebra, prime filters and maximal filters coincide. Maximal filters on a Boolean algebra are usually called *ultrafilters*. It is not hard to see that a filter F is an ultrafilter if and only if

$$(iv) \quad x \vee y \in F \implies x \in F \text{ or } y \in F.$$

Now let T be a set and $\mathbf{B} = \langle\, \mathrm{Sb}(T), \cap, \cup, ', \varnothing, T \,\rangle$. In this context, a filter on the Boolean algebra $\mathbf{B}$ is usually called a *filter over* T. If we simply translate the above conditions, a filter over T is a family $\mathcal{F}$ of subsets of T

such that

$$\begin{array}{ll} (i') & T \in \mathcal{F} \ \& \ \varnothing \notin \mathcal{F} \\ (ii') & X \in \mathcal{F} \ \& \ X \subseteq Y \implies Y \in \mathcal{F} \\ (iii') & X, Y \in \mathcal{F} \implies X \cap Y \in \mathcal{F}. \end{array}$$

And $\mathcal{F}$ is an ultrafilter if, further

$$(iv') \quad X \cup Y \in \mathcal{F} \implies X \in \mathcal{F} \text{ or } Y \in \mathcal{F}.$$

In particular, note that for any $s \in T$, the set $\hat{s} = \{ X \subseteq T : s \in X \}$ is an ultrafilter over T, called the *principal ultrafilter* at s. It is also useful to observe that condition (iv') can be rephrased as:

$$\text{for every } X \subseteq T, \text{ either } X \in \mathcal{F} \text{ or } T - X \in \mathcal{F}.$$

Here is a useful tool for building ultrafilters. It relies on Zorn's lemma.

Definition 5.1. Let T be a set. A family $\mathcal{F}$ of subsets has the *finite intersection property* (FIP) if for every finite subfamily $\mathcal{F}_0$ of $\mathcal{F}$ we have $\bigcap \mathcal{F}_0 \neq \varnothing$.

Proposition 5.2. *Let T be a set and $\mathcal{F}$ a family of subsets of T with the FIP. There is an ultrafilter $\mathcal{U}$ over T containing $\mathcal{F}$.*

Proof. Let

$$\overline{\mathcal{F}} = \left\{ X \subseteq T : X \supseteq \bigcap \mathcal{F}_0, \text{ some } \mathcal{F}_0 \subseteq_\omega \mathcal{F} \right\}$$

One easily checks that $\overline{\mathcal{F}}$ is a filter over T containing $\mathcal{F}$ and excluding $\varnothing$. Let $\mathcal{I} = \{\varnothing\}$. Then $\mathcal{I}$ is an ideal over T and $\mathcal{I}$ is disjoint from $\overline{\mathcal{F}}$. By the prime ideal theorem (Theorem 2.41), dualized if you like, there is an ultrafilter $\mathcal{U}$ separating $\mathcal{I}$ from $\overline{\mathcal{F}}$. This ultrafilter contains $\mathcal{F}$. □

Now, let $\mathbf{A}_i$ be an algebra for each $i \in I$, and $\mathbf{A} = \prod_{i \in I} \mathbf{A}_i$. For $a, b \in A$ we define

$$[\![a = b]\!] = \{ i \in I : a_i = b_i \}$$

called the *equalizer* of a and b. Actually, we use this notation much more generally. For example, if the $\mathbf{A}_i$s are all lattices, then the set

$$\{ i \in I : \mathbf{A}_i \text{ is lower bounded} \}$$

can be denoted $[\![(\exists x)(\forall y)\ x \leq y]\!]$.

For each $i \in I$ we have a congruence relation $\eta_i = \{ (a, b) : a_i = b_i \}$ on $\mathbf{A}$. (The kernel of the projection map.) We extend this notation as follows. For a subset J of I we define

$$\eta_J = \bigcap_{j \in J} \eta_j.$$

Notice that

$$a \,\eta_J\, b \iff J \subseteq [\![a = b]\!].$$

It is easy to check that $\mathbf{A}/\eta_J \cong \prod_{j\in J} \mathbf{A}_j$. Note also that for any subsets J and K, $\eta_{J\cup K} = \eta_J \cap \eta_K$.

Let $\mathcal{F}$ be a filter over I. We define

$$\eta_{\mathcal{F}} = \{\, (a,b) \in A^2 : [\![a = b]\!] \in \mathcal{F} \,\}.$$

It is straightforward to verify that $\eta_{\mathcal{F}}$ is a congruence relation on $\mathbf{A}$. If $\mathcal{F} = \hat{\jmath}$, we have $\eta_{\mathcal{F}} = \eta_j$.

The quotient structure $\mathbf{A}/\eta_{\mathcal{F}}$ is called a *reduced product.* In the special case that $\mathcal{F}$ is an ultrafilter, it is called an *ultraproduct.* If all the algebras $\mathbf{A}_i$ are isomorphic to a single algebra, say $\mathbf{B}$, then the ultraproduct $\mathbf{A}/\eta_{\mathcal{F}}$ is an *ultrapower,* and can be written $\mathbf{B}^I/\eta_{\mathcal{F}}$.

It turns out that for a great many properties φ and any ultrafilter $\mathcal{U}$, $\mathbf{A}/\eta_{\mathcal{U}}$ satisfies φ if and only if $[\![\varphi]\!] \in \mathcal{U}$. In particular, we shall prove in Theorem 5.21 that this is true when φ is any first-order formula. The exercises contain some specific examples.

We introduce two new class operators.

Definition 5.3. Let $\mathcal{K}$ be a class of algebras.

$\mathsf{P_r}(\mathcal{K})$ = the class of all algebras isomorphic to reduced products of members of $\mathcal{K}$

$\mathsf{P_u}(\mathcal{K})$ = the class of all algebras isomorphic to ultraproducts of members of $\mathcal{K}$.

It follows from the definitions that $\mathsf{P_u} \le \mathsf{P_r} \le \mathsf{HP}$. In particular, varieties are closed under the formation of ultraproducts. We present a few other relationships involving these new operators.

Theorem 5.4. (1) $\mathsf{P} \le \mathsf{P_r}$;

(2) $\mathsf{P_uS} \le \mathsf{SP_u}$;

(3) $\mathsf{P_uP} \le \mathsf{P_r}$;

(4) $\mathsf{P_r} \le \mathsf{SPP_u}$.

Proof. For the first claim observe that $\mathcal{F} = \{I\}$ is a filter over I and $\eta_{\mathcal{F}}$ is the identity congruence on $\prod_I \mathbf{A}_i$. The second claim is quite easy and is left to the reader.

Let us consider (3). Suppose $\mathbf{A} = (\prod_J \mathbf{B}_j)/\eta_{\mathcal{U}}$ in which $\mathcal{U}$ is an ultrafilter over J and for every $j \in J$, $\mathbf{B}_j = \prod_{I_j} \mathbf{C}_i$. Without loss of generality assume that the sets I_j are pairwise disjoint.

Let $I = \bigcup_{j\in J} I_j$ and define

$$\begin{aligned}\mathcal{F} &= \Big\{ X \subseteq I : \{ j \in J : I_j \subseteq X \} \in \mathcal{U} \Big\} \\ &= \Big\{ Y \subseteq I : Y \supseteq \bigcup_{j\in Z} I_j \text{ for some } Z \in \mathcal{U} \Big\}.\end{aligned}$$

It is easy to check that $\mathcal{F}$ is a filter on I. Consider the composite mapping

$$\prod_{j\in J} \mathbf{B}_j \xrightarrow{\ f\ } \prod_{i\in I} \mathbf{C}_i \xrightarrow{\ q\ } \Big(\prod_I \mathbf{C}_i\Big)/\eta_{\mathcal{F}}$$

where q is the natural surjection and f is an isomorphism between the products such that if $\mathbf{c} = f(\mathbf{b})$ then for all $j \in J$ and $i \in I_j$, $(b_j)_i = c_i$.

Suppose $\mathbf{b}$ and $\mathbf{b}'$ are members of $\prod_J \mathbf{B}_j$, $\mathbf{c} = f(\mathbf{b})$, $\mathbf{c}' = f(\mathbf{b}')$. Let $Y = [\![\mathbf{c} = \mathbf{c}']\!]$. We have

$$\begin{aligned}&(\mathbf{b}, \mathbf{b}') \in \ker(q \circ f) \iff (\mathbf{c}, \mathbf{c}') \in \eta_{\mathcal{F}} \iff \\ &\qquad [\![\mathbf{c} = \mathbf{c}']\!] \in \mathcal{F} \iff (\exists Z \in \mathcal{U})\ Y \supseteq \bigcup_{j\in Z} I_j \iff \\ &\qquad\qquad (\exists Z \in \mathcal{U})\ (\forall j \in Z)\ b_j = b'_j \iff [\![\mathbf{b} = \mathbf{b}']\!] \in \mathcal{U}.\end{aligned}$$

Thus by the fundamental homomorphism theorem, $\mathbf{A} \cong (\prod_I \mathbf{C}_i)/\eta_{\mathcal{F}}$.

To prove (4), let $\mathbf{A} = (\prod_J \mathbf{B}_j)/\eta_{\mathcal{F}}$ for some filter $\mathcal{F}$ over J. Let I denote the set of all ultrafilters over J containing $\mathcal{F}$. By Corollary 2.42 (dualized), $\mathcal{F} = \bigcap I$ and therefore $\eta_{\mathcal{F}} = \bigcap \{ \eta_{\mathcal{U}} : \mathcal{U} \in I \}$. From this it follows that there is an embedding of $\mathbf{A}$ into $\prod_{\mathcal{U}\in I}(\prod_J \mathbf{B}_j/\eta_{\mathcal{U}})$. □

The ultraproduct construction, when combined with various finiteness conditions, tends to degenerate. This can be quite important in practice. The following two theorems provide examples of this phenomenon.

Theorem 5.5. (1) *An algebra can be embedded into each of its ultrapowers.*

(2) *Every finite algebra is isomorphic to each of its ultrapowers.*

Proof. Let $\mathbf{A}$ be any algebra and $\mathcal{U}$ an ultrafilter over a set I. Let $d\colon A \to A^I$ be the "diagonal map" $d(x) = \langle x, x, x, \ldots \rangle$. The map d is clearly a homomorphism. Let q denote the natural map from A^I to $A^I/\eta_{\mathcal{U}}$.

$$\mathbf{A} \xrightarrow{\ d\ } \mathbf{A}^I \xrightarrow{\ q\ } \mathbf{A}^I/\eta_{\mathcal{U}}$$

Notice that $q \circ d$ is injective since

$$[\![d(a) = d(b)]\!] = \begin{cases} \varnothing & \text{if } a \neq b \\ I & \text{if } a = b \end{cases}$$

so

$$q \circ d(a) = q \circ d(b) \implies [\![d(a) = d(b)]\!] \in \mathcal{U} \implies a = b.$$

Thus (1) holds.

Now assume that A is finite, say $A = \{a_1, a_2, \ldots, a_n\}$. We claim that the map $q \circ d$ from the previous paragraph is surjective. A typical element of the ultraproduct is of the form $\mathbf{b}/\eta_{\mathcal{U}}$ for some $\mathbf{b} \in A^I$. For each $j \leq n$ let $I_j = \{\, i \in I : \mathbf{b}_i = a_j \,\}$. Then $I_1 \cup I_2 \cup \cdots \cup I_n = I \in \mathcal{U}$, so (using property *iv′*) some $I_j \in \mathcal{U}$. But then $[\![\mathbf{b} = d(a_j)]\!] = I_j \in \mathcal{U}$ so $q \circ d(a_j) = \mathbf{b}/\eta_{\mathcal{U}}$. □

Theorem 5.6. (1) *Let $\mathcal{K}_1, \mathcal{K}_2$ be classes of algebras. Then $\mathsf{P_u}(\mathcal{K}_1 \cup \mathcal{K}_2) = \mathsf{P_u}(\mathcal{K}_1) \cup \mathsf{P_u}(\mathcal{K}_2)$.*

(2) *Let $\mathcal{K}$ be a finite set of finite algebras. Then every member of $\mathsf{P_u}(\mathcal{K})$ is isomorphic to a member of $\mathcal{K}$.*

Proof. Let $\mathbf{B} = \prod_I \mathbf{A}_i/\eta_{\mathcal{U}}$ where $\mathcal{U}$ is an ultrafilter over I and for each $i \in I$, $\mathbf{A}_i \in \mathcal{K}_1 \cup \mathcal{K}_2$. For $t \in \{1, 2\}$ define $J_t = \{\, i \in I : \mathbf{A}_i \in \mathcal{K}_t \,\}$. Then $J_1 \cup J_2 = I \in \mathcal{U}$, so since $\mathcal{U}$ is an ultrafilter, either $J_1 \in \mathcal{U}$ or $J_2 \in \mathcal{U}$. Let us say $J = J_1 \in \mathcal{U}$.

Define $\mathcal{U}' = \{\, X \cap J : X \in \mathcal{U} \,\}$. It is straightforward to check that $\mathcal{U}'$ is an ultrafilter on J. Consider the composite map

$$\prod\nolimits_I \mathbf{A}_i \xrightarrow{\ p_J\ } \prod\nolimits_J \mathbf{A}_i \xrightarrow{\ q\ } \prod\nolimits_J \mathbf{A}_i \big/ \eta_{\mathcal{U}'}$$

where p_J projects an I-tuple onto its J-components and q is the natural map. Then $\ker(q \circ p_J) = \eta_{\mathcal{U}}$ since for any $\mathbf{a}, \mathbf{b} \in \prod_I A_i$

$$\begin{aligned} q \circ p_J(\mathbf{a}) = q \circ p_J(\mathbf{b}) &\iff \{\, i \in J : a_i = b_i \,\} \in \mathcal{U}' \\ &\iff \{\, i \in J : a_i = b_i \,\} \in \mathcal{U} \iff [\![\mathbf{a} = \mathbf{b}]\!] \cap J \in \mathcal{U} \\ &\iff [\![\mathbf{a} = \mathbf{b}]\!] \in \mathcal{U} \iff (\mathbf{a}, \mathbf{b}) \in \eta_{\mathcal{U}}. \end{aligned}$$

Thus $\mathbf{B} \cong \prod_J \mathbf{A}_i/\eta_{\mathcal{U}'} \in \mathsf{P_u}(\mathcal{K}_1)$.

For the second claim, let $\mathcal{K} = \{\mathbf{A}_1, \mathbf{A}_2, \ldots, \mathbf{A}_n\}$ be a finite set of finite algebras and let $\mathbf{B} \in \mathsf{P_u}(\mathcal{K})$. By part (1), there is some $j \leq n$ such that $\mathbf{B} \in \mathsf{P_u}(\mathbf{A}_j)$. By Theorem 5.5(2), $\mathbf{B} \cong \mathbf{A}_j$. □

R. Quackenbush discovered the following surprising result in 1971 [Qua72]. It was one of the first hints of the deep structure that governs the subdirectly irreducible algebras in a variety.

Theorem 5.7. *Let $\mathcal{V}$ be a locally finite variety with only finitely many finite subdirectly irreducible algebras. Then $\mathcal{V}$ has no infinite subdirectly irreducible algebras.*

Proof. Let $\mathcal{V}_{\mathrm{fsi}}$ denote the set of finite subdirectly irreducible algebras of $\mathcal{V}$. By assumption, $\mathcal{V}_{\mathrm{fsi}}$ is a finite set of finite algebras. Let $\mathbf{A}$ be a subdirectly irreducible member of $\mathcal{V}$, and let $\mathcal{K}$ be the set of finitely generated subalgebras of $\mathbf{A}$. By Exercise 5.8.3, $\mathbf{A} \in \mathsf{SP_u}(\mathcal{K})$.

By local finiteness, every member of $\mathcal{K}$ is finite. Thus, for $\mathbf{B} \in \mathcal{K}$ we have $\mathbf{B} \in \mathsf{P_s}(\mathcal{V}_{\mathrm{fsi}}) \subseteq \mathsf{SP}(\mathcal{V}_{\mathrm{fsi}})$. Combining this observation with Theorems 5.4 and 5.6

$$\mathsf{SP_u}(\mathcal{K}) \subseteq \mathsf{SP_uSP}(\mathcal{V}_{\mathrm{fsi}}) \subseteq \mathsf{SP_uP}(\mathcal{V}_{\mathrm{fsi}}) \subseteq \mathsf{SPP_u}(\mathcal{V}_{\mathrm{fsi}}) = \mathsf{SP}(\mathcal{V}_{\mathrm{fsi}}).$$

But since $\mathbf{A}$ is subdirectly irreducible, $\mathbf{A} \in \mathsf{SP}(\mathcal{V}_{\mathrm{fsi}})$ implies that $\mathbf{A} \in \mathsf{S}(\mathcal{V}_{\mathrm{fsi}})$, thus, A is finite. □

The converse of 5.7 has spawned a great deal of research in universal algebra. The question of whether the converse holds became known as the Quackenbush problem. It wasn't until 1975 that Baldwin and Berman [BB75] found an example that showed that the answer to the problem is "no." In 1996, McKenzie [McK96a] constructed a finite algebra that generated a counterexample to the Quackenbush problem. However, both the Baldwin-Berman and McKenzie examples have infinite similarity type. At the time of this writing, it is still unknown whether there is a finite algebra of finite similarity type generating a variety with arbitrarily large finite, but no infinite, subdirectly irreducible members. This version of the question is known as the *restricted Quackenbush problem.*

Exercise Set 5.8.

1. Show that each of the following classes is closed under ultraproducts, but is not a variety. Determine which of these classes are closed under reduced products.
   (a) The class of groups with trivial center.
   (b) The class of rings with no zero-divisors.
   (c) The class of self-dual lattices.
2. Let $\mathbf{L} = \langle \omega, \wedge, \vee \rangle$ be the usual lattice-ordering on ω, and let $\mathcal{U}$ be a nonprincipal ultrafilter on ω. Let $\mathbf{A} = \mathbf{L}^{\omega}/\eta_{\mathcal{U}}$.
   (a) Show that $\mathbf{A}$ is a chain.
   (b) Let $j\colon \mathbf{L} \to \mathbf{A}$ be the map $q \circ d$ of Theorem 5.5. Show that $\vec{\jmath}(L)$ is a proper ideal of $\mathbf{A}$.
   (c) Show that $A - \vec{\jmath}(L)$ has no least element.
   (d) Let $\theta = \left\{ (a,b) \in A^2 : \left|\mathbf{I}[a \wedge b, a \vee b]\right| < \infty \right\}$. Show that θ is a congruence on $\mathbf{A}$ and that $\mathbf{A}/\theta$ is a dense linear order with a lower bound.

3. Prove that every algebra can be embedded into an ultraproduct of its finitely generated subalgebras. (Hint: for an algebra $\mathbf{A}$, use an ultrafilter over $\mathrm{Sb}_\omega(A) - \{\varnothing\}$ containing all sets of the form $\widehat{X} = \{ Y : X \subseteq Y \}$.)

4. Show that if a locally finite variety has infinitely many finite subdirectly irreducible algebras then the cardinality of those algebras must be unbounded.

5.2 Jónsson's lemma

In the mid-1960s, B. Jónsson [Jón67] discovered a remarkable application of ultraproducts to the study of congruence-distributive varieties. This has evolved into the major tool for the study of these varieties.

Lemma 5.9 (Jónsson's Lemma). *Let $\mathbf{C}$ be a subalgebra of a product $\prod_I \mathbf{A}_i$ and suppose that $\mathbf{C}$ is congruence-distributive. Then for every meet-irreducible congruence θ on $\mathbf{C}$, there is an ultrafilter $\mathcal{U}$ over I such that $\eta_\mathcal{U}\restriction_C \subseteq \theta$.*

Proof. Let $\mathbf{C}$ and θ be as in the statement of the lemma. We seek an ultrafilter $\mathcal{U}$ such that $\eta_\mathcal{U}\restriction_C \subseteq \theta$. Equivalently, for all $\mathbf{a}, \mathbf{b} \in C$

$$(\mathbf{a}, \mathbf{b}) \notin \theta \implies (\mathbf{a}, \mathbf{b}) \notin \eta_\mathcal{U} \implies [\![\mathbf{a} = \mathbf{b}]\!] \notin \mathcal{U} \implies [\![\mathbf{a} \neq \mathbf{b}]\!] \in \mathcal{U}.$$

So let $\mathcal{F} = \{\, [\![\mathbf{a} \neq \mathbf{b}]\!] : (\mathbf{a}, \mathbf{b}) \in C^2 - \theta \,\}$.

We claim that $\mathcal{F}$ has the finite intersection property. For suppose that $[\![\mathbf{a}^1 \neq \mathbf{a}^2]\!] \cap \cdots \cap [\![\mathbf{a}^k \neq \mathbf{b}^k]\!] = \varnothing$. Taking complements,

$$[\![\mathbf{a}^1 = \mathbf{b}^1]\!] \cup \cdots \cup [\![\mathbf{a}^k = \mathbf{b}^k]\!] = I.$$

Let $J_m = [\![\mathbf{a}^m = \mathbf{b}^m]\!]$ for $m = 1, \ldots, k$. Then $J_1 \cup J_2 \cup \cdots \cup J_k = I$. From our earlier observations

$$0_C = \eta_I = \eta_{(J_1 \cup \cdots \cup J_k)} = \eta_{J_1} \cap \cdots \cap \eta_{J_k}.$$

By congruence-distributivity, $\theta = \theta \vee 0_C = \bigcap_m (\theta \vee \eta_{J_m})$. But by assumption, θ is meet-irreducible, so there is some $m \le k$ such that $\theta \ge \eta_{J_m}$. Thus $[\![\mathbf{a}^m \neq \mathbf{b}^m]\!] \notin \mathcal{F}$ showing that $\mathcal{F}$ has the FIP.

By Proposition 5.2, there is an ultrafilter $\mathcal{U}$ extending $\mathcal{F}$. From our observations at the beginning of the proof we have $\eta_\mathcal{U}\restriction_C \subseteq \theta$. □

Theorem 5.10 (Jónsson, 1967). *Let $\mathcal{V}$ be a congruence-distributive variety, and suppose that $\mathcal{V} = \mathsf{V}(\mathcal{K})$. Then $\mathcal{V}_{\mathrm{si}} \subseteq \mathsf{HSP_u}(\mathcal{K})$.*

$$\begin{array}{ccccc}
\prod \mathbf{C}_i & \twoheadrightarrow & \prod \mathbf{C}_i/\eta_{\mathcal{U}} & & \\
\uparrow & & \uparrow & & \\
\mathbf{C} & \twoheadrightarrow & \mathbf{C}/(\eta_{\mathcal{U}}\restriction_C) & \twoheadrightarrow & \mathbf{B}
\end{array}$$

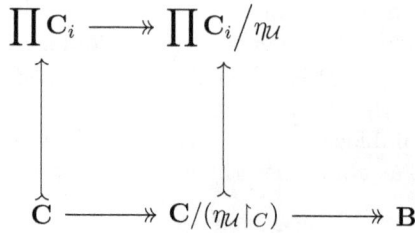

FIGURE 5.1: Diagram for Theorem 5.10

Proof. Let $\mathbf{B}$ be a subdirectly irreducible member of $\mathcal{V}$. Since $\mathcal{V} = \mathsf{HSP}(\mathcal{K})$, there are algebras $\mathbf{C}_i \in \mathcal{K}$, for $i \in I$, an algebra $\mathbf{C} \leq \prod_i \mathbf{C}_i$ and a congruence θ such that $\mathbf{B} \cong \mathbf{C}/\theta$. It follows from Theorem 3.23 that the congruence θ is meet-irreducible in $\mathrm{Con}(\mathbf{C})$. By the previous lemma, there is an ultrafilter $\mathcal{U}$ on I such that $\eta_{\mathcal{U}}\restriction_C \subseteq \theta$. (See Figure 5.1.)

Now $\mathbf{C}/(\eta_{\mathcal{U}}\restriction_C)$ can be embedded into $\prod_i \mathbf{C}_i/\eta_{\mathcal{U}}$ and this latter algebra is an ultraproduct of members of $\mathcal{K}$. Thus $\mathbf{B} \cong \mathbf{C}/\theta \in \mathsf{H}(\mathbf{C}/\eta_{\mathcal{U}}\restriction_C) \subseteq \mathsf{HSP}_{\mathsf{u}}(\mathcal{K})$. □

The importance of Jónsson's Lemma and the accompanying theorem to universal algebra can not be overstated. In the remainder of this section we present several easy consequences. The first is a strengthening of Theorem 3.50.

Corollary 5.11. *Let $\mathcal{K}$ be a finite set of finite algebras and suppose that $\mathcal{V} = \mathsf{V}(\mathcal{K})$ is congruence-distributive. Then $\mathcal{V}_{\mathrm{si}} \subseteq \mathsf{HS}(\mathcal{K})$.*

Proof. Combining Theorems 5.10 and 5.6(2) $\mathcal{V}_{\mathrm{si}} \subseteq \mathsf{HS}(\mathsf{P}_{\mathsf{u}}(\mathcal{K})) = \mathsf{HS}(\mathcal{K})$. □

Corollary 5.12. *Every finitely generated, congruence-distributive variety has only finitely many subvarieties.*

Proof. Let $\mathcal{K}$ be a finite set of finite algebras and $\mathcal{V} = \mathsf{V}(\mathcal{K})$. According to Corollary 3.45, the mapping $\mathcal{W} \mapsto \mathcal{W}_{\mathrm{si}}$ is injective on the subvarieties of $\mathcal{V}$. By the previous corollary $\mathcal{V}_{\mathrm{si}} = \mathsf{HS}(\mathcal{K})$ is (up to isomorphism) a finite set, consequently it has only finitely many subsets such as $\mathcal{W}_{\mathrm{si}}$. Therefore there can be only finitely many subvarieties. □

As an example, consider the variety $\mathcal{N}_5$ generated by the lattice $\mathbf{N}_5$. The members of $\mathsf{HS}(\mathbf{N}_5)$ are $\mathbf{N}_5$, $\mathbf{4}$, $\mathbf{2} \times \mathbf{2}$, $\mathbf{3}$, $\mathbf{2}$ and $\mathbf{1}$. Of these, only $\mathbf{N}_5$ and $\mathbf{2}$ are subdirectly irreducible. If $\mathcal{W}$ is a subvariety of $\mathcal{N}_5$ then there are four possibilities for $\mathcal{W}_{\mathrm{si}}$, namely $\{\mathbf{N}_5, \mathbf{2}\}$, $\{\mathbf{N}_5\}$, $\{\mathbf{2}\}$ and $\varnothing$. Each of these corresponds to a subvariety except for $\{\mathbf{N}_5\}$. Thus $\mathcal{N}_5$ has two proper subvarieties, namely the trivial variety and the variety of distributive lattices.

Corollary 5.13. *Let $\mathcal{V}$ be a congruence-distributive variety, and suppose that* $\mathbf{A}$ *and* $\mathbf{B}$ *are nonisomorphic members of $\mathcal{V}$ with* $\mathbf{A}$ *subdirectly irreducible and* $\mathbf{B}$ *finite. If $|A| \geq |B|$, then there is an identity that holds in* $\mathbf{B}$ *but not in* $\mathbf{A}$.

Proof. On cardinality grounds $\mathbf{A} \notin \mathsf{HS}(\mathbf{B}) \supseteq \mathsf{V}(\mathbf{B})_{\mathrm{si}}$. Since $\mathbf{A}$ is subdirectly irreducible, we conclude that $\mathbf{A} \notin \mathsf{V}(\mathbf{B}) = \mathrm{Mod}(\mathrm{Id}(\mathbf{B}))$. Consequently $\mathbf{A}$ fails to satisfy some identity holding in $\mathbf{B}$. □

Consider the lattices $\mathbf{A} = \mathbf{N}_5$ and $\mathbf{B} = \mathbf{M}_3$. In this case both $\mathbf{A}$ and $\mathbf{B}$ are finite and subdirectly irreducible. An identity holding in $\mathbf{B}$ but not $\mathbf{A}$ is, of course, the modular law. But suppose we reverse roles and take $\mathbf{A} = \mathbf{M}_3$, $\mathbf{B} = \mathbf{N}_5$. Now an identity holding in $\mathbf{B}$ but not $\mathbf{A}$ is harder to find. In fact, the reader can check that

$$x \wedge (y \vee (z \wedge (x \vee w))) \approx \Big(x \wedge (y \vee (x \wedge z))\Big) \vee \Big(x \wedge ((x \wedge y) \vee (z \wedge w))\Big)$$

is such an identity. (To see that $\mathbf{M}_3$ fails to satisfy, take x, y, z to be the three atoms and $w = y$.)

Definition 5.14. An algebra $\mathbf{A}$ has the *congruence extension property* (CEP) if for every $\mathbf{B} \leq \mathbf{A}$ and every $\theta \in \mathrm{Con}\,\mathbf{B}$ there is $\bar{\theta} \in \mathrm{Con}\,\mathbf{A}$ such that $\bar{\theta}\restriction_B = \theta$. A variety has the CEP if every member algebra has the CEP.

You showed in Exercise 3.51.3 that the variety of Abelian groups has CEP. Note that in a variety with CEP, every nontrivial subalgebra of a simple algebra must be simple. Thus the variety of groups can not have CEP. We seek a bit of a converse to this observation.

Definition 5.15. A variety $\mathcal{V}$ is called *semisimple* if every subdirectly irreducible member is simple. $\mathcal{V}$ is called *sub-semisimple* if every nontrivial subalgebra of a subdirectly irreducible algebra is simple.

Theorem 5.16. *Let $\mathcal{V}$ be finitely generated, congruence-distributive and sub-semisimple. Then $\mathcal{V}$ has the congruence extension property.*

Proof. Let $\mathbf{A} \leq \mathbf{B} \in \mathcal{V}$ and $\theta \in \mathrm{Con}(\mathbf{A})$. We wish to extend θ to $\mathbf{B}$. The algebra $\mathbf{B}$ has a subdirect representation by subdirectly irreducible algebras, say $\mathbf{B} \leq \prod_I \mathbf{B}_i$ with every $\mathbf{B}_i \in \mathcal{V}_{\mathrm{si}}$. By assumption, $\mathcal{V}$ is generated by a finite set $\mathcal{K}$ of finite algebras, so by Corollary 5.11, $\mathcal{V}_{\mathrm{si}} \subseteq \mathsf{HS}(\mathcal{K})$ is a finite set of finite simple algebras.

Let us first assume that θ is completely meet-irreducible. By Jónsson's Lemma there is an ultrafilter $\mathcal{U}$ on I such that $\eta_{\mathcal{U}}\restriction_A \subseteq \theta$. We have an embedding of $\mathbf{A}/(\eta_{\mathcal{U}}\restriction_A)$ into $\prod_I \mathbf{B}_i/\eta_{\mathcal{U}}$ (see Figure 5.2). But this latter algebra is a member of $\mathsf{P_u}(\mathcal{V}_{\mathrm{si}}) = \mathcal{V}_{\mathrm{si}}$ by Lemma 5.6, and is therefore simple. Hence by sub-semi-simplicity, $\mathbf{A}/(\eta_{\mathcal{U}}\restriction_A)$ is simple as well, so $\eta_{\mathcal{U}}\restriction_A$ is a maximal congruence on $\mathbf{A}$. But by assumption, this congruence is dominated by θ so we conclude that $\theta = \eta_{\mathcal{U}}\restriction_A$. Therefore θ extends to $\eta_{\mathcal{U}}\restriction_B$ on $\mathbf{B}$.

$$\begin{array}{ccccc}
\mathbf{A} & \rightarrowtail & \mathbf{B} & \rightarrowtail & \prod \mathbf{B}_i \\
\downarrow\!\!\!\downarrow & & \downarrow\!\!\!\downarrow & & \downarrow\!\!\!\downarrow \\
\mathbf{A}/(\eta_{\mathcal{U}}\restriction_A) & \rightarrowtail & \mathbf{B}/(\eta_{\mathcal{U}}\restriction_B) & \rightarrowtail & \prod \mathbf{B}_i\big/\eta_{\mathcal{U}}
\end{array}$$

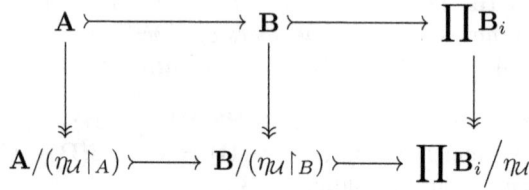

FIGURE 5.2: Diagram for Theorem 5.16

Finally, suppose that θ is an arbitrary congruence on $\mathbf{A}$. Since the congruence lattice is algebraic, $\theta = \bigcap_J \psi_j$ in which every ψ_j is completely meet-irreducible in $\mathrm{Con}(\mathbf{A})$. From the previous paragraph, every ψ_j extends to $\overline{\psi}_j$ on $\mathbf{B}$. But then $\bar{\theta} = \bigcap_J \overline{\psi}_j$ extends θ. □

As applications of the theorem, both the varieties of distributive lattices and implication algebras have the CEP since in each case, the only subdirectly irreducible algebra has two elements, so is obviously sub-semisimple. For any n, the variety $\mathcal{Cr}_n$ is congruence distributive (Exercise 2.22.4). It follows from Theorem 3.28 that $\mathcal{Cr}_n$ is sub-semisimple. Therefore it too has the CEP.

Exercise Set 5.17.

1. Let $\mathbf{A}$ denote the field of order 4, and $\mathbf{B}$ the subfield of order 2. Define $C = \{\, \mathbf{x} \in A^{\omega} : (\exists n)(\forall m > n)\; x_m \in B \,\}$. Clearly C is a subuniverse of $\mathbf{A}^{\omega}$.

   (a) Let $\mathcal{U}$ be an ultrafilter on ω. Show that if $\mathcal{U}$ is principal, then $\mathbf{C}/(\eta_{\mathcal{U}}\restriction_C) \cong \mathbf{A}$. If $\mathcal{U}$ is nonprincipal, then $\mathbf{C}/(\eta_{\mathcal{U}}\restriction_C) \cong \mathbf{B}$.

   (b) Let N denote the set of nonprincipal ultrafilters on ω. For every $j \in \omega$ let $\theta_j = \bigcap_{i \neq j} \eta_i$. Show that

   $$\bigvee_{j\in\omega} \theta_j = \bigwedge_{\mathcal{U}\in N} \eta_{\mathcal{U}} = \eta_{\mathcal{F}}$$

   where $\mathcal{F} = \{\, X \subseteq \omega : |\omega - X| < \infty \,\}$ is the *Fréchet filter* on ω.

   (c) Show that for all $\mathcal{U}, \mathcal{V} \in N, \mathcal{U} \neq \mathcal{V} \implies \eta_{\mathcal{U}}\restriction_C \neq \eta_{\mathcal{V}}\restriction_C$.

   (d) Let $D = \{\, \mathbf{x} \in C : (\exists n)(\forall m > n)\; x_m = x_n \,\}$. Show that, in contrast to part (c), $\mathcal{U}, \mathcal{V} \in N \implies \eta_{\mathcal{U}}\restriction_D = \eta_{\mathcal{V}}\restriction_D$.

2. Let $\mathcal{V}$ be a congruence-distributive variety, and let $\mathcal{W}_1, \mathcal{W}_2$ be subvarieties. Prove that $\mathcal{V} = \mathcal{W}_1 \vee \mathcal{W}_2 \iff \mathcal{V}_{\mathrm{si}} = (\mathcal{W}_1)_{\mathrm{si}} \cup (\mathcal{W}_2)_{\mathrm{si}}$.

3. For a cardinal κ, let $\mathbf{M}_\kappa$ denote the lattice of height 2 and cardinality $\kappa + 2$. Let $\mathcal{M}_\kappa = \mathbf{V}(\mathbf{M}_\kappa)$.

(a) Prove that an ultraproduct of lattices of height 2 also has height equal to 2.

(b) Prove that for any finite $n > 2$,

$$(\mathcal{M}_n)_{\mathrm{si}} = \{ \mathbf{M}_k : k = 0, 3, 4, \dots, n \} .$$

Consequently $\mathbf{L_V}(\mathcal{M}_n)$ is a chain of length $n - 1$.

(c) Prove that for any infinite κ, $\mathcal{M}_\kappa = \mathcal{M}_\omega = \bigvee \{ \mathcal{M}_n : n \in \omega \}$. Prove that $(\mathcal{M}_\omega)_{\mathrm{si}} = \{ \mathbf{M}_\kappa : \kappa \text{ a cardinal}, \kappa \neq 1, 2 \}$. (Hint: use Theorem 4.44.)

5.3 Model theory

Universal algebra, at least as we have construed it in this text, studies the relationship between algebraic structures and identities. First-order formulas are far more expressive than are identities, allowing us to define a much wider range of classes of algebras. But this is a double-edged sword. For the wider one casts one's net, the less powerful are the tools one can bring to bear.

And yet, the tools of model theory are considerable. First-order sentences behave nicely under ultraproducts (Theorem 5.21) and the elementary classes form an algebraic lattice (Theorem 5.22). These two properties will play a crucial role in our analysis of finitely based varieties in Section 5.5.

First-order logic also allows us to consider structures with basic relations in addition to operations. Thus graphs, posets and ordered fields become objects of study. However, as we broaden the scope, the subject becomes, well, less algebraic. For this reason and because logic is itself a huge subject worthy of an entire book of its own, we shall restrict our treatment to the bare minimum we need to achieve our universal algebraic ends. The interested reader can consult any number of texts, for example [Hin05] or [Men10].

Our definition of first-order formula can be seen as extending the notion of term as given in Definition 4.19. In addition to the operation symbols, we are provided with several *logical connectives*, namely "$\curlywedge$," "$\neg$," and "$\exists$." We use parentheses as additional punctuation when necessary.

Definition 5.18. Let $\rho\colon \mathcal{F} \to \omega$ be a similarity type. We define, by recursion on n, the sets Fm_n by

$$\begin{aligned} \mathrm{Fm}_0 &= \{ s \approx t : s, t \in T_\rho(X_\omega) \} \\ \mathrm{Fm}_{n+1} &= \mathrm{Fm}_n \cup \{ (\phi \curlywedge \psi),\ \neg\phi,\ \exists x\, \phi \ :\ \phi, \psi \in \mathrm{Fm}_n,\ x \in X_\omega \} . \end{aligned}$$

Finally, we define Fm_ρ to be $\bigcup_{n\in\omega} \mathrm{Fm}_n$, called the set of *first-order formulas of type ρ*.

Our definition of formula involves the minimum number of logical connectives that we can get away with. In practice it is often convenient to employ some additional notation. Thus

$$
\begin{array}{lll}
\phi \curlyvee \psi & \text{is shorthand for} & \neg(\neg\phi \curlywedge \neg\psi) \\
s \not\approx t & & \neg(s \approx t) \\
\phi \to \psi & & (\neg\phi) \curlyvee \psi \\
\forall x\ \phi & & \neg(\exists x\ \neg\phi).
\end{array}
$$

We also omit parentheses when no confusion will result.

Remark. In the literature it is customary to use "$\wedge$" and "$\vee$" as logical connectives. Since we are already using these symbols for our basic lattice operations, we shall use "curly" versions in the logic context.

Much as we did for terms, we often prove assertions about a formula by induction on its first appearance in the definition of Fm. Thus we define $|\phi|$ to be the least n such that $\phi \in \mathrm{Fm}_n$. We call $|\phi|$ the *height* of ϕ.

In order to clarify the way variables can be replaced in a formula, we need to distinguish between *free* and *bound* variables. The notion is defined by recursion on the height of the formula.

(i) If s and t are terms, then $\mathrm{Free}(s \approx t)$ is the set of all variables that actually appear in either s or t.

(ii) $\mathrm{Free}(\phi \curlywedge \psi) = \mathrm{Free}(\phi) \cup \mathrm{Free}(\psi)$.

(iii) $\mathrm{Free}(\neg\phi) = \mathrm{Free}(\phi)$.

(iv) $\mathrm{Free}(\exists x\ \phi) = \mathrm{Free}(\phi) - \{x\}$.

An appearance of a variable x in ψ is called *bound* if it is contained in a subformula of the form $\exists x\ \phi$.

Notice that a variable can be both free and bound in the same formula. For example, if ϕ is the formula $x + y \approx x \cdot y$ and ψ is the formula $\exists x\ (x + x \approx y)$ then $(\phi \curlyvee \psi)$ is a perfectly legal formula in which x is sometimes free and y is always free. We have $\mathrm{Free}(\phi \curlyvee \psi) = \{x, y\}$ in this case.

We write $\phi(x_1, \ldots, x_m)$ to indicate that $\mathrm{Free}(\phi) \subseteq \{x_1, \ldots, x_m\}$. A formula with no free variables is called a *sentence*. Ultimately it is the satisfaction of a sentence in an algebra that we are after. Because of the recursive nature of the definitions it is necessary to define the notion of satisfaction more generally for formulas.

Definition 5.19. Let $\mathbf{A}$ be an algebra of type ρ, ϕ and ψ formulas with free variables among $\{x_1, \ldots, x_m\}$, and $\mathbf{a} = \langle a_1, \ldots, a_m \rangle \in A^m$.

(i) $\mathbf{A} \models (s \approx t)(\mathbf{a})$ iff $s^{\mathbf{A}}(\mathbf{a}) = t^{\mathbf{A}}(\mathbf{a})$, for terms $s, t \in T(X_m)$;

(ii) $\mathbf{A} \models (\phi \curlywedge \psi)(\mathbf{a})$ iff $\mathbf{A} \models \phi(\mathbf{a})$ and $\mathbf{A} \models \psi(\mathbf{a})$;

(iii) $\mathbf{A} \models (\neg\phi)(\mathbf{a})$ iff $\mathbf{A} \not\models \phi(\mathbf{a})$;

(iv) $\mathbf{A} \models \exists x_i\, \phi(\mathbf{a})$ iff there is some $b \in A$ such that $\mathbf{A} \models \phi(a_1, \ldots, b, \ldots, a_m)$, where b appears in the i-th position;

One way to think about the above definition is that a formula with m free variables defines an m-ary relation on each algebra, namely, the set of those m-tuples for which the formula is true. For example, on the ring of real numbers the formula

$$\exists z (y \approx x + z \cdot z)$$

defines the usual ordering on the reals.

Notice that in our earlier definition of satisfaction of identities (Definition 4.35), there are implied universal quantifiers. That is, by

$$\mathbf{A} \models p(x_1, \ldots, x_m) \approx q(x_1, \ldots, x_m)$$

we mean, in the sense of Definition 5.19

$$\mathbf{A} \models \forall x_1 \cdots \forall x_m \, \big(p(x_1, \ldots, x_m) \approx q(x_1, \ldots, x_m)\big).$$

We extend this convention to formulas. Thus, $\mathbf{A} \models \phi(x_1, \ldots, x_m)$ is shorthand for $\mathbf{A} \models \forall x_1 \cdots \forall x_m \, \phi(x_1, \ldots, x_m)$.

What does Definition 5.19 mean for sentences? Since it has no free variables, a sentence is either true or false in a given algebra. (If you like, it is either true or false when applied to the unique 0-tuple of elements in the algebra.) For example, a commutative ring with identity satisfies

$$\forall x \forall y \, (x \cdot y \approx 0 \rightarrow (x \approx 0 \curlyvee y \approx 0))$$

if and only if it is an integral domain, while that same ring satisfies

$$\forall x \, (x \not\approx 0 \rightarrow \exists y \, (x \cdot y \approx 1))$$

if and only if it is a field.

Example 5.20. Let n be a positive integer. Define

$$\gamma_n = \exists x_1 \exists x_2 \cdots \exists x_n \, \big(\bigwedge_{i \neq j} x_i \not\approx x_j \big).$$

Then an algebra $\mathbf{A}$ satisfies γ_n if and only if it has at least n elements. Thus

$$\mathbf{A} \models (\gamma_n \curlywedge \neg\gamma_{n+1}) \iff |A| = n.$$

This shows that the class of all n-element algebras (of a fixed similarity type) is definable by a single first-order sentence, as is, say, the class of all n-element groups, rings, etc. These classes are obviously not closed under any of **H**, **S** or **P**, so they are not varieties.

We can juice this example up even more. Let $\mathbf{A}$ be any finite algebra of cardinality n and of finite similarity type. Then there is a single sentence that asserts that there are exactly n elements and that the operations on those elements behave exactly as they do on the corresponding elements of A. Any model of this sentence will be isomorphic to $\mathbf{A}$.

For example, if $\mathbf{A} = \langle\{0,1\}, \wedge\rangle$, then the sentence we have described above would be

$$\exists x_0 \exists x_1 \left(x_0 \not\approx x_1 \curlywedge (\forall y\,(x_0 \approx y \curlyvee x_1 \approx y)) \curlywedge \right.$$
$$\left.(x_0 \wedge x_0 \approx x_0) \curlywedge (x_0 \wedge x_1 \approx x_0) \curlywedge (x_1 \wedge x_0 \approx x_0) \curlywedge (x_1 \wedge x_1 \approx x_1)\right).$$

Not every interesting class can be described by first-order sentences. For example, the class of all finite groups has no first-order definition (Exercise 5.25.1), nor has the class of cyclic groups (Exercise 5.25.2). The intuitive idea is that with a first-order sentence, one can assert only finitely many things about an element. It is not possible to require (or deny) the existence of infinitely many elements, subsets or relations with a particular property.

One of the most useful tools for bringing algebraic techniques to the study of logic is the following theorem that relates the truth of a formula in an ultraproduct to its truth in the factors.

Theorem 5.21 (Łoś [Lo55b]). *Let I be a set, $\langle \mathbf{A}_i : i \in I\rangle$ a sequence of algebras, and $\mathcal{U}$ an ultrafilter on I. Then for any formula $\phi(x_1, \dots, x_m)$ and $\mathbf{b}^1, \dots, \mathbf{b}^m \in \prod_I \mathbf{A}_i$*

$$\Big(\prod_{i\in I} \mathbf{A}_i\Big)/\eta_{\mathcal{U}} \models \phi(\mathbf{b}^1/\eta_{\mathcal{U}}, \dots \mathbf{b}^m/\eta_{\mathcal{U}})$$
$$\text{if and only if} \tag{5–1}$$
$$\{\, i \in I : \mathbf{A}_i \models \phi(b_i^1, \dots, b_i^m)\,\} \in \mathcal{U}.$$

Proof. Let us extend our notation from Section 5.1 by writing

$$[\![\phi(\mathbf{b}^1, \dots \mathbf{b}^m)]\!] = \{\, i \in I : \mathbf{A}_i \models \phi(b_i^1, \dots, b_i^m)\,\}.$$

The proof is by induction on the height of ϕ.

If $|\phi| = 0$ then ϕ is of the form $p(x_1, \dots, x_m) \approx q(x_1, \dots, x_m)$. The correctness of (5–1) then follows from the fact that $\eta_{\mathcal{U}}$ is a congruence.

It is easy to see that

$$[\![\neg\psi(\mathbf{b}^1, \dots, \mathbf{b}^m)]\!] = I - [\![\psi(\mathbf{b}^1, \dots, \mathbf{b}^m)]\!] \quad \text{and}$$
$$[\![\psi_1(\mathbf{b}^1, \dots, \mathbf{b}^m) \curlywedge \psi_2(\mathbf{b}^1, \dots, \mathbf{b}^m)]\!] = [\![\psi_1(\mathbf{b}^1, \dots, \mathbf{b}^m)]\!] \cap [\![\psi_2(\mathbf{b}^1, \dots, \mathbf{b}^m)]\!].$$

For these formulas, (5–1) follows easily from the induction hypothesis and the usual properties of ultrafilters.

Now suppose that ϕ is $\exists y\, \psi(y, x_1, \dots, x_m)$. Let $J = [\![\phi(\mathbf{b}^1, \dots, \mathbf{b}^m)]\!]$. If ϕ is true in the ultraproduct, then there is $\mathbf{a} \in \prod_I \mathbf{A}_i$ such that

$$\prod_{i\in I} \mathbf{A}_i/\eta_{\mathcal{U}} \models \psi(\mathbf{a}/\eta_{\mathcal{U}}, \mathbf{b}^1/\eta_{\mathcal{U}}, \dots, \mathbf{b}^m/\eta_{\mathcal{U}}).$$

By the induction hypothesis $J \supseteq [\![\psi(\mathbf{a}, \mathbf{b}^1, \dots, \mathbf{b}^m)]\!] \in \mathcal{U}$, proving the left-to-right direction of (5–1). For the converse, for each $i \in J$ choose $a_i \in A_i$ so that $\mathbf{A}_i \models \psi(a_i, b_i^1, \dots, b_i^m)$ and for $i \in I - J$, choose $a_i \in A_i$ arbitrarily. Then (by the induction hypothesis) $\psi(\mathbf{a}/\eta_{\mathcal{U}}, \mathbf{b}^1/\eta_{\mathcal{U}}, \dots, \mathbf{b}^m/\eta_{\mathcal{U}})$ holds in the ultraproduct, hence ϕ holds in the ultraproduct as well. □

In Section 4.4 we investigated the Galois connection between algebras and identities. This has a natural extension to arbitrary first-order sentences. For a fixed similarity type ρ, let $\mathcal{S}_\rho$ denote the set of all sentences of type ρ. For a class $\mathcal{K}$ of algebras and a set Φ of sentences we write $\mathcal{K} \models \Phi$ to mean that for every $\mathbf{A} \in \mathcal{K}$ and $\phi \in \Phi$, $\mathbf{A} \models \phi$. The polarities of the Galois connection are

$$\mathrm{Th}(\mathcal{K}) = \{\, \phi \in \mathcal{S}_\rho : \mathcal{K} \models \phi \,\}$$
$$\mathrm{Mod}(\Phi) = \left\{\, \mathbf{A} \in \mathcal{Alg}_\rho : \mathbf{A} \models \Phi \,\right\}.$$

The "Mod" operator is the same one we have been using for identities and varieties. In the present context, classes of the form Mod(Φ) are called *elementary classes.* The "Th" operator is new. Th($\mathcal{K}$) is called the *elementary theory* of $\mathcal{K}$.

The polarities give rise to closure operators. Mod(Th($\mathcal{K}$)) is the elementary class generated by $\mathcal{K}$. Unfortunately, it is usually difficult to construct the members of Mod(Th($\mathcal{K}$)) starting from a knowledge of $\mathcal{K}$. On the other side, however, our understanding of Th(Mod(Φ)) is exceptionally rich. Briefly, the members of Th(Mod(Φ)) are precisely the sentences for which a proof (in the conventional sense of the word) exists, with Φ as the set of axioms.

The closed sets on each side of the Galois connection form a complete lattice, and the two lattices are dual to each other. Of fundamental importance is the fact that the lattice of elementary theories is algebraic. Recall that we proved in Theorem 4.61 that the lattice of equational theories is algebraic. Actually this followed immediately from the fact that the lattice is isomorphic to the congruence lattice of a certain algebra. The result for full first-order logic is not so easily achieved. We shall use an ultraproduct construction to accomplish our goal.

Theorem 5.22. *The lattice of elementary theories is algebraic.*

Proof. We shall prove that the closure operator Th ∘ Mod is algebraic. Suppose that Φ is a set of sentences, σ is another sentence and $\sigma \in \mathrm{Th}(\mathrm{Mod}(\Phi))$. We wish to show the existence of a finite subset, Φ_0, of Φ such that σ is a member of $\mathrm{Th}(\mathrm{Mod}(\Phi_0))$. Let us suppose that no such finite subset exists and derive a contradiction.

Let $I = \mathrm{Sb}_\omega(\Phi)$. For each $\Psi \in I$, since $\sigma \notin \mathrm{Th}(\mathrm{Mod}(\Psi))$, there is an algebra $\mathbf{A}_\Psi$ such that $\mathbf{A}_\Psi \models \Psi$ and $\mathbf{A}_\Psi \not\models \sigma$, which is to say $\mathbf{A}_\Psi \models \neg\sigma$.

For each $\phi \in \Phi$ let $J_\phi = \{\, \Psi \in I : \phi \in \Psi \,\}$. The family $\mathcal{F} = \{\, J_\phi : \phi \in \Phi \,\}$ has the finite intersection property since $J_{\phi_1} \cap J_{\phi_2} \cap \cdots \cap J_{\phi_n}$ contains the subset $\{\phi_1, \dots, \phi_n\}$. Thus, by Proposition 5.2, there is an ultrafilter $\mathcal{U}$ extending $\mathcal{F}$.

Let $\mathbf{B} = (\prod_{\Psi \in I} \mathbf{A}_\Psi)/\eta_{\mathcal{U}}$. For each $\phi \in \Phi$ we have $\mathbf{B} \models \phi$ since $[\![\phi]\!] = \{\Psi \in I : \mathbf{A}_\Psi \models \phi\} \supseteq J_\phi \in \mathcal{U}$. Thus $\mathbf{B} \models \Phi$. On the other hand, since σ fails in every factor, $\mathbf{B} \not\models \sigma$. This contradicts our assumption that $\sigma \in \mathrm{Th}(\mathrm{Mod}(\Phi))$. □

Theorem 5.22 has several equivalent formulations. In model theory, the most common is the following, which is usually called the "compactness theorem."

Corollary 5.23. *Let Φ be a set of sentences. If every finite subset of Φ has a model, then Φ has a model.*

Proof. Notice that

$$\text{“}\Phi \text{ has no models”} \iff \mathrm{Mod}(\Phi) = \varnothing \iff \mathrm{Th}(\mathrm{Mod}(\Phi)) = \mathcal{S}$$

where $\mathcal{S}$ denotes the set of all sentences. The second equivalence comes directly from the Galois connection. By taking Φ to be the single formula $x \not\approx x$ we deduce that $\mathcal{S} = \mathrm{Th}(\mathrm{Mod}(x \not\approx x))$. Since $\mathrm{Th} \circ \mathrm{Mod}$ is an algebraic closure operator, it follows that $\mathcal{S}$ is a compact theory.

Now if Φ has no models then

$$\mathcal{S} = \mathrm{Th}(\mathrm{Mod}(\Phi)) = \bigvee \{\, \mathrm{Th}(\mathrm{Mod}(\Phi_0)) : \Phi_0 \subseteq_\omega \Phi \,\}.$$

Since $\mathcal{S}$ is compact, it is dominated by a finite subjoin. Thus for some $\Psi \subseteq_\omega \Phi$, $\mathcal{S} = \mathrm{Th}(\mathrm{Mod}(\Psi))$. In other words, the finite subset Ψ has no models. □

As an application, we shall prove the "upwards Löwenheim-Skolem-Tarski theorem." The proof illustrates another useful technique: the expansion of the similarity type by additional operation symbols to produce an algebra with a particular property. At the end, the algebra is reducted to the original similarity type.

Theorem 5.24. *Let Φ be a set of first-order sentences and κ a cardinal. If Φ has an infinite model then it has a model of cardinality at least κ.*

Proof. Let $\rho \colon \mathcal{F} \to \omega$ be the similarity type and let $\mathbf{A}$ be an infinite model of Φ. Add to $\mathcal{F}$ new nullary operation symbols $\underline{c}_k$ for $k < \kappa$. Let $\Psi = \Phi \cup \{\, \underline{c}_j \not\approx \underline{c}_k : j \neq k \,\}$. We shall apply Corollary 5.23 to show that Ψ has a model.

Let $\Psi_0 \subseteq_\omega \Psi$. Then Ψ_0 mentions only finitely many of the new constants. Since A is infinite, we can choose distinct elements of A to represent the constant symbols in Ψ_0. Thus an appropriate expansion of $\mathbf{A}$ will be a model of Ψ_0.

Therefore by the corollary, Ψ has a model, $\mathbf{B}$. (Note that $\mathbf{B}$ is an algebra in the expanded language.) Since $\mathbf{B}$ satisfies every sentence $\underline{c}_j \not\approx \underline{c}_k$, it must contain a subset $\{\, \underline{c}_k^{\mathbf{B}} : k < \kappa \,\}$ of cardinality κ. Therefore the reduct of $\mathbf{B}$ to the original similarity type is a model of Φ of cardinality at least κ. □

The Löwenheim-Skolem-Tarski theorem has a tortured history, coming as it did at the dawn of model theory. The first complete proof of Theorem 5.24 was published by Maltsev [Mal36], who cites Skolem, who in turn, credits the proof to Tarski. It is said that Tarski had no recollection of proving the theorem. We discuss the "downwards" part of the theorem in Exercise 6 (which really is due to Skolem).

The compact elementary theories are precisely those that are finitely generated. The corresponding elementary classes are called *finitely axiomatizable* or *strictly elementary*. These classes are important for several reasons. For one thing, it is always possible to determine which finite algebras are members.

Note that in general, the complement of an elementary class is not elementary. However, by taking an appropriate conjunction, any finitely axiomatizable class can actually be axiomatized by a single first-order sentence. And therefore the complement of the class is axiomatized by the negation of that sentence. Thus in contrast to an arbitrary elementary class, the complement of a finitely axiomatizable class is again elementary, and is, in fact, finitely axiomatizable.

Exercise Set 5.25.

1. (a) Prove that the class of finite groups is not elementary. (Hint: If Φ were a set of defining equations, consider $\Phi \cup \{\, \gamma_n : n \geq 1 \,\}$, where γ_n is as in Example 5.20.)

   (b) Prove that the class of infinite groups is elementary but not strictly elementary.

2. Prove that the class of cyclic groups is not elementary.

3. Let $\mathcal{K}$ and $\mathcal{L}$ be elementary classes with $\mathcal{K} \subseteq \mathcal{L}$ and $\mathcal{L}$ strictly elementary. Prove that $\mathcal{K}$ is strictly elementary if and only if $\mathcal{L} - \mathcal{K}$ is closed under ultraproducts.

4. Let $\mathcal{V}$ be a variety and suppose that $\mathcal{V}$ is a strictly elementary class. Prove that $\mathcal{V}$ has a finite *equational* base.

5. Let $h = q \circ d$ be the embedding of the algebra $\mathbf{A}$ into an ultrapower $\mathbf{B} = \mathbf{A}^I/\eta_{\mathcal{U}}$ in Theorem 5.5. Show that for every formula $\phi(x_1, \dots, x_n)$ and every $a_1, \dots, a_n \in A$,

$$\mathbf{A} \models \phi(a_1, \dots, a_n) \iff \mathbf{B} \models \phi\big(h(a_1), \dots, h(a_n)\big). \qquad (5\text{–}2)$$

   Conclude that $\mathbf{A}$ and $\mathbf{B}$ generate the same elementary class. An embedding h satisfying (5–2) is called an *elementary embedding*.

6. *The downward Löwenheim-Skolem-Tarski theorem.* Let $\mathbf{A} = \langle A, F \rangle$ be an algebra. For every formula $\phi(x_1, \dots, x_n, y)$ define an n-ary operation f_ϕ on A so that for every $\mathbf{a} = \langle a_1, \dots, a_n \rangle$, $f_\phi(\mathbf{a})$ returns some b such

that $\mathbf{A} \models \phi(\mathbf{a}, b)$, if such a b exists. If there is no such b, $f_\phi(\mathbf{a})$ is arbitrary. Let $\mathbf{A}' = \langle A, F \cup \{ f_\phi : \phi \in \mathrm{Fm} \} \rangle$.

(a) For a subset X of A, let $\mathbf{B}' = \mathrm{Sg}^{(\mathbf{A}')}(X)$ and let $\mathbf{B}$ be the reduct of $\mathbf{B}'$ to the similarity type of $\mathbf{A}$. Prove that the embedding of $\mathbf{B}$ into $\mathbf{A}$ is elementary (see the previous exercise).

(b) Prove that $|B| = \max\{|X|, |F|, \aleph_0\}$.

(c) Prove the full Löwenheim-Skolem-Tarski theorem: Let Φ be a set of sentences of similarity type $\mathcal{F}$. If Φ has an infinite model, then it has a model of cardinality κ, where $\kappa \geq \max\{|F|, \aleph_0\}$.

5.4 Finitely based and nonfinitely based algebras

Recall that an *equational base* for a variety $\mathcal{V}$ is a set Σ of equations such that $\mathcal{V} = \mathrm{Mod}(\Sigma)$. We say that Σ is a base for an algebra $\mathbf{A}$ if Σ is a base for $\mathsf{V}(\mathbf{A})$. A variety (or algebra) is *finitely based* if it has a finite base. By Theorem 4.61, the lattice of equational theories is algebraic. Thus, the finitely based varieties correspond exactly to the compact equational theories. And therefore, if $\mathcal{V}$ is finitely based, every base for $\mathcal{V}$ contains a finite subbase.

It is desirable to have a finite base for a variety. Most familiar varieties are finitely based, such as the varieties of groups, rings, lattices and Boolean algebras. There are some obvious exceptions such as the variety of $\mathbb{R}$-vector spaces. But this variety has infinitely many operation symbols, so it is not really reasonable to talk about a finite base in this case.

Let $\mathbf{A}$ be a finite algebra of finite similarity type. Then $\mathbf{A}$ can be completely specified by a finite amount of information. In fact, we argued in Example 5.20 that the full first-order theory of $\mathbf{A}$ is finitely axiomatizable. However, in the context of the present discussion, we are asking about the *equational* theory of $\mathbf{A}$. It seems reasonable to hope that such an $\mathbf{A}$ will be finitely based. Birkhoff proved in 1935 (Theorem 5.27, just below) that one can come "close" to a finite basis for $\mathbf{A}$.

In 1951, Roger Lyndon [Lyn51, Ber80] proved that every 2-element algebra (of finite type) is indeed finitely based. But in 1954 Lyndon [Lyn54] found an example of a 7-element binar that is not finitely based. In 1965, V. L. Murskiĭ [Mur65] found a 3-element example.

In 1964 Oates and Powell [OP64] proved that every finite group is finitely based, and in 1973 Kruse and L′vov [Kru73, L′v73] did the same for rings. In 1977 Baker [Bak77] proved that if $\mathbf{A}$ generates a congruence-distributive variety then $\mathbf{A}$ is finitely based. It was hoped that one could somehow combine the arguments for groups and rings (which are congruence-permutable) and for Baker's theorem (congruence-distributive) to prove that if $\mathbf{A}$ generates a congruence-modular variety, then **A** is finitely based. It was subsequently discovered that S. V. Polin [Pol76] had constructed a counterexample in 1976. We shall prove Baker's theorem in Section 5.5. In this section we prove Birkhoff's theorem and present an example of a finite binar that is not finitely based.

Definition 5.26. Let $\mathcal{K}$ be a class of algebras and let n be a natural number. Then $\mathrm{Id}_n(\mathcal{K}) = \{ p \approx q : p, q \in T(X_n),\ \mathcal{K} \models p \approx q \}$.

Recall that $T(X_n)$ is the set of terms in the variables $\{x_1, x_2, \dots, x_n\}$. Note that $\mathrm{Id}_n(\mathcal{K})$ will not be an equational theory, since the variables that may appear are restricted to $\{x_1, \dots, x_n\}$. If Σ and Γ are sets of equations, we will use the notation $\Sigma \models \Gamma$ to denote the relation $\mathrm{Mod}(\Sigma) \subseteq \mathrm{Mod}(\Gamma)$. In other words, every model of Σ is a model of Γ.

Theorem 5.27 (Birkhoff [Bir35]). *Let* **A** *be a finite algebra of finite similarity type, and let n be a natural number. Then there is a finite subset Σ of* $\mathrm{Id}_n(\mathbf{A})$ *such that* $\Sigma \models \mathrm{Id}_n(\mathbf{A})$.

Proof. Recall from Lemma 4.27 that $\mathbf{T}(X_n)/\lambda_{\{\mathbf{A}\}} \in \mathbf{V}(\mathbf{A})$ and is generated by X_n. Since **A** is finite, it generates a locally finite variety (Theorem 3.49) so $T(X_n)/\lambda$ is finite. Fix a set $Q = \{q_1, q_2, \dots, q_k\}$ of terms from $T(X_n)$ to represent the λ-classes.

Let Σ be the set of identities $s \approx t$ in which $(s,t) \in \lambda$ and $s \approx t$ is of one of the forms

$$x_i \approx x_j$$
$$q_l \approx x_j$$
$$f(q_{i_1}, q_{i_2}, \dots, q_{i_m}) \approx q_l$$

where $i, j \leq n$, $i_1, i_2, \dots, i_m, l \leq k$ and f is a basic operation symbol of **A**. Note that the finiteness of Q and of the similarity type of **A** implies the finiteness of Σ.

Let $p \in T(X_n)$ and $q \in Q$. We claim that $\Sigma \models p \approx q$ if and only if $(p, q) \in \lambda$. For the left-to-right direction:

$$\Sigma \models p \approx q \implies \mathbf{T}(X_n)/\lambda \models p \approx q \implies (p,q) \in \lambda.$$

The proof of the converse is by induction on $|p|$. If $p = x_i$ then $p \approx q \in \Sigma$ by definition. So suppose that $p = f(p_1, \dots, p_m)$ and $(p, q) \in \lambda$. Then for each $i \leq m$ there is j_i such that $(p_i, q_{j_i}) \in \lambda$ (since Q is a set of representatives). By the induction hypothesis, $\Sigma \models p_i \approx q_{j_i}$, for $i \leq m$. Since λ is a congruence, $\big(f(p_1, \dots, p_m), f(q_{j_1}, \dots, q_{j_m})\big) = (p, f(q_{j_1}, \dots, q_{j_m})) \in \lambda$. Therefore, by transitivity, $(q, f(q_{j_1}, \dots, q_{j_m})) \in \lambda$. From the definition of Σ, $\Sigma \models f(q_{j_1}, \dots, q_{j_m}) \approx q$. Thus

$$\Sigma \models \{p_1 \approx q_{j_1}, \dots, p_m \approx q_{j_m}, f(q_{j_1}, \dots, q_{j_m}) \approx q\}$$

from which it follows that $\Sigma \models f(p_1, \dots, p_m) \approx q$.

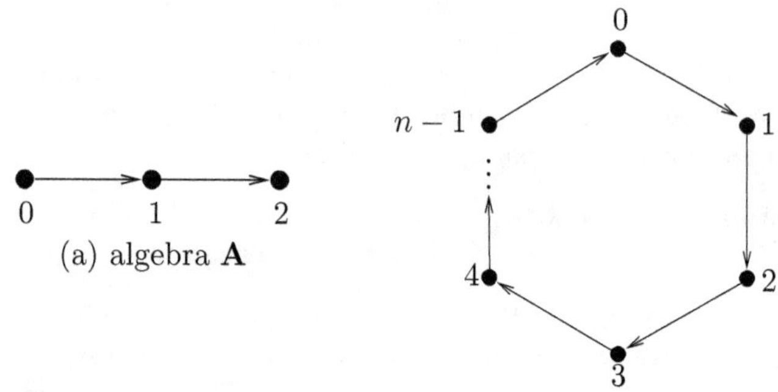

FIGURE 5.3: The graphs for the algebras **A** and **B** in Theorem 5.28

Now suppose that p and p' are arbitrary members of $T(X_n)$. Let q and q' be the representatives of p and p' in Q. Then

$$\Sigma \models p \approx p' \iff \Sigma \models \{p \approx q,\ p' \approx q',\ q \approx q'\} \iff (p,q),(p',q'),(q,q') \in \lambda \iff (p,p') \in \lambda \iff \mathbf{A} \models p \approx p'$$

with the last equivalence coming from Corollary 4.39. □

Since Lyndon's first example of a nonfinitely based algebra there has been considerable progress in understanding the nature of the property. We have already mentioned that every two-element algebra is finitely based but that the same is not true for 3-element algebras. Perkins found, in 1969, a 6-element semigroup that fails to be finitely based. In 1996 McKenzie proved that the problem of determining whether a given finite algebra is finitely based is undecidable. Additional discussion of the problem can be found in [JM].

As a counterbalance to Theorem 5.27, we present the following example due to Robert Park [Par80] of a finite binar which is not finitely based. Let $A = \{0, 1, 2, u\}$ and let "$\cdot$" be the following binary operation on A:

$\cdot$	0	1	2	u
0	0	1	u	u
1	1	1	2	u
2	u	2	2	u
u	u	u	u	u

Theorem 5.28. *The algebra* $\mathbf{A} = \langle A, \cdot \rangle$ *is not finitely based.*

Proof. Note that **A** is commutative, but not associative. When we write expressions without parentheses, we assume that they associate to the left. Thus, $x \cdot y \cdot z$ is shorthand for $(x \cdot y) \cdot z$.

It may be helpful to visualize the construction in the following way. Let σ be a reflexive, antisymmetric relation on a set X_0. The relation σ can be depicted as a directed graph as in Figure 5.3, in which we don't show the loops. Define a binary operation on the set $X = X_0 \cup \{u\}$, where $u \notin X_0$, by

$$x \cdot y = y \cdot x = y \quad \text{if } (x, y) \in \sigma$$
$$x \cdot y = u \qquad\qquad \text{if } (x, y) \notin \sigma \cup \sigma^{\smile}.$$

The binar **A** is defined in this way from the relation in Figure 5.3(a).

Suppose that Σ is a finite set of identities such that $\mathbf{A} \models \Sigma$. Choose an integer n such that all variables appearing in Σ lie in $\{x_1, \ldots, x_{n-1}\}$. Define two n-ary terms

$$s(x_1, \ldots, x_n) = x_1 \cdot x_2 \cdots x_{n-1} \cdot x_n \cdot x_1 \cdot x_2 \cdots x_n$$
$$t(x_1, \ldots, x_n) = x_2 \cdot x_3 \cdots x_n \cdot x_1 \cdot x_2 \cdot x_3 \cdots x_n \cdot x_1.$$

We shall show that $\mathbf{A} \models s \approx t$ but $\Sigma \not\models s \approx t$. Thus Σ can not be a base for **A**, proving the theorem.

To see that $\mathbf{A} \models s \approx t$ observe first that if any variable is replaced by u then $s = t = u$. So assume that each of $x_1, \ldots, x_n$ are replaced by one of 0, 1 or 2. Suppose that both 0 and 2 appear among the x_i's. Because each variable appears twice, each of s and t will look like

$$a_1 \cdot a_2 \cdots 2 \cdots 0 \cdots$$

As we evaluate this expression from left-to-right, we will eventually multiply on the right by 2. At that point the accumulated result is in $\{2, u\}$. Continuing to multiply, the result remains in $\{2, u\}$ until we reach the 0. At that point the result must be u, and from then on, it is stuck at u. Thus each of s and t will evaluate to u.

The only remaining possibility is that all variables come from $\{0, 1\}$ or from $\{1, 2\}$. But each of $\langle\{0, 1\}, \cdot\rangle$ and $\langle\{1, 2\}, \cdot\rangle$ forms a 2-element semilattice, which is both commutative and associative. On such a subset we must have $s \approx t$.

We now proceed to show that $\Sigma \not\models s \approx t$. For this, we use the following algebra **B**:

$\cdot$	0	1	2	3	4	$\cdots$	$n-2$	$n-1$	u
0	0	1	u	u	u	$\cdots$	u	0	u
1	1	1	2	u	u	$\cdots$	u	u	u
2	u	2	2	3	u	$\cdots$	u	u	u
3	u	u	3	3	4	$\cdots$	u	u	u
$\vdots$						$\vdots$			
$n-1$	0	u	u	u	u	$\cdots$	$n-1$	$n-1$	u
u	u	u	u	u	u	$\cdots$	u	u	u

We shall show that $\mathbf{B} \models \Sigma$ but $\mathbf{B} \not\models s \approx t$.

The binar $\mathbf{B}$ is obtained from the relation $\sigma = \{(i, i+1) : 0 \leq i < n\}$, with $i+1$ computed modulo n, see Figure 5.3(b). It is easy to see that

$$s^{\mathbf{B}}(0, 1, 2, \ldots, n-1) = n-1$$
$$t^{\mathbf{B}}(0, 1, 2, \ldots, n-1) = 0$$

showing that $\mathbf{B}$ fails to satisfy the identity $s \approx t$.

Let us suppose that $\mathbf{B} \not\models \Sigma$ and we shall derive a contradiction. There is an identity $(p \approx q) \in \Sigma$ and $b_1, \ldots, b_{n-1} \in B$ such that $p^{\mathbf{B}}(b_1, \ldots, b_{n-1}) \neq q^{\mathbf{B}}(b_1, \ldots, b_{n-1})$. On cardinality grounds there is some member of the set $\{0, \ldots, n-1\}$ that is not among $b_1, b_2, \ldots, b_{n-1}$. Because of the symmetry in the construction of $\mathbf{B}$, we can assume that $0 \notin \{b_1, \ldots, b_{n-1}\}$.

Let $C = B - \{0\}$. Then C is a subuniverse of $\mathbf{B}$ and it follows from above that $\mathbf{C} \not\models p \approx q$. For $i = 1, 2, \ldots, n-1$ we define elements $\mathbf{c}_i \in A^{n-3}$ by

$$\mathbf{c}_i(j) = \begin{cases} 0 & \text{for } j < n-1-i \\ 1 & \text{for } j = n-1-i \\ 2 & \text{for } j > n-1-i \end{cases}$$

Thus

$$\begin{aligned} \mathbf{c}_1 &= \langle 0, 0, \ldots, 0, 0, 0\rangle \\ \mathbf{c}_2 &= \langle 0, 0, \ldots, 0, 0, 1\rangle \\ \mathbf{c}_3 &= \langle 0, 0, \ldots, 0, 1, 2\rangle \\ \mathbf{c}_4 &= \langle 0, 0, \ldots, 1, 2, 2\rangle \\ &\vdots \\ \mathbf{c}_{n-1} &= \langle 2, 2, \ldots, 2, 2, 2\rangle. \end{aligned}$$

Let

$$D = \{\mathbf{c}_1, \mathbf{c}_2, \ldots, \mathbf{c}_{n-1}\} \cup \{\mathbf{y} \in A^{n-3} : \mathbf{y}(j) = u, \text{ some } j\}.$$

It is easy to check that D is a subuniverse of $\mathbf{A}^{n-3}$. Furthermore $\mathbf{C}$ is a homomorphic image of $\mathbf{D}$ under the mapping that sends each $\mathbf{c}_i$ to i and every other element to u. Thus $\mathbf{C} \in \mathsf{HSP}(\mathbf{A})$. But this is a contradiction since $\mathbf{C} \not\models \Sigma$ and we assumed at the outset that $\mathbf{A} \models \Sigma$. □

5.5 Definable principal (sub)congruences

It is not likely that there will ever be a truly satisfying characterization of those varieties that are finitely based. However, the search for promising sufficient conditions has zeroed in on two important criteria: (1) whether the class of subdirectly irreducible members is strictly elementary, and (2) whether the behavior of congruences on arbitrary algebras can be described using first-order formulas.

McKenzie considered versions of these properties first for varieties generated by a finite lattice [McK70] and then for varieties with definable principal congruences (see below). Baker generalized the lattice result to any finite algebra in a congruence-distributive variety. Numerous other results followed, including several new proofs of Baker's theorem. In 2002, Baker and Wang discovered a formulation that encompasses both Baker's original theorem as well as McKenzie's DPC result.

Definition 5.29. A variety $\mathcal{V}$ has *definable principal congruences* (DPC) if there is a first-order formula $\phi(u, v, x, y)$ such that for every $\mathbf{A} \in \mathcal{V}$ and $a, b, c, d \in A$,

$$(c, d) \in \mathrm{Cg}^{\mathbf{A}}(a, b) \iff \mathbf{A} \models \phi(c, d, a, b).$$

Example 5.30. (1) The variety of commutative rings with unit has DPC. For observe that $(c, d) \in \mathrm{Cg}(a, b)$ if and only if $c - d$ lies in the principal ideal generated by $a - b$. Thus for the defining formula we can take $\exists x\,(c - d \approx x \cdot (a - b))$.

(2) For any positive integer n, the variety $\mathcal{A}_n$ of Abelian groups of exponent n has DPC. In this case, using the fact that every subgroup is normal, $(c, d) \in \mathrm{Cg}(a, b)$ if and only if $c - d$ lies in the cyclic subgroup generated by $a - b$. Thus the quantifier-free formula

$$(c - d \approx a - b) \curlyvee (c - d \approx 2(a - b)) \curlyvee \cdots \curlyvee (c - d \approx n(a - b))$$

serves to define the congruence.

On the other hand, the variety of all Abelian groups lacks DPC. Let $\mathcal{U}$ be a nonprincipal ultrafilter on ω and take $\mathbf{A} = \mathbb{Z}^{\omega}/\eta_{\mathcal{U}}$. Let $\mathbf{a} = \langle 1, 1, 1, \ldots\rangle$ and $\mathbf{c} = \langle 0, 1, 2, \ldots\rangle$. Write $\bar{a} = \mathbf{a}/\eta_{\mathcal{U}}$ and $\bar{c} = \mathbf{c}/\eta_{\mathcal{U}}$. Observe that $(\bar{c}, \bar{0}) \notin \mathrm{Cg}^{\mathbf{A}}(\bar{a}, \bar{0})$ since for any natural number k, $[\![\mathbf{c} = k\mathbf{a}]\!] = \{k\} \notin \mathcal{U}$. Now suppose that ϕ defined principal congruences for the whole variety. Then $[\![\phi(\mathbf{c}, \mathbf{0}, \mathbf{a}, \mathbf{0})]\!] = \omega \in \mathcal{U}$ since, for every k, $\mathbb{Z} \models \phi(c_k, 0, a_k, 0)$. This implies that $\mathbf{A} \models \phi(\bar{c}, \bar{0}, \bar{a}, \bar{0})$, which we know to be false. □

Let us examine principal congruences in more detail. Have another look at what Theorem 4.17 says about the condition $(c, d) \in \mathrm{Cg}(a, b)$. There will be unary polynomials $f_0, f_1, \ldots, f_{n-1}$ such that

$$c = z_0, \{z_i, z_{i+1}\} = \{f_i(a), f_i(b)\}, \text{for } i = 0, \ldots, n-1, \text{ and } d = z_n.$$

Recalling that a polynomial is simply a term operation with some entries replaced by constants, we see that $(c, d) \in \mathrm{Cg}(a, b)$ if and only if $\phi(c, d, a, b)$

holds, where $\phi(u, v, x, y)$ is of the form

$$\exists w_1 \cdots w_m \left(u \approx p_0(s_0, \mathbf{w}) \curlywedge \bigcurlywedge_{i=0}^{n-1} \big(p_i(t_i, \mathbf{w}) \approx p_{i+1}(s_{i+1}, \mathbf{w})\big) \curlywedge p_n(t_n, \mathbf{w}) \approx v \right) \tag{5–3}$$

and $\{s_i, t_i\} = \{x, y\}$ for $i = 0, \dots, n$.

A formula of the form (5–3) is called a *strong congruence formula.* We can recast 4.17 as follows.

Lemma 5.31. *Let* $\mathbf{A}$ *be an algebra and* $a, b, c, d \in A$.

(1) *If* $(c, d) \in \mathrm{Cg}^{\mathbf{A}}(a, b)$ *then there exists a strong congruence formula* ϕ *such that* $\mathbf{A} \models \phi(c, d, a, b)$.

(2) *Conversely, if* ϕ *is a strong congruence formula, then* $\mathbf{A} \models \phi(c, d, a, b)$ *implies* $(c, d) \in \mathrm{Cg}^{\mathbf{A}}(a, b)$.

Since a finite algebra $\mathbf{A}$ has only finitely many quadruples $\langle a, b, c, d\rangle$, there is a finite disjunction, $\psi(u, v, x, y)$, of strong congruence formulas such that, for every $a, b, c, d \in A$, $(c, d) \in \mathrm{Cg}^{\mathbf{A}}(a, b)$ if and only if $\mathbf{A} \models \psi(c, d, a, b)$.

A finite disjunction of strong congruence formulas is called a *congruence formula.* Thus every finite algebra has a congruence formula that defines principal congruences on that algebra. More generally, if $\mathcal{K}$ is a finite set of finite algebras, then there is a congruence formula that works for every member of $\mathcal{K}$.

As the following lemma shows, in a variety with DPC, the defining formula can always be taken to be a congruence formula.

Lemma 5.32. *Let* $\mathcal{V}$ *be a variety and suppose that* ϕ *defines congruences on* $\mathcal{V}$. *There is a congruence formula* ψ *such that* $\mathcal{V} \models \phi \leftrightarrow \psi$.

Proof. We use a technique similar to the one used in Theorem 5.24. To the similarity type of $\mathcal{V}$ add four new nullary operation symbols, $\underline{a}$, $\underline{b}$, $\underline{c}$, $\underline{d}$. An algebra in the original similarity type will be denoted $\mathbf{A}$, $\mathbf{B}$, etc., while an expansion to the new type will be $\mathbf{A}', \mathbf{B}'$, etc.

Let Σ be an equational base for $\mathcal{V}$ and let Ψ be the set of all strong congruence formulas. Consider the set

$$\Omega = \Sigma \cup \{\, \neg\psi(\underline{c}, \underline{d}, \underline{a}, \underline{b}) : \psi \in \Psi \,\} \cup \{\phi(\underline{c}, \underline{d}, \underline{a}, \underline{b})\}.$$

Suppose that Ω has a model, $\mathbf{A}'$. Then since $\mathbf{A}' \models \phi(\underline{c}, \underline{d}, \underline{a}, \underline{b})$, we have, by assumption, $(\underline{c}^{\mathbf{A}'}, \underline{d}^{\mathbf{A}'}) \in \mathrm{Cg}^{\mathbf{A}'}(\underline{a}^{\mathbf{A}'}, \underline{b}^{\mathbf{A}'})$. But then from our discussion above, there is some strong congruence formula $\psi \in \Psi$ so that $\mathbf{A}' \models \psi(\underline{c}, \underline{d}, \underline{a}, \underline{b})$, which is false.

Thus Ω has no models. By the compactness theorem (5.23), there is a finite subset Ω_0 with no model. Let $\Psi_0 = \Omega_0 \cap \Psi$ and $\bar{\psi} = \bigcurlyvee \Psi_0$.

Now take any $\mathbf{A} \in V$. Then $\mathbf{A} \models \Sigma$. Suppose that $a, b, c, d \in A$ and

$\mathbf{A} \models \phi(c,d,a,b)$. Expand $\mathbf{A}$ to $\mathbf{A}'$ by defining $\underline{c}^{\mathbf{A}'} = c$, $\underline{d}^{\mathbf{A}'} = d$, etc. Then $\mathbf{A}' \models \phi(\underline{c},\underline{d},\underline{a},\underline{b})$. Since Ω_0 has no model, we must have $\mathbf{A}' \not\models \bigwedge_{\psi\in\Psi_0} \neg\psi(\underline{c},\underline{d},\underline{a},\underline{b})$. Hence $\mathbf{A} \models \bar{\psi}(c,d,a,b)$. This shows that $\mathcal{V} \models \phi \to \bar{\psi}$. The converse follows immediately from Lemma 5.31. □

Notice that a strong congruence formula has a very special form. It consists of several existential quantifiers followed by a conjunction of equations. A congruence formula, as a disjunction of strong congruence formulas, inherits some special properties.

Lemma 5.33. *Let ψ be a congruence formula.*

(1) *If $h\colon \mathbf{A} \to \mathbf{B}$ is a homomorphism, $a,b,c,d \in A$ and $\mathbf{A} \models \psi(c,d,a,b)$ then $\mathbf{B} \models \psi\big(h(c),h(d),h(a),h(b)\big)$.*

(2) *For every algebra $\mathbf{A}$, $\mathbf{A} \models \big(\psi(u,v,x,x) \to u \approx v\big)$.*

Proof. (1) follows from the aforementioned form of a strong congruence formula. (2) holds because of Lemma 5.31. □

Suppose that $\psi(u,v,x,y)$ is a congruence formula that defines principal congruences on a variety $\mathcal{V}$. Then each of the following formulas is true in $\mathcal{V}$.

- $\psi_R(x,y) = \forall u\ \psi(u,u,x,y)$;
- $\psi_S(x,y) = \forall u,v\ \big(\psi(u,v,x,y) \to \psi(v,u,x,y)\big)$;
- $\psi_T(x,y) = \forall u,v,w\ \big(\psi(u,v,x,y) \curlywedge \psi(v,w,x,y) \to \psi(u,w,x,y)$;
- $\psi_I(x,y) =$
  $$\bigwedge_{f\in\mathcal{F}} \forall u_1\cdots u_n \forall v_1\cdots v_n \Big(\bigwedge_{i=1}^{n} \psi(u_i,v_i,x,y) \to \psi(f(\mathbf{u}),f(\mathbf{v}),x,y)\Big);$$
  (here $\mathcal{F}$ is the set of operation symbols and n is the rank of f);
- $\psi_P(x,y) = \psi(x,y,x,y)$.

For any a,b, let $\theta = \{\,(c,d) : \psi(c,d,a,b)\,\}$. The validity of ψ_R, ψ_S and ψ_T means that θ is a reflexive, symmetric and transitive relation on any algebra in $\mathcal{V}$. ψ_I says that θ has the substitution property. ψ_P tells us that θ contains (a,b).

Let $\psi_C(x,y)$ be the conjunction of all of the above formulas. Thus if ψ defines congruences on $\mathcal{V}$, then $\mathcal{V} \models \psi_C$. Conversely, let $\mathcal{W}$ be a variety, $\mathbf{A} \in \mathcal{W}$ and $a,b \in A$. If $\mathbf{A} \models \psi_C(a,b)$ then $\theta = \{\,(c,d) : \psi(c,d,a,b)\,\} = \mathrm{Cg}^{\mathbf{A}}(a,b)$. To see this, note that we have just observed that θ will be a congruence on $\mathbf{A}$ containing (a,b). Thus $\mathrm{Cg}(a,b) \subseteq \theta$. Suppose conversely, that $(c,d) \in \theta$. Let $\mathbf{B} = \mathbf{A}/\mathrm{Cg}(a,b)$ and let $h\colon \mathbf{A} \to \mathbf{B}$ be the canonical map. By Lemmas 5.31 and 5.33,

$$\begin{aligned}(c,d) \in \theta &\implies \mathbf{A} \models \psi(c,d,a,b) \implies\\ &\mathbf{B} \models \psi\big(h(c),h(d),h(a),h(a)\big) \implies h(c) = h(d) \implies (c,d) \in \mathrm{Cg}(a,b).\end{aligned}$$

Thus $\theta = \mathrm{Cg}(a, b)$.

Unfortunately, DPC is a bit too strong to apply broadly. For example, no nondistributive variety of lattices has definable principal congruences. (For distributive lattices, see Exercise 5.41.1.) In [BW02], Baker and Wang introduced an interesting generalization of definable principal congruences and used it to excellent effect.

Definition 5.34. A variety $\mathcal{V}$ has *definable principal subcongruences* (DPSC) if there are congruence formulas π and ψ such that for all $\mathbf{B} \in \mathcal{V}$ and $a, b \in B$ with $a \neq b$, there are $c, d \in B$ with $c \neq d$ such that $\mathbf{B} \models \pi(c, d, a, b)$ and $\mathbf{B} \models \psi_C(c, d)$.

DPSC is a bit mysterious. It says that for every a and b, there is a principal congruence contained in $\mathrm{Cg}(a, b)$ that is defined by ψ, and we get to that principal subcongruence via π.

Observe that DPC implies DPSC. For if $\mathcal{V}$ has DPC with defining formula ψ, take $\pi = \psi$, $c = a$ and $d = b$ in Definition 5.34. As we now show, DPSC is strong enough to obtain a finite basis result.

Theorem 5.35 (Baker and Wang, 2002). *Let $\mathcal{V}$ be a variety with definable principal subcongruences such that $\mathcal{V}_{\mathrm{si}}$ is finitely axiomatizable. Then $\mathcal{V}$ is finitely based.*

In order to connect the finite axiomatizability of $\mathcal{V}_{\mathrm{si}}$ to that of $\mathcal{V}$, we need the following result of Jónsson's [Jón79].

Theorem 5.36. *Let $\mathcal{V}$ be a variety, $\mathcal{F}$ and $\mathcal{K}$ elementary classes. Suppose that*

(i) $\mathcal{V} \subseteq \mathcal{F}$

(ii) $\mathcal{F}_{\mathrm{si}} \subseteq \mathcal{K}$

(iii) *Both $\mathcal{F}$ and $\mathcal{V} \cap \mathcal{K}$ are strictly elementary.*

Then $\mathcal{V}$ is finitely based.

Proof. By Exercise 5.25.3, it is enough to show that $\mathcal{F} - \mathcal{V}$ is closed under ultraproducts. So let $\mathbf{B}_i \in \mathcal{F} - \mathcal{V}$, for $i \in I$, and suppose that $\mathbf{A} = (\prod_I \mathbf{B}_i)/\eta_{\mathcal{U}} \in \mathcal{V}$ for some ultrafilter $\mathcal{U}$ on I.

By Birkhoff's subdirect representation theorem, every $\mathbf{B}_i$ must have a subdirectly irreducible homomorphic image, $\mathbf{C}_i$, with $\mathbf{C}_i \notin \mathcal{V}$. Let $\mathbf{A}' = (\prod_I \mathbf{C}_i)/\eta_{\mathcal{U}}$. $\mathbf{A}'$ is, in a natural way, a homomorphic image of $\mathbf{A}$. Thus $\mathbf{A}' \in \mathcal{V} \subseteq \mathcal{F}$.

Since $\mathcal{K} \cap \mathcal{V}$ is strictly elementary, its complement is closed under ultraproducts. Thus $\mathbf{A}' \notin \mathcal{K} \cap \mathcal{V}$, so $\mathbf{A}' \notin \mathcal{K}$.

Let $J = \{\, i \in I : \mathbf{C}_i \in \mathcal{F} \,\}$. Since $\mathcal{F}$ is finitely axiomatizable and $\mathbf{A}' \in \mathcal{F}$ we must have $J \in \mathcal{U}$. Let $\mathcal{U}' = \{\, X \cap J : X \in \mathcal{U} \,\}$. Using the same argument as in Theorem 5.6(1), $\mathbf{A}' \cong (\prod_J \mathbf{C}_j)/\eta_{\mathcal{U}'}$.

Since for every $j \in J$, $\mathbf{C}_j \in \mathcal{F}_{\mathrm{si}} \subseteq \mathcal{K}$ and $\mathcal{K}$ is elementary, we conclude $\mathbf{A}' \in \mathcal{K}$, which is a contradiction. □

Proof of Theorem 5.35. Let π and ψ be the congruence formulas that witness DPSC. Let ϕ be the sentence

$$\forall a, b \left(a \not\approx b \rightarrow \exists c, d \left(c \not\approx d \curlywedge \pi(c,d,a,b) \curlywedge \psi_C(c,d)\right)\right)$$

and define $\mathcal{F} = \mathrm{Mod}(\phi)$. By assumption $\mathcal{V} \models \phi$ so $\mathcal{V} \subseteq \mathcal{F}$ and $\mathcal{F}$ is strictly elementary. Thus condition (i) and part of (iii) in Theorem 5.36 hold.

Now let σ be the sentence

$$\exists r, s \left(r \not\approx s \curlywedge \forall a, b \left(a \not\approx b \rightarrow \exists c, d \left(\pi(c,d,a,b) \curlywedge \psi(r,s,c,d)\right)\right)\right)$$

and define $\mathcal{K} = \mathrm{Mod}(\phi \curlywedge \sigma)$. Then $\mathcal{K} = \mathcal{F}_{\mathrm{si}}$. To see this, observe first that since $\mathcal{K} \models \phi$, $\mathcal{K} \subseteq \mathcal{F}$. Let $\mathbf{A} \in \mathcal{F}$. To have $\mathbf{A} \models \sigma$ is precisely for $\mathbf{A}$ to be (r,s)-irreducible (for the r,s given in σ). For if $a \neq b$ then $(c,d) \in \mathrm{Cg}(a,b)$ and $(r,s) \in \mathrm{Cg}(c,d)$. Thus $(r,s) \in \mathrm{Cg}(a,b)$.

Finally, $\mathcal{V} \cap \mathcal{K} = \mathcal{V} \cap \mathcal{F}_{\mathrm{si}} = \mathcal{V}_{\mathrm{si}}$ is strictly elementary, so the second half of condition (iii) in Theorem 5.36 holds. Therefore by that theorem, $\mathcal{V}$ is finitely based. □

Corollary 5.37 (McKenzie [McK78])**.** *Let $\mathcal{V}$ be a variety of finite type. If $\mathcal{V}$ has DPC and if $\mathcal{V}_{\mathrm{si}}$ is (up to isomorphism) a finite set of finite algebras, then $\mathcal{V}$ is finitely based.*

Proof. We have already observed that DPC implies DPSC. Suppose that, up to isomorphism, $\mathcal{V}_{\mathrm{si}} = \{\mathbf{C}_1, \dots, \mathbf{C}_k\}$ and every C_i is finite. Since the variety has finite type, there is a single sentence τ_i such that $\mathrm{Mod}(\tau_i) = \mathsf{I}\{\mathbf{C}_i\}$. Then $\mathcal{V}_{\mathrm{si}} = \mathrm{Mod}(\tau_1 \curlyvee \tau_2 \curlyvee \cdots \curlyvee \tau_k)$. Thus Theorem 5.35 applies. □

Example 5.38. Let n be a positive integer and $\mathcal{V}$ be a subvariety of $\mathcal{C}r_n$. Then $\mathcal{V}$ is finitely based. We have already observed that the variety of commutative rings with identity has DPC. We showed, in Theorem 3.28, that every member of $\mathcal{C}r_n$ has an identity and that $\mathcal{V}_{\mathrm{si}} \subseteq \{\, \mathbb{F}_d : (d-1) \mid (n-1) \,\}$. Thus the corollary applies.

The importance of DPSC is that it is a much weaker condition than DPC. In fact, it holds in any finitely generated, congruence-distributive variety.

Theorem 5.39 (Baker and Wang, 2002)**.** *Let $\mathbf{A}$ be a finite algebra of finite type that generates a congruence-distributive variety, $\mathcal{V}$. Then $\mathcal{V}$ has definable principal subcongruences.*

Proof. By Corollary 5.11 $\mathcal{V}_{\mathrm{si}} \subseteq \mathsf{HS}(\mathbf{A})$. Consequently, there is an integer $N \leq |A|$ bounding the cardinalities of the members of $\mathcal{V}_{\mathrm{si}}$. By local finiteness, for every positive integer k, the free $\mathcal{V}$-algebra on k generators, $\mathbf{F}(k)$, is finite. Since every k-generated member of $\mathcal{V}$ is a homomorphic image of $\mathbf{F}(k)$, there can be only (up to isomorphism) finitely many k-generated algebras. Therefore, there is a congruence formula π which defines congruences on every N-generated algebra, and a congruence formula ψ which defines congruences on every $(N+2)$-generated algebra. We shall show that π and ψ satisfy the requirements of Definition 5.34.

For this let $\mathbf{B} \in \mathcal{V}$ and $a, b \in B$ with $a \neq b$. Let $\mathbf{B} \leq \prod_I \mathbf{S}_i$ be a subdirect representation of $\mathbf{B}$ by subdirectly irreducible algebras. Let $J = \{\, i \in I : a_i \neq b_i \,\}$. Let m be an element of J such that $|S_m|$ is maximal in $\{\, |S_j| : j \in J \,\}$. Write p_i for the projection map from $\mathbf{B}$ to $\mathbf{S}_i$, for $i \in I$.

Let X consist of one preimage under p_m of each element of S_m, and ensure that $a, b \in X$. Thus $|X| = |S_m| \leq N$. Let $\mathbf{D} = \mathrm{Sg}^{\mathbf{B}}(X)$. Thus $\mathbf{D}$ is (at most) N-generated. For every $i \in I$, let $\eta_i = \ker(p_i{\restriction_D})$.

Since $\mathbf{D}/\eta_m \cong \mathbf{S}_m$, η_m is meet-irreducible. Since $\mathrm{Con}(\mathbf{D})$ is a finite distributive lattice

$$\begin{aligned}\eta_m \text{ meet irreducible} \implies (\eta_m] \text{ is a prime ideal} \implies&\\ \mathrm{Con}(\mathbf{D}) - (\eta_m] \text{ is a prime filter} \implies&\\ \alpha = \bigwedge\big(\mathrm{Con}(\mathbf{D}) - (\eta_m]\big) \text{ is join-irreducible.}&\end{aligned}$$

Certainly $\alpha = \bigvee\{\, \mathrm{Cg}^{\mathbf{D}}(x,y) : (x,y) \in \alpha \,\}$. Since α is join-irreducible and D is finite, there are $c, d \in D$ such that $\alpha = \mathrm{Cg}^{\mathbf{D}}(c,d)$.

Note that $\alpha \neq 0$ (since $\eta_m \notin \mathrm{Con}(\mathbf{D}) - (\eta_m]$), so $c \neq d$. Also, by the choice of m, $(a,b) \notin \eta_m$. Thus $\mathrm{Cg}(a,b) \in \mathrm{Con}(\mathbf{D}) - (\eta_m]$ and therefore $\alpha \leq \mathrm{Cg}(a,b)$. Put another way, $(c,d) \in \mathrm{Cg}^{\mathbf{D}}(a,b)$. Since $\mathbf{D}$ is N-generated, $\mathbf{D} \models \pi(c,d,a,b)$. Therefore by Lemma 5.33, $\mathbf{B} \models \pi(c,d,a,b)$, proving the first half of the DPSC criteria.

Now we must show that $\mathbf{B} \models \psi_C(c,d)$, i.e., for every $r, s \in B$,

$$(r,s) \in \mathrm{Cg}^{\mathbf{B}}(c,d) \implies \mathbf{B} \models \psi(r,s,c,d).$$

For this, assume that $(r,s) \in \mathrm{Cg}^{\mathbf{B}}(c,d)$ and let $\mathbf{C} = \mathrm{Sg}^{\mathbf{B}}(X \cup \{r,s\})$. Since $\mathbf{C}$ is $(N+2)$-generated, ψ does define principal congruences on $\mathbf{C}$. Thus if we can show that $(r,s) \in \mathrm{Cg}^{\mathbf{C}}(c,d)$, then $\mathbf{C}$, hence $\mathbf{B}$, satisfies $\psi(r,s,c,d)$.

Claim. For every $i \in I$, $\big(p_i(r), p_i(s)\big) \in \mathrm{Cg}^{\vec{p_i}(\mathbf{C})}\big(p_i(c), p_i(d)\big)$.

Suppose the claim holds. Let $\theta_i = \ker(p_i{\restriction_C})$. From Exercise 3.9.2,

$$\overleftarrow{p_i}\big(\mathrm{Cg}^{\vec{p_i}(\mathbf{C})}(p_i(c), p_i(d))\big) = \mathrm{Cg}^{\mathbf{C}}(c,d) \vee \theta_i.$$

Therefore, for every $i \in I$, $(r,s) \in \mathrm{Cg}^{\mathbf{C}}(c,d) \vee \theta_i$. Since C is finite, $\{\, \theta_i : i \in I \,\}$ is a finite set and $\bigcap_{i\in I} \theta_i = 0$. Thus

$$\mathrm{Cg}^{\mathbf{C}}(c,d) = \mathrm{Cg}^{\mathbf{C}}(c,d) \vee \bigwedge_{i\in I} \theta_i = \bigwedge_{i \in I}\big(\mathrm{Cg}^{\mathbf{C}}(c,d) \vee \theta_i\big)$$

by congruence-distributivity. Therefore $(r,s) \in \mathrm{Cg}^{\mathbf{C}}(c,d)$ as desired.

It remains to verify the claim. If $p_i(c) = p_i(d)$, then $\mathrm{Cg}^{\mathbf{B}}(c,d) \subseteq \ker p_i$ so $p_i(r) = p_i(s)$, and the claim holds for that i.

So suppose that $p_i(c) \neq p_i(d)$. Then $\alpha = \mathrm{Cg}^{\mathbf{D}}(c,d) \nsubseteq \eta_i$, so we must have $\eta_i \leq \eta_m$. Therefore $(a,b) \notin \eta_i$, so $i \in J$. There is a natural surjective homomorphism from $\mathbf{D}/\eta_i$ onto $\mathbf{D}/\eta_m \cong \mathbf{S}_m$. Thus $|S_i| \geq |D/\eta_i| \geq |S_m|$. By the maximal property of m, we must have $|S_i| = |S_m|$. Therefore, since C contains D, $\vec{p_i}(C) = S_i = \vec{p_i}(B)$. Finally, using Exercise 3.9.2

$$(r,s) \in \mathrm{Cg}^{\mathbf{B}}(c,d) \implies$$
$$\big(p_i(r), p_i(s)\big) \in \mathrm{Cg}^{\vec{p_i}(\mathbf{B})}\big(p_i(c), p_i(d)\big) = \mathrm{Cg}^{\vec{p_i}(\mathbf{C})}\big(p_i(c), p_i(d)\big)$$

as desired. □

Putting all of this together, we have a new proof of Baker's theorem on congruence-distributive varieties.

Corollary 5.40 (Baker, 1977). *Every finitely generated, congruence distributive variety of finite type is finitely based.*

Proof. The theorem shows that the variety has DPSC. The class of subdirectly irreducibles is (up to isomorphism) a finite set of finite algebras, so is strictly elementary. Therefore by Theorem 5.35, the variety is finitely based. □

Exercise Set 5.41.

1. Let $\mathbf{L}$ be a distributive lattice. Prove that $(c,d) \in \mathrm{Cg}^{\mathbf{L}}(a,b)$ iff

$$c \wedge (a \wedge b) = d \wedge (a \wedge b) \;\; \& \;\; c \vee (a \vee b) = d \vee (a \vee b).$$

   Thus the variety of distributive lattices has DPC.

2. Prove that the variety of semilattices has DPC. (Hint: there are not very many unary polynomials on a semilattice.)

3. Let $\mathcal{V}$ be a variety with definable principal congruences, and suppose that $\mathcal{V}$ is finitely based. Prove that $\mathcal{V}_{\mathrm{si}}$ is strictly elementary.

4. Let $\mathcal{V}$ be a locally finite variety and suppose that ψ is a congruence formula that defines principal congruences on the finite members of $\mathcal{V}$. Prove that ψ defines principal congruences on all of $\mathcal{V}$.

Chapter 6

Arithmetical Varieties

The previous chapter was concerned with the lattice-theoretic property of distributivity, and the influence it has on the structure of a variety. However, there is another property of a congruence lattice — permutability — that plays an important role. In this chapter, we study the combination of congruence-distributivity and -permutability and its effect on the clone of term operations.

6.1 Large clones

Universal algebra is at least partly a search for objects with an especially rich structure. Of course "rich structure" is in the eye of the beholder. Perhaps the congruence lattice has some important property or the identities that hold in the algebra are of a special form. In this section we consider algebras whose clone of term operations is as large as possible, subject to some imposed constraint. You might say we are looking for sets of basic operations that give us the most bang for the buck.

We begin with a beautiful theorem of Baker and Pixley [BP75] that elucidates the importance of a majority operation. While not the first result on the subject, it is one of the most general.

Recall from Definition 4.13 that $\mathcal{F}(\Theta)$ denotes the clone of all operations that preserve the relations in Θ and $\mathcal{R}(F)$ is the set of all relations invariant under the operations in F.

Theorem 6.1 (Baker-Pixley, 1975). *Let* $\mathbf{A}$ *be a finite algebra with a majority term. Then* $\operatorname{Clo}\mathbf{A} = \mathcal{F}\big(\operatorname{Sub}(\mathbf{A}^2)\big)$.

Before we begin the proof we introduce one additional piece of notation. For any set B define

$$\delta_B = \{\, (x,x) : x \in B \,\}.$$

The relation δ_B is sometimes called the *diagonal of* B. Among other things, δ_B is the graph of the identity function on B.

Proof of Theorem 6.1. For any algebra $\mathbf{A} = \langle A, F\rangle$, $\operatorname{Sub}(\mathbf{A}^2) \subseteq \mathcal{R}(F)$, thus

$$\mathcal{F}\big(\operatorname{Sub}(\mathbf{A}^2)\big) \supseteq \mathcal{F}\mathcal{R}(F) = \operatorname{Clo}(\mathbf{A}).$$

We must verify the converse.

Let f be an n-ary operation and assume that $f \mid\colon \mathrm{Sub}(\mathbf{A}^2)$. Using the finiteness of A, the following claim will suffice to prove that f is a term operation of $\mathbf{A}$.

Claim. For every $k \in \omega$ and $\mathbf{a}^1, \dots, \mathbf{a}^k \in A^n$, there is $p \in \mathrm{Clo}_n(\mathbf{A})$ such that $f(\mathbf{a}^i) = p(\mathbf{a}^i)$, for $i = 1, \dots, k$.

Proof. The key tool here is Theorem 4.32 which asserts that for any algebra $\mathbf{C}$ and $c_1, \dots, c_n \in C$

$$\mathrm{Sg}^{\mathbf{C}}(c_1, \dots, c_n) = \{\, p(c_1, \dots, c_n) : p \in \mathrm{Clo}_n(\mathbf{C}) \,\}. \qquad (6\text{–}1)$$

We prove the claim by induction on k. First assume that $k = 2$. Let $f(\mathbf{a}^1) = b_1$, $f(\mathbf{a}^2) = b_2$ and

$$\theta = \mathrm{Sg}^{\mathbf{A}^2}\big((a_1^1, a_1^2),\ (a_2^1, a_2^2), \dots, (a_n^1, a_n^2)\big).$$

Since f preserves θ, $(b_1, b_2) \in \theta$. Using (6–1) there is a term p such that $(b_1, b_2) = p^{\mathbf{A}^2}\big((a_1^1, a_1^2), \dots, (a_n^1, a_n^2)\big) = \big(p^{\mathbf{A}}(\mathbf{a}^1), p^{\mathbf{A}}(\mathbf{a}^2)\big)$ as required.

The case $k = 1$ follows from $k = 2$ by taking $\mathbf{a}^2 = \mathbf{a}^1$. So assume the claim holds for some $k \geq 2$ and we prove the claim for $k+1$. Let $b_i = f(\mathbf{a}^i)$, $i = 1, \dots, k+1$. By the induction hypothesis, there are terms p_j, for $j = 1, 2, 3$, such that for all $i \neq j$, $f(\mathbf{a}^i) = p_j(\mathbf{a}^i)$.

Let m denote the majority term operation assumed to exist for $\mathbf{A}$ and define $q = m[p_1, p_2, p_3]$. Note that $q \in \mathrm{Clo}(\mathbf{A})$. Then for every $i \leq k+1$ we have $q(\mathbf{a}^i) = f(\mathbf{a}^i)$, proving the claim and the theorem. ☐

Theorem 6.1 can be generalized. For $k > 2$ a k-ary *near-unanimity term* is a term $t(x_1, x_2, \dots, x_k)$ on an algebra such that the k identities

$$t(x, \dots, x, y, x, \dots, x) \approx x$$

hold, where the lone y can appear in any position. Note that a ternary near-unanimity term is exactly a majority term. Using the same argument we used above, Baker and Pixley show that if a finite algebra $\mathbf{A}$ has a k-ary near-unanimity term then $\mathrm{Clo}\,\mathbf{A} = \mathcal{F}\big(\mathrm{Sub}(\mathbf{A}^{k-1})\big)$.

The next two theorems provide useful tools for working with congruence-permutable varieties. The first of these is often referred to as "Fleischer's lemma." Suppose that η_1 and η_2 are congruences on an algebra $\mathbf{A}$. If $\eta_1 \cap \eta_2 = 0$ and $\eta_1 \circ \eta_2 = 1$ then $\mathbf{A}$ is a direct product $\mathbf{A}/\eta_1 \times \mathbf{A}/\eta_2$. If we weaken the first condition to $\eta_1 \cap \eta_2 = \alpha \neq 0$ then $\mathbf{A}/\alpha$ is a direct product. But what can we deduce about the structure of $\mathbf{A}$ if we keep the condition $\eta_1 \cap \eta_2 = 0$ but require only that $\eta_1 \circ \eta_2 = \eta_2 \circ \eta_1$? Fleischer's lemma provides the answer.

Theorem 6.2 (Fleischer, 1955). *Let $\mathbf{A} \leq \mathbf{B}_1 \times \mathbf{B}_2$ be a subdirect product, and suppose that $\eta_1 \circ \eta_2 = \eta_2 \circ \eta_1$, where η_i is the projection kernel of $\mathbf{A}$ onto $\mathbf{B}_i$ for $i = 1, 2$. Then there is an algebra $\mathbf{C}$ and surjective homomorphisms $g_i \colon \mathbf{B}_i \to \mathbf{C}$ such that $A = \{\, (x_1, x_2) \in B_1 \times B_2 : g_1(x_1) = g_2(x_2) \,\}$.*

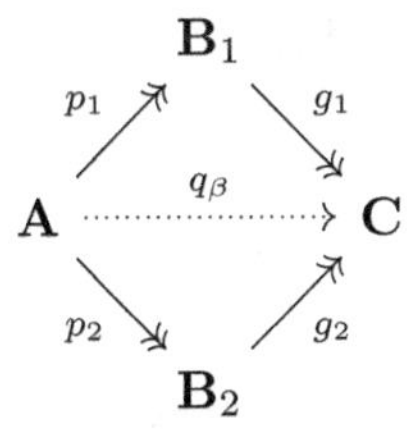

FIGURE 6.1: Fleischer's Lemma

Proof. Because of the permutability assumption, $\beta = \eta_1 \circ \eta_2$ is a congruence on $\mathbf{A}$. Let $\mathbf{C} = \mathbf{A}/\beta$. Since $\eta_i \subseteq \beta$ there are surjective homomorphisms g_i mapping $\mathbf{B}_i$ onto $\mathbf{C}$, for $i = 1, 2$. We have a commuting diagram as in Figure 6.1. Thus for every (x_1, x_2) in A, $q_\beta(x_1, x_2) = g_i p_i(x_1, x_2) = g_i(x_i)$. Since this is true for $i = 1$ and 2 we obtain $g_1(x_1) = g_2(x_2)$.

Conversely, suppose that $x_1 \in B_1$, $x_2 \in B_2$ and $g_1(x_1) = g_2(x_2)$. Since $\mathbf{A}$ is a subdirect product, there are elements y_1 and y_2 such that (x_1, y_2) and (y_1, x_2) are members of A. Then

$$q_\beta(x_1, y_2) = g_1 p_1(x_1, y_2) = g_1(x_1) = g_2(x_2) = g_2 p_2(y_1, x_2) = q_\beta(y_1, x_2)$$

from which it follows that $(x_1, y_2)\ \beta\ (y_1, x_2)$. Since $\beta = \eta_1 \circ \eta_2$ in $\mathbf{A}$ we conclude that $(x_1, y_2)\ \eta_1\ (x_1, x_2)\ \eta_2\ (y_1, x_2)$, hence $(x_1, x_2) \in A$. ☐

We comment that the conclusion of Theorem 6.2 asserts that $\mathbf{A}$ is the equalizer, $E(g_1p_1, g_2p_2)$, of g_1p_1 and g_2p_2 in $\mathbf{B}_1 \times \mathbf{B}_2$ (see Exercise 1.16.6).

Theorem 6.3. *Let* $\mathbf{A}$ *be an algebra with permuting congruences, and suppose that* $\mathbf{A}$ *is a subdirect product of finitely many simple algebras. Then* $\mathbf{A}$ *is isomorphic to a direct product of some of those simple algebras.*

Proof. Assume $\mathbf{A}$ is a subdirect product of $\prod_{i=1}^{n} \mathbf{B}_i$ in which every $\mathbf{B}_i$ is simple. Let η_i be the kernel of the projection of $\mathbf{A}$ onto $\mathbf{B}_i$, for $i \leq n$. Since $\mathbf{A}$ is a subdirect product, $\mathbf{A}/\eta_i \cong \mathbf{B}_i$ is simple, so η_i is a maximal proper congruence. Furthermore $\bigcap_{i=1}^{n} \eta_i = 0_A$.

Pick a minimal subset J of $\{1, 2, 3, \ldots, n\}$ such that $\bigcap_{j \in J} \eta_j = 0_A$. Then for each $j \in J$ we must have $\eta_j \not\geq \bigcap_{k \neq j} \eta_k$. From this, the maximality of η_j, and congruence-permutability, we deduce $\eta_j \circ \bigcap_{k \neq j} \eta_k = 1_A$. Applying Exercise 3.15.5, $\mathbf{A} \cong \prod_{j \in J} \mathbf{B}_j$. ☐

If we are to talk about algebras with large clones, we should probably start with algebras whose clone contains all possible operations. It is well-known that the two-element Boolean algebra has this property. In propositional logic this is the observation that every truth table can be built from the propositional variables using the logical connectives of "and," "or," and "not".

We shall call such algebras *primal.* As in Section 4.8, we prefer to blur the distinction between nullary and constant unary operations. Our definition of primality reflects this preference.

Definition 6.4. A finite algebra $\mathbf{P}$ is called *primal* if for every $n > 0$, $\mathrm{Clo}_n(\mathbf{P}) = \mathrm{Op}_n(P)$.

In Theorem 6.5, we *do* distinguish between nullary and constant unary operations, so we don't use the word "primal." The formal characterization of primal algebra is in the corollary following the theorem.

Theorem 6.5. *Let* $\mathbf{P}$ *be a finite algebra. Then* $\mathrm{Clo}(\mathbf{P}) = \mathrm{Op}(P)$ *if and only if* $\mathsf{V}(\mathbf{P})$ *is arithmetical and* $\mathbf{P}$ *is simple, rigid, and has no proper subuniverses.*

Proof. Let $\mathrm{Clo}(\mathbf{P}) = \mathrm{Op}(P)$. Then $\mathbf{P}$ certainly has a Pixley term so $\mathsf{V}(\mathbf{P})$ is arithmetical. Since every constant operation is a term, $\mathbf{P}$ has no proper subalgebras or nontrivial automorphisms. Since it has nullary operations, the empty set is not a subuniverse. To see that $\mathbf{P}$ is simple, let θ be a nonzero congruence. Choose $(a, b) \in \theta - 0_P$ and $(c, d) \in P^2$. There is a unary operation f such that $f(a) = c$ and $f(b) = d$. By assumption, f is a term of $\mathbf{P}$, so $(c, d) \in \theta$, proving that $\mathbf{P}$ is simple.

So now assume that $\mathbf{P}$ has all of the properties listed in the statement of the theorem. Since $\mathsf{V}(\mathbf{P})$ is arithmetical, by Corollary 4.71 $\mathbf{P}$ has a majority term, m. Therefore by Theorem 6.1 $\mathrm{Clo}(\mathbf{P}) = \mathcal{F}(\mathrm{Sub}(\mathbf{P}^2))$. We shall show that $\mathrm{Sub}(\mathbf{P}^2)$ contains only two members: P^2 itself and $\delta_P = \{ (x, x) : x \in P \}$. From this it follows easily that $\mathrm{Clo}(\mathbf{P}) = \mathcal{F}(P^2, \delta_P) = \mathrm{Op}(P)$.

So suppose that A is a subuniverse of $\mathbf{P}^2$. Note that if A is empty then $\varnothing$ is a proper subuniverse of $\mathbf{P}$, contrary to our assumption. For $i = 1, 2$, let $p_i \colon \mathbf{A} \to \mathbf{P}$ be the coordinate projection. Since $\mathbf{P}$ has no proper subalgebra, each p_i must be surjective. Therefore by Fleischer's lemma there is an algebra $\mathbf{B}$ and surjective homomorphisms $g_i \colon \mathbf{P} \to \mathbf{B}$, for $i = 1, 2$, so that $A = E(g_1 \circ p_1, g_2 \circ p_2)$. Since $\mathbf{P}$ is simple, there are only two possibilities for $\mathbf{B}$. Either $\mathbf{B}$ is trivial or $\mathbf{B} \cong \mathbf{P}$.

In the first case, for all $(x, y) \in P^2$, $g_1(x) = g_2(y)$ from which it follows that $A = P^2$. Suppose on the other hand that $\mathbf{B} \cong \mathbf{P}$, which is to say, both g_1 and g_2 are isomorphisms. Then $g_2^{-1} \circ g_1$ is an automorphism of $\mathbf{P}$. Since $\mathbf{P}$ is assumed rigid, we conclude that $g_1 = g_2$ and therefore $A = \delta_P$. ☐

Corollary 6.6 (Foster-Pixley, 1964). *Let* $\mathbf{P}$ *be a finite algebra. Then* $\mathbf{P}$ *is primal if and only if* $\mathsf{V}(\mathbf{P})$ *is arithmetical and* $\mathbf{P}$ *is simple, rigid, and has no proper subalgebras.*

Proof. If $\mathbf{P}$ has at least one nullary operation then, since it has every constant unary operation, it has every nullary operation. This is the case covered by the previous theorem. Suppose, on the other hand, that $\mathbf{P}$ has no nullary operations. This corresponds precisely to the presence of the empty subuniverse. Allowing for this difference, the proof of 6.5 goes through just as before. ☐

Example 6.7. For any prime integer p, the finite field $\mathbb{F}_p$ satisfies the conditions of Theorem 6.5 so it is primal. For this example we must assume that the basic operations of $\mathbb{F}_p$ contain a nullary constant 1. We can see the primality directly using the following multivariable version of the Lagrange interpolation theorem. Let $d(x) = 1 - x^{p-1}$. Then $d(0) = 1$ and $d(x) = 0$ for $x \neq 0$. Let $f(x_1, x_2, \dots, x_n)$ be any n-ary operation on F_p. For each $\mathbf{a} \in (F_p)^n$ define $b_{\mathbf{a}}$ to be $f(\mathbf{a})$. Then one easily checks that the term operation

$$g(x_1, \dots, x_n) = \sum_{\mathbf{a} \in F^n} b_{\mathbf{a}} \cdot \prod_{i=1}^{n} d(x_i - a_i)$$

coincides with f.

In particular, the two-element Boolean algebra, which has the same clone as $\mathbb{F}_2$, is primal. On the other hand, for $k > 1$, the field $\mathbb{F}_{p^k}$ has both proper subalgebras and automorphisms, so according to Theorem 6.6 is not primal. It is instructive to find an operation that fails to lie in the clone. Why does the above construction of g fail for the operation you find?

Corollary 6.8. *Let* $\mathbf{P}$ *be a primal algebra. Then the finite members of* $\mathsf{V}(\mathbf{P})$ *are (up to isomorphism) precisely the algebras* $\mathbf{P}^n$*, for* $n \in \omega$.

Proof. Let $\mathcal{V} = \mathsf{V}(\mathbf{P})$. By Theorem 6.6, $\mathcal{V}$ is congruence-distributive and congruence-permutable, so by Jónsson's lemma $\mathcal{V}_{\mathrm{si}} = \mathsf{HSP_u}(\mathbf{P}) = \{\mathbf{P}\}$ since $\mathbf{P}$ is finite, simple, and has no proper subalgebras. Therefore if $\mathbf{A}$ is a finite member of $\mathcal{V}$ then $\mathbf{A}$ is a finite subdirect power of $\mathbf{P}$. Then by Theorem 6.3 $\mathbf{A}$ is a power of $\mathbf{P}$. □

Following the discovery of Theorem 6.6, Pixley found numerous generalizations of the notion of primality. Eventually he came upon the following definition and theorem which seemed to be a good compromise between generality and utility.

Definition 6.9. Let A be a set. The *discriminator* on A is the ternary operation d_A defined by

$$d_A(x, y, z) = \begin{cases} z & \text{if } x = y, \\ x & \text{if } x \neq y. \end{cases}$$

A *discriminator algebra* is an algebra $\mathbf{A}$ such that $d_A \in \operatorname{Clo} \mathbf{A}$. A finite discriminator algebra is called *quasiprimal.*

Note that the discriminator function is a Pixley function. But unlike the definition of Pixley function (or Maltsev or majority function), there is only one discriminator on a set A.

Suppose that $\mathbf{A}$ is an algebra, $\mathbf{B}_1, \mathbf{B}_2 \leq \mathbf{A}$ and $h\colon \mathbf{B}_1 \to \mathbf{B}_2$ is an isomorphism. Then $h^{\square} = \{\, (x, h(x)) : x \in B_1 \,\}$ is an invariant binary relation on $\mathbf{A}$, called an *internal isomorphism.* We let $\operatorname{Iso}(\mathbf{A})$ denote the set of internal isomorphisms of $\mathbf{A}$.

Theorem 6.10. *Let* **A** *be a finite algebra. The following are equivalent.*

(a) **A** *is quasiprimal;*

(b) *Every subalgebra of* **A** *is simple or trivial, and* $\mathbf{V}(\mathbf{A})$ *is arithmetical;*

(c) $\operatorname{Clo}\mathbf{A} = \mathcal{F}\big(\operatorname{Iso}(\mathbf{A})\big)$.

Proof. Let **A** be quasiprimal. Then there is a term p such that $p^{\mathbf{A}} = d_A$. Since p is a Pixley term, by Theorem 4.70, $\mathbf{V}(\mathbf{A})$ is arithmetical. Let **B** be a nontrivial subalgebra of **A** and θ a nonzero congruence on **B**. Pick $(a,b) \in \theta - 0_B$. For any $c \in B$, $p^{\mathbf{A}^2}\big((a,a),(a,b),(c,c)\big) = (c,a) \in \theta$. Since c was arbitrary, we have $a/\theta = B$. Thus $\theta = 1_B$, so **B** is simple. This shows (a) $\Rightarrow$ (b).

Now assume (b). To prove (c) we first show that

$$\operatorname{Sub}(\mathbf{A}^2) = \operatorname{Iso}(\mathbf{A}) \cup \{\, A_1 \times A_2 : A_1, A_2 \in \operatorname{Sub}(\mathbf{A})\,\}. \tag{6–2}$$

Suppose that **C** is a subalgebra of $\mathbf{A}^2$. Let $A_i = \vec{p_i}(C)$, for $i = 1,2$. Then by assumption, $\mathbf{A}_i$ is either simple or trivial. Since $\mathbf{V}(\mathbf{A})$ is arithmetical, it is certainly congruence-permutable, so we can apply Fleischer's lemma. Thus there is an algebra **B** and homomorphisms $g_i \colon \mathbf{A}_i \to \mathbf{B}$ for $i = 1,2$ such that $C = E(g_1 \circ p_1, g_2 \circ p_2)$. Since $\mathbf{A}_1$, $\mathbf{A}_2$ are simple or trivial, either $|B| = 1$ or g_1 and g_2 are isomorphisms. In the former case $C = A_1 \times A_2$. In the latter, C is the graph of the isomorphism $g_2^{-1} \circ g_1 \in \operatorname{Iso}(\mathbf{A})$. Thus Equation (6–2) holds.

Now, by Corollary 4.71, **A** has a majority term, so by Theorem 6.1, $\operatorname{Clo}(\mathbf{A}) = \mathcal{F}(\operatorname{Sub}(\mathbf{A}^2))$. In light of Equation (6–2), to prove (c) it is enough to show that for any $\mathbf{A}_1, \mathbf{A}_2 \leq \mathbf{A}$, $f \mid\colon \operatorname{Iso}(\mathbf{A}) \implies f \mid\colon A_1 \times A_2$. So suppose that f is an n-ary operation preserving $\operatorname{Iso}(\mathbf{A})$. We shall show that f preserves $A_1 \times A_2$. For this let $(a_1,b_1), (a_2,b_2), \dots (a_n,b_n) \in A_1 \times A_2$. Since $\mathbf{A}_i$ is a subalgebra of **A**, $\delta_{A_i} \in \operatorname{Iso}(\mathbf{A})$, so $f \mid\colon \delta_{A_i}$. Since $(a_1,a_1), \dots, (a_n,a_n) \in \delta_{A_1}$ we deduce that $\big(f(\mathbf{a}), f(\mathbf{a})\big) \in \delta_{A_1}$, hence $f(\mathbf{a}) \in A_1$. Similarly $f(\mathbf{b}) \in A_2$. Therefore $\big(f(\mathbf{a}), f(\mathbf{b})\big) \in A_1 \times A_2$ as required.

Finally, (c) $\Rightarrow$ (a) since d_A always preserves the members of $\operatorname{Iso}(\mathbf{A})$. □

Every finite field is quasiprimal. One can see this by applying condition 6.10(b). For a field of order n, a discriminator term is

$$d(x,y,z) = (x-y)^{n-1} \cdot (x-z) + z.$$

In this case, the discriminator term is easy to find and the conditions of Theorem 6.10 easy to verify. For contrast, consider the following example.

Example 6.11. Let $\mathbf{Q} = \langle Q, \cdot \rangle$ be the binar with Cayley table

$\cdot$	1	2	3	4	5
1	1	2	3	4	5
2	2	3	4	5	1
3	3	5	2	1	4
4	4	1	5	2	3
5	5	4	1	3	2

$\mathbf{Q}$ is a Latin square, so according to Exercise 4.80.6, $\mathbf{Q}$ has the clone of a quasigroup. Therefore it generates a Maltsev variety. Since the cardinality of Q is 5, $\mathbf{Q}$ must be simple (Exercise 1.26.6). By inspection we see that the only subbinar of $\mathbf{Q}$ is $\{1\}$. Thus every subalgebra of $\mathbf{Q}$ is simple or trivial. Finally, we will show in Example 7.64 that $\mathbf{Q}$ generates a congruence distributive variety. Thus we can apply condition (b) of Theorem 6.10 and deduce that $\mathbf{Q}$ is quasiprimal. Therefore, it has a discriminator term. However, finding such a term is not so easy. In [Bur95], Burris describes a computer-search for such a term. The smallest discriminator term found occupied two pages of single-spaced text.

In Theorem 6.1 and the remarks following it, we characterized those finite algebras $\mathbf{A}$ for which $\mathrm{Clo}(\mathbf{A}) = \mathcal{F}(\mathrm{Sub}(\mathbf{A}^k))$, for $k > 1$. It is natural to wonder about the case $k = 1$. A finite algebra $\mathbf{A}$ is called *subalgebra-primal* if $\mathrm{Clo}(\mathbf{A}) = \mathcal{F}(\mathrm{Sub}(\mathbf{A}))$.

Corollary 6.12. *Let $\mathbf{A}$ be a finite algebra. Then $\mathbf{A}$ is subalgebra-primal if and only if $\mathbf{A}$ is quasiprimal, distinct nontrivial subalgebras of $\mathbf{A}$ are nonisomorphic, and no subalgebra of $\mathbf{A}$ has a nontrivial automorphism.*

Proof. ($\Leftarrow$) Let $\mathrm{Sub}_1(\mathbf{A})$ denote the set of trivial subuniverses of $\mathbf{A}$. The conditions in the corollary statement imply that $\mathrm{Iso}(\mathbf{A}) = \{\, \delta_B : \mathbf{B} \le \mathbf{A} \,\} \cup \mathrm{Sub}_1(\mathbf{A}^2)$. It is easy to see that the operations on A that preserve this set of binary relations are the same ones that preserve the subalgebras of $\mathbf{A}$. Thus by Theorem 6.10, $\mathrm{Clo}(\mathbf{A}) = \mathcal{F}\big(\mathrm{Sub}(\mathbf{A})\big)$.

($\Rightarrow$) Assume that $\mathbf{A}$ is subalgebra-primal. Since d_A preserves $\mathrm{Sub}(\mathbf{A})$, the algebra $\mathbf{A}$ is quasiprimal. Suppose that $\mathbf{B}_1$ and $\mathbf{B}_2$ are nontrivial subalgebras and $h\colon \mathbf{B}_1 \to \mathbf{B}_2$ is an isomorphism. We must show that $B_1 = B_2$ and that h is the identity map. Let $a \in B_1$ and $b = h(a)$. Since $\mathbf{B}_1$ is assumed nontrivial, there is $z \in B_1$ with $z \neq a$. Let $w = h(z)$. Since h is injective, $w \neq b$.

Define the binary operation f on A by

$$f(x,y) = \begin{cases} y & \text{if } (x,y) = (a,z) \\ x & \text{otherwise.} \end{cases}$$

It is easy to see that f preserves every subalgebra of $\mathbf{A}$, thus $f \in \mathrm{Clo}(\mathbf{A})$. Since h is a homomorphism, $h\big(f(a,z)\big) = f\big(h(a), h(z)\big)$. But $h\big(f(a,z)\big) = h(z) = w$. Thus $w = f\big(h(a), h(z)\big) = f(b,w)$. Since $w \neq b$ the definition of f implies that $b = a$. This shows that $B_1 = B_2$ and h is the identity. $\square$

Example 6.13. For a positive integer n, define $W_n = \{0, \frac{1}{n}, \frac{2}{n}, \ldots, \frac{n-1}{n}, 1\}$ and

$$\mathbf{W}_n = \langle W_n, \rightarrow, \neg, 1 \rangle$$

to be the algebra of similarity type $\langle 2, 1, 0 \rangle$ in which

$$x \rightarrow y = \min\{1, 1 - x + y\} \quad \text{and} \quad \neg x = 1 - x.$$

The algebra $\mathbf{W}_n$ is called a *Wajsberg algebra* of cardinality $n + 1$. These algebras were among the earliest generalizations of classical propositional logic to multiple truth values. $\mathbf{W}_n$ is subalgebra-primal for all $n \geq 1$. We sketch the proof and leave the details to the interested reader.

First, by defining $x \vee y = (x \rightarrow y) \rightarrow y$ and $x \wedge y = \neg(\neg x \vee \neg y)$ we see that the ordinary lattice operations on W_n lie in the clone. (Consider the cases $x \leq y$ and $x > y$ separately.) Next, the term $(\neg x) \rightarrow x$ is equal to $\min\{1, 2x\}$. Iterating this construction, we can define a "scalar product" by $1 \cdot x = x$ and $(k+1) \cdot x = (\neg x) \rightarrow (k \cdot x)$. Using these terms we see that the subalgebras of $\mathbf{W}_n$ are the algebras $\mathbf{W}_k$ for k a divisor of n, generated by the single element $\frac{1}{k}$. Thus $\mathbf{W}_n$ satisfies all of the conditions of Corollary 6.12 except perhaps the existence of a discriminator term.

To find the discriminator, first define

$$g(x, y) = n \cdot \Big(\neg\big((x \vee y) \rightarrow (x \wedge y)\big)\Big).$$

It is easy to see that

$$g(x, y) = \begin{cases} 0 & \text{if } x = y \\ 1 & \text{if } x \neq y \end{cases}$$

Finally the discriminator term is obtained as

$$d(x, y, z) = \big(x \wedge g(x, y)\big) \vee \big(z \wedge g(g(x, y), 1)\big).$$

Example 6.14. Continuing Example 6.7, the fields $\mathbb{F}_{p^k}$ for $k > 1$ are not subalgebra-primal since the subalgebras (including $\mathbb{F}_{p^k}$ itself) are not rigid. They are however, *automorphism-primal*, i.e., $\operatorname{Clo}(\mathbb{F}_{p^k}) = \mathcal{F}\big(\operatorname{Aut}(\mathbb{F}_{p^k})\big)$, see [BB96a, Example 5.10].

Exercise Set 6.15.

1. Let $k > 2$ and let t be a k-ary near unanimity term for a variety $\mathcal{V}$ (page 170). Define, for $j = 1, \ldots, k-1$, terms

$$\begin{aligned} m_j(x, y, z) &= t(x, \ldots, x, y, z, \ldots, z) \\ p_{2j-1}(x, y, z) &= t(x, m_j(x, y, z), \ldots, m_j(x, y, z), y, z) \\ p_{2j}(x, y, z) &= t(x, m_j(x, z, z), \ldots, m_j(x, z, z), y, z) \end{aligned}$$

where the lone y in m_j appears in position $k - j$. Use these terms and Theorem 4.66 to show that $\mathcal{V}$ is congruence-distributive, in fact, the variety $\mathcal{CD}_{2(k-2)}$ is interpretable in $\mathcal{V}$. (Mitschke [Mit78])

2. Let $\mathbf{A}$ be a finite algebra with a majority term. Show that $\mathbf{A}$ is term-equivalent to an algebra of finite similarity type.

3. Use Fleischer's Lemma (Theorem 6.2) to give another proof of Theorem 4.65.1.

4. Let $\mathbf{P}$ be a finite algebra of cardinality k and $\mathcal{V} = \mathsf{V}(\mathbf{P})$. Prove that the following are equivalent.

   (a) $\mathbf{P}$ is primal

   (b) For all $n > 0$, $\mathbf{F}_{\mathcal{V}}(n) \cong \mathbf{P}^{(k^n)}$

   (c) $|\mathbf{F}_{\mathcal{V}}(3)| = k^{(k^3)}$.

   (Sioson, [Sio61])

5. Let $\mathbf{A}$ and $\mathbf{B}$ be quasiprimal algebras and assume that $\mathcal{V} = \mathsf{V}(\mathbf{A}, \mathbf{B})$ is a Maltsev variety. Suppose further that $\mathsf{V}(\mathbf{A}) \cap \mathsf{V}(\mathbf{B})$ is trivial.

   (a) Prove that for any set X, $F_{\mathcal{V}}(X) \cong F_{\mathsf{V}(\mathbf{A})}(X) \times F_{\mathsf{V}(\mathbf{B})}(X)$. (See Exercise 4.75.8.)

   (b) Prove that there is a single ternary term t inducing the discriminator on both $\mathbf{A}$ and $\mathbf{B}$.

6. Let $\mathbf{A}$ be a discriminator algebra with discriminator term d and let $\mathcal{V} = \mathsf{V}(\mathbf{A})$.

   (a) Prove that every subdirectly irreducible member of $\mathcal{V}$ is a discriminator algebra (under the same discriminator term). (Jónsson's lemma.)

   (b) Let $\mathbf{B} \in \mathcal{V}$, and let $a, b \in B$. Prove that

   $$\mathrm{Cg}^{\mathbf{B}}(a, b) = \{\, (x, y) \in B^2 : d(a, b, x) = d(a, b, y) \,\}.$$

   Thus $\mathcal{V}$ has definable principal congruences.

7. An algebra $\mathbf{A}$ is called *idemprimal* if $\mathrm{Clo}(\mathbf{A})$ contains every idempotent operation on A. Let $\mathbf{A}$ be a finite algebra. Prove that the following are equivalent.

   (a) $\mathbf{A}$ is idemprimal;

   (b) $\mathrm{Clo}(\mathbf{A}) = \mathcal{F}(\mathrm{Sub}_1(\mathbf{A}))$;

   (c) $\mathbf{A}$ is simple, rigid, has no proper, nontrivial subalgebras, and generates an arithmetical variety.

8. Let $\mathbf{2}_r$ denote the two-element r-lattice (see Example 4.72).

   (a) Show that $\mathbf{2}_r$ is quasiprimal, in fact, idemprimal.

(b) Show that $\mathbf{2}_r$ generates the variety of all distributive r-lattices.

9. Suppose we modify the definition of $\mathbf{W}_n$ (Example 6.13) to obtain the algebra $\langle W_n, \to, \neg, \frac{1}{n}\rangle$. I.e., we replace the constant 1 with the constant $\frac{1}{n}$. Prove that the resulting algebra is primal. Thus for every $m > 1$ there is a primal algebra of cardinality m with finite similarity type.

10. Let A be a set. The *dual-discriminator* is the ternary operation u_A on A given by

$$u_A(x, y, z) = \begin{cases} x & \text{if } x = y \\ z & \text{if } x \neq y \end{cases}$$

(compare to Definition 6.9). A binary relation ρ on A is called *p-rectangular* if

$$\begin{aligned} (x, y_1), (x, y_2), (u, v) \in \rho \;\&\; y_1 \neq y_2 &\implies (x, v) \in \rho \text{ and} \\ (x_1, y), (x_2, y), (u, v) \in \rho \;\&\; x_1 \neq x_2 &\implies (u, y) \in \rho. \end{aligned}$$

Let $P(\mathbf{A})$ denote the set of p-rectangular subalgebras of $\mathbf{A}^2$. Prove that a finite algebra $\mathbf{A}$ has u_A in its clone if and only if $\text{Clo}(\mathbf{A}) = \mathcal{F}(P(\mathbf{A}))$. (Fried-Pixley [FP79])

11. A finite algebra $\mathbf{A}$ is a *dual-discriminator algebra* if $u_A \in \text{Clo}(\mathbf{A})$ (see Exercise 10). Let $\mathbf{A}$ be a dual-discriminator algebra and $\mathcal{V} = \mathbf{V}(\mathbf{A})$.

(a) Prove that $\mathbf{A}$ is simple.

(b) Prove that $\mathcal{V}$ is congruence-distributive and that every member of $\mathcal{V}_{\text{si}}$ is a dual-discriminator algebra.

(c) Prove that $\mathcal{V}$ has the congruence extension property.

6.2 How rare are primal algebras?

The theorems of the previous section would seem to suggest that primal algebras are very special. In this section, we will prove a theorem of Murskiĭ's that shows that in some sense, far from being unusual, primal algebras are ubiquitous.

Murskiĭ asked: What is the probability that a randomly chosen algebra will be primal?[1] In order to make this question precise, let us proceed as

[1] Actually, Murskiĭ was looking at the probability that a finite algebra is finitely based. But, in fact, he proved the stronger statement that we discuss here.

follows. Let P be a property of finite algebras and let ρ be a fixed similarity type. For a positive integer n let

$$\mathcal{Alg}_{\rho,n} = \{\, \mathbf{A} : \mathbf{A} \text{ has type } \rho \text{ and } A = \{1, 2, \ldots, n\}\,\}$$
$$\mathcal{Alg}_{\rho,n}[P] = \{\, \mathbf{A} \in \mathcal{Alg}_{\rho,n} : \mathbf{A} \text{ has } P\,\}.$$

And now define the probability that a finite algebra of type ρ has P by

$$\Pr_\rho(P) = \lim_{n\to\infty} \frac{\left|\mathcal{Alg}_{\rho,n}[P]\right|}{\left|\mathcal{Alg}_{\rho,n}\right|}$$

(assuming the limit exists). It is straightforward to argue that this yields a finitely additive probability measure on the set $\bigcup_{n\in\omega}\mathcal{Alg}_{\rho,n}$. For a detailed discussion of this, and a related, notion of probability, see [Fre90].

Murskiĭ proved that when P is taken to be the property of primality, then with a mild assumption on ρ we have $\Pr_\rho(P) = 1$. In other words, almost every finite algebra is primal. We begin the analysis by considering a simpler property of an algebra, namely the presence of idempotent elements.

Theorem 6.16. *Let E be the property that an algebra has no trivial subalgebras.*

(1) *Let $\rho = \langle k\rangle$, with $k > 1$. Then $\Pr_\rho(E) = 1/e \approx 0.368$.*

(2) *If ρ contains a least two operation symbols, at least one of which is nonunary, then $\Pr_\rho(E) = 1$.*

Proof. Let's first consider the case of a single binary operation on $\{1, 2, \ldots, n\}$. The operation is uniquely specified by its Cayley table. Since there are n^2 spaces in the table, and there are n choices for each space, we see that $\left|\mathcal{Alg}_n\right| = n^{(n^2)}$.

Now, in order for the algebra to have no trivial subalgebras, we must have $1 \cdot 1 \neq 1$, $2 \cdot 2 \neq 2$, etc. This means that each diagonal entry of the Cayley table has only $n-1$ choices instead of n. Since there are n diagonal entries, we obtain $\mathcal{Alg}_n[E] = n^{(n^2-n)} \cdot (n-1)^n$. Therefore

$$\Pr(E) = \lim_{n\to\infty} \frac{n^{(n^2-n)} \cdot (n-1)^n}{n^{n^2}} = \lim_{n\to\infty} \left(\frac{n-1}{n}\right)^n = \frac{1}{e}.$$

There is nothing special about a binary operation. For a k-ary operation with $k > 1$ the above equation is still valid, but with n^k replacing n^2 in the exponents.

For the second statement of the theorem, let us assume that $\rho = \langle 2, 1\rangle$. A more general similarity type follows the same argument, but the notation is messier. Call the unary operation symbol f. Its operation table is a sequence of length n. Arguing as above, the total number of spaces in the two tables is

n^2+n, so $\left|\mathcal{Alg}_{\rho,n}\right| = n^{(n^2+n)}$. Now, in order to have no 1-element subalgebras, we must have either $1\cdot 1 \neq 1$ or $f(1)\neq 1$ and either $2\cdot 2\neq 2$ or $f(2)\neq 2$, etc. There are n pairs $\big(j\cdot j,\ f(j)\big)$, $j=1,2,\ldots,n$ and for each pair, there are n^2-1 ways of filling it in. Thus $\left|\mathcal{Alg}_{\rho,n}[E]\right| = (n^2-1)^n\cdot n^{(n^2-n)}$. Consequently

$$\mathrm{Pr}_\rho(E) = \lim_{n\to\infty}\frac{(n^2-1)^n\cdot n^{n^2-n}}{n^{n^2+n}} = \lim_{n\to\infty}\left(\frac{n-1}{n}\cdot\frac{n+1}{n}\right)^n = \frac{1}{e}\cdot e = 1.$$

□

Recall, that an algebra is *idemprimal* (see Exercise 6.15.7) if its clone contains all idempotent operations. Let I be the property of being an idemprimal algebra.

Theorem 6.17. *Let ρ be a similarity type containing a single k-ary operation symbol, with $k>1$. Then*

(1) $\mathrm{Pr}_\rho(P) = 1/e$;

(2) $\mathrm{Pr}_\rho(I) = 1$.

If ρ contains at least two operation symbols, at least one of which is nonunary, then $\mathrm{Pr}_\rho(P)=1$.

Theorem 6.17(1) is due to R. O. Davies, 1968. The remainder of the theorem is due to V. L. Murskiĭ, [Mur75, Mur79]. Our proof follows unpublished treatments by R. Quackenbush and R. McKenzie. The entire theorem follows easily once we have proved part (2) in the case $k=2$. The next several pages (up through page 188) are devoted to a proof of Theorem 6.23 that asserts this fact.

Let A be a finite set, $m>1$ and denote by $q_1,\ldots,q_m$ the projection maps from A^m to A. Let us (for the remainder of this section) define a subset B of A^m to be *reduced* if

$$\begin{array}{ll}\text{(i)} & \forall i\leq m\quad |\vec{q_i}(B)|>1 \text{ and}\\ \text{(ii)} & \forall i\neq j\quad q_i{\restriction_B}\neq q_j{\restriction_B}.\end{array}$$

Lemma 6.18. *Let* **A** *be a finite algebra.* **A** *is idemprimal if and only if for every $m\geq 1$,* $\mathbf{A}^m$ *has no proper, reduced subalgebra.*

Proof. Suppose that **A** is idemprimal and **B** is a reduced subalgebra of $\mathbf{A}^m$. We want to show $B=A^m$. Let q_i be the ith projection map and $\eta_i=\ker(q_i)$, for $i\leq m$. For every i, $\vec{q_i}(B)$ is a subalgebra of **A**, and nontrivial since B is reduced. Applying the results of Exercise 6.15.7 we deduce that $\vec{q_i}(B)=A$ and **A** is simple. Since $\mathbf{B}/\eta_i\cong\vec{q_i}(B)$ it follows that η_i is a maximal proper congruence of **B**.

Suppose that for some $i \neq j$ we have $\eta_i = \eta_j$. Then by Exercise 1.26.8 there is an isomorphism $f\colon \mathbf{A} \to \mathbf{A}$ such that $f \circ q_i = q_j$.

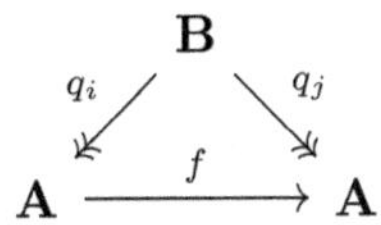

But again by Exercise 6.15.7, $\mathbf{A}$ is rigid, so f is the identity map. Consequently $q_i = q_j$ which contradicts the fact that $\mathbf{B}$ is reduced.

Thus the η_i's are pairwise distinct, and since they are maximal, $\eta_i \vee \eta_j = 1_B$ whenever $i \neq j$. Now $\mathbf{V}(\mathbf{A})$ is arithmetical (Exercise 6.15.7 again) so

$$\eta_i \circ \bigcap_{j \neq i} \eta_j = \eta_i \vee \bigcap_{j \neq i} \eta_j = \bigcap_{j \neq i} (\eta_i \vee \eta_j) = 1$$

and $\bigcap_{i=1}^{m} \eta_i = 0_B$. Therefore $\mathbf{B} \cong \mathbf{A}^m$ by Exercise 3.15.5.

Now, for the converse, suppose that $\mathbf{A}$ is not idemprimal. Then there is an operation f with $f \in \mathcal{F}(\mathrm{Sub}_1(\mathbf{A})) - \mathrm{Clo}(\mathbf{A})$. Let $n = \operatorname{rank} f$. Recall that $\mathrm{Clo}_n(\mathbf{A})$ is a subalgebra of $\mathbf{A}^{(A^n)}$ (Theorem 4.9). Because it omits f, it is a proper subalgebra. Unfortunately, it is not quite reduced, so we have to throw away several coordinates.

Let us define

$$I = A^n - \{\, (u, u, \dots, u) : \{u\} \in \mathrm{Sub}_1(\mathbf{A}) \,\}$$
$$B = \{\, g{\restriction}_I : g \in \mathrm{Clo}_n(\mathbf{A}) \,\}.$$

Then $\mathbf{B} \leq \mathbf{A}^I$. We claim that $f{\restriction}_I \notin B$. To see this, observe that if $f{\restriction}_I = g{\restriction}_I$ for some $g \in \mathrm{Clo}_n(\mathbf{A})$, then, since $g(u, u, \dots, u) = u$ for each $\{u\} \in \mathrm{Sub}_1(\mathbf{A})$, $f = g \in \mathrm{Clo}(\mathbf{A})$ which is false.

Thus $\mathbf{B}$ is a proper subalgebra of $\mathbf{A}^I$. We show it is reduced. Keep in mind that the "coordinates" of a member of B are the elements of I, which are n-tuples from A. For the second condition, suppose that $\mathbf{x}$ and $\mathbf{y}$ are distinct elements of I. Then for some $j \leq n$, $x_j \neq y_j$. Let $p_j^n \in \mathrm{Clo}_n(\mathbf{A})$ be the projection operation. Thus $q_{\mathbf{x}}(p_j^n) = p_j^n(\mathbf{x}) = x_j \neq y_j = q_{\mathbf{y}}(p_j^n)$. So condition (ii) holds.

Finally, suppose that for some $\mathbf{x} \in I$, $q_{\mathbf{x}}(B) = \{u\}$. Since for every $j \leq n$, $x_j = q_{\mathbf{x}}(p_j^n)$, we must have that $\mathbf{x} = (u, u, \dots, u)$. Since $\mathbf{B}$ is an algebra, u must be an idempotent of $\mathbf{A}$, contradicting the fact that $\mathbf{x} \in I$. ☐

In order to prove Theorem 6.23, we divide the set of nonidemprimal binars into 10 subsets, and prove that the probability that each one occurs is 0. In the next lemma, X and Y range over arbitrary subsets of A, a, b, c over elements of A and α and β over permutations of A.

Lemma 6.19. *Let $\mathbf{A} = \langle A, \cdot \rangle$ be a binar of cardinality n. If $\mathbf{A}$ is not idemprimal, then (at least) one of the following ten conditions must hold.*

(1) $\exists X\ (2 \le |X| \le n-1\ \&\ X \cdot X \subseteq X)$;

(2) $\exists X\ (3 \le |X| \le n-1\ \&\ |X \cdot X| \le |X|)$;

(3) $\exists X\ (|X| = 2\ \&\ |X \cdot X| = 1)$;

(4) $\exists X, Y\ (|X| = |Y| = 2\ \&\ X \cdot X = Y\ \&\ |Y \cdot Y| = 2)$;

(5) $A \cdot A \neq A$;

(6) $\exists a, b\ (a \neq b\ \&\ a \cdot a = a \cdot b = b \cdot a = a)$;

(7) $\exists X\ (1 \le |X \cdot A| \le |X| \le n-1)$;

(8) $\exists X\ (1 \le |A \cdot X| \le |X| \le n-1)$;

(9) $\exists a, b\ (a \neq b\ \&\ (\forall c)(a \cdot c = b \cdot c))$;

(10) $\exists \alpha, \beta\ (\alpha \neq \iota\ \&\ (\forall a, b)(\alpha(a) \cdot \alpha(b) = \beta(a \cdot b)))$.

Proof. Assume that $\langle A, \cdot \rangle$ is not idemprimal. Then by Lemma 6.18, for some $m > 0$, $\mathbf{A}^m$ has a proper, reduced subalgebra $\mathbf{B}$. Choose $\mathbf{B}$ so as to minimize m. Since the projection of B onto any $m-1$ coordinates will still be reduced, by the minimality of m, these projections must all be isomorphic to $\mathbf{A}^{m-1}$.

Suppose $m = 1$. Then $\mathbf{B}$ is a proper nontrivial subalgebra of $\mathbf{A}$, so (1) holds. So from now on assume that $m > 1$. For every $\mathbf{a} = \langle a_1, a_2, \ldots, a_{m-1} \rangle$ in A^{m-1} let

$$B(\mathbf{a}) = \{\, b \in A : \langle a_1, a_2, \ldots, a_{m-1}, b \rangle \in B \,\}.$$

Note that

$$B(\mathbf{a}) \cdot B(\mathbf{a}') \subseteq B(\mathbf{a} \cdot \mathbf{a}'). \tag{6–3}$$

Also, for every $\mathbf{a} \in A^{m-1}$, $B(\mathbf{a}) \neq \varnothing$ since B projects onto A^{m-1}. Finally, let $k = \max\{\, |B(\mathbf{a})| : \mathbf{a} \in A^{m-1} \,\}$. Let $\mathbf{a}$ give rise to the maximum value of k. The argument breaks into cases depending on the value of k.

Case 1. $3 \le k \le n-1$.
Then $X = B(\mathbf{a})$ satisfies (2) since $B(\mathbf{a}) \cdot B(\mathbf{a}) \subseteq B(\mathbf{a} \cdot \mathbf{a})$ by (6–3).

Case 2. $k = 2$.
Let $X = B(\mathbf{a})$ and let $Y = X \cdot X$. Note that $|Y| \le |X| = 2$. If $|Y| = 1$ or $|Y \cdot Y| = 1$ then (3) holds. Otherwise, $|Y \cdot Y| = 2$, so (4) holds.

Case 3. $k = n$.
Let $C = \{\, \mathbf{c} \in A^{m-1} : B(\mathbf{c}) = A \,\}$. By assumption, C is nonempty. If $C = A^{m-1}$ then $B = A^m$ which is false. So C is a proper nonempty subset of A^{m-1}. If $A \cdot A \neq A$ then (5) holds. So assume $A \cdot A = A$.

It follows that C is a subuniverse of $\mathbf{A}^{m-1}$. For suppose that $\mathbf{c}, \mathbf{c}' \in C$. For any $x \in A$ there are $b, b' \in A$ such that $x = b \cdot b'$ (since $A \cdot A = A$). Thus

$$\langle c_1 \cdot c_1', c_2 \cdot c_2', \ldots, c_{m-1} \cdot c_{m-1}', x \rangle =$$
$$\langle c_1, c_2, \ldots, c_{m-1}, b \rangle \cdot \langle c_1', c_2', \ldots, c_{m-1}', b' \rangle \in B \cdot B = B.$$

From the previous two paragraphs we conclude that $\mathbf{C}$ is a proper subalgebra of $\mathbf{A}^{m-1}$. By the minimality of m, $\mathbf{C}$ is not reduced. One of the two conditions in the definition of reduced must fail. Suppose first that condition (i) fails, i.e., for some $i \leq m-1$ we have $\vec{q}_i(C) = \{a\}$. Note that since C is a subuniverse, a must be an idempotent element. Choose $\mathbf{b} \in A^{m-1}$ with $b_i \neq a$ and $B(\mathbf{b})$ as large as possible. If $a \cdot b_i = b_i \cdot a = a$ then (6) holds.

On the other hand, if $b_i \cdot a \neq a$ then (7) holds and if $a \cdot b_i \neq a$ then (8) holds. To see this, assume that $b_i \cdot a \neq a$ and let $X = B(\mathbf{b})$. Since $b_i \neq a$, $\mathbf{b} \notin C$, so $B(\mathbf{b}) \neq A$, and therefore $|X| < n$. Also, $X \cdot A = B(\mathbf{b}) \cdot B(\mathbf{a}) \subseteq B(\mathbf{b} \cdot \mathbf{a})$. By assumption, $b_i \cdot a \neq a$, so $\mathbf{b} \cdot \mathbf{a} \notin C$. Therefore, by the maximality of $\mathbf{b}$, $|X \cdot A| \leq |B(\mathbf{b} \cdot \mathbf{a})| \leq |B(\mathbf{b})| = |X|$. This is (7). The argument for (8) is dual.

Now assume that $\mathbf{C}$ satisfies condition (i) but fails to be reduced because of condition (ii). Thus, there are $i \neq j$ such that for all $\mathbf{c} \in C$, $c_i = c_j$. Since C satisfies condition (i), $|\vec{q}_i(C)| > 1$. If $|\vec{q}_i(C)| < n$ then (1) holds (C is a subuniverse of $\mathbf{A}^{m-1}$ so $X = \vec{q}_i(C)$ is a subuniverse of $\mathbf{A}$). So we assume that $\vec{q}_i(C) = \vec{q}_j(C) = A$.

Pick $\mathbf{u} \in A^{m-1}$ with $u_i \neq u_j$ and $|B(\mathbf{u})|$ as large as possible. Let $a = u_i$ and $b = u_j$. If (9) fails, there is $c \in A$ with $a \cdot c \neq b \cdot c$. Pick $\mathbf{x} \in C$ with $x_i = c$. Then $x_j = c$ as well. Now $\mathbf{v} = \mathbf{u} \cdot \mathbf{x}$ has $v_i = a \cdot c \neq b \cdot c = v_j$. And therefore

$$1 \leq |B(\mathbf{u}) \cdot B(\mathbf{x})| \leq |B(\mathbf{u} \cdot \mathbf{x})| = |B(\mathbf{v})| \leq |B(\mathbf{u})|,$$

the last inequality following from the maximality of $\mathbf{u}$. Therefore (7) holds with $X = B(\mathbf{u})$ and $A = \mathbf{B}(\mathbf{x})$.

Case 4. $k = 1$.
Let B' be the result of transposing the last two coordinates of each element of B. If any of cases 1–3 apply to B', we are done. So we may assume that for all $\mathbf{a} \in A^{m-1}$, $|B'(\mathbf{a})| = |B(\mathbf{a})| = 1$. Put more explicitly

$$(\forall a_1, a_2, \ldots, a_{m-1})\,(\exists! b)\,\langle a_1, a_2, \ldots, a_{m-1}, b\rangle \in B \tag{6–4}$$

and for this b we have

$$\langle a_1, \ldots, a_{m-2}, b, a_{m-1}\rangle \in B'. \tag{6–5}$$

Fix $\mathbf{c} = \langle c_1, c_2, \ldots, c_{m-2}\rangle \in A^{m-2}$. Because of (6–4), there is a map $f_{\mathbf{c}}\colon A \to A$ such that for every x, $\langle c_1, \ldots, c_{m-2}, x, f_{\mathbf{c}}(x)\rangle \in B$. (6–5) says that $f_{\mathbf{c}}$ is injective. Therefore by the finiteness of A, $f_{\mathbf{c}}$ is a permutation.

Recall that B is reduced. Thus $q_{m-1}\restriction_B \neq q_m\restriction_B$. Therefore there is $\langle u_1, u_2, \ldots, u_m\rangle \in B$ with $u_{m-1} \neq u_m$. Let $\mathbf{u} = \langle u_1, \ldots, u_{m-2}\rangle$. Then $f_{\mathbf{u}}$ is not the identity. Let $\mathbf{y} = \mathbf{u} \cdot \mathbf{u}$, $\alpha = f_{\mathbf{u}}$, and $\beta = f_{\mathbf{y}}$. Then for every $a, b \in A$, $\langle \mathbf{u}, a, \alpha(a)\rangle \in B$ and $\langle \mathbf{u}, b, \alpha(b)\rangle \in B$, so (since B is a subuniverse) $\langle \mathbf{y},\, a \cdot b,\, \alpha(a) \cdot \alpha(b)\rangle \in B$. From this last relationship we obtain $\beta(a \cdot b) = \alpha(a) \cdot \alpha(b)$, so (10) holds. □

It turns out that except for the second one, all of the cases in the above lemma are easy to handle.

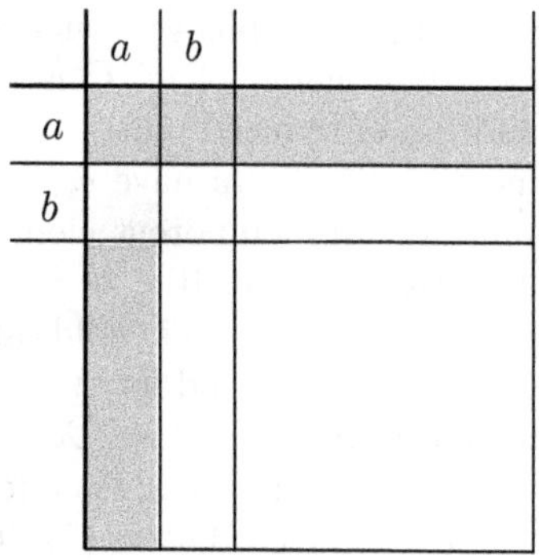

FIGURE 6.2: Cayley table if condition 10 holds. The shaded entries are determined by the unshaded ones.

Lemma 6.20. *For $c = 1, \ldots, 10$, the probability that a random binar satisfies condition* (c) *and not condition* (2) *of Lemma 6.19 is* 0.

Proof. Suppose first that **A** satisfies condition (10) of Lemma 6.19. Since α is not the identity, there are $a, b \in A$ with $\alpha(b) = a$ and $a \neq b$. Then for all $x \in A$

$$a \cdot x = \alpha(b) \cdot \alpha(y) = \beta(b \cdot y)$$
$$x \cdot a = \alpha(y) \cdot \alpha(b) = \beta(y \cdot b)$$

with $y = \alpha^{-1}(x)$. This means that in the Cayley table, the row for a is completely determined by the row for b, and similarly for the column of a. (Actually, there is one possible exception. If $\alpha(a) = b$ then $b \cdot a = \beta(a \cdot b)$. In this case one of the entries in the a-column is not determined by the b-column. See Figure 6.2.)

There are $n!$ choices for each of α and β. Note that $n! = 2 \cdot (3 \cdot 4 \cdots n) \leq 2 \cdot n^{n-2}$. There are at most $(n-1)^2 + 1$ unshaded squares in the Cayley table of Figure 6.2. Thus we compute

$$\frac{|\mathcal{Alg}_n[\text{cond.}(10)]|}{|\mathcal{Alg}_n|} \leq \frac{(n!)^2 \cdot n^{(n-1)^2+1}}{n^{n^2}} \leq \frac{4 \cdot n^{2n-4} \cdot n^{n^2-2n+2}}{n^{n^2}} = \frac{4}{n^2} \to 0.$$

Now assume that **A** satisfies (9). Thus there are $a, b \in A$ such that for all $c \in A$, $a \cdot c = b \cdot c$. This means the a-row and the b-row are equal. Then

$$\frac{|\mathcal{Alg}_n[\text{cond.}(9)]|}{|\mathcal{Alg}_n|} = \frac{\binom{n}{2} \cdot n^{n^2-n}}{n^{n^2}} \leq \frac{n^2}{2n^n} \to 0.$$

Suppose **A** satisfies condition (7) but not condition (2). Then

$$(\exists X)\ (1 \leq |X \cdot A| \leq |X| \leq n-1) \qquad \text{but not}$$
$$(\exists X)\ (3 \leq |X| \leq n-1 \ \&\ |X \cdot X| \leq |X|).$$

So the X that satisfies the first condition must have cardinality at most 2, and so must $X \cdot A$. Thus there are $\binom{n}{2} < n^2$ choices for each of those two sets. If $a \in X$ then the a-row of the Cayley table contains only two distinct values. There are 2^n choices for such a row. Therefore

$$\frac{|\mathcal{Alg}_n[\text{cond.}(7) \text{ not } (2)]|}{|\mathcal{Alg}_n|} \leq \frac{n^2 \cdot n^2 \cdot 2^n \cdot 2^n \cdot n^{n^2-2n}}{n^{n^2}} = \frac{4^n}{n^{2n-4}} = \left(\frac{4}{n}\right)^n \cdot \frac{1}{n^{n-4}} \to 0.$$

Condition (6) is similar to (9) and (3) and (4) are similar to (7). Of course (8) is dual to (7). If **A** satisfies (5) but not (2) then $n \leq 3$ (otherwise choose $a \in A - A \cdot A$ and take $X = A - \{a\}$ in condition (2).)

Finally, suppose that **A** satisfies (1) but not (2). Then for some subset X

$$(2 \leq |X| \leq n-1 \ \& \ X \cdot X \subseteq X) \qquad \text{but not}$$
$$(3 \leq |X| \leq n-1 \ \& \ |X \cdot X| \leq |X|).$$

Thus X must be a two-element subuniverse of **A**. There are 16 possible Cayley tables for each X. We compute

$$\frac{|\mathcal{Alg}_n[\text{cond.}(1) \text{ not } (2)]|}{|\mathcal{Alg}_n|} \leq \frac{\binom{n}{2} \cdot 16 \cdot n^{n^2-4}}{n^{n^2}} = \frac{8n(n-1)}{n^4} \to 0. \qquad \square$$

Unfortunately, condition (2) is much more delicate than the others. Let

$$\psi_n(k) = \binom{n}{k}^2 \cdot \left(\frac{k}{n}\right)^{(k^2)}$$

Lemma 6.21. *The probability that a random binar of cardinality n satisfies condition (2) is at most $\sum_{k=3}^{n-1} \psi_n(k)$.*

Proof. Condition (2) is equivalent to the assertion

$$(\exists X, Y \subseteq A) \ (3 \leq |X| = |Y| \leq n-1 \ \& \ X \cdot X \subseteq Y).$$

Let $|A| = n$, $X, Y \subseteq A$, $|X| = |Y| = k$ with $3 \leq k \leq n-1$. What is the probability that $X \cdot X \subseteq Y$? There are k^2 positions in the Cayley table for the products that make up $X \cdot X$ and the probability that each falls into Y is k/n. Thus the desired probability is $(k/n)^{k^2}$. Since there are $\binom{n}{k}$ choices for each of X and Y, the result follows. $\square$

Lemma 6.22. *There are real numbers c and d such that each of the three sums:*

$$\sum_{3 \leq k \leq cn} \psi_n(k) \qquad \sum_{cn \leq k \leq dn} \psi_n(k) \qquad \sum_{dn \leq k \leq n-1} \psi_n(k)$$

converges to 0 as n tends to infinity.

Each of these three sums requires a different argument. They form the content of the next three lemmas.

Lemma 6.22.A. *If* $0 < c < d < 1$ *then* $\lim_{n\to\infty} \sum_{k=cn}^{dn} \psi_n(k) = 0$.

Proof. Let $cn \le k \le dn$. Then

$$\psi_n(k) = \binom{n}{k}^2 \cdot \left(\frac{k}{n}\right)^{k^2} \le (2^n)^2 \cdot d^{k^2} \le 4^n d^{(cn)^2} = (4d^{c^2n})^n.$$

(The first inequality follows from the fact that $2^n = \sum_{i=0}^{n} \binom{n}{i} > \binom{n}{k}$ for any k. The second uses the fact that for $d < 1$, d^x is a decreasing function of x.) Therefore

$$\sum_{k=cn}^{dn} \psi_n(k) \le \sum_{k=cn}^{dn} (4d^{c^2n})^n \le (d-c)n(4d^{c^2n})^n \le n \cdot (4d^{c^2n})^n \to 0$$

since for large n, $d^{c^2n} < \frac{1}{4}$. □

Lemma 6.22.B. *There is a constant* c *with* $0 < c < 1$ *such that*

$$3 \le k \le cn \implies \frac{\psi_n(k+1)}{\psi_n(k)} < 1. \textit{ Thus}$$

$$\sum_{k=3}^{cn} \psi_n(k) \le \frac{cn \cdot 3^6}{n^3} \to 0$$

Proof.

$$\frac{\psi_n(k+1)}{\psi_n(k)} = \left(\frac{n-k}{k+1}\right)^2 \left(\frac{k+1}{n}\right)^{2k+1} \left(\frac{k+1}{k}\right)^{k^2} =$$

$$\left(\frac{n-k}{n}\right)^2 \left(\frac{k+1}{n}\right)^{2k-1} \left(\frac{k+1}{k}\right)^{k\cdot k} \le \left(\frac{k+1}{n}\right)^{2k-1} \cdot e^k \le \left(\frac{k+1}{n}\right)^k \cdot e^k.$$

Let $c = 1/4$ and $3 \le k \le n/4$, which is vacuous unless $n \ge 12$. Then

$$\frac{k+1}{n} = \frac{k}{n} + \frac{1}{n} \le \frac{1}{4} + \frac{1}{n} \le \frac{1}{3}$$

so $\frac{\psi_n(k+1)}{\psi_n(k)} \le (\frac{e}{3})^k < 1$.

It follows that $\psi_n(k)$ is decreasing in k. Also

$$\psi_n(3) = \left[\frac{n(n-1)(n-2)}{6}\right]^2 \cdot \frac{3^9}{n^9} \le \frac{n^6}{3^2 \cdot 4} \cdot \frac{3^9}{n^9} \le \frac{3^6}{n^3}.$$

Therefore

$$\sum_{k=3}^{cn} \psi_n(k) \le cn\psi_n(3) \le \frac{3^6 c}{n^2} \to 0. \quad \square$$

OK, now it's time to roll out the big guns. Here is a sharp version of Stirling's formula [Rob55].

$$(\forall x \geq 1)\ \sqrt{2\pi}x^{x+1/2}e^{-x}e^{(12x+1)^{-1}} \leq x! \leq \sqrt{2\pi}x^{x+1/2}e^{-x}e^{(12x)^{-1}} \tag{6–6}$$

As a consequence, for integers $a > b \geq 1$ we have

$$\binom{a}{b} < \sqrt{a}\left[\left(\frac{b}{a}\right)^{b/a} \cdot \left(\frac{a-b}{a}\right)^{(a-b)/a}\right]^{-a} \tag{6–7}$$

Lemma 6.22.C. *There is a real number d with $1/2 < d < 1$ such that* $\sum_{k=dn}^{n-1} \psi_n(k) \to 0$.

Proof. Assume that $1/2 < d < 1$ and let $u = \lfloor dn \rfloor$. It follows from inequality (6–7) that

$$\binom{n}{u}^2 < n \cdot \left[\left(\frac{u}{n}\right)^{u/n} \cdot \left(\frac{n-u}{n}\right)^{(n-u)/n}\right]^{-2n} \tag{6–8}$$

Notice that $0 \leq dn - u < 1 \implies 0 \leq d - u/n < 1/n \implies d - 1/n < u/n \leq d < 1$. Consequently

$$\left(\frac{u}{n}\right)^{u/n} > \left(d - \frac{1}{n}\right)^{u/n} \geq \left(d - \frac{1}{n}\right)^{d}. \tag{6–9}$$

Similarly, $(n-u)/n = 1 - u/n \geq 1 - d$ and $1 - u/n < 1 - (d - 1/n) = 1 - d + 1/n$. Thus

$$\left(\frac{n-u}{n}\right)^{(n-u)/n} \geq (1-d)^{(n-u)/n} \geq$$
$$(1-d)^{1-d+1/n} = (1-d)^{1-d} \cdot (1-d)^{1/n}. \tag{6–10}$$

Combining the inequalities in (6–8), (6–9) and (6–10) we obtain

$$\binom{n}{u}^2 < n\left[\left(d - 1/n\right)^d (1-d)^{1-d}(1-d)^{1/n}\right]^{-2n}. \tag{6–11}$$

Furthermore

$$\left(\frac{n-1}{n}\right)^{d^2n^2} = \left[\left(\frac{n-1}{n}\right)^n\right]^{d^2n} < \left(\frac{1}{e}\right)^{d^2n}. \tag{6–12}$$

Since $d > 1/2$, $k > dn > n/2$, so $\binom{n}{k}$ decreases as k increases. Therefore $\binom{n}{k} \leq \binom{n}{u}$. Also, $k/n < 1$ implies that $\left(\frac{k}{n}\right)^x$ is decreasing in x. Therefore

$$\left(\frac{k}{n}\right)^{(k^2)} \leq \left(\frac{n-1}{n}\right)^{(k^2)} \leq \left(\frac{n-1}{n}\right)^{(dn)^2}.$$

Applying (6–11) and (6–12) to the definition of $\psi_n(k)$, we obtain, for $dn \leq k \leq n-1$,

$$\psi_n(k) \leq n\left[\left(\frac{1}{(d-1/n)^d(1-d)^{1-d}}\right)^2 \cdot \left(\frac{1}{e}\right)^{d^2}\right]^n \cdot (1-d)^{-2}. \tag{6–13}$$

Now assume that $n > 4$. Then

$$\lim_{d\to 1^-} de^{d^2}\left[(d-1/n)^d(1-d)^{(1-d)}\right]^2 \geq 1.$$

Choose d large enough so that this quantity is larger than 1. (Casual calculations indicate that d should be about 0.965.) For this d

$$d \geq \left(\frac{1}{\left(d-\frac{1}{n}\right)^d(1-d)^{(1-d)}}\right)^2 \cdot \left(\frac{1}{e}\right)^{d^2}. \tag{6–14}$$

From (6–13) and (6–14) we deduce

$$\psi_n(k) \leq nd^n(1-d)^{-2}$$

and therefore

$$\sum_{dn\leq k\leq n-1} \psi_n(k) \leq n^2d^n(1-d)^{-2} \to 0.$$

□

Combining Lemmas 6.19–6.22, we have proved

Theorem 6.23. *The probability that a random finite binar is idemprimal is 1.*

Proof of Theorem 6.17. It follows from Definition 6.4 and Exercise 6.15.7 that a finite algebra is primal iff it is idemprimal and has no trivial subalgebras. Put another way, if we view the properties P, E and I as subsets of $\mathcal{Alg}_\rho$, then $P = E \cap I$. Let $\overline{I}$ denote the complement of the set I.

Now let $\rho = \langle k\rangle$ with $k > 2$ and let $\mathbf{A} = \langle A, f\rangle$ be a finite algebra of type ρ. Define $x \cdot y = f(x, y, y, \dots, y)$ and $\mathbf{A}_0 = \langle A, \cdot\rangle$. Then $\mathrm{Clo}(\mathbf{A}_0) \subseteq \mathrm{Clo}(\mathbf{A})$. Since with probability 1, a randomly chosen $\mathbf{A}_0$ is idemprimal (i.e., $\mathrm{Clo}(\mathbf{A}_0)$ contains every idempotent operation on A), the same is true for a random $\mathbf{A}$. This proves 6.17(2).

Since by 6.16(1), $\Pr_\rho(E) = 1/e$ and $E = (I \cap E) \uplus (\overline{I} \cap E) = P \uplus (\overline{I} \cap E)$, we obtain

$$1/e = \Pr(P) + \Pr(\overline{I} \cap E) = \Pr(P) \tag{6–15}$$

since $\overline{I} \cap E \subseteq \overline{I}$ which occurs with probability 0 by the previous paragraph.

Finally, if the similarity type contains at least one more operation symbol, then $1/e$ can be replaced with 1 in (6–15) according to Theorem 6.16(2). □

Chapter 7

Maltsev Varieties

In this chapter we drop our assumption of congruence-distributivity, but retain congruence-permutability. Structures now will be much more group-like. We lose the power of Jónsson's lemma, so the analysis of subdirectly irreducible algebras becomes more delicate. The study of the congruence lattices of products becomes essential. One might say that the central goal of this chapter is an understanding of skew congruences.

7.1 Directly representable varieties

Any conception of "highly structured variety" should surely include the varieties of Boolean algebras and Abelian groups. Among their attributes, the finite algebras in each of these varieties decompose nicely. In fact:

- every finite Boolean algebra is a direct power of $\mathbb{B}_1$;
- every finite Abelian group is a product of cyclic groups of prime-power order.

This is the inspiration for the following definition.

Definition 7.1. A finitely generated variety is called *directly representable* if and only if it has only finitely many finite directly indecomposable algebras.

Now we should acknowledge that the variety of all Abelian groups does not qualify as directly representable since it is not finitely generated (and, in fact, has infinitely many finite directly indecomposable algebras). However, every proper subvariety of Abelian groups is directly representable.

Several mathematicians contributed to a characterization of directly representable varieties, including D. Clark, P. Krauss, R. Quackenbush and S. Burris. The project was largely completed in 1982 by Ralph McKenzie [McK82], who reduced the question to finding those finite rings whose variety of modules is directly representable. This characterization is one of the milestones of modern universal algebra. The problem for rings is still open.

The reader will not be surprised to learn that congruence-permutability plays a role in the analysis of direct representability. McKenzie's first achievement was to prove necessity: every directly representable variety is Maltsev.

We derive that result in this section. Using that result we settle the direct representability question for congruence-distributive varieties. In Section 7.5 we shall complete McKenzie's characterization.

First, we develop a few more examples of directly representable varieties.

Theorem 7.2. *Let $\mathcal{V}$ be a finitely generated, arithmetical and semisimple variety. Then $\mathcal{V}$ is directly representable.*

Proof. Assume $\mathcal{V}$ is generated by a finite set $\mathcal{K}$ of finite algebras. Since $\mathcal{V}$ is congruence-distributive, $\mathcal{V}_{\mathrm{si}} \subseteq \mathbf{HS}(\mathcal{K})$ by Jónsson's lemma, hence all subdirectly irreducible algebras are finite and there are only finitely many of them.

Let $\mathbf{A}$ be a finite directly indecomposable algebra. We can write $\mathbf{A}$ as a subdirect product $\prod_{i=1}^{n} \mathbf{B}_i$ of finite subdirectly irreducible algebras. By assumption, each $\mathbf{B}_i$ is simple. Therefore by Theorem 6.3, $\mathbf{A}$ is a direct product of some of the $\mathbf{B}_i$'s. But $\mathbf{A}$ is directly indecomposable. Thus $\mathbf{A} \cong \mathbf{B}_i \in \mathcal{V}_{\mathrm{si}}$. So there are only finitely many finite directly indecomposable algebras. □

That the variety of Boolean algebras is directly representable follows immediately from this theorem. More generally, the varieties $\mathcal{Cr}_n$ of commutative rings satisfying $x^n \approx x$ as well as the variety generated by any quasiprimal algebra will be directly representable.

As another example consider the 6-element r-lattice, $\mathbf{MO}_2$, pictured to the right, in which a_0 and b_0 are complements of each other, as are a_1 and b_1. We showed in Example 4.72 that any r-lattice will generate an arithmetical variety. Using Jónsson's lemma, it is easy to see that the only subdirectly irreducible members of the variety are $\mathbf{MO}_2$ itself as well as the two-element r-lattice, which is polynomially equivalent to $\mathbb{B}_1$. Thus Theorem 7.2 implies that $\mathbf{MO}_2$ generates a directly representable variety. (In the literature, this algebra is called a *modular ortholattice.*)

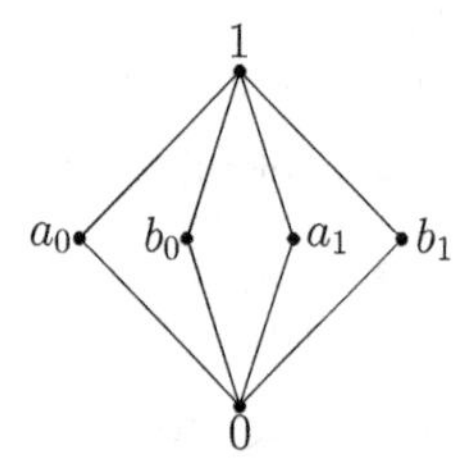

Now we turn to consequences of direct representability. We begin with a definition. In it we write $\mathcal{V}_{\mathrm{fin}}$ to denote the class of finite members of the variety $\mathcal{V}$.

Definition 7.3. (1) Let S be a set of positive integers. S is *narrow* if the elements of S collectively have only finitely many prime divisors.

(2) A finitely generated variety $\mathcal{V}$ is *narrow* if the set $\{\, |A| : \mathbf{A} \in \mathcal{V}_{\mathrm{fin}} \,\}$ is a narrow set of integers.

What is the relevance of narrowness to the subject at hand? Suppose that $\mathcal{V}$ is a directly representable variety. Let $\{\mathbf{D}_1, \ldots, \mathbf{D}_t\}$ be the finite directly indecomposable members of $\mathcal{V}$. There are only finitely many primes $p_1, \ldots, p_m$

dividing the cardinality of any of the D's. Now let $\mathbf{A}$ be a finite member of $\mathcal{V}$. We must have $\mathbf{A} \cong \mathbf{D}_1^{k_1} \times \mathbf{D}_2^{k_2} \times \cdots \times \mathbf{D}_t^{k_t}$ for some natural numbers $k_1, \ldots, k_t$. But then the prime divisors of $|A|$ must lie among $p_1, \ldots, p_m$. Thus

Proposition 7.4. *Every directly representable variety is narrow.*

Lemma 7.5 (Pólya). *Let $c_1, \ldots, c_t$ be a finite sequence of positive integers, not all equal. Then the set $\{ c_1^n + c_2^n + \cdots + c_t^n : n > 0 \}$ is not narrow.*

Proof. Let $s_n = c_1^n + \cdots + c_t^n$ and $S = \{ s_n : n > 0 \}$. Suppose that S is narrow. We shall derive a contradiction.

Since S is narrow, there is a finite set $\{p_1, \ldots, p_r\}$ of primes so that every s_n is a product of $p_1, \ldots, p_r$. Let d be the greatest common divisor of $\{c_1, \ldots, c_t\}$. If $d > 1$ then $\frac{c_1}{d}, \frac{c_2}{d}, \ldots, \frac{c_t}{d}$ would yield a set at least as narrow as S. So without loss of generality, assume that $d = 1$. Furthermore, since $c_1, \ldots, c_t$ are not all equal to 1, the set S must be unbounded.

Recall the Euler totient function, $\phi(n)$, from elementary number theory. For any prime p and $k \geq 1$ we have $\phi(p^{k+1}) = p^k(p-1)$. Then for any integer x

$$\begin{aligned} p \nmid x &\implies x^{\phi(p^{k+1})} \equiv 1 \pmod{p^{k+1}} \\ p \mid x &\implies x^{\phi(p^{k+1})} \equiv 0 \pmod{p^{k+1}}. \end{aligned} \tag{7–1}$$

The first implication follows from Euler's theorem. For the second, if $p \mid x$ then $p^{k+1} \mid x^{k+1}$. Since $\phi(p^{k+1}) = p^k(p-1) \geq 2^k \geq k+1$ the implication follows.

Claim. Let p be a prime, $k \geq 1$, and assume that $p^{k+1} > t$. Then for any positive integer ℓ, $\phi(p^{k+1}) \mid \ell \implies p^{k+1} \nmid s_\ell$.

Proof. By renumbering, there is $m \leq t$ so that $p \nmid c_1$, $p \nmid c_2, \ldots, p \nmid c_m$, $p \mid c_{m+1}, \ldots, p \mid c_t$. Since $\gcd(c_1, \ldots, c_t) = 1$, $m \geq 1$. By assumption there is an integer n such that $\ell = n \cdot \phi(p^{k+1})$. Then using the implications in (7–1)

$$\begin{aligned} s_\ell &= c_1^{n\phi(p^{k+1})} + c_2^{n\phi(p^{k+1})} + \cdots + c_t^{n\phi(p^{k+1})} \\ &\equiv 1 + 1 + \cdots + 0 + \cdots + 0 = m \pmod{p^{k+1}}. \end{aligned}$$

Since $1 \leq m \leq t < p^{k+1}$ we conclude that $p^{k+1} \nmid s_\ell$. □

Now choose k large enough so that $p_j^{k+1} > t$ for $j = 1, \ldots, r$ and set $m = \phi(p_1^{k+1}) \cdots \phi(p_r^{k+1}) = \phi((p_1 \cdots p_r)^{k+1})$. For all $j \leq r$ and for all $n \geq 1$, take $\ell = nm = \phi(p_j^{k+1}) \cdot (m/\phi(p_j)^{k+1}) \cdot n$. Then the claim implies that $p_j^{k+1} \nmid s_{nm}$.

However, the only allowable prime divisors of s_{nm} are $p_1, \ldots, p_r$. So we must have $s_{nm} \mid (p_1^k p_2^k \cdots p_r^k)$. It follows that S is bounded, which is a contradiction. □

McKenzie rediscovered Lemma 7.5. He used it to construct a key link in the chain of consequences.

Definition 7.6. An algebra $\mathbf{A}$ is *congruence-uniform* if for every congruence θ and every $a, b \in A$, $|a/\theta| = |b/\theta|$.

We learn that groups are congruence-uniform in our first class in abstract algebra. More generally, quasigroups are congruence-uniform (Exercise 1.26.6). Lattices, by contrast, are not congruence-uniform.

We shall have several opportunities to make use of the following construction. Let $\mathbf{A}$ be an algebra and θ a congruence on $\mathbf{A}$. For a positive integer n define

$$\mathbf{A}_n(\theta) = \{\, \mathbf{x} \in A^n : 1 \leq i < j \leq n \implies (x_i, x_j) \in \theta \,\}. \qquad (7\text{–}2)$$

It is easy to check that $\mathbf{A}_n(\theta)$ is always a subalgebra of $\mathbf{A}^n$. We can also write $A_n(\theta) = \bigcup\{\, (a/\theta)^n : a \in A \,\}$.

Proposition 7.7. *In a narrow variety, every finite member is congruence-uniform.*

Proof. Assume that $\mathbf{A}$ is a finite algebra in a narrow variety and that θ is a congruence on $\mathbf{A}$. Denote the θ-classes by $C_1, C_2, \ldots, C_t$ and let $c_i = |C_i|$, for $i \leq t$. We must show that $c_1 = c_2 = \cdots = c_t$.

From the definition in Equation (7–2) we see that for every $n > 0$, $A_n(\theta) = C_1^n \cup C_2^n \cup \ldots C_t^n$. Since these sets are pairwise disjoint,

$$|A_n(\theta)| = c_1^n + c_2^n + \cdots + c_t^n.$$

By assumption the variety is narrow, thus by Lemma 7.5 we must have $c_1 = c_2 = \cdots = c_t$. □

Proposition 7.8. *Let* $\mathbf{A}$ *be a finite algebra and suppose that each member of* $\mathrm{Sub}(\mathbf{A}^2)$ *is congruence-uniform. Then* $\mathbf{A}$ *has permuting congruences.*

Proof. Let α and β be congruences on $\mathbf{A}$. We shall show that $\beta \circ \alpha \subseteq \alpha \circ \beta$. Let $B = \alpha \circ \beta$. Note that $\mathbf{B}$ is a subalgebra of $\mathbf{A}^2$ (Exercise 1.26.1). By assumption $\mathbf{B}$ is congruence-uniform. Also $\mathbf{A}$ is congruence-uniform since it is isomorphic to the subalgebra δ_A of $\mathbf{A}^2$.

Let θ be the restriction of the congruence $\beta \times \beta \in \mathrm{Con}(\mathbf{A}^2)$ to B. In other words,

$$(x_1, x_2) \equiv_\theta (y_1, y_2) \iff (x_1, x_2) \in B,\ (y_1, y_2) \in B,\ x_1 \equiv_\beta y_1, \text{ and } x_2 \equiv_\beta y_2.$$

Since both α and β are reflexive it follows that B is reflexive as well. We claim that for any $a \in A$

$$(a, a)/\theta = a/\beta \times a/\beta. \qquad (7\text{–}3)$$

The left-to-right inclusion is immediate. For the converse, suppose (b, c) is a member of $a/\beta \times a/\beta$. Then $b \mathrel{\beta} a \mathrel{\beta} c$ so $(b, c) \in \beta \subseteq B$. Thus $(b, c) \mathrel{\theta} (a, a)$.

Utilizing congruence-uniformity, let r denote the size of any β-class on $\mathbf{A}$

and s the size of a θ-class on $\mathbf{B}$. Equation (7–3) implies that $s = r^2$. But now, for any $(x_1, x_2) \in B$ we have

$$(x_1, x_2)/\theta = (x_1/\beta \times x_2/\beta) \cap B \subseteq x_1/\beta \times x_2/\beta.$$

Since $|(x_1, x_2)/\theta| = s$ and $|x_1/\beta \times x_2/\beta| = r^2$ and $s = r^2$ the above inclusion must be an equality. Thus for any $(x_1, x_2) \in B$, $(x_1/\beta \times x_2/\beta) \subseteq B$.

Finally, we prove the proposition. Let $(a, c) \in \beta \circ \alpha$. Then for some $b \in A$, $a \ \beta \ b \ \alpha \ c$. We have $(b, c) \in \alpha \subseteq B$ so from the equality in the previous paragraph $(a, c) \in b/\beta \times c/\beta \subseteq B = \alpha \circ \beta$. □

We can now combine the results of the last three propositions to achieve our goal.

Theorem 7.9 (McKenzie, 1982). *Every directly representable variety is Maltsev.*

Proof. Let $\mathcal{V}$ be a directly representable variety. As we have observed, $\mathcal{V}$ is narrow, hence every finite member is congruence-uniform. Let $\mathbf{F}$ denote the $\mathcal{V}$-free algebra on 3 generators. Since $\mathcal{V}$ is locally finite, F^2 is finite. Therefore every subalgebra of $\mathbf{F}^2$ is congruence-uniform. By Proposition 7.8, $\mathbf{F}$ has permuting congruences. But now by Theorem 4.64, $\mathcal{V}$ has permuting congruences, i.e., it is Maltsev. □

We shall use Theorem 7.9 to characterize those congruence-distributive varieties that are directly representable. The proof hinges on the following observation.

Proposition 7.10. *Let* $\mathbf{A}$ *be a subdirectly irreducible algebra in a congruence-distributive variety. If* μ *is the monolith of* $\mathbf{A}$ *and* $\mu \neq 1_A$ *then for every* $n > 0$ *the algebra* $\mathbf{A}_n(\mu)$ *is directly indecomposable.*

Proof. The algebra $\mathbf{A}_n(\mu)$ is defined in Equation (7–2). For every $i \leq n$, let $p_i \colon \mathbf{A}_n(\mu) \to \mathbf{A}$ be the projection onto the ith coordinate. Note that every p_i is surjective. (Every constant n-tuple is a member of $A_n(\mu)$.) As always, denote by η_i the kernel of p_i. We have $\bigcap_i \eta_i = 0$ and $\mathbf{A}_n(\mu)/\eta_i \cong \mathbf{A}$. Thus every η_i is completely meet-irreducible.

Let $\mu_i = \bar{p}_i(\mu)$. By the correspondence theorem, $1 > \mu_i \succ \eta_i$. Also, from the definition of $\mathbf{A}_n(\mu)$ we have $\mu_1 = \mu_2 = \cdots = \mu_n$.

Let us suppose that $\mathbf{A}_n(\mu)$ is directly decomposable. Then there is a nontrivial pair $\{\alpha, \beta\}$ of factor congruences on $\mathbf{A}_n(\mu)$, i.e.,

$$\alpha \cap \beta = 0, \quad \alpha \circ \beta = 1, \quad 0 < \alpha, \beta < 1.$$

For every $i \leq n$ define $\alpha_i = \alpha \vee \eta_i$ and $\beta_i = \beta \vee \eta_i$. By distributivity we obtain $\alpha = \bigcap_i \alpha_i$ and $\beta = \bigcap_i \beta_i$. Therefore

$$0 = \alpha \cap \beta = \bigcap_{i=1}^{n} (\alpha_i \cap \beta_i). \tag{7–4}$$

Claim. For every $i \leq n$, $\alpha_i \cap \beta_i = \eta_i$.
Proof. Suppose that for some $j \leq n$, $\alpha_j \cap \beta_j > \eta_j$. Then $\alpha_j \cap \beta_j \geq \mu_j$. Pick $(a, b) \in \mu - 0_A$. Let $\mathbf{x} = \langle a, a, a, \ldots, a \rangle$ and $\mathbf{y} = \langle a, a, \ldots, b, \ldots, a \rangle$, with the lone b appearing in the jth place. Notice that both $\mathbf{x}$ and $\mathbf{y}$ are members of $A_n(\mu)$. Therefore, for every $i \leq n$, $\mathbf{x} \equiv \mathbf{y} \pmod{\alpha_i \cap \beta_i}$ which contradicts equation (7–4).

Since every η_i is completely meet-irreducible, it follows from the claim that for every $i \leq n$ either $\alpha_i = \eta_i$ or $\beta_i = \eta_i$. On the other hand, using distributivity

$$1_{A_n(\mu)} = \alpha \vee \beta = \bigcap_{i=1}^{n} \alpha_i \vee \bigcap_{j=1}^{n} \beta_j = \bigcap_{i,j=1}^{n} (\alpha_i \vee \beta_j)$$

from which it follows that $\alpha_i \vee \beta_j = 1$ for all i, j.

Now, since $\alpha \neq 0$, there is some $j \leq n$ such that $\alpha_j > \eta_j$, so from the previous paragraph, $\beta_j = \eta_j$. Similarly, there is $i \leq n$ such that $\alpha_i = \eta_i$. But then

$$1 = \alpha_i \vee \beta_j = \eta_i \vee \eta_j < \mu_i \vee \mu_j = \mu_1$$

since all the μ_i's are equal. This contradiction completes the proof. □

Theorem 7.11 (Burris, 1978). *Let $\mathcal{V}$ be a finitely generated and congruence-distributive variety. Then $\mathcal{V}$ is directly representable if and only if it is semi-simple and arithmetical.*

Proof. The right-to-left direction is Theorem 7.2. So assume that $\mathcal{V}$ is directly representable. By Theorem 7.9, $\mathcal{V}$ is Maltsev, hence arithmetical. Suppose that $\mathbf{A}$ is subdirectly irreducible but not simple. Since $\mathcal{V}$ is congruence-distributive and finitely generated, $\mathbf{A}$ is finite by Jónsson's lemma. By Proposition 7.10 every $\mathbf{A}_n(\mu)$ is directly indecomposable. Then $\mathcal{V}$ has infinitely many finite directly indecomposable algebras which contradicts direct representability. □

As part of his effort to understand the structure of directly representable varieties, McKenzie had earlier proved that these varieties have definable principal congruences. As a result, they are all finitely based. The proof does not invoke Theorem 7.9.

Lemma 7.12. *Let $\mathcal{V}$ be a directly representable variety. Then $\mathcal{V}$ has only finitely many subdirectly irreducible algebras, all finite.*

Proof. By assumption, $\mathcal{V}$ is locally finite. Since every subdirectly irreducible algebra is directly indecomposable, direct representability implies that there are only finitely many finite subdirectly irreducible algebras in $\mathcal{V}$. It follows from Theorem 5.7, that $\mathcal{V}$ has no infinite subdirectly irreducible algebras. □

Theorem 7.13. *Every directly representable variety has definable principal congruences.*

Proof. Let $\mathcal{V}$ be directly representable and let $\mathcal{K}$ be the set of finite, directly indecomposable member of $\mathcal{V}$ (one isomorphic copy of each). By Exercise 5.41.4, it is enough to show that $\mathcal{V}_{\text{fin}}$ has DPC. Suppose not. Then for each congruence formula, ψ,

$$\text{there is a finite algebra } \mathbf{A} \text{ and } a, b, c, d \in A \text{ such that } (c, d) \in \mathrm{Cg}^{\mathbf{A}}(a, b) \text{ but } \mathbf{A} \models \neg\psi(c, d, a, b). \tag{7–5}$$

Choose a congruence formula ψ_1. Invoke (7–5) to obtain a finite algebra $\mathbf{A}_1$ and elements a_1, b_1, c_1, d_1 in A_1 such that $(c_1, d_1) \in \mathrm{Cg}^{\mathbf{A}_1}(a_1, b_1)$, while $\mathbf{A}_1 \models \neg\psi_1(c_1, d_1, a_1, b_1)$. There is a congruence formula ϕ_1 which defines principal congruences on $\mathbf{A}_1$ (see the remarks following Lemma 5.31). Let $\psi_2 = \psi_1 \curlyvee \phi_1$. Then ψ_2 is a congruence formula, so invoking (7–5) again, there is a finite algebra $\mathbf{A}_2$ and $a_2, b_2, c_2, d_2 \in A_2$ such that $(c_2, d_2) \in \mathrm{Cg}^{\mathbf{A}_2}(a_2, b_2)$ but $\mathbf{A}_2 \models \neg\psi_2(c_2, d_2, a_2, b_2)$. By iterating this process, we obtain congruence formulas $\psi_1, \psi_2, \ldots$, finite algebras $\mathbf{A}_1, \mathbf{A}_2, \ldots$, and, for each n, $a_n, b_n, c_n, d_n \in A_n$ such that

$$\begin{gathered} \mathcal{V} \models \psi_n \to \psi_{n+1}, \quad \mathbf{A}_n \models \neg\psi_n(c_n, d_n, a_n, b_n), \\ \mathbf{A}_n \models \psi_{n+1}(c_n, d_n, a_n, b_n). \end{gathered} \tag{7–6}$$

Since the variety is directly representable and every $\mathbf{A}_n$ is finite, we can write $\mathbf{A}_n \cong \prod_{i=1}^{m_n} \mathbf{B}_{n,i}$ in which every $\mathbf{B}_{n,i} \in \mathcal{K}$ and m_n is a natural number. Let us define

$$Y = \{\, (a, b, c, d, \mathbf{B}) : \mathbf{B} \in \mathcal{K} \text{ and } a, b, c, d \in B \,\}$$

and, for every positive integer n

$$Y_n = \{\, (a_{n,i}, b_{n,i}, c_{n,i}, d_{n,i}, \mathbf{B}_{n,i}) : i \le m_n \,\}.$$

Note that every $Y_n \subseteq Y$. By $a_{n,i}$ we mean the ith coordinate of $a_n \in \prod \mathbf{B}_{n,j}$.

Suppose that for some n,

$$(a_{n,1}, b_{n,1}, c_{n,1}, d_{n,1}, \mathbf{B}_{n,1}) = (a_{n,2}, b_{n,2}, c_{n,2}, d_{n,2}, \mathbf{B}_{n,2}).$$

Let $\mathbf{A}_n'$ be the projection of $\mathbf{A}_n$ onto all coordinates except the first. Since $\mathbf{B}_{n,1} = \mathbf{B}_{n,2}$, $\mathbf{A}_n'$ is not only a homomorphic image of $\mathbf{A}_n$, but isomorphic to a subalgebra of $\mathbf{A}_n$ as well, obtained by mapping $(x_2, x_3, \ldots, x_{m_n})$ to $(x_2, x_2, x_3, \ldots, x_{m_n})$. Therefore, by Lemma 5.33, we can replace $\mathbf{A}_n$ with $\mathbf{A}_n'$ and maintain the validity of (7–6).

Obviously, there is nothing special about the indices 1 and 2. Furthermore, we can apply this reduction repeatedly to eliminate any redundant coordinates in the representation of $\mathbf{A}_n$. Thus, without loss of generality, we can assume that in Y_n, no quintuple appears more than once, i.e., $|Y_n| = m_n$.

Since $\mathcal{K}$ is a finite set of finite algebras, the set Y is finite. Therefore, there are integers $n < m$ such that $Y_n = Y_m$. Because we have eliminated redundant coordinates, it follows that $(a_n, b_n, c_n, d_n, \mathbf{A}_n) = (a_m, b_m, c_m, d_m, \mathbf{A}_m)$ (after a reordering of coordinates). However, from (7–6),

$$\mathbf{A}_n \models \psi_{n+1}(c_n, d_n, a_n, b_n), \implies \mathbf{A}_n \models \psi_m(c_n, d_n, a_n, b_n) \implies \mathbf{A}_m \models \psi_m(c_m, d_m, a_m, b_m)$$

which is a contradiction. □

Corollary 7.14. *Every directly representable variety of finite type is finitely based.*

Proof. We wish to apply Corollary 5.37. Let $\mathcal{V}$ be directly representable. By Theorem 7.13, $\mathcal{V}$ has definable principal congruences. By Lemma 7.12, $\mathcal{V}_{\text{si}}$ is (up to isomorphism) a finite set of finite algebras. □

Exercise Set 7.15.

1. Is the variety of rectangular bands directly representable?

2. The purpose of this exercise is to prove the following result from [CK76]. Let $\mathcal{V}$ be a locally finite variety in which every finite algebra is congruence uniform (as in Proposition 7.7). Then $\mathcal{V}$ is congruence-uniform. For this, let $\mathbf{A} \in \mathcal{V}$, $\theta \in \text{Con}(\mathbf{A})$, $a, b \in A$ and suppose that $|a/\theta| < |b/\theta|$.

   (a) Suppose $|a/\theta| = n < \infty$. Pick distinct elements $b_1, b_2, \ldots, b_{n+1}$ in b/θ with $b = b_1$. Let $B = \text{Sg}^{\mathbf{A}}\{a, b_1, b_2, \ldots, b_{n+1}\}$ and $\psi = \theta\restriction_B$. Then $|a/\psi| < |b/\psi|$ which is impossible.

   (b) Now suppose that a/θ is infinite. We can choose $c \in (b/\theta) - \text{Sg}^{\mathbf{A}}(a/\theta \cup \{b\})$. Let $B_1 = \text{Sg}^{\mathbf{A}}\{a, b, c\}$ and $\psi_1 = \theta\restriction_{B_1}$. B_1 is certainly finite. Now let $B_2 = \text{Sg}^{\mathbf{B}_1}(a/\psi_1 \cup \{b\})$ and $\psi_2 = \psi_1\restriction_{B_2}$. Then

   $$|b/\psi_2| = |a/\psi_2| = |a/\psi_1| = |b/\psi_1|.$$

   But $c \in b/\psi_1$ and $c \notin b/\psi_2$ which is a contradiction.

3. An algebra is *congruence-regular* if, for any two congruences α, β and every element a, $a/\alpha = a/\beta \implies \alpha = \beta$. (Cf. Exercise 4.75.9) A variety is congruence-regular (resp. uniform) if every member is congruence-regular (resp. uniform). Prove that every congruence-uniform variety is congruence-regular.

7.2 The centralizer congruence

In Chapter 5 we developed tools that allowed us to make a detailed analysis of congruence-distributive varieties. This gives us great insight into the structure of the subdirectly irreducible algebras, the subvarieties, and the equations that hold in such a variety. These tools are the key to our characterization of primal and quasi-primal algebras in Section 6.1.

In the 1970s, researchers sought analogous insight into the structure of congruence-permutable varieties. The work of R. McKenzie, P. Krauss, D. Clark and J. D. H. Smith collectively developed a structure theory that has become known as "commutator theory." Although the behavior of Maltsev varieties is quite different from that of congruence-distributive varieties, the commutator too gives us information about the fine structure of varieties. In particular, it illuminates the boundary between Maltsev varieties in general and the more special arithmetical varieties. (Have another look at the schematic on page 138.)

Very shortly after Smith nailed down the precise definition of the commutator in [Smi76], the lattice theorists H. P. Gumm, J. Hagemann and C. Herrmann showed that the entire framework of commutator theory could be defined, albeit with considerably more effort, in the more general context of congruence-modular varieties. This has allowed us to unite the commutator construction with the insights of Jónsson's lemma into a single theory that gives us a very clear view of congruence-modularity. Along the way it clarifies the role of distributivity and permutability.

In the next three sections we shall develop the basic notions of commutator theory but only in the congruence-permutable context. This will already allow us to derive some of the most striking consequences of the theory. The motivated reader can consult either [FM87] or [Gum83] for the full treatment.

Definition 7.16. Let $\mathbf{A}$ be an arbitrary algebra and α a congruence on $\mathbf{A}$. We define a binary relation α^* called the *centralizer of* α by

$$\alpha^* = \{ (a,b) \in A^2 : \forall t \in \mathrm{Clo}_{n+1}(\mathbf{A})\ \forall(\mathbf{c},\mathbf{d}) \in \alpha, \\ t(a,\mathbf{c}) = t(a,\mathbf{d}) \iff t(b,\mathbf{c}) = t(b,\mathbf{d}) \}.$$

We emphasize that despite the comments that introduced this section, the centralizer is defined for any congruence on any algebra whatsoever. Let us derive what we can in this completely general setting.

Proposition 7.17. *For any algebra* $\mathbf{A}$ *and congruence* α, α^* *is a congruence on* $\mathbf{A}$.

Proof. It is obvious from the form of the definition that α^* will be an equivalence relation. To show it is a congruence it is enough, by Theorem 4.16, to check that $\mathrm{Pol}_1(\mathbf{A}) \models \alpha^*$. So let $f(x) = g(x, e_1, e_2, \ldots, e_k) \in \mathrm{Pol}_1(\mathbf{A})$ and $(a, b) \in \alpha^*$. We must show that $\big(f(a), f(b)\big) \in \alpha^*$.

For this, let $t \in \mathrm{Clo}_{n+1}(\mathbf{A})$ and $(\mathbf{c}, \mathbf{d}) \in \alpha$. Suppose that $t\big(f(a), \mathbf{c}\big) = t\big(f(a), \mathbf{d}\big)$. We wish to show that $t\big(f(b), \mathbf{c}\big) = t\big(f(b), \mathbf{d}\big)$. Define the term $s(x, y_1, \ldots, y_{k+n}) = t\big(g(x, y_1, \ldots, y_k), y_{k+1}, \ldots, y_{k+n}\big)$. Let $\mathbf{c}'$ denote the concatenation of the vector $e_1, \ldots, e_k$ with the vector $\mathbf{c}$, and $\mathbf{d}'$ the concatenation of $e_1, \ldots, e_k$ with $\mathbf{d}$. Then

$$s(a, \mathbf{c}') = t(g(a, \mathbf{e}), \mathbf{c}) = t(f(a), \mathbf{c}) = t(f(a), \mathbf{d}) = s(a, \mathbf{d}').$$

Since $(a, b) \in \alpha^*$ we obtain $s(b, \mathbf{c}') = s(b, \mathbf{d}')$, i.e., $t(f(b), \mathbf{c}) = t(f(b), \mathbf{d})$ as desired. □

Thus the centralizer of a congruence is again a congruence. Naturally we begin by investigating this concept in the context of a few familiar algebraic structures.

Example 7.18. Suppose $\mathbf{A}$ is a group with identity e, α a congruence on $\mathbf{A}$ and $N = e/\alpha$. Then $e/(\alpha^*) = \{\, a \in A : (\forall x \in N)\ ax = xa \,\}$. The set $e/(\alpha^*)$ is what is generally called the centralizer, $C(N)$, of N in group theory.

Proof. Let $M = e/(\alpha^*)$. Take any $a \in M$ and $d \in N$. Then $(e, a) \in \alpha^*$ and $(e, d) \in \alpha$. Consider the term $t(x, y) = yxy^{-1}$. Since $t(e, e) = t(e, d)$, the definition of α^* yields $t(a, e) = t(a, d)$. Thus $eae^{-1} = dad^{-1}$, i.e., a and d commute. this shows $M \subseteq C(N)$.

Conversely, let $a \in C(N)$. We wish to show that $(a, e) \in \alpha^*$. We must apply Definition 7.16. Let $t \in \mathrm{Clo}_{n+1}(\mathbf{A})$ and $\mathbf{c}\ \alpha\ \mathbf{d}$. We must show

$$t(e, \mathbf{c}) = t(e, \mathbf{d}) \iff t(a, \mathbf{c}) = t(a, \mathbf{d}). \tag{7–7}$$

Instead we shall verify the following claim from which Equation (7–7) easily follows.

Claim. For every term operation t, $a \in C(N)$, and $\mathbf{c}\ \alpha\ \mathbf{d}$,

$$t(e, \mathbf{c})^{-1} \cdot t(e, \mathbf{d}) \cdot t(a, \mathbf{d})^{-1} \cdot t(a, \mathbf{c}) = e \tag{7–8}$$

Proof. The term operation $t(x, \mathbf{y})$ is of the form $z_1 \cdot z_2 \cdots z_k$ in which k is a natural number and every z_i is a member of $\{x^{\pm 1}, y_1^{\pm 1}, \ldots, y_n^{\pm 1}\}$. We prove the claim by induction on k.

If $k = 0$ then $t(x, \mathbf{y}) = e$ and the above equality obviously holds. If $k = 1$ then $t(x, \mathbf{y})$ is equal to either $x^{\pm 1}$ or $y_i^{\pm 1}$. In either case the claim is easily seen to be true.

So assume that $k > 1$ and the claim holds for all terms of length less than k. First suppose that $t(x, \mathbf{y}) = x^{\pm 1} \cdot s(x, \mathbf{y})$. Then the expression in Equation (7–8) becomes

$$s(e, \mathbf{c})^{-1} \cdot s(e, \mathbf{d}) \cdot s(a, \mathbf{d})^{-1} \cdot a^{\mp 1} \cdot a^{\pm 1} \cdot s(a, \mathbf{c}) = e$$

by the induction hypothesis applied to s.

Finally suppose that $t(x, \mathbf{y}) = y_i^{\pm 1} \cdot s(x, \mathbf{y})$. Let $r_i = c_i^{-1} \cdot d_i$, for $i = 1, \ldots, n$. Note that every $r_i \in N$. Then

$$\begin{aligned}
&t(e, \mathbf{c})^{-1} \cdot t(e, \mathbf{d}) \cdot t(a, \mathbf{d})^{-1} \cdot t(a, \mathbf{c}) = \\
&s(e, \mathbf{c})^{-1} \cdot (c_i^{\mp 1} \cdot d_i^{\pm 1}) \cdot (s(e, \mathbf{d}) \cdot s(a, \mathbf{d})^{-1}) \cdot (d_i^{\mp 1} \cdot c_i^{\pm 1}) \cdot s(a, \mathbf{c}) = \\
&s(e, \mathbf{c})^{-1} \cdot r_i^{\pm 1} \cdot w \cdot r_i^{\mp 1} \cdot s(a, \mathbf{c})
\end{aligned} \tag{7–9}$$

where $w = s(e, \mathbf{d}) \cdot s(a, \mathbf{d})^{-1}$. Since $C(N)$ is a normal subgroup, it corresponds to a congruence θ. So

$$a \in C(N) \implies e \;\theta\; a \implies s(e, \mathbf{d}) \;\theta\; s(a, \mathbf{d}) \implies w \in C(N).$$

Since $r_i \in N$ it follows that $r_i w = w r_i$. Applying this relationship to the last expression in (7–9) yields

$$s(e, \mathbf{c})^{-1} \cdot s(e, \mathbf{d}) \cdot s(a, \mathbf{d})^{-1} \cdot s(a, \mathbf{c}) = e$$

from the induction hypothesis. □

Example 7.19. Let $\mathbf{A}$ be a ring and $I = 0/\alpha$. Then $0/\alpha^*$ is equal to $\{\, a \in A : aI = Ia = (0) \,\}$. Thus, the centralizer of a congruence on a ring corresponds to the annihilator of the associated ideal.

Proof. Let $J = 0/\alpha^*$. Since α^* is a congruence, J is a two-sided ideal. Let $a \in J$ and $c \in I$. Take $t(x, y) = x \cdot y$. Then according to Definition 7.16 since $(a, 0) \in \alpha^*$

$$t(0, 0) = t(0, c) \iff t(a, 0) = t(a, c).$$

Thus $ac = 0$. A similar argument shows that $ca = 0$. Thus $J \subseteq \operatorname{Ann}(I)$.

Conversely, assume that a annihilates I. We wish to show that $a \in J$, i.e., $(a, 0) \in \alpha^*$. So let $t \in \operatorname{Clo}_{n+1}(\mathbf{A})$ and $\mathbf{c} \;\alpha\; \mathbf{d}$. We shall show that

$$t(0, \mathbf{c}) - t(0, \mathbf{d}) = t(a, \mathbf{c}) - t(a, \mathbf{d}).$$

Assume first that $t(x, \mathbf{y})$ is a monomial. We once again argue by induction on the length of t. The base case is left to the reader. Suppose that $t(x, \mathbf{y}) = r(x, \mathbf{y}) s(x, \mathbf{y})$. Then

$$\begin{aligned}
&t(0, \mathbf{c}) - t(0, \mathbf{d}) = r(0, \mathbf{c}) s(0, \mathbf{c}) - r(0, \mathbf{d}) s(0, \mathbf{d}) = \\
&r(0, \mathbf{c}) s(0, \mathbf{c}) - r(0, \mathbf{c}) s(0, \mathbf{d}) + r(0, \mathbf{c}) s(0, \mathbf{d}) - r(0, \mathbf{d}) s(0, \mathbf{d}) = \\
&r(0, \mathbf{c}) \big(s(0, \mathbf{c}) - s(0, \mathbf{d})\big) + \big(r(0, \mathbf{c}) - r(0, \mathbf{d})\big) s(0, \mathbf{d}) =
\end{aligned}$$

(by the induction hypothesis)

$$\begin{aligned}
&r(0, \mathbf{c}) \big(s(a, \mathbf{c}) - s(a, \mathbf{d})\big) + \big(r(a, \mathbf{c}) - r(a, \mathbf{d})\big) s(0, \mathbf{d}) \overset{*}{=} \\
&r(a, \mathbf{c}) \big(s(a, \mathbf{c}) - s(a, \mathbf{d})\big) + \big(r(a, \mathbf{c}) - r(a, \mathbf{d})\big) s(a, \mathbf{d}) = t(a, \mathbf{c}) - t(a, \mathbf{d}).
\end{aligned}$$

To see why the equality marked with a "$*$" holds, observe that the difference of the two sides is

$$\big(r(0,\mathbf{c}) - r(a,\mathbf{c})\big)\big(s(a,\mathbf{c}) - s(a,\mathbf{d})\big) + \big(r(a,\mathbf{c}) - r(a,\mathbf{d})\big)\big(s(0,\mathbf{d}) - s(a,\mathbf{d})\big).$$

Now $\mathbf{c}\ \alpha\ \mathbf{d}$, so each of $\big(s(a,\mathbf{c}) - s(a,\mathbf{d})\big)$ and $\big(r(a,\mathbf{c}) - r(a,\mathbf{d})\big)$ are members of I, while $a \in \mathrm{Ann}(I)$ so each of $\big(r(0,\mathbf{c}) - r(a,\mathbf{c})\big)$ and $\big(s(0,\mathbf{d}) - s(a,\mathbf{d})\big)$ are members of $\mathrm{Ann}(I)$. Therefore the above difference is 0.

To complete the proof, suppose that $t(x,\mathbf{y}) = \sum_j t_j(x,\mathbf{y})$ is a sum of monomials. The equation $t(0,\mathbf{c}) - t(0,\mathbf{d}) = t(a,\mathbf{c}) - t(a,\mathbf{d})$ follows immediately from the monomial case. □

Example 7.20. Let $\mathbf{A}$ be a lattice. Then α^* is the pseudocomplement of α in $\mathrm{Con}(\mathbf{A})$.

This is Exercise 7.31.2.

It is of course with an eye towards group theory that α^* is called the centralizer of α. Continuing the analogy, we make the following definitions.

Definition 7.21. Let $\mathbf{A}$ be an algebra. The *center* of $\mathbf{A}$ is the congruence $\zeta_{\mathbf{A}} = (1_A)^*$. The algebra $\mathbf{A}$ is *Abelian* if $\zeta_{\mathbf{A}} = 1_A$.

The proof of the following proposition is a straightforward application of Definition 7.16.

Proposition 7.22. (1) *If* $\mathbf{B}$ *is a subalgebra of* $\mathbf{A}$ *and* $\alpha \in \mathrm{Con}(\mathbf{A})$, *then* $(\alpha{\restriction_B})^* \supseteq \alpha^*{\restriction_B}$.

(2) *If* $\mathbf{A} = \prod_{i\in I} \mathbf{A}_i$, *and for each* $i \in I$, $\alpha_i \in \mathrm{Con}(\mathbf{A}_i)$, *then* $(\prod \alpha_i)^* = \prod \alpha_i^*$.

It follows from this proposition that every subalgebra of an Abelian algebra is Abelian, and every product of Abelian algebras is Abelian. There is not much else that can be said about centralizers in this completely general context. In particular, it is not clear how centralizers behave under homomorphism.

However, much more is true if we assume congruence-permutability. In order to examine this phenomenon, we need to define certain congruences on subdirect squares of an arbitrary algebra. In order to keep things readable we shall denote members of the square as vertical ordered pairs. Thus, $A^2 = \big\{\binom{x}{y} : x, y \in A\big\}$. Furthermore, an ordered pair of these pairs can be written as a 2×2 matrix. Thus we write $\left(\begin{smallmatrix} x & y \\ z & w\end{smallmatrix}\right)$ instead of $\big(\binom{x}{z}, \binom{y}{w}\big)$.

Definition 7.23. Let $\mathbf{A}$ be an algebra and $\alpha, \beta \in \mathrm{Con}(\mathbf{A})$.

(1) $\mathbf{A}(\beta)$ is the subalgebra of $\mathbf{A}^2$ with universe β. In other words, $\mathbf{A}(\beta) = \big\{\binom{x}{y} : (x,y) \in \beta\big\}$.

(2) $\Delta_\alpha^\beta = \mathrm{Cg}^{\mathbf{A}(\beta)}\{\left(\begin{smallmatrix} x & y \\ x & y\end{smallmatrix}\right) : x\ \alpha\ y\}$.

Notice that the algebra $\mathbf{A}(\beta)$ defined here is identical to $\mathbf{A}_2(\beta)$ as defined in Equation (7–2). For that matter, the algebra $\mathbf{A}(\beta)$ is nothing but the relation β viewed as a subalgebra of $\mathbf{A}^2$. We introduce this new notation to emphasize the role of $\mathbf{A}(\beta)$ as a subalgebra rather than a congruence.

As an immediate consequence of these definition we have:

$$\alpha_1 \subseteq \alpha_2 \implies \Delta^{\beta}_{\alpha_1} \subseteq \Delta^{\beta}_{\alpha_2}.$$

Lemma 7.24. *Let* $\mathbf{A}$ *be a member of a Maltsev variety,* $\alpha, \beta \in \operatorname{Con}(\mathbf{A})$*. Then*

$$\Delta^{\beta}_{\alpha} = \left\{ \begin{pmatrix} t(\mathbf{a},\mathbf{c}) & t(\mathbf{b},\mathbf{c}) \\ t(\mathbf{a},\mathbf{d}) & t(\mathbf{b},\mathbf{d}) \end{pmatrix} : t \in \operatorname{Clo}_{m+n}(\mathbf{A}),\ (\mathbf{a},\mathbf{b}) \in \alpha,\ (\mathbf{c},\mathbf{d}) \in \beta \right\}.$$

Proof. Let $\theta = \{ \left(\begin{smallmatrix} x & y \\ x & y \end{smallmatrix}\right) : x \mathrel{\alpha} y \}$ be the generating set of Δ^{β}_{α} given in 7.23(2). According to Theorem 4.65, a typical element of Δ^{β}_{α} is a pair $\big(t(\mathbf{u},\mathbf{w}),\, t(\mathbf{v},\mathbf{w})\big)$ in which t is an $(n+m)$-ary term, $\mathbf{u} \mathrel{\theta} \mathbf{v}$ and $\mathbf{w} \in \mathbf{A}(\beta)^m$. But note that each component, u_i, of $\mathbf{u}$ is of the form $\binom{a_i}{a_i}$, and each component of $\mathbf{v}$ is $\binom{b_i}{b_i}$. The condition $\mathbf{u} \mathrel{\theta} \mathbf{v}$ is equivalent to $a_i \mathrel{\alpha} b_i$ for $i = 1, \dots, n$. Similarly, each component of $\mathbf{w}$ is of the form $\binom{c_i}{d_i}$ with $c_i \mathrel{\beta} d_i$. Making all of these substitutions yields the description of Δ^{β}_{α} in the lemma. □

For the remainder of this section, we shall be working in a Maltsev variety. Recall from Theorem 4.64 that such a variety has a term q satisfying the identities $q(x,x,y) \approx q(y,x,x) \approx y$. q is called a Maltsev term for the variety. Using Lemma 7.24 and the definition of the centralizer, we obtain the following relationships between α, β^* and Δ^{β}_{α}.

Theorem 7.25. *Let* $\mathbf{A}$ *be a member of a Maltsev variety with Maltsev term* q*, and let* $\alpha, \beta, \gamma \in \operatorname{Con}(\mathbf{A})$*. Then*

(1) $\alpha \subseteq \beta^*$ *if and only if* $\Big(\left(\begin{smallmatrix} u & v \\ u & w \end{smallmatrix}\right) \in \Delta^{\beta}_{\alpha} \implies v = w \Big)$.

Furthermore, if $\alpha \subseteq \beta^*$ *then*

(2) $\left(\begin{smallmatrix} a & b \\ c & d \end{smallmatrix}\right) \in \Delta^{\beta}_{\alpha} \implies q(a,b,d) = c.$

(3) $\left(\begin{smallmatrix} a & b \\ c & d \end{smallmatrix}\right) \in \Delta^{\beta}_{\alpha}$ *and* $b \mathrel{\gamma} d \implies a \mathrel{\gamma} c.$

Proof. Let us first assume that $\alpha \subseteq \beta^*$ and $\left(\begin{smallmatrix} u & v \\ u & w \end{smallmatrix}\right) \in \Delta^{\beta}_{\alpha}$. Lemma 7.24 tells us that $u = t(\mathbf{a},\mathbf{c}) = t(\mathbf{a},\mathbf{d})$, $v = t(\mathbf{b},\mathbf{c})$, and $w = t(\mathbf{b},\mathbf{d})$, for some term t, $(\mathbf{a},\mathbf{b}) \in \alpha$ and $(\mathbf{c},\mathbf{d}) \in \beta$. Now using the assumption that $\alpha \subseteq \beta^*$ and repeatedly applying Definition 7.16,

$$\begin{aligned}
u = t(\underline{a_1}, a_2, a_3 \dots, a_n, \mathbf{c}) &= t(\underline{a_1}, a_2, a_3 \dots, a_n, \mathbf{d}) \implies \\
t(b_1, \underline{a_2}, a_3 \dots, a_n, \mathbf{c}) &= t(b_1, \underline{a_2}, a_3 \dots, a_n, \mathbf{d}) \implies \\
t(b_1, b_2, \underline{a_3} \dots, a_n, \mathbf{c}) &= t(b_1, b_2, \underline{a_3} \dots, a_n, \mathbf{d}) \implies \\
&\ \ \vdots \\
t(b_1, b_2, b_3, \dots, \underline{a_n}, \mathbf{c}) &= t(b_1, b_2, b_3, \dots, \underline{a_n}, \mathbf{d}) \implies \\
v = t(b_1, b_2, b_3, \dots, b_n, \mathbf{c}) &= t(b_1, b_2, b_3, \dots, b_n, \mathbf{d}) = w.
\end{aligned}$$

To see that each of the above implications holds, observe that at each step the underlined variable a_i is β^*-equivalent to b_i while all remaining variables are β-equivalent. Thus with a suitable reordering of the variables in the term t we can apply Definition 7.16 at each step.

For the converse of part (1) let $(a, b) \in \alpha$, $t \in \mathrm{Clo}_{n+1}(\mathbf{A})$ and $\mathbf{c}\ \beta\ \mathbf{d}$. Then $\left(\begin{smallmatrix} t(a,\mathbf{c}) & t(b,\mathbf{c}) \\ t(a,\mathbf{d}) & t(b,\mathbf{d}) \end{smallmatrix}\right) \in \Delta_\alpha^\beta$. Applying the implication in part (1), $t(a, \mathbf{c}) = t(a, \mathbf{d}) \implies t(b, \mathbf{c}) = t(b, \mathbf{d})$. Thus $(a, b) \in \beta^*$.

Now we turn to the second statement of the theorem. Assume that $\alpha \subseteq \beta^*$. Let $\left(\begin{smallmatrix} a & b \\ c & d \end{smallmatrix}\right) \in \Delta_\alpha^\beta$. Note that $\binom{a}{c}$ and $\binom{b}{d}$ are members of $\mathbf{A}(\beta)$. Then

$$\begin{pmatrix} d \\ d \end{pmatrix} = q^{\mathbf{A}(\beta)}\left(\begin{pmatrix} b \\ d \end{pmatrix}, \begin{pmatrix} b \\ d \end{pmatrix}, \begin{pmatrix} d \\ d \end{pmatrix}\right) \equiv_{\Delta_\alpha^\beta}$$
$$q^{\mathbf{A}(\beta)}\left(\begin{pmatrix} a \\ c \end{pmatrix}, \begin{pmatrix} b \\ d \end{pmatrix}, \begin{pmatrix} d \\ d \end{pmatrix}\right) = \begin{pmatrix} q^{\mathbf{A}}(a, b, d) \\ c \end{pmatrix}.$$

The first and last equalities follow from the fact that q is a Maltsev term. Writing the above congruence in another form, $\left(\begin{smallmatrix} d & q(abd) \\ d & c \end{smallmatrix}\right) \in \Delta_\alpha^\beta$. Then from statement (1), $q(a, b, d) = c$.

Finally consider statement (3). Let $\left(\begin{smallmatrix} a & b \\ c & d \end{smallmatrix}\right) \in \Delta_\alpha^\beta$ and $b\ \gamma\ d$. Then by (2), $c = q(a, b, d)\ \gamma\ q(a, d, d) = a$. □

Lemma 7.24 is only true in a Maltsev variety. Although we won't do so here, it turns out that with some extra effort, one can prove Theorem 7.25(1) in the more general context of a congruence-modular variety. Let us continue with another important relationship. For a binary relation $\Gamma \subseteq A^2 \times A^2$, let us write

$$\Gamma^{\mathrm{t}} = \left\{ \begin{pmatrix} x & y \\ z & w \end{pmatrix} : \begin{pmatrix} x & z \\ y & w \end{pmatrix} \in \Gamma \right\}.$$

In other words, Γ^{t} consists of the "transposes" of the matrices in Γ.

Theorem 7.26. *Let* $\mathbf{A}$ *lie in a Maltsev variety,* $\alpha, \beta \in \mathrm{Con}(\mathbf{A})$. *Then* $(\Delta_\alpha^\beta)^t = \Delta_\beta^\alpha$. *Consequently,* $\alpha \subseteq \beta^* \iff \beta \subseteq \alpha^*$.

Proof. The equality follows immediately from Lemma 7.24. Assume that $\alpha \subseteq \beta^*$. Let $\left(\begin{smallmatrix} u & v \\ u & w \end{smallmatrix}\right) \in \Delta_\beta^\alpha$. Taking the transpose, $\left(\begin{smallmatrix} u & u \\ v & w \end{smallmatrix}\right) \in \Delta_\alpha^\beta \subseteq A(\beta)^2$. So in particular, $u\ \beta\ v$. Also by Theorem 7.25(2), $v = q(u, u, w) = w$. Hence by 7.25(1), $\beta \subseteq \alpha^*$. □

Since Δ_α^β (and Δ_β^α) are equivalence relations, they are transitive on the columns. One consequence of Theorem 7.26 is that they are also transitive on the rows. That is: $\left(\begin{smallmatrix} a & b \\ c & d \end{smallmatrix}\right) \in \Delta_\alpha^\beta$ & $\left(\begin{smallmatrix} c & d \\ e & f \end{smallmatrix}\right) \in \Delta_\alpha^\beta \implies \left(\begin{smallmatrix} a & b \\ e & f \end{smallmatrix}\right) \in \Delta_\alpha^\beta$. In the literature, the phrase Δ_α^β "respects transitivity" is used for this implication.

We say "α centralizes β" whenever $\alpha \subseteq \beta^*$. The theorem tells us that in a Maltsev variety, this relationship is symmetric. Moreover, using Theorem 7.25, we can finally derive a relationship between centralizers and homomorphisms.

Theorem 7.27. *Let* **A** *and* **B** *be members of a Maltsev variety, and suppose that* $h\colon \mathbf{A} \twoheadrightarrow \mathbf{B}$ *is a surjective homomorphism. Then for every* $\gamma \in \mathrm{Con}(\mathbf{B})$ *we have* $\overleftarrow{h}(\gamma^*) \supseteq \overleftarrow{h}(\gamma)^*$.

Proof. Let $\beta = \overleftarrow{h}(\gamma)$ and $\alpha = \ker h$. We want $\beta^* \subseteq \overleftarrow{h}(\gamma^*)$. Let $(a, b) \in \beta^*$. We must show $(h(a), h(b)) \in \gamma^*$. For this let $\mathbf{c}\ \gamma\ \mathbf{d}$ and let t be an n-ary term. We must show

$$t^{\mathbf{B}}(ha, \mathbf{c}) = t^{\mathbf{B}}(ha, \mathbf{d}) \iff t^{\mathbf{B}}(hb, \mathbf{c}) = t^{\mathbf{B}}(hb, \mathbf{d}).$$

Since h is surjective there are $\mathbf{c}', \mathbf{d}' \in A^n$ such that $h(c_i') = c_i$ and $h(d_i') = d_i$ for $i = 1, \ldots, n$. Therefore $\mathbf{c}'\ \beta\ \mathbf{d}'$. It follows that

$$\begin{pmatrix} t(a, \mathbf{c}') & t(b, \mathbf{c}') \\ t(a, \mathbf{d}') & t(b, \mathbf{d}') \end{pmatrix} \in \Delta^{\beta}_{\beta^*}.$$

Suppose that $t^{\mathbf{B}}(ha, \mathbf{c}) = t^{\mathbf{B}}(ha, \mathbf{d})$. Then $h(t^{\mathbf{A}}(a, \mathbf{c}')) = h(t^{\mathbf{A}}(a, \mathbf{d}'))$, i.e., $t^{\mathbf{A}}(a, \mathbf{c}') \equiv_\alpha t^{\mathbf{A}}(a, \mathbf{d}')$. Then by Theorem 7.25(3), $t^{\mathbf{A}}(b, \mathbf{c}') \equiv_\alpha t^{\mathbf{A}}(b, \mathbf{d}')$. Thus $t^{\mathbf{B}}(b, \mathbf{c}) = t^{\mathbf{B}}(b, \mathbf{d})$ as desired. □

If $h\colon \mathbf{A} \twoheadrightarrow \mathbf{B}$ is surjective then $\overleftarrow{h}(1_B) = 1_A$ always. Thus Theorem 7.27 tells us that $\overleftarrow{h}(1_B^*) \supseteq 1_A^*$. Expressed in terms of the center, $\overleftarrow{h}(\zeta_{\mathbf{B}}) \supseteq \zeta_{\mathbf{A}}$.

Corollary 7.28. *Let* $\mathcal{V}$ *be a Maltsev variety. A homomorphic image of an Abelian algebra in* $\mathcal{V}$ *is Abelian. The Abelian members of* $\mathcal{V}$ *form a subvariety, denoted* $\mathcal{V}_{\mathrm{ab}}$.

Proof. Suppose that **A** is Abelian and $h\colon \mathbf{A} \twoheadrightarrow \mathbf{B}$ is a surjective homomorphism. As we just observed, $1_A = \zeta_{\mathbf{A}} \subseteq \overleftarrow{h}(\zeta_{\mathbf{B}})$. Applying $\overrightarrow{h}$, $1_B \subseteq \zeta_{\mathbf{B}}$, i.e., **B** is Abelian. The subvariety assertion follows from this together with Proposition 7.22. □

To conclude this section we shall develop some characterizations of Abelian algebras in Maltsev varieties. By taking $\alpha = \alpha^* = 1$ in Definition 7.16 we see that an algebra, **A**, is Abelian if and only if

$$\forall t \in \mathrm{Clo}(\mathbf{A})\ \forall a, b \in A\ \forall \mathbf{c}, \mathbf{d} \in A^n\ \ t(a, \mathbf{c}) = t(a, \mathbf{d}) \iff t(b, \mathbf{c}) = t(b, \mathbf{d}) \tag{7–10}$$

where the rank of t is $n+1$. Note also that Definition 7.23 specializes to $\Delta_1^1 = \mathrm{Cg}^{\mathbf{A}^2}\left\{ \left(\begin{smallmatrix} x & y \\ x & y \end{smallmatrix}\right) : x, y \in A \right\}$. Of course neither of these definitions are predicated on the assumption that **A** lies in a Maltsev variety. In fact, as we shall see in Chapter 8, the Abelian property has consequences even for arbitrary algebras. By contrast, the congruence Δ_1^1 seems to be useful primarily in the congruence-permutable context.

Theorem 7.29. *Let* **A** *be an algebra in a Maltsev variety. Suppose that there are congruences* α, β *and* γ *on* **A** *such that* $\{0_A, \alpha, \beta, \gamma, 1_A\}$ *forms a sublattice isomorphic to* $\mathbf{M}_3$. *Then* **A** *is Abelian.*

Proof. By the symmetry of the situation, it is enough to show that $\zeta_{\mathbf{A}} \supseteq \alpha$. So let $(a, b) \in \alpha$ and let t be a term operation on $\mathbf{A}$. Suppose that $t(a, \mathbf{c}) = t(a, \mathbf{d})$ for some vectors $\mathbf{c}$ and $\mathbf{d}$ of A.

Since $\beta \circ \gamma = 1_A$, there is a vector $\mathbf{e}$ such that $\mathbf{c}\ \beta\ \mathbf{e}\ \gamma\ \mathbf{d}$. Thus

$$t(a, \mathbf{c})\ \beta\ t(a, \mathbf{e})\ \gamma\ t(a, \mathbf{d}).$$

But $t(a, \mathbf{c}) = t(a, \mathbf{d})$ so $t(a, \mathbf{c})\ (\beta \wedge \gamma)\ t(a, \mathbf{e})$. Therefore, since $\beta \wedge \gamma = 0$, $t(a, \mathbf{c}) = t(a, \mathbf{d}) = t(a, \mathbf{e})$.

Now

$$t(b, \mathbf{c})\ \alpha\ t(a, \mathbf{c}) = t(a, \mathbf{e})\ \alpha\ t(b, \mathbf{e}).$$

And $t(b, \mathbf{c})\ \beta\ t(b, \mathbf{e})$. Therefore, since $\alpha \wedge \beta = 0$, $t(b, \mathbf{c}) = t(b, \mathbf{e})$. Similarly, $t(b, \mathbf{d}) = t(b, \mathbf{e})$. Combining, we conclude that $t(b, \mathbf{c}) = t(b, \mathbf{d})$, which shows that $(a, b) \in 1_A^* = \zeta_{\mathbf{A}}$. □

Note that there are simple Abelian algebras (for example the groups $\mathbb{Z}_p$ for p a prime) and these obviously fail to satisfy the condition of the above theorem. However, when one moves to the square of an algebra, the existence of a "spanning $\mathbf{M}_3$" does indeed characterize Abelianness.

Theorem 7.30. *Let* $\mathbf{A}$ *be a nontrivial algebra in a Maltsev variety. The following are equivalent.*

(a) $\mathbf{A}$ *is Abelian.*

(b) $\left\{ \binom{a}{a} : a \in A \right\}$ *is the coset of a congruence on* $\mathbf{A}^2$.

(c) $\{0_{A^2}, \eta_1, \Delta_1^1, \eta_2, 1_{A^2}\}$ *forms a sublattice of* $\operatorname{Con}(\mathbf{A}^2)$ *isomorphic to* $\mathbf{M}_3$.

Proof. By taking $\alpha = \beta = 1_A$ in Theorem 7.25(1), we see that (a) $\Leftrightarrow$ (b) with Δ_1^1 as the congruence in (b). Also (c) together with Theorem 7.29 implies that $\mathbf{A}^2$ is Abelian, hence $\mathbf{A}$ is Abelian by Corollary 7.28.

Finally if $1^* = 1$ on $\mathbf{A}$, then by 7.25(2), $\eta_1 \cap \Delta_1^1 = 0_A$. Let $\binom{a}{b}$ and $\binom{c}{d}$ be arbitrary pairs in A^2 and let $u = q^{\mathbf{A}}(a, c, d)$. Then

$$\binom{a}{u} = q\left(\binom{a}{a}, \binom{d}{c}, \binom{d}{d}\right) \equiv_{\Delta_1^1} q\left(\binom{c}{c}, \binom{d}{c}, \binom{d}{d}\right) = \binom{c}{d}.$$

So $\binom{a}{b}\ \eta_1\ \binom{a}{u}\ \Delta_1^1\ \binom{c}{d}$, i.e., $\eta_1 \circ \Delta_1^1 = 1_A$. A similar argument holds with η_2 replacing η_1. Hence (a) $\Rightarrow$ (c). □

Exercise Set 7.31.

1. Let α and β be congruences on an arbitrary algebra $\mathbf{A}$. Prove that if $\beta \cap \alpha = 0$ then $\beta \subseteq \alpha^*$.

2. Let **A** be an algebra with a majority term. (**A** is not assumed to lie in a Maltsev variety.) Prove that for every congruence α, the centralizer α^* coincides with the pseudocomplement of α. (See the previous exercise and Example 3.31.6.) Extra credit: generalize to the case that **A** lies in a congruence-distributive variety.

3. Prove that every algebra whose basic operations are all essentially unary is Abelian.

4. Prove that every rectangular band (see Exercise 3.15.6) is Abelian.

5. Let **A** be an algebra in a Maltsev variety, $\alpha, \beta \in \operatorname{Con} \mathbf{A}$ and $\alpha \subseteq \beta \cap \beta^*$. Show that $\operatorname{Con}\big(\mathbf{A}(\beta)\big)$ has, as a sublattice, a homomorphic image of the following lattice.

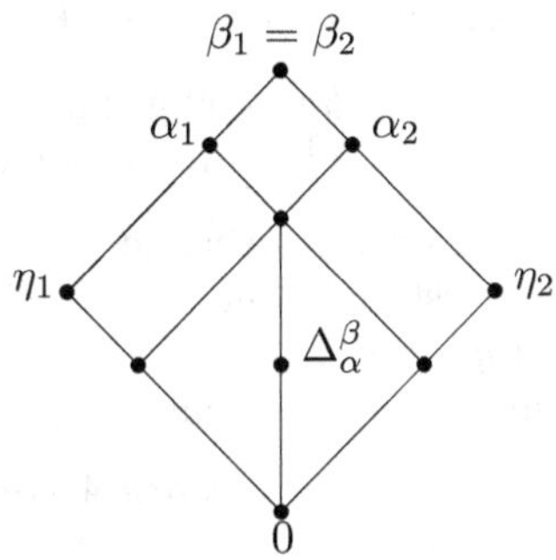

6. Let **A** be an arbitrary algebra.

   (a) Prove that $\alpha_1 \subseteq \alpha_2 \implies \alpha_1^* \supseteq \alpha_2^*$ and that $\zeta_{\mathbf{A}} \subseteq \zeta_{\mathbf{A}}^*$.

   (b) Suppose that **A** lies in a Maltsev variety. Prove that the algebra $\mathbf{A}(\zeta)$ is Abelian. (Hint: use the previous exercise.)

7.3 Abelian varieties

It follows easily from Theorem 7.30 that for every ring **R**, every **R**-module is Abelian. There is a strong sense in which the converse is true. In this section we develop this idea. To begin with, recall that for a module **M**, a Maltsev term is given by $q(x, y, z) = x - y + z$. Observe further that viewed as a function, $q\colon \mathbf{M}^3 \to \mathbf{M}$ is a homomorphism. This turns out to be true for any Abelian algebra in a Maltsev variety.

Theorem 7.32. *Let* **A** *be an Abelian algebra with Maltsev term q. Then* $q\colon \mathbf{A}^3 \to \mathbf{A}$ *is a homomorphism.*

Proof. Let f be an n-ary operation symbol. We must show that

$$\forall (a_1, b_1, c_1), (a_2, b_2, c_2), \ldots, (a_n, b_n, c_n) \in A^3,$$
$$f^{\mathbf{A}}\big(q(a_1, b_1, c_1), \ldots, q(a_n, b_n, c_n)\big) = q\big(f^{\mathbf{A}^3}(\,(a_1, b_1, c_1), \ldots, (a_n, b_n, c_n)\,)\big).$$

Let $u_i = q(a_i, b_i, c_i)$, for $i = 1, \ldots, n$. Then the equality we wish to verify can be rewritten as $f(\mathbf{u}) = q\big(f(\mathbf{a}),\ f(\mathbf{b}),\ f(\mathbf{c})\big)$. For each $i \leq n$,

$$\begin{pmatrix} a_i \\ u_i \end{pmatrix} = \begin{pmatrix} q(a_i, c_i, c_i) \\ q(a_i, b_i, c_i) \end{pmatrix} \equiv_{\Delta_1^1} \begin{pmatrix} q(b_i, c_i, c_i) \\ q(b_i, b_i, c_i) \end{pmatrix} = \begin{pmatrix} b_i \\ c_i \end{pmatrix}.$$

Therefore $\begin{pmatrix} f(\mathbf{a}) & f(\mathbf{b}) \\ f(\mathbf{u}) & f(\mathbf{c}) \end{pmatrix} \in \Delta_1^1$. Then $f(\mathbf{u}) = q\big(f(\mathbf{a}), f(\mathbf{b}), f(\mathbf{c})\big)$ follows from Theorem 7.25(2). ☐

H. P. Gumm introduced the elegant concept of a ternary Abelian group.

Definition 7.33. A *ternary Abelian group* is an algebra $\mathbf{A} = \langle A, q\rangle$ in which q is a Maltsev operation and $q\colon \mathbf{A}^3 \to \mathbf{A}$ is a homomorphism.

Notice that the class of ternary Abelian groups forms a variety, defined by three identities: two to establish that q is Maltsev and a single 9-variable identity that asserts that q is a homomorphism. What do these algebras have to do with Abelian groups? The following theorem provides the answer.

Theorem 7.34. *Let $\mathbf{A} = \langle A, q\rangle$ be a ternary Abelian group and fix an element $0 \in A$. Define $x + y = q(x, 0, y)$ and $-x = q(0, x, 0)$. Then $\langle A, +, -, 0\rangle$ is an Abelian group. Furthermore, $q(x, y, z) = x - y + z$.*

Proof. Consider a 3×3 array of variables. That q is a homomorphism with respect to itself is equivalent to the assertion that applying q across the rows of the matrix and then applying q to the result is the same as first applying q to the columns followed by q on the result.

$$\begin{array}{ccccc} \begin{bmatrix} x_1 & x_2 & x_3 \\ y_1 & y_2 & y_3 \\ z_1 & z_2 & z_3 \end{bmatrix} & & & \begin{array}{c}\longrightarrow\\ \longrightarrow\\ \longrightarrow\end{array} & \begin{array}{c} q(x_1, x_2, x_3) \\ q(y_1, y_2, y_3) \\ q(z_1, z_2, z_3)\end{array} \\ \downarrow\quad\quad\downarrow\quad\quad\downarrow & & & & \downarrow \\ q(x_1, y_1, z_1)\ \ q(x_2, y_2, z_2)\ \ q(x_3, y_3, z_3) & & & \longrightarrow & \star \end{array}$$

Each of the identities defining an Abelian group can now be verified with a quick look at an appropriate matrix. Thus

$$(x + y) + z = x + (y + z) \qquad \begin{bmatrix} x & 0 & y \\ 0 & 0 & 0 \\ 0 & 0 & z \end{bmatrix}$$

$$x + y = y + x \qquad \begin{bmatrix} 0 & 0 & x \\ 0 & 0 & 0 \\ y & 0 & 0 \end{bmatrix}$$

$$x + (-y) = q(x, y, 0) \qquad \begin{bmatrix} x & 0 & 0 \\ 0 & 0 & 0 \\ 0 & y & 0 \end{bmatrix}$$

$$q(x, y, z) = (x - y) + z \qquad \begin{bmatrix} x & 0 & 0 \\ y & 0 & 0 \\ 0 & 0 & z \end{bmatrix}$$

It follows from the third identity that $x + (-x) = 0$ and of course $x + 0 = q(x, 0, 0) = x$. ☐

The operations "+" and "−" defined in this theorem are not term operations of **A** because they involve the constant 0, which is not necessarily a basic operation. The best we can say is that "+" and "−" are polynomials of **A**. Thus every ternary Abelian group is polynomially equivalent to an Abelian group. (See Definition 4.79.)

Combining the ideas we have developed, we arrive at the following very striking theorem about Abelian varieties.

Theorem 7.35. *Let $\mathcal{W}$ be an Abelian, Maltsev variety. There is a ring* **R** *with identity such that every member of $\mathcal{W}$ is polynomially equivalent to a unital* **R***-module.*

Proof. Let $\mathbf{F} = \mathbf{F}_{\mathcal{W}}(x, y)$ be the free $\mathcal{W}$-algebra on two generators and define $R = \{\, t(x, y) \in F : t^{\mathbf{F}}(y, y) = y \,\}$. Let q denote the Maltsev term for $\mathcal{W}$.

R is closed under q since the composition of idempotent operations is idempotent. Thus, since **F** is Abelian, $\langle R, q \rangle$ is a ternary Abelian group. We take y to be the zero of R. From 7.34

$$t_1 + t_2 = q(t_1, y, t_2) \text{ and } -t = q(y, t, y)$$

define an additive group structure on R. We define a multiplication on R by

$$t_1(x, y) \cdot t_2(x, y) = t_1(t_2(x, y), y).$$

Multiplication is obviously associative. We must check the distributive laws.

$$s \cdot (t_1 + t_2) = s(t_1 + t_2, y) = s\big(q(t_1, y, t_2), q(y, y, y)\big) = \\ q\big(s(t_1, y), s(y, y), s(t_2, y)\big) = q\big(s(t_1, y), y, s(t_2, y)\big) = s \cdot t_1 + s \cdot t_2$$

using Theorem 7.32, and

$$(t_1 + t_2) \cdot s = q(t_1, y, t_2) \cdot s = q\big(t_1(s, y), y, t_2(s, y)\big) = t_1 \cdot s + t_2 \cdot s.$$

Finally, the term x serves as the multiplicative identity of R by definition of multiplication.

We have established that $\mathbf{R} = \langle R, +, -, \cdot, y, x \rangle$ is a ring. Let $\mathbf{A} \in \mathcal{W}$. Then by assumption, $\langle A, q \rangle$ is a ternary abelian group. Fix $0 \in A$ and define addition and negation as in Theorem 7.34. To turn $\langle A, +, -, 0 \rangle$ into an **R**-module, define a scalar multiplication by $t \cdot a = t^{\mathbf{A}}(a, 0)$. The module axioms are easy to verify. For example

$$t \cdot (a + b) = t(q(a, 0, b), 0) = t(q(a, 0, b),\, q(0, 0, 0)) = \\ q(t(a, 0),\, t(0, 0),\, t(b, 0)) = t \cdot a + t \cdot b$$

using 7.32 again. We denote the resulting **R**-module by A_R.

It remains to show that $\text{Pol}(A_R) = \text{Pol}(\mathbf{A})$. Recall that the clone of polynomial operations is generated by the basic operations of the algebra together with all nullary operations. Since the basic operations of A_R (namely addition, subtraction and each scalar multiplication) are clearly polynomials of $\mathbf{A}$, $\text{Pol}(A_R) \subseteq \text{Pol}(\mathbf{A})$. Let us consider the converse. Let $f(z_1, \ldots, z_n)$ be an n-ary term operation of $\mathbf{A}$. We shall show that there are ring elements $r_1, \ldots, r_n$ and a constant $d \in A$ such that

$$f(z_1, \ldots, z_n) = \sum_{i=1}^{n} r_i \cdot z_i + d. \tag{7–11}$$

From this it will follow that each term operation of $\mathbf{A}$ is a polynomial of A_R.

Since $\mathbf{A}$ is Abelian, by Theorem 7.32 for any $\mathbf{a}, \mathbf{b}, \mathbf{c} \in A^n$

$$q(f(\mathbf{a}), f(\mathbf{b}), f(\mathbf{c})) = f(q(a_1, b_1, c_1), \ldots, q(a_n, b_n, c_n))$$

which, by Theorem 7.34 can be rewritten

$$f(\mathbf{a}) - f(\mathbf{b}) + f(\mathbf{c}) = f(\mathbf{a} - \mathbf{b} + \mathbf{c}).$$

Taking $\mathbf{b} = \mathbf{0}$

$$f(\mathbf{a}) + f(\mathbf{c}) = f(\mathbf{a} + \mathbf{c}) + f(\mathbf{0}). \tag{7–12}$$

For $i = 1, \ldots, n$ let

$$\begin{aligned} r_i(x, y) = q\big(f(y, y, \ldots, x, y, \ldots),\ f(y, \ldots, y),\ y\big) \text{ and} \\ d = f(\mathbf{0}) \end{aligned} \tag{7–13}$$

with the lone x in the ith position of the first appearance of f. Note that each r_i is an element of R. Then for each $i \leq n$,

$$r_i \cdot z_i = q\big(f(0, 0, \ldots, z_i, 0, \ldots, 0),\ f(\mathbf{0}),\ 0\big) = f(0, \ldots, z_i, \ldots, 0) - d.$$

Therefore by repeated application of equation (7–12)

$$\begin{aligned} &\sum_{i=1}^{n} r_i \cdot z_i = \\ &\quad f(z_1, 0, \ldots, 0) + f(0, z_2, \ldots, 0) + \cdots + f(0, \ldots, 0, z_n) - nd = \\ &\qquad f(z_1, z_2, \ldots, z_n) + (n-1)d - nd = f(z_1, \ldots, z_n) - d \end{aligned}$$

from which equation (7–11) follows. □

Example 7.36. Recall the variety of quasigroups from Example 1.6. This is a very big and complicated variety. (For one thing, it contains a subvariety term-equivalent to the variety of groups.) Let us consider the subvariety $\mathcal{C}$ defined by the three additional identities

$$\begin{aligned} &\text{(c)} & x \cdot y &\approx y \cdot x \\ &\text{(i)} & x \cdot x &\approx x \\ &\text{(e)} & (x \cdot y) \cdot (z \cdot w) &\approx (x \cdot z) \cdot (y \cdot w) \end{aligned}$$

These are the commutative, idempotent and *entropic* laws. Originally, the identity (e) was called the Abelian law, and for this reason the variety $\mathcal{C}$ is known as the variety of *CIA-quasigroups.*

To justify this curious terminology, let us show that every entropic quasigroup is Abelian. Let $\mathbf{A}$ be an entropic quasigroup. Then in $\mathbf{A}$, $x \cdot y = ((x/z) \cdot z) \cdot ((y/w) \cdot w) = ((x/z) \cdot (y/w)) \cdot (z \cdot w)$. Dividing on the right by $z \cdot w$ yields

$$\mathbf{A} \models (x \cdot y)/(z \cdot w) \approx (x/z) \cdot (y/w).$$

A dual identity holds with right division replaced by left division. Now consider the map $h\colon \mathbf{A} \times \mathbf{A} \to \mathbf{A}$ given by $h(x, y) = x \cdot y$. It follows from (e) and the above identity (and its dual) that h is a homomorphism. Let $\theta = \ker h = \{\, \left(\begin{smallmatrix} x & z \\ y & w \end{smallmatrix}\right) : x \cdot y = z \cdot w \,\}$. Then $\{0_A, \eta_1, \theta, \eta_2, 1_A\}$ forms a spanning M_3. Therefore by Theorem 7.29, $\mathbf{A}^2$, hence $\mathbf{A}$, is Abelian.

Thus the variety $\mathcal{C}$ consists of Abelian algebras. One easily shows that in $\mathcal{C}$

$$\begin{aligned} x/y &\approx y\backslash x \\ x/x &\approx x \\ (x \cdot y)/z &\approx (x/z) \cdot (y/z). \end{aligned} \tag{7–14}$$

Recall from Exercise 4.75.1 that $q(x, y, z) = (x/(y\backslash y)) \cdot (y\backslash z)$ is a Maltsev term for the variety of quasigroups. Using the above identities we see that in $\mathcal{C}$, $q(x, y, z) = (x \cdot z)/y$.

Now let us apply the construction described in Theorem 7.35 and determine the ring associated with $\mathcal{C}$. Let $\mathbf{A} \in \mathcal{C}$. Fix $z \in A$ to be the zero. Then $a + b = q(a, z, b) = (a \cdot b)/z$ and $-a = q(z, a, z) = z/a$ are the group operations. Note also that $(a \cdot z) + (a \cdot z) = ((a \cdot z) \cdot (a \cdot z))/z = a$ from which it follows that $a \cdot z = \frac{1}{2}a$. More generally, for an arbitrary $b \in A$, $a \cdot b = \frac{1}{2}(a + b)$.

What are the elements of the ring? Every term is idempotent (this follows from (7–14)) so the ring consists of all terms in x, y. What terms can we build? We have

$$\begin{gathered} y = 0, \quad x = 1, \\ x \cdot x = x/x = x\backslash x = x = 1, \\ y \cdot x = x \cdot y = \tfrac{1}{2}x = \tfrac{1}{2}, \\ x/y = 2, \ y/x = -1, \ldots \end{gathered}$$

So the ring contains $\mathbb{Z}[\frac{1}{2}]$.

Conversely, given a $\mathbb{Z}[\frac{1}{2}]$-module $\mathbf{M}$, we can define operations

$$a \cdot b = \tfrac{1}{2}(a+b), \quad a/b = b\backslash a = 2a - b$$

and obtain an algebra $\langle M, \cdot, /, \backslash \rangle$. It is straightforward to check that this is a CIA-quasigroup. Thus, the ring associated with this variety is precisely $\mathbb{Z}[\frac{1}{2}]$, every member of $\mathcal{C}$ is polynomially equivalent to a module over this ring and every module is polynomially equivalent to a CIA-quasigroup. □

Note that, despite the last paragraph of the previous example, the transformation from modules to algebras is not covered by Theorem 7.35. In fact, as the next example shows, the structure of the ring alone does not contain enough information to reconstruct the original variety.

Example 7.37. For this example we will use two basic operation symbols: "+" and "g", of ranks 2 and 1, respectively. It will be important to distinguish between the basic operation symbol "+" and the module addition, which will be written "$+_c$" to indicate that the zero-element is c.

Let $\mathbf{B} = \langle \mathbb{Z}_6, +, g^{\mathbf{B}} \rangle$ with $g^{\mathbf{B}}(x) = x+1$, and addition computed modulo 6, and let $\mathbf{A} = \langle \mathbb{Z}_6, +, g^{\mathbf{A}} \rangle$, with $g^{\mathbf{A}}(x) = x$. Obviously, $\mathbf{A}$ is term-equivalent to the group $\langle \mathbb{Z}_6, + \rangle$ since the operation $g^{\mathbf{A}}$ is nothing but a projection operation. $\mathbf{B}$, on the other hand, is term-equivalent to $\langle \mathbb{Z}_6, +, 1 \rangle$. As usual, we shall write kx in place of x added to itself k times and $x - y$ in place of $x + 5y$.

Let $\mathcal{V} = \mathbf{V}(\mathbf{B})$ and $\mathcal{W} = \mathbf{V}(\mathbf{A})$. There is a homomorphism from $\mathbf{B}^2$ onto $\mathbf{A}$, mapping (a, b) to $a - b$. This is similar to Exercise 3.15.7. The kernel of this map is Δ_1^1. Thus both $\mathcal{V}$ and $\mathcal{W}$ are Abelian varieties. The Maltsev operation is, of course, $q(x, y, z) = x - y + z$. Since $\mathbf{A} \models g(x) \approx x$ while $\mathbf{B}$ fails to satisfy this identity, $\mathcal{W}$ is a proper subvariety of $\mathcal{V}$.

Our objective is to understand the modules corresponding to both $\mathcal{V}$ and $\mathcal{W}$. Let us begin with $\mathcal{W}$. The terms on an Abelian group are the linear maps. Thus $\mathbf{F}_{\mathcal{W}}(x, y) = \{ kx + ly : k, l \in \mathbb{Z}_6 \}$. Suppose that $r(x, y) = kx + ly$. Then $r(y, y) = y$ is equivalent to $k + l = 1$. (Scalars are computed mod 6.) Thus the ring obtained in 7.35 is

$$R_{\mathcal{W}} = \{ kx + (1-k)y : k \in \mathbb{Z}_6 \}.$$

Let $r = kx + (1-k)y$ and $s = lx + (1-l)y$ be two elements of the ring. Addition is defined on R by choosing y as the zero-element. Thus

$$r +_y s = q(r, y, s) = (kx + (1-k)y) - y + (lx + (1-l)y) = (k+l)x + (1-(k+l))y$$

and we multiply by

$$r \cdot_y s = r(s, y) = k(lx + (1-l)y) + (1-k)y = (kl)x + (1-kl)y.$$

Thus, although the initial form of the elements of R may have been a bit unexpected, we ultimately determine that $\mathbf{R}$ is isomorphic to the ring $\mathbb{Z}_6$. The reader will not be surprised by this result, nor by the conclusion that every member of $\mathcal{W}$ is polynomially equivalent to a $\mathbb{Z}_6$-module. For future reference, here is the computation. Let $\mathbf{C} \in \mathcal{W}$ and take $c \in C$ to be the zero. Then

$$\begin{aligned} a +_c b &= q(a,c,b) = a + b - c \\ ra &= r(a,c) = ka + (1-k)c. \end{aligned} \tag{7–15}$$

Now consider the larger variety $\mathcal{V}$. Because of the constant operation 1, we have $\mathbf{F}_{\mathcal{V}}(x,y) = \{\, kx + ly + m : k,l,m \in \mathbb{Z}_6 \,\}$. The condition $r(y,y) = y$ forces $k + l = 1$ and $m = 0$. Thus $R_{\mathcal{V}} = R_{\mathcal{W}}$. Furthermore, the equations in (7–15) apply equally well to $\mathcal{V}$. We see that both $\mathcal{V}$ and its proper subvariety $\mathcal{W}$ yield the same variety of $\mathbf{R}$-modules (for the same ring $\mathbf{R}$)!

We shall resolve the above dilemma illustrated in Example 7.39 shortly. But first we need to describe a method of augmenting a variety with some additional information.

The variety of algebras under J

It is sometimes useful to consider classes of algebras that are anchored to a fixed algebra via a homomorphism. For example, in commutative ring theory, the phrase "**R**-algebra" refers to a commutative ring **S** together with a ring homomorphism from **R** to **S**. **R**-algebra homomorphisms are ring homomorphisms that make the triangle commute, as in Figure 7.1.

In category theory, classes of the above kind are called "coslice categories" (or, more generally, "comma categories"). It turns out that this construction can be performed entirely within the framework of universal algebra. There are a great many easy details to be verified, so we only provide a sketch.

Let $\rho\colon \mathcal{F} \to \omega$ be a similarity type, $\mathcal{V}$ a variety of algebras of type ρ, and $\mathbf{J}$ an algebra of type ρ. Additionally, let Σ be an equational base for $\mathcal{V}$ and C a generating set for $\mathbf{J}$. Expand ρ by adding a new nullary operation symbol, $\underset{\sim}{c}$, for every $c \in C$, to obtain a new similarity type ρ_C.

Let $\mathbf{T} = \mathbf{T}_\rho(C)$ be the term algebra of type ρ over C. The identity map on C extends to a homomorphism $\bar{\iota}\colon \mathbf{T} \to \mathbf{J}$. This simply means that $\bar{\iota}(p^{\mathbf{T}}(c_1,\dots,c_n)) = p^{\mathbf{J}}(c_1,\dots,c_n)$ for every term p. Define

$$\Theta = \bigl\{\, p(\underset{\sim}{c}_1,\dots,\underset{\sim}{c}_n) \approx q(\underset{\sim}{c}_1,\dots,\underset{\sim}{c}_n) : \bigl(p(c_1,\dots,c_n),\, q(c_1,\dots,c_n)\bigr) \in \ker(\bar{\iota}) \,\bigr\}.$$

Note that Θ is a set of equations in the new similarity type, ρ_C. Finally, we define the variety $(\mathbf{J} \downarrow \mathcal{V})$ of *$\mathcal{V}$-algebras under* $\mathbf{J}$ to be $\mathrm{Mod}(\Sigma \cup \Theta)$. The algebras, of course, have similarity type ρ_C.

Let $\mathbf{A} \in \mathcal{V}$ and let $h\colon \mathbf{J} \to \mathbf{A}$ be a homomorphism. Then h induces a natural ρ_C-expansion of $\mathbf{A}$ by defining $\underset{\sim}{c}^{\mathbf{A}} = h(c)$ for each $c \in C$. The expanded algebra, call it $\mathbf{A}_C$, will satisfy the equations in Θ because

$$\begin{aligned} p^{\mathbf{A}_C}(\underset{\sim}{c}_1,\dots,\underset{\sim}{c}_n) &= p^{\mathbf{A}}(h(c_1),\dots,h(c_n)) = h(p^{\mathbf{J}}(c_1,\dots,c_n)) = \\ &h\bar{\iota}(p^{\mathbf{T}}(c_1,\dots,c_n)) = h\bar{\iota}(q^{\mathbf{T}}(c_1,\dots,c_n)) = h(q^{\mathbf{J}}(c_1,\dots,c_n)) = \\ &q^{\mathbf{A}}(h(c_1),\dots,h(c_n)) = q^{\mathbf{A}_C}(\underset{\sim}{c}_1,\dots,\underset{\sim}{c}_n). \end{aligned}$$

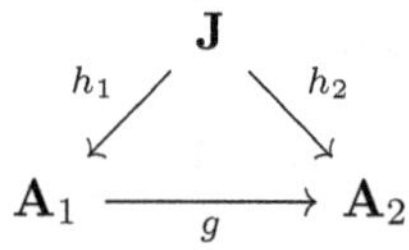

FIGURE 7.1: A coslice homomorphism

Thus from the pair $(\mathbf{A}, h)$ we obtain the algebra $\mathbf{A}_C \in (\mathbf{J} \downarrow \mathcal{V})$.

Conversely, let $\mathbf{A}_C$ be a member of $(\mathbf{J} \downarrow \mathcal{V})$. Then the ρ-reduct, $\mathbf{A}$, is a member of $\mathcal{V}$ since $\mathbf{A}_C \models \Sigma$ (so $\mathbf{A} \models \Sigma$). The mapping $h_0 \colon c \mapsto \underline{c}^{\mathbf{A}_C}$ extends to a homomorphism $h \colon \mathbf{J} \to \mathbf{A}$ precisely because $\mathbf{A}_C \models \Theta$. (To be more precise, h_0 extends to a homomorphism $\bar{h}_0 \colon \mathbf{T}(C) \to \mathbf{A}$. The condition $\mathbf{A}_C \models \Theta$ is equivalent to $\ker(\bar{h}_0) \supseteq \ker(\bar{\iota})$. Apply Exercise 1.26.8 to obtain h.)

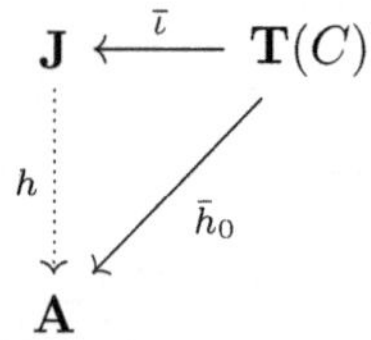

In summary, we can think of the members of $(\mathbf{J} \downarrow \mathcal{V})$ interchangeably as algebras $\mathbf{A}_C$ or as pairs $(\mathbf{A}, h)$. It is also important to understand the homomorphisms in $(\mathbf{J} \downarrow \mathcal{V})$. Given two pairs $(\mathbf{A}_1, h_1)$ and $(\mathbf{A}_2, h_2)$ in $(\mathbf{J} \downarrow \mathcal{V})$, a ρ-homomorphism $g \colon A_1 \to A_2$ will be a homomorphism of ρ_C-algebras if and only if $g \circ h_1 = h_2$. See Figure 7.1.

One consequence of this alternate characterization is that it becomes clear that, despite the definition, the variety $(\mathbf{J} \downarrow \mathcal{V})$ is actually independent of the generating set C. Specifically, generating sets C_1 and C_2 will produce term-equivalent varieties.

We might also point out that since every member of $(\mathbf{J} \downarrow \mathcal{V})$ is polynomially equivalent to a member of $\mathcal{V}$, the congruence lattices that appear in the coslice variety already appear in $\mathcal{V}$. Thus, if $\mathcal{V}$ is congruence-distributive or permutable, the same will hold for $(\mathbf{J} \downarrow \mathcal{V})$.

We now present our strengthening of Theorem 7.35. Of course what we would really like is a term-equivalence between the Abelian variety and a variety of modules. Unfortunately, that is not quite the case. However, we obtain an equivalence by resorting to coslice varieties.

Theorem 7.38. *Let $\mathcal{W}$ be an Abelian, Maltsev variety. There is a ring $\mathbf{R}$, an algebra $\mathbf{J} \in \mathcal{W}$ and an $\mathbf{R}$-module $\widetilde{J}$ such that the varieties $(\mathbf{J} \downarrow \mathcal{W})$ and $(\widetilde{J} \downarrow {}_{\mathbf{R}}\mathcal{M}od)$ are term-equivalent.*

Proof. The ring $\mathbf{R}$ is unchanged from 7.35. Let $\mathbf{J} = \mathbf{F}_{\mathcal{W}}(u)$ be the free $\mathcal{W}$-algebra on one generator. Recall that in the earlier theorem, for each $\mathbf{A} \in \mathcal{W}$, we chose an arbitrary element to be the zero of the group. But observe that there is a one-to-one correspondence between elements of **A** and homomorphisms from **J** to **A**. Thus the module constructed in 7.35 depends precisely on an algebra **A** and a homomorphism $h\colon \mathbf{J} \to \mathbf{A}$. We shall denote this module $\mathcal{M}(\mathbf{A}, h)$.

Let $\widetilde{J} = \mathcal{M}(\mathbf{J}, \iota_J)$. We wish to argue that $\mathcal{M}$ constitutes an interpretation of $(\widetilde{J} \downarrow {}_{\mathbf{R}}\mathcal{Mod})$ into $(\mathbf{J} \downarrow \mathcal{W})$. We know from 7.35 that all of the module operations are in the domain of the interpretation. Defining the constant symbols corresponding to elements of $\widetilde{J}$ is equivalent to checking that for all $(\mathbf{A}, h) \in (\mathbf{J} \downarrow \mathcal{W})$, the map $h\colon \widetilde{J} \to \mathcal{M}(\mathbf{A}, h)$ is a module homomorphism. But this is easy:

$$h(s+t) = h(q^{\mathbf{J}}(s,u,t)) = q^{\mathbf{A}}(h(s), 0, h(t)) = h(s) + h(t),$$
$$h(r \cdot t) = h(r^{\mathbf{J}}(t,u)) = r^{\mathbf{A}}\big(h(t(u)), 0\big) = r \cdot h(t).$$

In the above derivation, we used the facts that h is a $\mathcal{W}$-homomorphism and $h(u) = 0$ is the zero element of the module $\mathcal{M}(\mathbf{A}, h)$.

Now we must go the other way, that is, interpret each basic operation symbol of $(\mathbf{J} \downarrow \mathcal{W})$ as a term of $(\widetilde{J} \downarrow {}_{\mathbf{R}}\mathcal{Mod})$. The recipe for this is already implicit in equations (7–11) and (7–13). Let f be a basic n-ary operation symbol of $\mathcal{W}$. Define

$$r_{f,i}(x,y) = q\big(f(y,y,\ldots,x,y,\ldots),\, f(y,y,\ldots),\, y\big), \quad \text{for } 1 \le i \le n,$$
$$d_f = f^{\mathbf{J}}(u,u,\ldots,u).$$

Note that $r_{f,i}$ is the same $\mathcal{W}$-term as in (7–13), so we already know that it is an element of the ring. d_f is an element of $\mathbf{J} = \widetilde{J}$. We treat it as one of the basic nullary operation symbols of the variety $(\widetilde{J} \downarrow {}_{\mathbf{R}}\mathcal{Mod})$.

Given a module $(\widetilde{M}, h) \in (\widetilde{J} \downarrow {}_{\mathbf{R}}\mathcal{Mod})$, we define an algebra $\mathbf{M} = \mathcal{A}(\widetilde{M}, h)$ with basic operations

$$f^{\mathbf{M}}(z_1,\ldots,z_n) = \sum_{i=1}^{n} r_{f,i} \cdot z_i + h(d_f).$$

We must show that $(\mathbf{M}, h)$ is a member of $(\mathbf{J} \downarrow \mathcal{W})$. That is: $\mathbf{M} \in \mathcal{W}$ and $h\colon \mathbf{J} \to \mathbf{M}$ is a $\mathcal{W}$-homomorphism. The second of these is easy. Since h is assumed to be an ${}_{\mathbf{R}}\mathcal{Mod}$-homomorphism,

$$h(f^{\mathbf{J}}(t_1,\ldots,t_n)) = h\big(\textstyle\sum r_{f,i} \cdot t_i + d_f\big) =$$
$$\textstyle\sum r_i \cdot h(t_i) + h(d_f) = f^{\mathbf{M}}\big(h(t_1),\ldots,h(t_n)\big).$$

The equality $f^{\mathbf{J}}(z_1,\ldots,z_n) = \sum r_{f,i} \cdot z_i + d_f$ comes from 7.35.

How are we to show that **M** is a member of $\mathcal{W}$? By Theorem 4.44 it is enough to show that every finitely generated subalgebra of **M** lies in $\mathcal{W}$. So let

$a_1, \ldots, a_m \in M$ and $N = \mathrm{Sg}^{\mathbf{M}}\{a_1, \ldots, a_m\}$. Let $\mathbf{F}$ denote the free $\mathcal{W}$-algebra on generators $\{z_1, \ldots, z_m, u\}$ and $j\colon \mathbf{J} \to \mathbf{F}$ the homomorphism taking u to itself. We shall show that $\mathbf{N} \in \mathsf{SH}(\mathbf{F}) \subseteq \mathcal{W}$.

Claim. $\mathcal{M}(\mathbf{F}, j) \cong \mathbf{F}_{\mathbf{R}\mathcal{Mod}}(z_1, \ldots, z_m) \oplus \widetilde{J}$.

In this claim, "$\oplus$" denotes the direct sum of modules. Assume for the moment that the claim is true. Since $\widetilde{M}$ is an $\mathbf{R}$-module and h an $\mathbf{R}$-module homomorphism, there is an $\mathbf{R}$-module homomorphism

$$g\colon \mathbf{F}_{\mathbf{R}\mathcal{Mod}}(z_1, \ldots, z_m) \oplus \widetilde{J} \to \widetilde{M} \text{ such that } g(z_i) = a_i, \text{ and } g{\restriction}_{\widetilde{J}} = h.$$

Then $g\colon \mathbf{F} \to \mathbf{M}$ is a $\mathcal{W}$-homomorphism since

$$g\big(f^{\mathbf{F}}(t_1, \ldots, t_n)\big) = g\big(\sum r_i \cdot t_i + j(d)\big) = \\ \sum r_i \cdot g(t_i) + h(d) = f^{\mathbf{M}}\big(g(t_1), \ldots, g(t_n)\big).$$

Since $\mathbf{N}$ is a subalgebra of the image of g, it lies in $\mathcal{W}$.

Our final task is to prove the claim. The free $\mathbf{R}$-module $\mathbf{F}_{\mathbf{R}\mathcal{Mod}}(z_1, \ldots, z_m)$ is the direct sum $Rz_1 \oplus Rz_2 \oplus Rz_m$. A typical element of $\mathbf{F}$ is $t(z_1, \ldots, z_m, u)$ for some $\mathcal{W}$-term t. From equation (7–11),

$$t(z_1, \ldots, z_m, u) = \sum_{i=1}^{m} r_i \cdot z_i + (r_{m+1} \cdot u + t(u, \ldots, u)) \in \sum_{i=1}^{m} Rz_i + mJ.$$

To see that the sum is direct, recall that u is the zero of $\mathcal{M}(\mathbf{F}, j)$. Suppose that $r_1 z_1 + \ldots r_m z_m + s(u) = u$, for some $r_1, \ldots, r_m \in R$ and $s(u) \in J$. (So s is a $\mathcal{W}$-term.) By evaluating all of the z_i's at u we obtain $s(u) = u$. Now by evaluating every z_i at u for $i \neq k$ we obtain $r_k \cdot z_k = u$, or, by definition of scalar product, $r_k^{\mathbf{F}}(z_k, u) = u$. Since $\mathbf{F}$ is free in $\mathcal{W}$, we obtain $\mathcal{W} \models r_k(z, u) \approx u$. But this means that $r_k = 0$ in $\mathbf{R}$. □

Example 7.39. Let us revisit Example 7.37. Recall that the ring, and the varieties of modules obtained from $\mathcal{W}$ and from $\mathcal{V}$, are the same. Since $\mathcal{W}$ is a proper subvariety of $\mathcal{V}$, we expect the more detailed construction of Theorem 7.38 to reflect this relationship. The key is, of course, in the objects $\mathbf{J}$ and $\widetilde{J}$.

According to the definitions

$$J_{\mathcal{W}} = \mathbf{F}_{\mathcal{W}}(u) = \{\, ku : k \in \mathbb{Z}_6 \,\}$$
$$J_{\mathcal{V}} = \mathbf{F}_{\mathcal{V}}(u) \ = \{\, ku + m : k, m \in \mathbb{Z}_6 \,\}.$$

Thus $\widetilde{J}_{\mathcal{W}}$ is the module $\mathbb{Z}_6$, while $\widetilde{J}_{\mathcal{V}} \cong \widetilde{J}_{\mathcal{W}} \oplus \widetilde{N}$, where $\widetilde{N} \cong \mathbb{Z}_6$. The upshot is that there are many more module homomorphisms from $\widetilde{J}_{\mathcal{V}}$ to an $\mathbf{R}$-module $\widetilde{M}$ than there are from $\widetilde{J}_{\mathcal{W}}$. In fact, if $h\colon \widetilde{J}_{\mathcal{V}} \to \widetilde{M}$, then $(\widetilde{M}, h) \in (\widetilde{J}_{\mathcal{W}} \downarrow {}_{\mathbf{R}}\mathcal{Mod})$ if and only if $\vec{h}(N) = 0$.

Let us make one more observation on this subject. Consider the maps

$h_i\colon \mathbf{J}_{\mathcal{W}} \to \mathbf{A}$ given by $h_i(u) = i$. The algebras $(\mathbf{A}, h_1)$ and $(\mathbf{A}, h_3)$ are not isomorphic in $(\mathbf{J}_{\mathcal{W}} \downarrow \mathcal{W})$ since 1 and 3 have different additive orders in $\mathbf{A}$. And yet they produce isomorphic modules, $\widetilde{M}$. What, in the structure of $(\widetilde{J}_{\mathcal{W}} \downarrow {}_{\mathbf{R}}\mathcal{Mod})$, causes $(\widetilde{M}, h_1) \not\cong (\widetilde{M}, h_3)$? The answer lies in the constant d of equation (7–13). By definition, $d_+ = u + u$. Thus $h_1(d_+) = 1 + 1 = 2$, while $h_3(d_+) = 3 + 3 = 0$. Since there is no module automorphism f such that $f \circ h_3 = h_1$, the structures $(\widetilde{M}, h_1)$ and $(\widetilde{M}, h_3)$ are nonisomorphic.

Exercise Set 7.40.

1. Prove the converse to Theorem 7.32, that is, if q is a Maltsev term for an algebra $\mathbf{A}$, and if $q\colon \mathbf{A}^3 \to \mathbf{A}$ is a homomorphism, then $\mathbf{A}$ is Abelian.

2. Let $\mathbf{A}$ be an Abelian algebra in a Maltsev variety.

   (a) Prove that $\mathbf{A}$ has uniform congruences.

   (b) Suppose that $\mathbf{A}$ is subdirectly irreducible. Prove that every nontrivial subalgebra of $\mathbf{A}$ is subdirectly irreducible.

   (c) Suppose that $\mathbf{A}$ is simple. Prove that $\mathbf{A}$ has no proper nontrivial subalgebras.

3. Let $\mathbf{A}$ be an Abelian algebra in a Maltsev variety. Suppose that q_1 and q_2 are both Maltsev terms for $\mathbf{A}$. Prove that $\mathbf{A}$ satisfies the identity $q_1(x, y, z) \approx q_2(x, y, z)$.

4. Prove that an Abelian loop is an (Abelian) group.
(Remark: It is not true that every commutative loop is a group. Consider, for example, the operation $x * y = |x - y|$ on the set of nonnegative real numbers.)

5. In our definition of coslice variety, we did not require that the algebra $\mathbf{J}$ be a member of the variety $\mathcal{V}$. Prove that the varieties $(\mathbf{J} \downarrow \mathcal{V})$ and $(\mathbf{J}/\lambda^{\mathbf{J}}_{\mathcal{V}} \downarrow \mathcal{V})$ are term-equivalent. Thus, we may as well assume that $\mathbf{J} \in \mathcal{V}$. (The congruence $\lambda^{\mathbf{J}}_{\mathcal{V}}$ is defined in 4.26.)

6. Let $\mathcal{V}$ be the variety of idempotent and entropic binars satisfying the additional identities $(x \cdot y) \cdot y \approx (y \cdot x) \cdot y \approx x$.

   (a) Show that $q(x, y, z) = (x \cdot z) \cdot y$ is a Maltsev term for $\mathcal{V}$.

   (b) Show that $\mathcal{V}$ is Abelian and that the ring associated with $\mathcal{V}$ is $\mathbb{Z}_3$. Describe explicitly the term-equivalence between modules and $\mathcal{V}$-algebras. (This is an example of a *symmetric binary mode*, see Pilitowska, Romanowska and Smith, [PRS95].)

7. Let $\mathcal{W}$ be the variety of binars considered in Exercise 4.34.5(b).

   (a) Show that $q(x, y, z) = (y \cdot x) \cdot (z \cdot y)$ is a Maltsev term for $\mathcal{W}$. Show that in fact, every member of $\mathcal{W}$ is term-equivalent to a quasigroup.

   (b) Since $\mathcal{W}$ is entropic, it is Abelian. Determine the ring of $\mathcal{W}$. Show that in this case, the variety of modules is term-equivalent to the variety of "pointed $\mathcal{W}$-algebras." (Mitschke and Werner, [MW73])

7.4 Commutators

We continue our discussion from the previous sections by considering the centralizer notion "relative" to a third congruence. Let **A** be an arbitrary algebra, and let α and β be congruences on **A**. We define

$$C(\alpha, \beta) = \{\, \gamma \in \operatorname{Con}(\mathbf{A}) : \gamma \subseteq \alpha \wedge \beta \ \ \& \ \ (\alpha/\gamma) \subseteq (\beta/\gamma)^* \,\}.$$

Note that in this description, the relationship $(\alpha/\gamma) \subseteq (\beta/\gamma)^*$ is computed in the algebra $\mathbf{A}/\gamma$. We only consider congruences γ contained in $\alpha \wedge \beta$ since otherwise α/γ and β/γ don't make sense. It follows from Exercise 7.31.1 that $\alpha \wedge \beta \in C(\alpha, \beta)$, so this set is always nonempty. In the literature, the relationship $\gamma \in C(\alpha, \beta)$ is expressed as α *centralizes* β *modulo* γ.

Theorem 7.41. *Let* $\mathbf{A}, \alpha, \beta$ *be as above, and let* $\gamma = \bigcap C(\alpha, \beta)$. *Then* γ *lies in* $C(\alpha, \beta)$.

Proof. Let us write C in place of $C(\alpha, \beta)$. We have an embedding

$$\mathbf{A}/\gamma \rightarrowtail \prod_{\nu \in C} \mathbf{A}/\nu$$

Let $\bar{\alpha} = \prod_{\nu \in C} \alpha/\nu$ and $\bar{\beta} = \prod \beta/\nu$. By Proposition 7.22(2) $\bar{\alpha} = \prod \alpha/\nu \leq \prod (\beta/\nu)^* = \bar{\beta}^*$. Then by part (1) of the same proposition

$$\alpha/\gamma = \bar{\alpha}{\restriction}_{A/\gamma} \leq (\bar{\beta}{\restriction}_{A/\gamma})^* = (\beta/\gamma)^*. \qquad \square$$

Thus there is always a smallest congruence γ such that α centralizes β modulo γ.

Definition 7.42. Let **A** be an algebra, $\alpha, \beta \in \operatorname{Con} \mathbf{A}$. We define $[\alpha, \beta]$ to be $\bigcap C(\alpha, \beta)$, the *commutator of* (α,β).

Let us consider this notion in the context of groups. Let **G** be a group, $\alpha, \beta, \gamma \in \operatorname{Con}(\mathbf{G})$. Let A, B and D be the normal subgroups corresponding to α, β and γ respectively. From Example 7.18, $\gamma \in C(\alpha, \beta)$ if and only if $D \subseteq A \cap B$ and A/D is contained in the centralizer of B/D. In other words, for all $a \in A$ and $b \in B$, $(aD)(bD) = (bD)(aD)$, equivalently, $a^{-1}b^{-1}ab \in D$. Thus Definition 7.42 tells us that $[\alpha, \beta]$ corresponds to the smallest normal subgroup containing $\{ a^{-1}b^{-1}ab : a \in A, b \in B \}$. Among group theorists, this subgroup is denoted $[A, B]$.

Based on its definition, the relationship $\gamma \in C(\alpha, \beta)$ seems to involve the algebra $\mathbf{A}/\gamma$. However, using the correspondence theorem, we can describe it directly within the algebra $\mathbf{A}$ in a manner similar to Definition 7.16 as follows.

Lemma 7.43. *Let* $\mathbf{A}$ *be an algebra,* $\alpha, \beta, \gamma \in \operatorname{Con} \mathbf{A}$. *Then* α *centralizes* β *modulo* γ *if and only if* $\gamma \leq \alpha \wedge \beta$ *and*

$$\forall t \in \operatorname{Clo}_{n+1}(\mathbf{A}),\ \forall (a, b) \in \alpha,\ \forall (\mathbf{c}, \mathbf{d}) \in \beta,$$
$$t(a, \mathbf{c})\ \gamma\ t(a, \mathbf{d}) \implies t(b, \mathbf{c})\ \gamma\ t(b, \mathbf{d}). \quad (7\text{–}16)$$

Proof. Suppose first that α centralizes β mod γ. Then by definition, $\gamma \leq \alpha \wedge \beta$ and for every term t, $(a, b) \in \alpha$, and $(\mathbf{c}, \mathbf{d}) \in \beta$

$$t(a/\gamma, \mathbf{c}/\gamma) = t(a/\gamma, \mathbf{d}/\gamma) \iff t(b/\gamma, \mathbf{c}/\gamma) = t(b/\gamma, \mathbf{d}/\gamma), \quad (7\text{–}17)$$

with this computation taking place, of course, in $\mathbf{A}/\gamma$. The implication in (7–16) clearly follows from this.

Conversely, assume that $\gamma \leq \alpha \wedge \beta$ and (7–16) holds. Then the left-to-right implication in (7–17) holds. Since α is a symmetric relation, we can interchange a and b in (7–16) and obtain the right-to-left implication in the definition of the centralizer. □

We say that γ *has the* α,β*-term condition* if the implication in (7–16) holds. This turns out to be all we need to obtain the commutator.

Theorem 7.44. *Let* $\mathbf{A}$ *be an algebra,* $\alpha, \beta \in \operatorname{Con} \mathbf{A}$.

(1) *If a congruence* γ *has the* α,β*-term condition, then* $\gamma \wedge \alpha \wedge \beta$ *has the* α,β*-term condition.*

(2) $[\alpha, \beta] = \bigcap \{ \gamma \in \operatorname{Con} \mathbf{A} : \gamma$ *has the* α,β*-term condition* $\}$.

Proof. Let $\gamma' = \gamma \wedge \alpha \wedge \beta$. Suppose that t is a term operation, $a\ \alpha\ b$ and $\mathbf{c}\ \beta\ \mathbf{d}$. Suppose further that $t(a, \mathbf{c})\ \gamma'\ t(a, \mathbf{d})$. Then $t(a, \mathbf{c})\ \gamma\ t(a, \mathbf{d})$ so by assumption $t(b, \mathbf{c})\ \gamma\ t(b, \mathbf{d})$. Also $\mathbf{c}\ \beta\ \mathbf{d} \implies t(b, \mathbf{c})\ \beta\ t(b, \mathbf{d})$. Thirdly, $\gamma' \leq \alpha$ and $t(a, \mathbf{c})\ \gamma'\ t(a, \mathbf{d})$. Thus

$$t(b, \mathbf{c})\ \alpha\ t(a, \mathbf{c})\ \alpha\ t(a, \mathbf{d})\ \alpha\ t(b, \mathbf{d}).$$

Thus γ' has the α,β-term condition.

The second claim follows from the first together with Definition 7.42 and Lemma 7.43. □

In this completely general setting, we do not even have $[\alpha,\beta]=[\beta,\alpha]$. We can make only the following observations. The second of these says that the commutator is monotone in each variable.

Proposition 7.45. *Let* $\mathbf{A}$ *be an algebra,* $\alpha,\alpha',\beta,\beta'\in \operatorname{Con}\mathbf{A}$.

(1) $[\alpha,\beta]\leq \alpha\wedge\beta$.

(2) $\alpha\leq\alpha'$ *and* $\beta\leq\beta'$ *implies* $[\alpha,\beta]\leq[\alpha',\beta']$.

Proof. The first inequality follows from the fact that $\alpha\wedge\beta\in C(\alpha,\beta)$. We turn to the second. Assume that $\alpha\leq\alpha'$ and $\beta\leq\beta'$ and define $\gamma'=[\alpha',\beta']$. let $\gamma=\alpha\wedge\beta\wedge\gamma'$. By part (2) of Theorem 7.44, γ' has the α',β'-term condition, so it certainly has the α,β-term condition. By part (1) of the theorem γ has the α,β-term condition. Thus $[\alpha,\beta]\leq\gamma\leq\gamma'=[\alpha',\beta']$. □

To get any useful work done, we must return to the realm of Maltsev varieties. We first obtain commutativity of the commutator.

Theorem 7.46. *Let* $\mathbf{A}$ *lie in a Maltsev variety,* $\alpha,\beta\in\operatorname{Con}\mathbf{A}$. *Then* $[\alpha,\beta]=[\beta,\alpha]$.

Proof. For any $\gamma\leq\alpha\wedge\beta$, we have $\alpha/\gamma\leq(\beta/\gamma)^*$ iff $\beta/\gamma\leq(\alpha/\gamma)^*$ by Theorem 7.26. Thus $C(\alpha,\beta)=C(\beta,\alpha)$. □

One of the most important properties of the commutator is Theorem 7.48 which asserts that it is completely additive. This will allow us, in some circumstances, to replace the meet operation with the commutator and achieve a certain amount of distributivity. To prove this we need a lemma that asserts that the congruences with the α,β-term condition form a principal filter of the congruence lattice.

Lemma 7.47. *Let* $\mathbf{A}$ *lie in a Maltsev variety,* $\alpha,\beta,\gamma\in\operatorname{Con}\mathbf{A}$. *Suppose that* $\gamma\geq[\alpha,\beta]$. *Then* γ *has the* α,β*-term condition.*

Proof. Let $\nu=[\alpha,\beta]$. Then $\alpha/\nu\leq(\beta/\nu)^*$. Therefore, for any term operation t, $(a,b)\in\alpha$, and $(\mathbf{c},\mathbf{d})\in\beta$,

$$\begin{pmatrix} t(a,\mathbf{c})/\nu & t(b,\mathbf{c})/\nu \\ t(a,\mathbf{d})/\nu & t(b,\mathbf{d})/\nu \end{pmatrix}\in\Delta^{\beta/\nu}_{\alpha/\nu}.$$

By Theorem 7.25(3)

$$t(a,\mathbf{c})/\nu\equiv t(a,\mathbf{d})/\nu \implies t(b,\mathbf{c})/\nu\equiv t(b,\mathbf{d})/\nu \pmod{\gamma/\nu}$$

which is equivalent to γ possessing the α,β-term condition. □

Theorem 7.48. *Let* $\mathbf{A}$ *lie in a Maltsev variety,* $\alpha\in\operatorname{Con}\mathbf{A}$, $\Gamma\subseteq\operatorname{Con}\mathbf{A}$. *Then* $[\alpha,\bigvee\Gamma]=\bigvee_{\gamma\in\Gamma}[\alpha,\gamma]$.

Proof. $[\alpha, \bigvee\Gamma] \geq \bigvee_\gamma[\alpha,\gamma]$ holds by Proposition 7.45. Let $\beta = \bigvee_\gamma[\alpha,\gamma]$. Thus for every $\gamma \in \Gamma$, β has the γ,α-term condition. We can complete the proof by showing that β has the $(\bigvee\Gamma),\alpha$-term condition.

So let $(a,b) \in \bigvee\Gamma$, $(\mathbf{c},\mathbf{d}) \in \alpha$, and t a term operation. By Proposition 2.16 there are $a_0,\dots,a_k \in A$ and $\gamma_1,\dots,\gamma_k \in \Gamma$ such that

$$a = a_0\;\gamma_1\;a_1\;\gamma_2\cdots\gamma_k\;a_k = b.$$

Now using the fact that β has the γ_i,α-term conditions we deduce

$$\begin{aligned} t(a,\mathbf{c})\;\beta\;t(a,\mathbf{d}) &\implies t(a_0,\mathbf{c})\;\beta\;t(a_0,\mathbf{d}) \implies t(a_1,\mathbf{c})\;\beta\;t(a_1,\mathbf{d}) \implies \cdots \\ &\implies t(a_k,\mathbf{c})\;\beta\;t(a_k,\mathbf{d}) \implies t(b,\mathbf{c})\;\beta\;t(b,\mathbf{d}). \end{aligned}$$

Thus β has the $\bigvee\Gamma,\alpha$-term condition. □

As an illustration, consider a group $\mathbf{G}$ with normal subgroups A, B and D. Let α, β and γ be the corresponding congruences. Suppose that each element of A commutes with every element of $B \cup D$. This means that $[\alpha,\beta] = [\alpha,\gamma] = 0$, from which we deduce, using Theorem 7.48, that $[\alpha, \beta\vee\gamma] = 0$. This is equivalent to the obvious group-theoretic conclusion that A centralizes the normal subgroup BD.

The properties expressed in Theorems 7.46 and 7.48 are the ones that make the commutator such a powerful tool. The properties expressed in Theorems 7.22 and 7.27 can be stated much more generally in terms of the commutator.

Theorem 7.49. *Let $\mathcal{V}$ be a Maltsev variety, $\mathbf{A},\mathbf{B},\mathbf{A}_1,\mathbf{A}_2 \in \mathcal{V}$.*

(1) *If $\mathbf{B} \leq \mathbf{A}$ and $\alpha,\beta \in \operatorname{Con}\mathbf{A}$ then $[\alpha{\restriction_B}, \beta{\restriction_B}] \leq [\alpha,\beta]{\restriction_B}$.*

(2) *Let $h\colon \mathbf{A} \twoheadrightarrow \mathbf{B}$ be a surjective homomorphism, and let $\alpha,\beta \in \operatorname{Con}\mathbf{B}$. Then $\overleftarrow{h}[\alpha,\beta] = [\overleftarrow{h}(\alpha), \overleftarrow{h}(\beta)] \vee \ker h$.*

(3) *Let $\alpha_1,\beta_1 \in \operatorname{Con}\mathbf{A}_1$ and $\alpha_2,\beta_2 \in \operatorname{Con}\mathbf{A}_2$. Then $[\alpha_1\times\alpha_2, \beta_1\times\beta_2] = [\alpha_1,\beta_1]\times[\alpha_2,\beta_2]$, with the left-hand side computed in $\operatorname{Con}(\mathbf{A}_1\times\mathbf{A}_2)$ and the right-hand side in $\operatorname{Con}(\mathbf{A}_1)\times\operatorname{Con}(\mathbf{A}_2)$.*

Proof. (1) Let $\gamma = [\alpha,\beta]$. That $\gamma{\restriction_B} \in C(\alpha{\restriction_B},\beta{\restriction_B})$ is a straightforward verification.

For (2), let $\gamma = \overleftarrow{h}[\alpha,\beta]$. Then $\gamma \geq \ker h$ by the correspondence theorem. It is straightforward to verify that γ has the $\overleftarrow{h}(\alpha),\overleftarrow{h}(\beta)$-term condition. Hence $\gamma \geq [\overleftarrow{h}(\alpha),\overleftarrow{h}(\beta)] \vee \ker h$.

Conversely, let $\nu = [\overleftarrow{h}(\alpha),\overleftarrow{h}(\beta)] \vee \ker h$. By Lemma 7.47, ν has the $\overleftarrow{h}(\alpha),\overleftarrow{h}(\beta)$-term condition. Using this (together with the surjectivity of h) we easily check that $\overrightarrow{h}(\nu)$ has the α,β-term condition. Hence $\overrightarrow{h}(\nu) \geq [\alpha,\beta]$. By the correspondence theorem again $\nu \geq \overleftarrow{h}[\alpha,\beta] = \gamma$.

We turn to (3), which is not quite so straightforward. Let

$$\gamma = [\alpha_1 \times \alpha_2,\, \beta_1 \times \beta_2], \text{ and } \gamma_i = [\alpha_i, \beta_i], \text{ for } i = 1, 2.$$

We must show that $\gamma = \gamma_1 \times \gamma_2$. One direction is easy: $\gamma_1 \times \gamma_2$ has the $(\alpha_1 \times \alpha_2), (\beta_1 \times \beta_2)$-term condition. Thus $\gamma_1 \times \gamma_2 \geq \gamma$.

For $i = 1, 2$ let $p_i \colon \mathbf{A}_1 \times \mathbf{A}_2 \to \mathbf{A}_i$ be the projection map as usual, with kernel denoted η_i. For a congruence $\theta_i \in \mathrm{Con}(\mathbf{A}_i)$ we shall denote by $\overleftarrow{\theta}_i$ the congruence $\overleftarrow{p}_i(\theta_i)$ on the product algebra. Recall from Section 3.2 that the congruence $\alpha_1 \times \alpha_2$ coincides with $\overleftarrow{\alpha}_1 \wedge \overleftarrow{\alpha}_2$. From part (2) above

$$\overleftarrow{\gamma}_i = \overleftarrow{p}_i[\alpha_i, \beta_i] = [\overleftarrow{p}_i(\alpha_i), \overleftarrow{p}_i(\beta_i)] \vee \ker(p_i) = [\overleftarrow{\alpha}_i, \overleftarrow{\beta}_i] \vee \eta_i.$$

Now, $\gamma_1 \times \gamma_2 = (\overleftarrow{\gamma}_1 \wedge \eta_2) \circ (\overleftarrow{\gamma}_2 \wedge \eta_1) = (\overleftarrow{\gamma}_1 \wedge \eta_2) \vee (\overleftarrow{\gamma}_2 \wedge \eta_1)$ because

$$\begin{aligned}(a_1, a_2)\ (\gamma_1 \times \gamma_2)\ (b_1, b_2) \iff (a_1\ \gamma_1\ b_1)\ \ \&\ \ (a_2\ \gamma_2\ b_2) \iff \\ (a_1, a_2)\ (\overleftarrow{\gamma}_1 \wedge \eta_2)\ (b_1, a_2)\ (\eta_1 \wedge \overleftarrow{\gamma}_2)\ (b_1, b_2)\end{aligned}$$

together with the fact that we are in a Maltsev variety. Therefore, to complete the proof, it is enough to show that $\gamma \geq \overleftarrow{\gamma}_1 \wedge \eta_2$.

Again using congruence permutability, $\overleftarrow{\alpha}_1 = (\overleftarrow{\alpha}_1 \wedge \eta_2) \vee \eta_1$, and similarly for $\overleftarrow{\beta}_1$. Thus

$$\begin{aligned}\overleftarrow{\gamma}_1 = [\overleftarrow{\alpha}_1, \overleftarrow{\beta}_1] \vee \eta_1 = [(\overleftarrow{\alpha}_1 \wedge \eta_2) \vee \eta_1,\, (\overleftarrow{\beta}_1 \wedge \eta_2) \vee \eta_1] \vee \eta_1 = \\ [\overleftarrow{\alpha}_1 \wedge \eta_2, \overleftarrow{\beta}_1 \wedge \eta_2] \vee [\eta_1, \overleftarrow{\beta}_1 \wedge \eta_2] \vee [\overleftarrow{\alpha}_1 \wedge \eta_2, \eta_1] \vee [\eta_1, \eta_1] \vee \eta_1 = \\ [\overleftarrow{\alpha}_1 \wedge \eta_2, \overleftarrow{\beta}_1 \wedge \eta_2] \vee \eta_1\end{aligned}$$

using Proposition 7.45 and Theorem 7.48. Finally, another application of congruence-permutability and the monotonicity of the commutator yields

$$\begin{aligned}\overleftarrow{\gamma}_1 \wedge \eta_2 = \big([\overleftarrow{\alpha}_1 \wedge \eta_2, \overleftarrow{\beta}_1 \wedge \eta_2] \circ \eta_1\big) \wedge \eta_2 = [\overleftarrow{\alpha}_1 \wedge \eta_2, \overleftarrow{\beta}_1 \wedge \eta_2] \leq \\ [\overleftarrow{\alpha}_1 \wedge \overleftarrow{\alpha}_2, \overleftarrow{\beta}_1 \wedge \overleftarrow{\beta}_2] = \gamma\end{aligned}$$

completing the proof. □

Unfortunately, part (3) of this theorem is not true for infinite products. The first paragraph of that argument holds, showing that

$$\Big[\prod_{i \in I} \alpha_i, \prod_{i \in I} \beta_i\Big] \leq \prod_{i \in I} [\alpha_i, \beta_i].$$

The converse however, can fail. See [FM87, Remark 4.6] for an example.

Theorem 7.49(2) is particularly useful since it tells us that the commutator on a homomorphic image is entirely determined by its value in the originating algebra. To facilitate computations, it is convenient to introduce an additional notation. Let α, β, θ be congruences on an algebra $\mathbf{A}$. We write

$$[\alpha, \beta]_\theta = [\alpha, \beta] \vee \theta$$

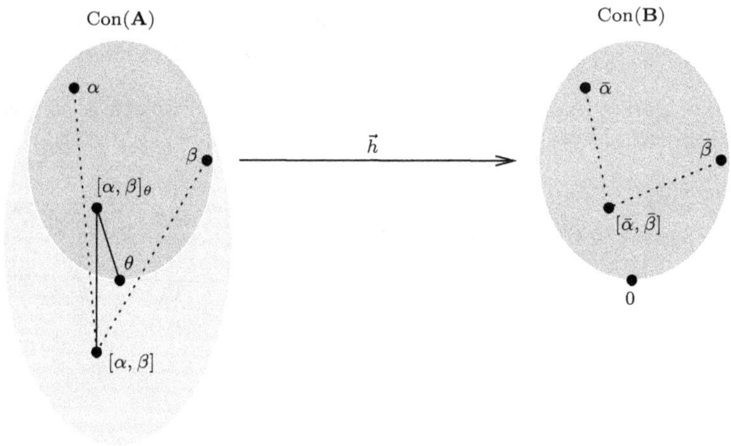

FIGURE 7.2: The relative commutator

and call $[\alpha, \beta]_\theta$ a *relative commutator.* Let us elucidate this definition.

Suppose that $\mathbf{B} = \mathbf{A}/\theta$ and $h \colon \mathbf{A} \to \mathbf{B}$ is the canonical homomorphism. Let $\bar{\alpha}$ and $\bar{\beta}$ be congruences on $\mathbf{B}$. Then $\alpha = \overleftarrow{h}(\bar{\alpha})$ and $\beta = \overleftarrow{h}(\bar{\beta})$ are congruences on $\mathbf{A}$. Theorem 7.49(2) asserts that $\overleftarrow{h}[\bar{\alpha}, \bar{\beta}] = [\overleftarrow{h}(\bar{\alpha}), \overleftarrow{h}(\bar{\beta})] \vee \ker h = [\alpha, \beta]_\theta$. Thus, if α and β are the congruences corresponding to $\bar{\alpha}$ and $\bar{\beta}$, then $[\alpha, \beta]_\theta$ is the congruence corresponding to $[\bar{\alpha}, \bar{\beta}]$, see Figure 7.2.

Because of their construction, the congruences α and β that appear in the previous paragraph must dominate θ. But observe that if α and β are *any* two congruences on $\mathbf{A}$ then

$$[\alpha \vee \theta,\, \beta \vee \theta]_\theta = [\alpha, \beta] \vee [\alpha, \theta] \vee [\theta, \beta] \vee [\theta, \theta] \vee \theta = [\alpha, \beta]_\theta \tag{7–18}$$

by the additivity and monotonicity of the commutator. As we just observed, the first congruence in this chain corresponds to a commutator on $\mathbf{B}$. Thus, properties of the commutator on $\mathbf{A}/\theta$ can often be used to deduce properties on $\mathbf{A}$. It is (7–18), together with Exercise 7.52.4 that forms the argument in 7.49(3).

From the definitions, if α^* denotes the centralizer of α then $[\alpha^*, \alpha] = 0$. In fact, α^* is the largest congruence β such that $[\beta, \alpha] = 0$. An algebra is Abelian if $1^* = 1$, equivalently, $[1, 1] = 0$. It follows from monotonicity that $\mathbf{A}$ is Abelian if and only if $[\alpha, \beta] = 0$ for all $\alpha, \beta \in \mathrm{Con}(\mathbf{A})$, i.e., $[\alpha, \beta]$ is always as small as possible. At the opposite extreme, we shall call an algebra $\mathbf{A}$ *neutral* if for all congruences α and β, we have $[\alpha, \beta] = \alpha \wedge \beta$, which is, by Proposition 7.45(1) its largest possible value.

From Theorem 7.48 we see that every neutral algebra in a Maltsev variety is congruence-distributive. The converse is false (consider the group $\mathbb{Z}_p$ for a prime p). However, at the level of an entire (Maltsev) variety we do have the following.

Theorem 7.50 (Hagemann and Herrmann [HH79]). *Let $\mathcal{V}$ be a Maltsev variety. Then $\mathcal{V}$ is congruence-distributive if and only if $\mathcal{V}$ is neutral.*

Proof. We have already observed that every neutral variety is congruence-distributive. Suppose that $\mathcal{V}$ is not neutral, say $\mathbf{A} \in \mathcal{V}$, $\alpha, \beta \in \mathrm{Con}(\mathbf{A})$ and $[\alpha, \beta] < \alpha \wedge \beta$.

Let $\bar{\gamma} = \alpha \wedge \beta$. By monotonicity, $[\bar{\gamma}, \bar{\gamma}] \leq [\alpha, \beta] < \bar{\gamma}$. Take $\mathbf{B} = \mathbf{A}/[\alpha, \beta]$ and $\gamma = \bar{\gamma}/[\alpha, \beta]$. Then $\gamma > 0$ and $[\gamma, \gamma] = 0$ by Theorem 7.49(2).

Now $\mathrm{Con}(\mathbf{B}(\gamma))$ contains the following sublattice, which shows that $\mathcal{V}$ is not congruence-distributive. (Here Δ is Δ^{γ}_{γ}.)

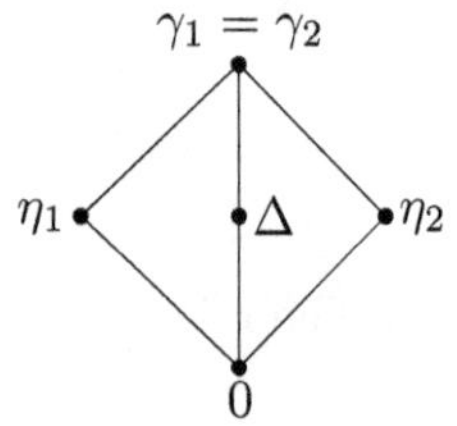

To see this observe that $[\gamma, \gamma] = 0$ implies $\gamma \leq \gamma^*$. Thus by Theorem 7.25 $\left(\begin{smallmatrix} a & c \\ b & d \end{smallmatrix}\right) \in \eta_1 \wedge \Delta$ implies that $a = c$ and $b = q(a, c, d) = q(a, a, d) = d$. Thus $\eta_1 \wedge \Delta = 0$. On the other hand, if $a \mathrel{\gamma} c$ then

$$\begin{pmatrix} a \\ b \end{pmatrix} \eta_1 \begin{pmatrix} a \\ a \end{pmatrix} \Delta \begin{pmatrix} c \\ c \end{pmatrix} \eta_1 \begin{pmatrix} c \\ d \end{pmatrix}$$

showing that $\eta_1 \vee \Delta = \gamma_1$. □

As an application, we can give another proof of Lemma 4.74. Suppose that $\mathcal{V}$ is a congruence-distributive variety of commutative rings. Let $\mathbf{F} = \mathbf{F}_{\mathcal{V}}(x)$ be the free algebra on one generator. Since $\mathcal{V}$ is congruence-distributive, it is neutral. Thus $[1_F, 1_F] = 1_F$. Then according to Exercise 7.52.1, $F \cdot F = F$ (as ideals). Since $F = x\mathbb{Z}[x]$, there are polynomials $g_1(x), g_2(x) \in x\mathbb{Z}[x]$ such that $g_1(x) \cdot g_2(x) = x$ in $\mathbf{F}$. By factoring out an x from g_1, we get $\mathcal{V} \models xf(x) \approx x$ for some $f(x) \in x\mathbb{Z}[x]$.

Theorem 7.51. *Let $\mathcal{V}$ be a Maltsev variety. A subdirect product of finitely many neutral algebras from $\mathcal{V}$ is neutral.*

Proof. Let $\mathbf{A}_1, \mathbf{A}_2$ be neutral algebras in $\mathcal{V}$ and $\mathbf{C}$ a subdirect product of $\mathbf{A}_1 \times \mathbf{A}_2$. It suffices to show that $\mathbf{C}$ is neutral and apply induction. Let $p_i \colon \mathbf{C} \twoheadrightarrow \mathbf{A}_i$ be the canonical projection and $\eta_i = \ker p_i$. We claim that

$$(\forall \theta, \psi \in \mathrm{Con}(\mathbf{C}))\ \ \theta \wedge \psi = [\theta, \psi] \vee (\eta_i \wedge \theta \wedge \psi), \quad \text{for } i = 1, 2. \tag{7–19}$$

First, since $\mathbf{A}_i$ is neutral,

$$[\theta \vee \eta_i,\, \psi \vee \eta_i]_{\eta_i} = (\theta \vee \eta_i) \wedge (\psi \vee \eta_i) \geq \theta \wedge \psi.$$

On the other hand,

$$[\theta \vee \eta_i,\, \psi \vee \eta_i]_{\eta_i} = [\theta, \psi] \vee \eta_i$$

by (7–18). Putting these two together $\theta \wedge \psi \leq [\theta, \psi] \vee \eta_i$. Finally, using congruence-permutability,

$$\theta \wedge \psi = \big([\theta, \psi] \vee \eta_i\big) \wedge (\theta \wedge \psi) = \big([\theta, \psi] \circ \eta_i\big) \wedge (\theta \wedge \psi) = [\theta, \psi] \vee (\eta_i \wedge \theta \wedge \psi).$$

This verifies Equation (7–19).

Now, to show $\mathbf{C}$ neutral, let $\alpha, \beta \in \operatorname{Con}(\mathbf{C})$. Applying (7–19),

$$\alpha \wedge \beta = [\alpha, \beta] \vee (\eta_2 \wedge \alpha \wedge \beta).$$

Applying (7–19) again, this time with $\theta = \alpha \wedge \eta_2$, $\psi = \beta \wedge \eta_2$ and $i = 1$,

$$\alpha \wedge \beta \wedge \eta_2 = [\alpha \wedge \eta_2,\, \beta \wedge \eta_2] \vee (\eta_1 \wedge \alpha \wedge \beta \wedge \eta_2) = [\alpha \wedge \eta_2,\, \beta \wedge \eta_2]$$

(since $\eta_1 \wedge \eta_2 = 0$). Thus $\alpha \wedge \beta = [\alpha, \beta] \vee [\alpha \wedge \eta_2,\, \beta \wedge \eta_2] = [\alpha, \beta]$. □

Exercise Set 7.52.

1. Let $\mathbf{R}$ be a ring, $\alpha, \beta \in \operatorname{Con}(\mathbf{R})$ and let I and J be the ideals corresponding to α and β. Show that the ideal corresponding to $[\alpha, \beta]$ is $IJ + JI$.

2. Let $\mathcal{V}$ be a Maltsev variety. An algebra in $\mathcal{V}$ is called *nil-2* if $[1, 1] \leq \zeta$. Prove that the class of nil-2 members of $\mathcal{V}$ forms a subvariety.

3. Let $\mathcal{W}$ be a Maltsev variety.

   (a) Suppose that $\mathcal{V}_1$ and $\mathcal{V}_2$ are subvarieties of $\mathcal{W}$. Prove that if both $\mathcal{V}_1$ and $\mathcal{V}_2$ are congruence-distributive, then so is $\mathcal{V}_1 \vee \mathcal{V}_2$.

   (b) Prove that if $\mathcal{W}$ is locally finite, then it has a largest congruence-distributive subvariety.

4. Let $\mathbf{A}$ lie in a Maltsev variety. Suppose that $\alpha, \beta, \gamma, \phi \in \operatorname{Con}(\mathbf{A})$ and $\mathbf{I}[\alpha, \beta] \nearrow \mathbf{I}[\gamma, \phi]$. Recall (Exercise 2.12.5) that this means that $\alpha = \beta \wedge \gamma$ and $\phi = \beta \vee \gamma$. Consequently the mappings $h(\theta) = \theta \vee \gamma$ and $k(\theta) = \theta \wedge \beta$ are lattice isomorphisms. Prove that

   (a) for all $\theta, \psi \in \mathbf{I}[\alpha, \beta]$, $\;h\big([\theta, \psi]_\alpha\big) = \big[h(\theta),\, h(\psi)\big]_\gamma$ and

   (b) for all $\theta, \psi \in \mathbf{I}[\gamma, \psi]$, $\;k\big([\theta, \psi]_\gamma\big) = \big[k(\theta),\, k(\psi)\big]_\alpha$

5. Let $\mathbf{B}_1, \dots, \mathbf{B}_n$ be simple, nonabelian algebras in a Maltsev variety, and $\mathbf{C} = \mathbf{B}_1 \times \dots \mathbf{B}_n$. Prove that for any $\psi \in \operatorname{Con}(\mathbf{C})$, $\mathbf{C}/\psi \cong \prod_{i \in I} \mathbf{B}_i$ for some $I \subseteq \{1, 2, \dots, n\}$.

6. Let $\mathbf{A}$ be an algebra in a Maltsev variety, $\alpha, \beta \in \operatorname{Con}\mathbf{A}$. Our goal in this problem is to prove that for all $a, b \in A$,
$a \equiv b \pmod{[\alpha, \beta]} \iff \left(\begin{smallmatrix} a & b \\ b & b \end{smallmatrix}\right) \in \Delta_\alpha^\beta$.
   (a) Let $\gamma = \{\, (a, b) \in A^2 : \left(\begin{smallmatrix} a & b \\ b & b \end{smallmatrix}\right) \in \Delta_\alpha^\beta \,\}$. Show that γ is a congruence on $\mathbf{A}$. (Hint: see the comments after 7.26.)
   (b) Show that γ has the α, β-term condition.
   (c) Show that $\gamma \subseteq [\alpha, \beta]$. (One way to do this is to use 7.24 and the fact that $[\alpha, \beta]$ has the α, β-term condition.)
   (d) Prove the theorem. (Gumm, 1980)

7. Let $\mathbf{A}$ be an algebra in a Maltsev variety and suppose that $\mathbf{A}^2$ has the congruence extension property (see Definition 5.14).
   (a) Prove that for all congruences α, β on $\mathbf{A}$, $[\alpha, \beta] = \beta \wedge [1, \alpha] = \alpha \wedge \beta \wedge [1, 1]$. [Hint: first show that $\Delta_\alpha^1 \restriction_{\mathbf{A}(\beta)} = \Delta_\alpha^\beta$. Then use Exercise 6.]
   (b) Prove that for any subalgebra, $\mathbf{B}$, of $\mathbf{A}$, $[1_A, 1_A]\restriction_B = [1_B, 1_B]$. (Kiss, 1985)

7.5 Directly representable varieties revisited

Earlier, we considered directly representable varieties and proved that they are always Maltsev. It turns out that every directly representable variety is a join of two contrasting subvarieties: one that is semisimple and arithmetical (hence neutral), and one that resembles the variety of modules over a ring. We derived a complete classification of the first sort in Theorem 7.11: a congruence-distributive variety is directly representable if and only if it is semisimple and arithmetical.

We now have the tools to consider the opposite extreme: Abelian algebras. It is not true that every finitely generated Abelian variety is directly representable. More specifically, a characterization of those finite rings $\mathbf{R}$ such that the variety of $\mathbf{R}$-modules is directly representable is still unknown. But modulo that difficulty (which properly belongs in the domain of ring theory, not universal algebra) we can completely characterize directly representable varieties. The proof is somewhat involved. The reader may wish to skip the proofs of Theorems 7.55 and 7.56 on first reading.

As we have observed throughout this text, the congruence lattices of the members of a variety $\mathcal{V}$ have a profound influence on the properties of $\mathcal{V}$. Within the realm of Maltsev varieties we can extend our menu of congruence properties to include those that involve the commutator operation. For example $\mathcal{V}$ is neutral if and only if

$$[\alpha, \beta] = \alpha \wedge \beta.$$

As a congruence identity on $\mathcal{V}$ this expression is intended to apply to every algebra $\mathbf{A} \in \mathcal{V}$ and every $\alpha, \beta \in \operatorname{Con}(\mathbf{A})$. As another example, $\mathcal{V}$ is Abelian if it satisfies the congruence identity $[\alpha, \beta] = 0$. Also, in Exercise 7.52.7 you showed that a variety with the congruence extension property satisfies the congruence identity $[\alpha, \beta] = \alpha \wedge \beta \wedge [1, 1]$. We now introduce a new congruence identity.

Definition 7.53. A Maltsev variety $\mathcal{V}$ satisfies (C1) if, for all $\mathbf{A} \in \mathcal{V}$ and $\alpha, \beta \in \operatorname{Con}(\mathbf{A})$

$$\alpha \wedge [\beta, \beta] \leq [\alpha, \beta]. \tag{C1}$$

It turns out that (C1) is closely related to bounds on the sizes of subdirectly irreducible algebras. We shall show one-half of this relationship in Theorem 7.55. A full analysis appears in [FM81]. First we provide several conditions equivalent to (C1).

Lemma 7.54. *Let $\mathcal{V}$ be a Maltsev variety. The following are equivalent.*

(a) *$\mathcal{V}$ satisfies* (C1)*;*

(b) $(\forall \mathbf{A} \in \mathcal{V}_{\mathrm{si}})$ $(\forall \beta \in \operatorname{Con}(\mathbf{A}))$ $[\mu, \beta] = 0 \implies [\beta, \beta] = 0$*;*

(c) *$\mathcal{V}$ satisfies the congruence implication* $\alpha \leq [\beta, \beta] \implies [\alpha, \beta] = \alpha$.

Proof. We shall prove the equivalence of (a) and (b). Condition (c) is left for the exercises. Suppose that $\mathcal{V}$ satisfies (C1) and $\mathbf{A} \in \mathcal{V}_{\mathrm{si}}$. Then $\mu \wedge [\beta, \beta] \leq [\mu, \beta]$. If $[\mu, \beta] = 0$ then $[\beta, \beta] \not\geq \mu$, so, since μ is the monolith, we must have $[\beta, \beta] = 0$.

Now we show that the second condition implies (C1). Suppose that there is an algebra $\mathbf{B} \in \mathcal{V}$ and $\alpha, \beta \in \operatorname{Con}(\mathbf{B})$ such that $\alpha \wedge [\beta, \beta] \not\leq [\alpha, \beta]$ (i.e., (C1) fails). Since every congruence is a meet of completely meet-irreducible congruences, there is a completely meet-irreducible congruence θ such that $[\alpha, \beta] \leq \theta$ but $\alpha \wedge [\beta, \beta] \not\leq \theta$. Let ψ be the upper cover of θ. Since $\alpha \not\leq \theta$, we must have $\alpha \vee \theta \geq \psi$ and similarly $[\beta, \beta] \vee \theta \geq \psi$.

Using the monotonicity and additivity of the commutator we compute

$$\theta \leq [\psi, \theta \vee \beta] \vee \theta \leq [\theta \vee \alpha,\, \theta \vee \beta] \vee \theta = [\theta, \theta] \vee [\theta, \beta] \vee [\alpha, \theta] \vee [\alpha, \beta] \vee \theta = \theta$$

hence $[\psi, \theta \vee \beta] \vee \theta = \theta$. On the other hand

$$[\theta \vee \beta,\, \theta \vee \beta] \vee \theta = [\theta, \theta] \vee [\theta, \beta] \vee \big([\beta, \beta] \vee \theta\big) \geq \psi.$$

Now let $\mathbf{A} = \mathbf{B}/\theta$ and $h\colon \mathbf{B} \twoheadrightarrow \mathbf{A}$ the natural map. By Theorem 7.49(2), with $\bar{\beta} = \vec{h}(\theta \vee \beta)$ and $\mu = \vec{h}(\psi)$, $[\mu, \bar{\beta}] = 0$ but $[\bar{\beta}, \bar{\beta}] \geq \mu$. Thus (b) fails. ☐

Theorem 7.55 (Freese-McKenzie). *Let $\mathcal{V}$ be a Maltsev variety and suppose that there is an integer n such that every subdirectly irreducible member of $\mathcal{V}$ has cardinality less than n. Then $\mathcal{V}$ satisfies* (C1).

Proof. Assume that $\mathcal{V}$ fails (C1). We will derive the existence of arbitrarily large finite subdirectly irreducible members of $\mathcal{V}$. In fact, we shall show that for every positive n, there is $\mathbf{C} \in \mathcal{V}_{\mathrm{si}}$ such that $|\mathrm{Con}(\mathbf{C})| > n$.

According to Lemma 7.54, since (C1) fails there is $\mathbf{A} \in \mathcal{V}_{\mathrm{si}}$ with monolith μ and congruence γ such that $[\gamma, \gamma] > 0$ but $[\mu, \gamma] = 0$. Let $\Delta = \Delta_\mu^\gamma \in \mathrm{Con}(\mathbf{A}(\gamma))$.

Let $\mathbf{B} = \mathbf{A}_n(\gamma)$. (See Equation (7–2) for the definition.) For each $i \leq n$ let $p_i \colon \mathbf{B} \to \mathbf{A}$ be the coordinate projection. If we write $\gamma_i = \overleftarrow{p}_i(\gamma)$, then for each $i \neq j$ we have $\gamma_i = \gamma_j$. Let us call this congruence $\bar{\gamma}$. Furthermore, for $i \neq j$, $\eta_i \vee \eta_j = \bar{\gamma}$. We also write μ_i in place of $\overleftarrow{p}_i(\mu)$.

We define the following additional congruences on $\mathbf{B}$:

$$\kappa_j = \left\{ (\mathbf{a}, \mathbf{b}) \in B^2 : \left(\begin{smallmatrix} a_1 & b_1 \\ a_j & b_j \end{smallmatrix} \right) \in \Delta_\mu^\gamma \ \& \ i \notin \{1, j\} \implies a_i = b_i \right\}, \quad 1 < j \leq n,$$

$$\theta_j = \{ (\mathbf{a}, \mathbf{b}) \in B^2 : a_j \mathrel{\mu} b_j \ \& \ i \neq j \implies a_i = a_j \}, \quad 1 \leq j \leq n,$$

$$\eta'_j = \bigwedge_{i \neq j}^{n} \eta_i, \quad \theta = \bigvee_{j=1}^{n} \theta_j, \quad \kappa = \bigvee_{j=2}^{n} \kappa_j.$$

Note that for every $j \geq 1$, $\theta_j = \mu_j \wedge \eta'_j$.

Claim 1. For every $j > 1$, $\theta_1 \leq \eta'_j \vee \kappa_j$.

Proof. Suppose that $\mathbf{a} \mathrel{\theta_1} \mathbf{b}$. Then $a_1 \mathrel{\mu} b_1$, $a_2 = b_2, \ldots, a_n = b_n$. Therefore

$$\begin{aligned} \mathbf{a} =& (a_1, a_2, \ldots, a_j, \ldots, a_n) \equiv_{\eta'_j} (a_1, a_2, \ldots, a_1, \ldots, a_n) \equiv_{\kappa_j} \\ & (b_1, a_2, \ldots, b_1, \ldots, a_n) \equiv_{\eta'_j} (b_1, a_2, \ldots, a_j, \ldots, a_n) = \mathbf{b}. \end{aligned}$$

Claim 2. For every $j > 1$, $\theta_1 \leq \theta_j \vee \kappa_j$.

Proof. $\theta_1 \wedge \mu_j \leq (\eta'_j \vee \kappa_j) \wedge \mu_j$ from Claim 1. But $\theta_1 \leq \mu_j$ (since $\mathbf{a} \mathrel{\theta_1} \mathbf{b} \implies a_j = b_j \implies a_j \mathrel{\mu} b_j$). Also $\kappa_j \leq \mu_j$ (since $\binom{a_1}{a_j} \ \Delta_\mu^\gamma \ \binom{b_1}{b_j} \implies a_j \mathrel{\mu} b_j$). Therefore $\theta_1 \leq (\eta'_j \wedge \mu_j) \vee \kappa_j = \theta_j \vee \kappa_j$ by modularity.

Claim 3. Similarly, for all $j > 1$, $\theta_j \leq \theta_1 \vee \kappa_j$.

Claim 4. For every $i, j \neq 1$, $\kappa \vee \theta_j \geq \theta_i \vee \theta_1$.

Proof. Using Claims 2 and 3 at the indicated equalities,

$$\begin{aligned} \kappa \vee \theta_j = \kappa \vee \kappa_i \vee (\kappa_j \vee \theta_j) &\stackrel{2}{=} \kappa \vee \kappa_i \vee \kappa_j \vee \theta_j \vee \theta_1 \\ &= \kappa \vee \kappa_i \vee \theta_1 \vee \theta_j \stackrel{3}{=} \kappa \vee \kappa_i \vee \theta_i \vee \theta_1 \vee \theta_j \geq \theta_i \vee \theta_1. \end{aligned}$$

Claim 5. For every $j \le n$, $\kappa \vee \theta_j = \theta$.

Proof. Certainly $\theta = \bigvee_i \theta_i \ge \theta_j$. Also $\kappa_j \le \theta_1 \vee \theta_j \le \theta$ since

$$\mathbf{a}\,\kappa_j\,\mathbf{b} \implies \begin{pmatrix} a_1 & b_1 \\ a_j & b_j \end{pmatrix} \in \Delta_\mu^\gamma \implies a_1\,\mu\,b_1 \;\&\; a_j\,\mu\,b_j \implies$$
$$\mathbf{a}\,\theta_1\,(b_1, a_2, \dots, a_n)\,\theta_j\,\mathbf{b}.$$

So $\kappa \vee \theta_j \le \theta$. But, by claim 4, $\kappa \vee \theta_j \ge \theta_i$, for all $i \le n$. Therefore $\kappa \vee \theta_j \ge \theta$.

Claim 6. $\theta_1 \not\le \kappa$.

Proof. Pick $(x, y) \in \mu$ with $x \ne y$ and $a_2, \dots, a_n \in A$. Let $\mathbf{a}^1 = (x, a_2, \dots, a_n)$ and $\mathbf{a}^n = (y, a_2, \dots, a_n)$. Then $\mathbf{a}^1\,\theta_1\,\mathbf{a}^n$. Suppose that $(\mathbf{a}^1, \mathbf{a}^n) \in \kappa$. Since we are in a Maltsev variety, there are $\mathbf{a}^2, \dots, \mathbf{a}^{n-1}$ such that

$$\mathbf{a}^1\,\kappa_2\,\mathbf{a}^2\,\kappa_3\,\mathbf{a}^3\,\kappa_4 \cdots \kappa_n\,\mathbf{a}^n.$$

Write $\mathbf{a}^i = (a_1^i, a_2^i, \dots, a_n^i)$ for $i = 1, \dots, n$. Since $a_1^1 = x \ne y = a_1^n$, there is some index i with $1 < i < n$ such that $a_1^{i-1} \ne a_1^i$. But $\mathbf{a}^{i-1}\,\kappa_i\,\mathbf{a}^i$ implies that $\begin{pmatrix} a_1^{i-1} & a_1^i \\ a_i^{i-1} & a_i^i \end{pmatrix} \in \Delta_\mu^\gamma$. By our initial assumption, $[\mu, \gamma] = 0$, i.e., γ centralizes μ. Thus by Theorems 7.25 and 7.26 we must have $a_i^{i-1} \ne a_i^i$. On the other hand, by the definition of κ_j for $j \ne i$, $a_i^1 = a_i^2 = \cdots = a_i^{i-1}$ and $a_i^i = a_i^{i+1} = \cdots = a_i^n$, from which it follows that $a_i^1 \ne a_i^n$, which is a contradiction.

Claim 7. $\kappa < \theta$.

Proof. From Claim 5, $\kappa \le \kappa \vee \theta_j \le \theta$. Since $\theta_1 \le \theta$ we conclude from Claim 6 that $\kappa \ne \theta$.

Since every congruence is a meet of completely meet-irreducible congruences, by Claim 7 there is a completely meet-irreducible congruence λ such that $\lambda \ge \kappa$ but $\lambda \not\ge \theta$. Let $\mathbf{C} = \mathbf{B}/\lambda$. Then $\mathbf{C}$ is subdirectly irreducible.

Claim 8. For every $1 \le i < j \le n$, $\lambda \vee \eta_i \ne \lambda \vee \eta_j$.

Proof. First $\mathbf{I}[\eta_j, \bar\gamma] \searrow \mathbf{I}[0, \eta_j']$ (since $\eta_i \vee \eta_j = \bar\gamma$ and $\eta_i \le \eta_j'$, so $\eta_j' \vee \eta_j = \bar\gamma$). In $\mathrm{Con}(\mathbf{A})$ we have $0 \prec \mu \le \gamma$, so in $\mathrm{Con}(\mathbf{B})$, $\eta_j \prec \mu_j \le \gamma_j = \bar\gamma$. Therefore (Exercise 2.12.5) $0 \prec \theta_j \le \eta_j'$, i.e., θ_j is the upper cover of 0 in $\mathbf{I}[0, \eta_j']$.

Now $\lambda \not\ge \theta$ and by Claim 5, $\theta = \kappa \vee \theta_j$. Since $\lambda \ge \kappa$ we conclude that $\lambda \not\ge \theta_j$. Therefore, from the previous paragraph $\lambda \wedge \eta_j' = 0$.

Recall that $\eta_j \vee \eta_j' = \bar\gamma$. Suppose that for some $j \le n$, $\lambda \vee \eta_j \ge \bar\gamma$. Then

$$[\bar\gamma, \bar\gamma] \le [\lambda \vee \eta_j, \eta_j \vee \eta_j'] = [\lambda, \eta_j] \vee [\lambda, \eta_j'] \vee [\eta_j, \eta_j] \vee [\eta_j, \eta_j'] \le \eta_j$$

since $[\lambda, \eta_j'] \le \lambda \wedge \eta_j' = 0$. But then in $\mathrm{Con}(\mathbf{A})$, $[\gamma, \gamma] = 0$ which is a contradiction. So we conclude that for every $j \le n$

$$\lambda \vee \eta_j \not\ge \bar\gamma = \eta_i \vee \eta_j, \quad \text{for } i \ne j.$$

Therefore $\lambda \vee \eta_j \not\ge \eta_i$ so $\lambda \vee \eta_j \ne \lambda \vee \eta_i$. This proves Claim 8.

Finally, since $\mathrm{Con}(\mathbf{C}) \cong \mathbf{I}[\lambda, 1]$, claim 8 implies that $|\mathrm{Con}(\mathbf{C})| \ge n$, proving the theorem. ☐

The next theorem is a generalization of Lemma 7.10. The proof is similar, although the replacement of the meet operation (in the distributive case) with the commutator (in the permutable case) requires a more complex argument.

Theorem 7.56 (McKenzie). *In a directly representable variety, every subdirectly irreducible algebra is either simple or Abelian.*

Proof. Let $\mathcal{V}$ be directly representable. By Lemma 7.12 there is a finite upper bound on the cardinality of the members of $\mathcal{V}_{\mathrm{si}}$. Therefore by Theorem 7.55, $\mathcal{V}$ satisfies (C1).

Suppose that there is a subdirectly irreducible algebra $\mathbf{A}$ in $\mathcal{V}$ such that $\mathbf{A}$ is neither simple nor Abelian. Denoting the monolith of $\mathbf{A}$ by μ, we have $[1,1] > 0$, so $[1,1] \geq \mu$ and also $1 > \mu$. Applying Lemma 7.54(c), $[\mu, 1] = \mu$.

For a positive integer n let $\mathbf{B} = \mathbf{A}_n(\mu)$ (see equation (7–2)). Since $\mathbf{A}$ is finite, so is $\mathbf{B}$. We shall show that $\mathbf{B}$ is directly indecomposable. This will provide an unbounded sequence of finite directly indecomposable algebras, contradicting the direct representability of $\mathcal{V}$.

For $i \leq n$, let $p_i \colon \mathbf{B} \to \mathbf{A}$ be the coordinate projection and define $\mu_i = \overleftarrow{p_i}(\mu)$. Then $\mu_1 = \mu_2 = \cdots = \mu_n$. We shall denote this congruence $\bar{\mu}$. Furthermore, for $i \leq n$, let $\eta_i = \ker(p_i)$ and

$$\eta_i' = \bigwedge_{j \neq i} \eta_j.$$

Then one easily checks that

$$\eta_i \wedge \eta_i' = 0 \text{ and } \eta_i \vee \eta_i' = \eta_i \vee \eta_j = \bigvee_{k=1}^{n} \eta_k' = \bar{\mu}, \quad \text{for } i \neq j. \tag{7–20}$$

Let us assume that Con($\mathbf{B}$) contains a pair $\{\alpha, \beta\}$ of nontrivial factor congruences and derive a contradiction.

Claim. For every $i \leq n$ either

$$\big([\eta_i', \beta] = \eta_i' \;\&\; [\eta_i, \alpha] = 0\big) \text{ or } \big([\eta_i', \alpha] = \eta_i' \;\&\; [\eta_i, \beta] = 0\big).$$

Proof. Since $[\mu, 1] = \mu$ on $\mathbf{A}$, $[\bar{\mu}, 1]_{\eta_i} = \bar{\mu}$ on $\mathbf{B}$. But

$$\begin{aligned}
&[\bar{\mu}, 1]_{\eta_i} = [\eta_i \vee \eta_i', \alpha \vee \beta] \vee \eta_i = && \text{since } \alpha \vee \beta = 1 \\
&[\eta_i, \alpha] \vee [\eta_i, \beta] \vee [\eta_i', \alpha] \vee [\eta_i', \beta] \vee \eta_i = && \\
&[\eta_i', \alpha] \vee [\eta_i', \beta] \vee \eta_i && \text{by Prop. 7.45.}
\end{aligned} \tag{7–21}$$

Now $\mathbf{I}[0, \mu] \cong \mathbf{I}[\eta_i, \bar{\mu}] \searrow \mathbf{I}[0, \eta_i']$. (The first of these intervals is in Con($\mathbf{A}$), the other two in Con($\mathbf{B}$). The transpose follows from Equation (7–20).) Since μ covers 0 on $\mathbf{A}$, η_i' must cover 0 on $\mathbf{B}$. Because $0 \leq [\eta_i', \alpha] \leq \eta_i'$, we must have $[\eta_i', \alpha] \in \{0, \eta_i'\}$. The same holds for $[\eta_i', \beta]$. If $[\eta_i', \alpha] = [\eta_i', \beta] = 0$ then

Equation (7–21) implies that $\bar{\mu} = \eta_i$, which is false. Therefore one of these two commutators must be equal to η_i'. If both equal η_i' then

$$\eta_i' = [\eta_i', \alpha] \wedge [\eta_i', \beta] \leq \alpha \wedge \beta = 0$$

which is false. Thus one commutator is η_i' and the other is 0, which proves the claim.

Now define

$$S_\alpha = \{\, i : [\eta_i', \alpha] = \eta_i' \,\}, \quad S_\beta = \{\, i : [\eta_i', \beta] = \eta_i' \,\}$$
$$\widehat{\alpha} = \bigvee_{i \in S_\alpha} \eta_i', \qquad \widehat{\beta} = \bigvee_{i \in S_\beta} \eta_i'.$$

It follows from the claim and Equation (7–20) that

$$S_\alpha \cup S_\beta = \{1, 2, \ldots, n\}, \quad S_\alpha \cap S_\beta = \varnothing, \quad \widehat{\alpha} \vee \widehat{\beta} = \bar{\mu}.$$

Furthermore, $j \in S_\alpha \implies \eta_j' = [\eta_j', \alpha] \leq \alpha$. Consequently, $\widehat{\alpha} \leq \alpha$. Of course analogous relationships apply to β.

Claim. $i \in S_\alpha \implies \beta \leq \bar{\mu}$. $i \in S_\beta \implies \alpha \leq \bar{\mu}$.

Proof. By the symmetry of the situation it is only necessary to verify the first implication. Suppose that $i \in S_\alpha$ but $\beta \not\leq \bar{\mu}$. Since $\bar{\mu} = \mu_i > \eta_i$ we certainly have $\beta \not\leq \eta_i$, i.e., $\eta_i \vee \beta > \eta_i$. Since μ_i covers η_i we obtain $\eta_i \vee \beta \geq \mu_i$. Rewriting and applying modularity

$$\bar{\mu} = \bar{\mu} \wedge (\eta_i \vee \beta) = \eta_i \vee (\bar{\mu} \wedge \beta) \overset{*}{=} \eta_i \vee \widehat{\beta}. \tag{7–22}$$

To see the indicated equality, observe that $\widehat{\alpha} \wedge \beta \leq \alpha \wedge \beta = 0$, hence by modularity

$$\bar{\mu} \wedge \beta = (\widehat{\alpha} \vee \widehat{\beta}) \wedge \beta = (\widehat{\alpha} \wedge \beta) \vee \widehat{\beta} = \widehat{\beta}.$$

Also, if $j \in S_\beta$ then $j \neq i$, so $\eta_j' = \bigwedge_{k \neq j} \eta_k \leq \eta_i$. Therefore $\widehat{\beta} \leq \eta_i$. Applying this observation to (7–22), we conclude $\bar{\mu} = \eta_i$ which is a contradiction, proving the claim.

To finish the proof of the theorem, we note that since $\mu < 1$ on $\mathbf{A}$, $\bar{\mu} < 1$ on $\mathbf{B}$. On the other hand, $\alpha \vee \beta = 1$ on $\mathbf{B}$. Therefore we must have (without loss of generality) that $\beta \not\leq \bar{\mu}$. From the above claim, $S_\alpha = \varnothing$, so $S_\beta = \{1, 2, \ldots, n\}$. And therefore again by the claim

$$\alpha \leq \bar{\mu} = \bigvee_{i=1}^{n} \eta_i' = \widehat{\beta} \leq \beta$$

which contradicts the nontriviality of the pair $\{\alpha, \beta\}$ of factor congruences. ☐

Theorem 7.57. *Let* $\mathbf{C}$ *be an algebra in a Maltsev variety. Suppose that* $\mathbf{C}$ *is a subdirect product of* $\mathbf{A}_1 \times \cdots \times \mathbf{A}_m \times \mathbf{B}_1 \times \cdots \times \mathbf{B}_n$ *in which every* $\mathbf{A}_i$ *is Abelian and every* $\mathbf{B}_i$ *is simple and nonabelian. Then* $\mathbf{C} \cong \mathbf{C}/[1,1] \times \mathbf{C}/\zeta$, $\mathbf{C}/[1,1]$ *is Abelian, and* $\mathbf{C}/\zeta \cong \prod_{i \in I} \mathbf{B}_i$, *for some subset* I *of* $\{1, 2, \ldots, n\}$.

Proof. Let $h\colon \mathbf{C} \to \prod_{i=1}^{m} \mathbf{A}_i$ be the natural map, $A = \vec{h}(C)$ and $\alpha = \ker(h)$. For $i \leq n$, let η_i be the kernel of the projection of $\mathbf{C}$ onto $\mathbf{B}_i$. Define $\theta = \bigcap_i \eta_i$. From our assumptions, $\theta \wedge \alpha = 0_C$.

Since $\mathbf{C}/\alpha \cong \mathbf{A}$ which is Abelian, $[1,1] \leq \alpha$. Let $\mathbf{C}' = \mathbf{C}/\theta$ which is a subdirect product of $\prod_i \mathbf{B}_i$. By assumption, $\mathbf{C}'$ is congruence permutable and each $\mathbf{B}_i$ is simple, so by Theorem 6.3, $\mathbf{C}'$ is a direct product of some of the $\mathbf{B}_i$'s. By Theorem 7.51, $\mathbf{C}'$ is neutral. Therefore in $\mathrm{Con}(\mathbf{C})$, $1 = [1,1]_\theta = [1,1] \vee \theta$. Since $\alpha \geq [1,1]$, we have, using modularity

$$\alpha = \alpha \wedge \big([1,1] \vee \theta\big) = [1,1] \vee (\alpha \wedge \theta) = [1,1].$$

Thus, by congruence permutability, $\mathbf{C} \cong \mathbf{C}/[1,1] \times \mathbf{C}/\theta$.

From the neutrality of $\mathbf{C}'$, $0 = [1^*,1] = 1^*$ (in $\mathbf{C}'$). Hence by Proposition 7.22

$$\zeta_{\mathbf{C}} = 1^*_{\mathbf{C}} = 1^*_{\mathbf{C}/[1,1]} \times 1^*_{\mathbf{C}/\theta} = 1_{\mathbf{C}/[1,1]} \times 0_{\mathbf{C}/\theta} = \theta. \qquad \square$$

We are now ready to complete McKenzie's characterization of directly representable varieties.

Theorem 7.58 (McKenzie [McK82]). *Let $\mathcal{K}$ be a finite set of finite algebras and let $\mathcal{V} = \mathsf{V}(\mathcal{K})$. Then $\mathcal{V}$ is directly representable if and only if*

(1) *$\mathcal{V}$ is Maltsev,*

(2) *every member of $\mathsf{S}(\mathcal{K})$ is a product of simple and Abelian algebras, and*

(3) *$\mathcal{V}_{\mathrm{ab}}$ is directly representable.*

Proof. First assume that $\mathcal{V}$ is directly representable. Since direct representability is inherited by subvarieties, (3) holds. By Theorem 7.9, $\mathcal{V}$ is Maltsev and by 7.56, every subdirectly irreducible algebra is either simple or Abelian. Let $\mathbf{C} \in \mathsf{S}(\mathcal{K})$. $\mathbf{C}$ is a finite subdirect product of subdirectly irreducible algebras, so the previous theorem applies. Thus $\mathbf{C}$ is a product of simple and Abelian algebras.

For the converse, assume that all three conditions hold. Let $\mathbf{C}$ be a finite, directly indecomposable algebra in $\mathcal{V}$. It suffices to show that either $\mathbf{C} \in \mathcal{V}_{\mathrm{ab}}$ or $\mathbf{C} \in \mathsf{HS}(\mathcal{K})$, since each of these two classes has only finitely many finite directly indecomposable algebras.

Of course, $\mathbf{C} \in \mathsf{HSP}(\mathcal{K})$. Therefore there is an algebra $\mathbf{D} \in \mathsf{SP}(\mathcal{K})$ and $\psi \in \mathrm{Con}(\mathbf{D})$ with $\mathbf{C} \cong \mathbf{D}/\psi$. Since C is finite and $\mathcal{V}$ is locally finite we can assume that D is finite as well. Thus $\mathbf{D}$ is a subdirect product of algebras $\mathbf{D}_1, \dots, \mathbf{D}_t$ in which each $\mathbf{D}_i \in \mathsf{S}(\mathcal{K})$. By assumption, each $\mathbf{D}_i$ is a product of simple and Abelian algebras. Thus we can write $\mathbf{D}$ as a subdirect product of $\mathbf{A}_1 \times \cdots \times \mathbf{A}_m \times \mathbf{B}_1 \times \cdots \times \mathbf{B}_n$ in which every $\mathbf{A}_i$ is Abelian, and every $\mathbf{B}_i$ is simple, nonabelian and is a member of $\mathsf{HS}(\mathcal{K})$. By Theorem 7.57, $\mathbf{D} \cong \mathbf{D}/[1,1] \times \mathbf{D}/\zeta$. Put another way, $[1,1]$ and ζ form a pair of complementary factor congruences on $\mathbf{D}$.

Now $\mathbf{D}/\zeta \cong \mathbf{B}_1 \times \cdots \times \mathbf{B}_n$ is neutral. Therefore, in $\mathrm{Con}(\mathbf{D})$, $[1, \psi \vee \zeta]_\zeta = \psi \vee \zeta$. Thus

$$\psi \vee \zeta = [1, \psi \vee \zeta]_\zeta = [1, \psi] \vee [1, \zeta] \vee \zeta = [1, \psi] \vee \zeta$$

since $\zeta = 1^*$ centralizes 1. Therefore, by modularity

$$[1,1] \wedge (\psi \vee \zeta) = [1,1] \wedge ([1,\psi] \vee \zeta) = [1,\psi] \vee ([1,1] \wedge \zeta) = [1,\psi] \leq \psi$$

using the fact that $[1,1] \wedge \zeta = 0$ (because they are complementary factor congruences). Thus

$$\psi = \psi \vee \big([1,1] \wedge (\psi \vee \zeta)\big) = (\psi \vee [1,1]) \wedge (\psi \vee \zeta).$$

Since $[1,1] \circ \zeta = 1$ we conclude that

$$\mathbf{C} \cong \mathbf{D}/\psi \cong \mathbf{D}/(\psi \vee [1,1]) \times \mathbf{D}/(\psi \vee \zeta).$$

But $\mathbf{C}$ is directly indecomposable, so either $\psi \vee \zeta = 1$, in which case $\mathbf{C} \in \mathsf{H}(\mathbf{D}/[1,1]) \subseteq \mathcal{V}_{\mathrm{ab}}$ or $\psi \vee [1,1] = 1$ which implies that $\mathbf{C} \cong \mathbf{D}/(\psi \vee \zeta)$ is a member of $\mathsf{H}(\mathbf{D}/\zeta)$. In this latter case, by Exercise 7.52.5, $\mathbf{C} \cong \prod_{i \in I} \mathbf{B}_i$, for some subset I. Finally, since $\mathbf{C}$ is directly indecomposable, we must have $\mathbf{C} \cong \mathbf{B}_i \in \mathsf{HS}(\mathcal{K})$ for some i, completing the proof. □

We can apply Theorem 7.38 to the subvariety $\mathcal{V}_{\mathrm{ab}}$ in the previous theorem. According to 7.38, for some ring $\mathbf{R}$, there is an algebra $\mathbf{J} \in \mathcal{V}_{\mathrm{ab}}$ and an $\mathbf{R}$-module $\widetilde{J}$ such that $(\mathbf{J} \downarrow \mathcal{V}_{\mathrm{ab}})$ is term-equivalent to $(\widetilde{J} \downarrow {}_{\mathbf{R}}\mathcal{M}od)$. We claim that if any one of the four varieties

$$\mathcal{V}_{\mathrm{ab}}, \quad (\mathbf{J} \downarrow \mathcal{V}_{\mathrm{ab}}), \quad (\widetilde{J} \downarrow {}_{\mathbf{R}}\mathcal{M}od), \quad {}_{\mathbf{R}}\mathcal{M}od$$

is directly representable, then so are they all.

To see this, let $\mathbf{A}$ be a finite member of $\mathcal{V}_{\mathrm{ab}}$. Each $h\colon \mathbf{J} \to \mathbf{A}$ gives rise to $(\mathbf{A}, h) \in (\mathbf{J} \downarrow \mathcal{V}_{\mathrm{ab}})$. Since the algebras $\mathbf{A}$ and $(\mathbf{A}, h)$ are polynomially equivalent, they have the same congruence lattices. Hence $\mathbf{A}$ is directly indecomposable if and only if $(\mathbf{A}, h)$ is directly indecomposable.

Both A and J are finite, so there are only finitely many maps, h, from J to A. Therefore, for each directly indecomposable algebra $\mathbf{A}$, there are only finitely many directly indecomposable algebras $(\mathbf{A}, h)$. Thus, $\mathcal{V}_{\mathrm{ab}}$ is directly representable if and only if $(\mathbf{J} \downarrow \mathcal{V}_{\mathrm{ab}})$ is directly representable.

The same argument applies to the varieties ${}_{\mathbf{R}}\mathcal{M}od$ and $(\widetilde{J} \downarrow {}_{\mathbf{R}}\mathcal{M}od)$. Finally, under a term-equivalence, direct indecomposability is preserved. This proves our claim.

We conclude that condition (3) of Theorem 7.58 is equivalent to a condition on a certain finite ring $\mathbf{R}$. If the variety ${}_{\mathbf{R}}\mathcal{M}od$ is directly representable, then $\mathbf{R}$ is said to be of *finite representation type.* The characterization of these rings is a long-standing open problem in ring theory. It is not hard to see that any semisimple ring has finite representation type. On the other hand, Colby proved in[Col66] (see also [Pie82, Theorem 6.7]) that if an Artinian ring $\mathbf{R}$ has finite representation type, then $\mathrm{Con}(\mathbf{R})$ is distributive. So rings of finite representation type are quite rare.

Example 7.59. Let $\mathbf{B} = \langle B, \cdot \rangle$ be the binar with Cayley table

$\cdot$	0	1	2	3	4
0	0	1	2	3	4
1	1	0	3	4	2
2	2	3	4	0	1
3	3	4	1	2	0
4	4	2	0	1	3

and let $\mathcal{V} = \mathbf{V}(\mathbf{B})$. We shall apply Theorem 7.58 to show that $\mathcal{V}$ is directly representable. A glance at the Cayley table shows that $\mathbf{B}$ is a finite Latin square. Therefore by Exercise 4.80.6, $\mathbf{B}$ is term-equivalent to a quasigroup, hence $\mathcal{V}$ is a Maltsev variety (condition 1). Having prime cardinality, $\mathbf{B}$ must be simple. The only proper nonempty subuniverses of $\mathbf{B}$ are $\{0\}$ and $A = \{0,1\}$. The quasigroup $\mathbf{A}$ is isomorphic to the group of order 2, so it is Abelian (condition 2). Finally by Exercise 7.60.2 below, $\mathcal{V}_{\text{ab}} = \mathbf{V}(\mathbf{A})$ is term-equivalent to ${}_{\mathbb{F}_2}\mathcal{M}od$, the variety of vector spaces over the field $\mathbb{F}_2$ which is directly representable.

Exercise Set 7.60.

1. Prove that each of the following congruence identities (or implications) is equivalent to (C1).
   (a) $\alpha \wedge [\beta, \beta] = [\alpha \wedge \beta,\, \beta]$;
   (b) $[\alpha, \beta] = (\alpha \wedge [\beta, \beta]) \vee (\beta \wedge [\alpha, \alpha])$;
   (c) $\alpha \leq [\beta, \beta] \implies [\alpha, \beta] = \alpha$.

2. Continuation of Example 7.59. You are to prove that $\mathcal{V}_{\text{ab}} = \mathbf{V}(\mathbf{A})$. For this, let $\mathbf{D}$ be an Abelian member of $\mathcal{V}$. We wish to show that $\mathbf{D} \in \mathbf{V}(\mathbf{A})$.
   (a) $\mathbf{B}$ is simple and nonabelian.
   (b) It is enough to show that every finitely generated subalgebra of $\mathbf{D}$ lies in $\mathbf{V}(\mathbf{A})$. So without loss of generality, assume D is finite.
   (c) There is a finite algebra $\mathbf{C}$ and congruence θ such that $\mathbf{D} \cong \mathbf{C}/\theta$ and $\mathbf{C} \leq \mathbf{B}^t$ for some positive integer t.
   (d) $\mathbf{C}$ is a subdirect product of $\mathbf{B}^n \times \mathbf{A}^m$.
   (e) $\mathbf{C}/[1,1] \leq \mathbf{A}^m$ and $\theta \geq [1,1]$, so $\mathbf{D} \in \mathbf{V}(\mathbf{A})$.

3. Let n be a positive integer and let $\mathcal{V}$ be the variety of commutative rings satisfying the identities $x^2y \approx xy$ and $nx \approx 0$ (see Exercise 3.39.1). Prove that $\mathcal{V}$ is directly representable.

4. The congruence identity $[\alpha, \beta] = \alpha \wedge \beta \wedge [1,1]$ is called (C2). Prove that every directly representable variety satisfies (C2). (Hint: use Theorem 7.56.)

7.6 Minimal varieties

A variety is *minimal* if it is nontrivial and has no proper nontrivial subvarieties. Familiar examples include the varieties of distributive lattices, Boolean algebras, Abelian groups of prime exponent, and **F**-vector spaces over any field **F**.

Recall the duality between varieties and equational theories. Under this duality, minimal varieties correspond to maximal proper equational theories. The smallest variety — the variety of trivial algebras—corresponds to the theory containing every possible equation. Thus an equational theory is maximal and proper if and only if the addition of any additional identity allows one to deduce every equation. In analogy with similar notions in first-order logic, both maximal equational theories and minimal varieties are often called *equationally complete.*

The first order of business is to prove the existence of minimal varieties. The following proposition shows that, in particular, it is not possible to have an infinite descending chain of varieties whose limit is trivial.

Proposition 7.61. *Every nontrivial variety contains a minimal subvariety.*

Proof. Let $\mathcal{V}$ be a nontrivial variety of similarity type ρ, and let $\mathbf{T} = \mathbf{T}_\rho(X_\omega)$ be the term algebra. Under the duality we recalled just above, every variety of type ρ corresponds to an equational theory, and the equational theories are the fully invariant congruences on $\mathbf{T}$. The variety $\mathcal{V}$ corresponds to a congruence $\theta \in \mathrm{Con}_{\mathrm{fi}}(\mathbf{T})$ and we seek a maximal member of $\{\, \psi \in \mathrm{Con}_{\mathrm{fi}}(\mathbf{T}) : \theta \leq \psi < 1 \,\}$. Note that a fully invariant congruence is proper if and only if it excludes the pair (x_1, x_2). Using this observation, an easy Zorn's lemma argument shows the existence of the desired congruence. □

Obviously, a minimal variety is generated by any one of its nontrivial members. Furthermore, let $\mathcal{V}$ be any variety containing at least one finite, nontrivial member and let **A** be a finite, nontrivial member of $\mathcal{V}$ of smallest cardinality. Then **A** is *strictly simple,* that is, **A** is finite, simple and has no proper, nontrivial subalgebras. Strictly simple algebras have also called been called *plain.*

Thus any locally finite, minimal variety is generated by a strictly simple algebra. It is reasonable to wonder about the converse: which strictly simple algebras generate minimal varieties? We shall give a complete answer for those algebras that lie in either a congruence-distributive or a congruence-permutable variety.

Proposition 7.62. *Let* $\mathbf{A}$ *be a finite, strictly simple algebra in a congruence-distributive variety. Then* $\mathsf{V}(\mathbf{A})$ *is minimal.*

Proof. By Jónsson's lemma, the subdirectly irreducible algebras in $\mathsf{V}(\mathbf{A})$ are members of $\mathsf{HSP_u}(\mathbf{A})$. Since $\mathbf{A}$ is strictly simple, the only nontrivial member of $\mathsf{HSP_u}(\mathbf{A})$ is $\mathbf{A}$ itself. Thus $\mathsf{V}(\mathbf{A})$ is minimal. □

As examples, both the varieties of distributive lattices and implication algebras (Exercise 4.75.7) are minimal since each is congruence distributive and generated by a two-element algebra, which is obviously strictly simple. Moreover, since every idemprimal algebra satisfies the conditions of the above proposition, it follows from Theorem 6.17 that almost every finite algebra generates a minimal variety.

Now we turn to congruence-permutable varieties. Suppose that $\mathbf{A}$ is a strictly simple algebra lying in such a variety. Then $\mathbf{A}$ is either Abelian or nonabelian. We consider the two alternatives separately.

Theorem 7.63. *Let* $\mathcal{V}$ *be a Maltsev variety and let* $\mathbf{A}$ *be a finite, strictly simple, nonabelian member of* $\mathcal{V}$. *Then* $\mathbf{A}$ *is quasiprimal and* $\mathsf{V}(\mathbf{A})$ *is minimal.*

Proof. Let $\mathcal{W} = \mathsf{V}(\mathbf{A})$ and $\mathbf{F}$ be the free $\mathcal{W}$-algebra on 3 generators. Then $\mathbf{F}$ is a subalgebra of $\mathbf{A}^n$ for some integer n (Exercise 4.34.3). Since $\mathbf{A}$ has no proper nontrivial subalgebras, the embedding of $\mathbf{F}$ into $\mathbf{A}^n$ can be assumed to be subdirect. Since $\mathbf{A}$ is simple and nonabelian, it is neutral. By Theorem 7.51 $\mathbf{F}$ is neutral, hence is congruence-distributive. Therefore, by Theorem 4.66, $\mathcal{W}$ is congruence-distributive, and minimal by Proposition 7.62. Finally, $\mathbf{A}$ is quasi-primal by Theorem 6.10. □

Example 7.64. Let us revisit the binar $\mathbf{Q}$ considered in Example 6.11. At that time we observed that $\mathbf{Q}$ generates a Maltsev variety and is strictly simple. Furthermore, 1 is an identity element. The binar $\mathbf{Q}$ is polynomially equivalent to a simple loop $\langle Q, \cdot, /, \backslash, 1\rangle$. According to Exercise 7.40.4, if this loop were Abelian, then the multiplication would be commutative, which it obviously is not. So the loop, and hence the binar $\mathbf{Q}$, is nonabelian. Therefore by Theorem 7.63, $\mathbf{Q}$ is quasiprimal, and generates a congruence-distributive variety which is minimal.

In order to address strictly simple, Abelian algebras, we need a bit of modular lattice theory. Let $\mathbf{L}$ be a lattice with 1. A *coatom* of $\mathbf{L}$ is an element c such that $c \prec 1$.

Proposition 7.65. *Let* $\mathbf{L}$ *be a finite modular lattice and suppose that* $\bigwedge C = 0$*, where* C *is the set of coatoms of* $\mathbf{L}$*. Then every member of* $\mathbf{L}$ *is a meet of coatoms.*

Proof. Since L is finite, it has both a 0 and a 1. We first show that $\mathbf{L}$ is complemented. Obviously 0 and 1 are complements. Let $x \in L - \{0, 1\}$. Pick $u \in L$ minimal with respect to the property that $x \vee u = 1$. (Such a u exists because the lattice is finite and $x \vee 1 = 1$.) To show that $\mathbf{L}$ is complemented, it is enough to show that $x \wedge u = 0$.

Suppose to the contrary that $x \wedge u > 0$. Since $\bigwedge C = 0$, there is $c \in C$ with $c \not\geq x \wedge u$. We have $\mathbf{I}[c, 1] \searrow \mathbf{I}[u \wedge c, u]$. Since c is a coatom, it follows that u covers $u \wedge c$ (Exercise 2.12.5). By the minimality of u, $x \vee (u \wedge c) < 1$. Since L is finite, every element (except 1) lies below a coatom. Thus there is $c' \in C$ with $x \vee (u \wedge c) \leq c'$. Joining both sides with u: $1 = x \vee u \leq c' \vee u$. From this it follows that $u \not\leq c'$. Now

$$u \wedge c \leq c' \implies u \wedge c \leq u \wedge c' < u$$

and $u \wedge c \prec u$, hence $u \wedge c = u \wedge c'$. But

$$c' \geq x \implies c' \wedge u \geq x \wedge u \implies c \geq c \wedge u \geq x \wedge u$$

which contradicts the choice of c.

Now, to prove the proposition, let $a \in L$, $a \notin \{0, 1\}$. We wish to show that a is a meet of coatoms. Take $b = \bigwedge \{c \in C : a \leq c\}$. Suppose that $a < b$. Let b' denote a complement of b and $w = a \vee b'$. By modularity, $w \wedge b = a$ and $w \vee b = 1$. Since $a \neq b$ we have $w \neq 1$. Hence there is $c \in C$ with $w \leq c$. Finally

$$c \geq w \geq a \implies c \geq b \implies c = 1$$

which is false. Thus $a = b$ as desired. □

Definition 7.66. Let $\mathbf{A}$, $\mathbf{B}$ and $\mathbf{C}$ be algebras. We say that $\mathbf{A}$ and $\mathbf{B}$ are *isotopic via* $\mathbf{C}$ if there is an isomorphism $f\colon \mathbf{C} \times \mathbf{A} \to \mathbf{C} \times \mathbf{B}$ such that $p_1 \circ f = p_1$. In other words, for all $a \in A$ and $c \in C$ there is $b \in B$ such that $f(c, a) = (c, b)$.

Suppose that $\mathbf{A}$ and $\mathbf{B}$ are isotopic via the algebra $\mathbf{C}$. Letting η_1 and η_2 be the projection congruences on $\mathbf{C} \times \mathbf{A}$ and $\theta = \ker(p_2 \circ f)$, we see that

$$\eta_1 \wedge \eta_2 = \eta_1 \wedge \theta = 0 \qquad \text{and} \qquad \eta_1 \circ \eta_2 = \eta_1 \circ \theta = 1 \tag{7–23}$$

(see Figure 7.3). Conversely, if θ is a congruence on $\mathbf{C} \times \mathbf{A}$ satisfying (7–23), then $\mathbf{A}$ is isotopic to $(\mathbf{C} \times \mathbf{A})/\theta$ via $\mathbf{C}$. We can summarize this situation with the phrase "η_1 is a common complement to θ and η_2."

Furthermore, if $\mathrm{Con}(\mathbf{C} \times \mathbf{A})$ is modular, then $\mathbf{I}[\theta, 1] \searrow \mathbf{I}[0, \eta_1] \nearrow \mathbf{I}[\eta_2, 1]$. Consequently, $\mathrm{Con}(\mathbf{A}) \cong \mathrm{Con}(\mathbf{B})$. In particular, if $\mathbf{A}$ is subdirectly irreducible or simple, then so is $\mathbf{B}$.

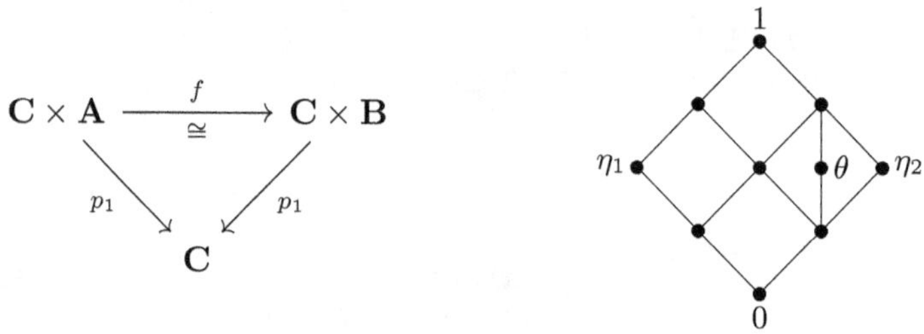

FIGURE 7.3: Isotopy and Con($\mathbf{C} \times \mathbf{A}$)

Observe also that if $\mathbf{A}$ is isotopic to $\mathbf{B}$ via $\mathbf{C}$, and if $\mathbf{C}$ has an idempotent element e, then the mapping $f(e, \_)$ is an isomorphism of $\mathbf{A}$ with $\mathbf{B}$.

Now suppose that $\mathbf{A}$ is an Abelian algebra. Recall that the congruences $\{0, \eta_1, \Delta_1^1, \eta_2, 1\}$ form a sublattice of Con($\mathbf{A}^2$) isomorphic to $\mathbf{M}_3$. Therefore, Δ_1^1 can play the role of θ in the equations (7–23). Writing $\mathbf{A}_\nabla$ for the algebra $\mathbf{A}^2/\Delta_1^1$, we conclude that $\mathbf{A}$ is isotopic to $\mathbf{A}_\nabla$ via $\mathbf{A}$.

Note that the elements of $\mathbf{A}_\nabla$ are the congruence classes of Δ_1^1. One of those classes is the set $\delta_A = \{ (a, a) : a \in A \}$. It is easy to see that this class is always an idempotent element of $\mathbf{A}_\nabla$. Thus we conclude that $\mathbf{A}_\nabla$ always has an idempotent element and furthermore, $\mathbf{A}$ has an idempotent element if and only if $\mathbf{A} \cong \mathbf{A}_\nabla$.

We are now ready to discuss the variety generated by a strictly simple Abelian algebra. First, it follows from Exercise 7.40.2 that every finite simple Abelian algebra is strictly simple. That observation allows us to state our theorem in a stronger form.

Theorem 7.67. *Let $\mathcal{V}$ be a Maltsev variety, $\mathbf{A}$ a finite, simple, Abelian algebra in $\mathcal{V}$, and $\mathcal{W} = \mathsf{V}(\mathbf{A})$.*

(1) *If $\mathbf{A}$ has an idempotent element, then $\mathcal{W}_{\text{fin}} = \{ \mathbf{A}^n : n \in \omega \}$, and $\mathcal{W}$ is a minimal variety.*

(2) *If $\mathbf{A}$ has no idempotent elements, then $\mathcal{W}_{\text{fin}} = \{ \mathbf{A}^n, (\mathbf{A}_\nabla)^n : n \in \omega \}$ and $\mathcal{W}$ has one proper nontrivial subvariety, namely $\mathsf{V}(\mathbf{A}_\nabla)$.*

Proof. We first show that if $\mathbf{B}$ is a finite, subdirectly irreducible member of $\mathcal{W}$ then $\mathbf{B} \in \{\mathbf{A}, \mathbf{A}_\nabla\}$. Since $\mathbf{B} \in \mathsf{HSP}(\mathbf{A})$ and A is finite, there is a positive integer k, a subalgebra $\mathbf{F}$ of $\mathbf{A}^k$, and a congruence θ such that $\mathbf{F}/\theta \cong \mathbf{B}$. Choose $\mathbf{F}$ in order to minimize k. Then since $\mathbf{A}$ is strictly simple, we actually have $\mathbf{F} = \mathbf{A}^k$ (Theorem 6.3).

We proceed by induction on k. If $k = 1$ then $\mathbf{F} = \mathbf{A}$ is simple, thus $\mathbf{B} = \mathbf{F} \in \{\mathbf{A}, \mathbf{A}_\nabla\}$. So assume that $k \geq 2$ and the claim is true for any subdirectly irreducible quotient of $\mathbf{A}^l$ when $l < k$.

Among the coatoms of Con($\mathbf{A}^k$) are the kernels of the projection maps, which must meet to 0. Therefore, by Proposition 7.65, θ is a meet of coatoms. But θ is completely meet-irreducible, so θ must itself be a coatom.

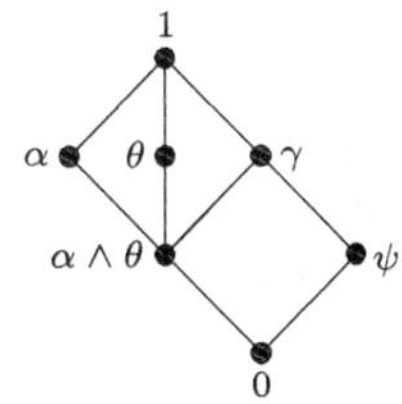

Since $k \geq 2$, $\mathbf{F} \cong \mathbf{A}_\nabla \times \mathbf{A}^{k-1}$. Let α denote the kernel of the projection of $\mathbf{F}$ onto $\mathbf{A}_\nabla$ and let ψ be the kernel onto $\mathbf{A}^{k-1}$. If $\theta = \alpha$ then our claim is certainly true. So assume that $\theta \neq \alpha$. Since both α and θ are coatoms, $\mathbf{I}[\theta, 1] \searrow \mathbf{I}[\alpha \wedge \theta, \alpha]$, so $\alpha \wedge \theta \prec \alpha$. Therefore, since $\mathbf{I}[0, \alpha] \nearrow \mathbf{I}[\psi, 1]$, we deduce that $\gamma = (\alpha \wedge \theta) \vee \psi$ is a coatom of $\mathbf{I}[\psi, 1]$. We conclude that $\{\alpha \wedge \theta, \gamma, \theta, \alpha, 1\}$ forms a sublattice isomorphic to $\mathbf{M}_3$.

Working in the lattice $\mathrm{Con}(\mathbf{F}/(\alpha \wedge \theta))$ this says that $\bar{\alpha}$ is a common complement of $\bar{\theta}$ and $\bar{\gamma}$, where we write $\bar{\alpha}$ in place of $\alpha/(\alpha \wedge \theta)$, etc. Thus $\mathbf{F}/\theta$ is isotopic to $\mathbf{F}/\gamma$ via $\mathbf{F}/\alpha$. But $\mathbf{F}/\alpha \cong \mathbf{A}_\nabla$ which has an idempotent. Therefore $\mathbf{B} \cong \mathbf{F}/\theta \cong \mathbf{F}/\gamma$.

Since γ is a coatom, $\mathbf{F}/\gamma$ is a simple homomorphic image of $\mathbf{F}/\psi \cong \mathbf{A}^{k-1}$. By the induction hypothesis, $\mathbf{B} \cong \mathbf{F}/\gamma \in \{\mathbf{A}, \mathbf{A}_\nabla\}$, proving the claim.

Now we move on to verify the statements in the theorem. It follows from our conclusion above and Theorem 5.7 that $\mathcal{W}_{\mathrm{si}} = \{\mathbf{A}, \mathbf{A}_\nabla\}$. Therefore, by Corollary 3.45, $\mathcal{W}$ has at most 4 possible subvarieties, namely $\mathsf{V}(\varnothing)$ (the trivial variety), $\mathsf{V}(\mathbf{A})$, $\mathsf{V}(\mathbf{A}_\nabla)$ and $\mathsf{V}(\mathbf{A}, \mathbf{A}_\nabla)$. Since $\mathbf{A}_\nabla = \mathbf{A}^2/\Delta_1^1$, we see that $\mathsf{V}(\mathbf{A}_\nabla) \subseteq \mathsf{V}(\mathbf{A}, \mathbf{A}_\nabla) = \mathsf{V}(\mathbf{A}) = \mathcal{W}$.

Thus, if $\mathbf{A}$ has an idempotent, then $\mathbf{A} = \mathbf{A}_\nabla$ and $\mathcal{W}$ is minimal. Every finite algebra must be a finite subdirect power of $\mathbf{A}$. Since $\mathbf{A}$ is strictly simple, a finite subdirect power is in fact direct. This proves (1).

Finally, if $\mathbf{A}$ has no idempotent, then $\mathbf{A} \ncong \mathbf{A}_\nabla$ (since $\mathbf{A}_\nabla$ *does* have an idempotent). In this case $\mathsf{V}(\mathbf{A}_\nabla)$ is the unique proper nontrivial subvariety of $\mathcal{W}$. Any finite algebra is a finite subdirect, hence direct, product of copies of $\mathbf{A}$ and $\mathbf{A}_\nabla$. Using the fact that $\mathbf{A}_\nabla \times \mathbf{A} \cong \mathbf{A} \times \mathbf{A}$ we see that any finite algebra not of the form $(\mathbf{A}_\nabla)^n$ must be isomorphic to $\mathbf{A}^n$ for some n. □

Remarks.

(1) With a bit of ring theory together with Theorem 7.35, one can prove in the above theorem that $|A|$ must be a prime-power. Consequently, the same applies to the cardinality of every finite member of $\mathcal{W}$. See [FM87, Theorem 12.4].

(2) It is immediate from Theorem 7.67 that every finite, simple, Abelian algebra generates a directly representable variety.

Corollary 7.68. *Every minimal, locally finite, Maltsev variety is either arithmetical or Abelian.*

Proof. Apply Theorems 7.63 and 7.67 to the strictly simple generator. □

Without the assumption of congruence-permutability, things are more complex. An assertion similar to Theorem 7.67 for Abelian algebras still holds, when the "Maltsev" assumption is dropped. However, the nonabelian case seems to be quite hard. There is no straightforward characterization of those strictly simple, nonabelian algebras that generate a minimal variety. A thorough analysis of the problem is presented by Kearnes and Szendrei in [KS97].

Of course not every minimal variety is locally finite, although concrete examples are not so easy to come by. The variety of vector spaces over an infinite field is minimal and not locally finite. However, it has infinitely many basic operation symbols, which makes the example less than satisfying. A more interesting example is the "Jónsson-Tarski variety," discussed in Exercise 4 below (although minimality was proved by Maltsev). Not only is $\mathcal{JT}$ not locally finite, it has no finite nontrivial members at all.

Exercise Set 7.69.

1. Another proof of Theorem 7.61. Let $\mathcal{V}$ be a nontrivial variety. Define $\mathbf{P} = \{\, \mathcal{W} \subseteq \mathcal{V} : \mathcal{W} \text{ nontrivial} \,\}$ under inclusion. Apply Zorn's lemma to $\mathbf{P}^\partial$ to obtain a minimal variety. (Hint: suppose that $\langle \mathcal{W}_i : i \in I \rangle$ is a chain in $\mathbf{P}^\partial$. Let $\mathbf{A}_i \in \mathcal{W}_i$ be nontrivial. Consider a nonprincipal ultraproduct of the $\mathbf{A}_i$'s.)

2. Prove that a locally finite variety has only finitely many minimal subvarieties. (Scott, 1956)

3. Revisit Examples 7.37 and 7.39. Show that (in the notation of those exercises) $\mathbf{A} \cong \mathbf{B}_\nabla$. Determine the lattice of subvarieties of $\mathcal{V}$.

4. In this exercise we show that the variety $\mathcal{JT}$ of Exercise 4.34.10 is equationally complete. We follow a proof presented in [AS68].

   For an identity ϵ of similarity type $\langle 2, 1, 1 \rangle$, let $L(\epsilon)$ denote the number of appearances of the binary operation in ϵ. Suppose that $\mathcal{W}$ is a proper subvariety of $\mathcal{JT}$. We shall show that $\mathcal{W}$ is trivial.

   (a) There is an identity ϵ such that $\mathcal{W} \models \epsilon$ and $\mathcal{JT} \not\models \epsilon$. Choose ϵ with $L(\epsilon)$ as small as possible.

   (b) Suppose $L(\epsilon) = 0$. Then ϵ is of the form

   $$f_1 f_2 \cdots f_k(x_i) \approx g_1 g_2 \cdots g_m(x_j)$$

   with $k \geq m$. Assume ϵ is chosen to minimize k. By replacing x_i with $x_i \cdot x_i$ and x_j with $x_j \cdot x_j$ show that $m = 0$.

   (c) Suppose that $i = j$. Let $f' \neq f_k$ and $i' \neq i$. Apply f' to both sides of ϵ and replace x_i with $x_i \cdot x_{i'}$ and we obtain a new identity that holds in $\mathcal{W}$.

   (d) Thus we may assume that $i \neq j$ in ϵ. Substitute $x_i \cdot x_i$ for x_i to conclude that ϵ is the identity $x_i \approx x_j$.

(e) Now suppose that $L(\epsilon) > 0$. Thus ϵ is of the form $s_1 \cdot s_2 \approx t$ for some terms s_1, s_2, t. Then both of the identities $s_1 \approx \ell(t)$ and $s_2 \approx r(t)$ hold in $\mathcal{W}$. If either identity fails to hold in $\mathcal{JT}$, we contradict the minimality of $L(\epsilon)$. On the other hand, if both identities hold, then in $\mathcal{JT}$ we can deduce

$$s_1 \cdot s_2 \approx \ell(t) \cdot r(t) \approx t$$

contradicting our assumption that $\mathcal{JT} \not\models \epsilon$.

5. Suppose that **A**, **B** and **C** lie in a Maltsev variety. Show that if **A** is isotopic to **B** via **C**, then **A** is isotopic to **B** via **C**′, where **C**′ is Abelian.

7.7 Functionally complete algebras

Recall from Definition 6.4 that a finite algebra **A** is primal if every operation on A is induced by a term. As we saw in the subsequent theorem, primal algebras have very strong properties. One consequence of the definition is that a primal algebra must have every possible constant (unary) operation. One might wonder what happens if we bend the rules a bit and throw in a constant symbol for each element of an algebra. This is equivalent to the following.

Definition 7.70. A finite algebra **A** is called *functionally complete* if $\operatorname{Pol} \mathbf{A} = \operatorname{Op}(A)$.

In other words, instead of requiring that every operation be a term, we only require that each be a polynomial. As we shall see, functional completeness is noticeably weaker than primality.

For any algebra $\mathbf{A} = \langle A, F \rangle$, let $\mathbf{A}^+$ denote the algebra with universe A and with basic operations consisting of F together with a constant nullary operation for each member of A. Observe that regardless of the properties of the algebra **A**, $\mathbf{A}^+$ is rigid and has no proper subalgebras. On the other hand since the addition of constant operations does not affect congruences, we have $\operatorname{Con}(\mathbf{A}) = \operatorname{Con}(\mathbf{A}^+)$.

It is immediate from the definition that **A** is functionally complete if and only if $\mathbf{A}^+$ is primal. Since every primal algebra is simple, it follows from the previous paragraph that every functionally complete algebra must be simple as well. However, it is entirely possible for a functionally complete algebra to have subalgebras and automorphisms. It is also possible for a functionally complete algebra to generate a variety that is not even congruence modular. This has made them much harder to characterize than primal algebras.

A useful tool in the study of primal algebras is the discriminator operation, introduced in Definition 6.9. It turns out to play a decisive role here as well.

Proposition 7.71 (Werner, 1970). *A finite algebra is functionally complete if and only if the discriminator operation is a polynomial of the algebra.*

Proof. Let d_A denote the discriminator operation on a set A. If an algebra $\mathbf{A}$ is functionally complete then $\mathbf{A}^+$ is primal, hence $d_A \in \mathrm{Clo}(\mathbf{A}^+) = \mathrm{Pol}(\mathbf{A})$. Conversely, suppose that $d_A \in \mathrm{Clo}(\mathbf{A}^+)$. Then (since A is finite), $\mathbf{A}^+$ is quasiprimal and has no proper subalgebras or nontrivial automorphisms. Therefore by Theorems 6.10 and 6.6, $\mathbf{A}^+$ is primal, hence $\mathbf{A}$ is functionally complete. □

As an obvious application of this proposition, every quasiprimal algebra is functionally complete. However, as we now show, there are numerous other examples.

Within Maltsev varieties, functionally complete algebras are easily recognized. This was first discovered in the case of groups by Maurer and Rhodes (1965) and generalized to an arbitrary Maltsev variety by Werner [Wer74]. In its present form, the result is due to McKenzie.

Theorem 7.72. *Let $\mathcal{V}$ be a Maltsev variety. A finite member of $\mathcal{V}$ is functionally complete if and only if it is simple and nonabelian.*

Proof. For any algebra $\mathbf{A}$, $\mathrm{Con}(\mathbf{A}^2) = \mathrm{Con}((\mathbf{A}^+)^2)$. Thus by Theorem 7.30

$$\mathbf{A} \text{ is Abelian} \iff \mathrm{Con}(\mathbf{A}^2) \text{ has a spanning } \mathbf{M}_3 \iff$$
$$\mathrm{Con}((\mathbf{A}^+)^2) \text{ has a spanning } \mathbf{M}_3 \iff \mathbf{A}^+ \text{ is Abelian.}$$

Now suppose that $\mathbf{A} \in \mathcal{V}$ and $\mathbf{A}$ is functionally complete. Then as we have observed, $\mathbf{A}$ is simple and $\mathbf{A}^+$ is primal. Therefore $\mathrm{Con}((\mathbf{A}^+)^2)$ is a distributive lattice, so it surely has no spanning $\mathbf{M}_3$. Hence $\mathbf{A}$ is nonabelian.

Conversely let $\mathbf{A} \in \mathcal{V}$ be finite, simple and nonabelian. Then $\mathbf{A}^+$ is strictly simple. Therefore by Theorem 7.63, $\mathbf{A}^+$ is quasiprimal, so by Proposition 7.71, $\mathbf{A}$ is functionally complete. □

Theorem 7.72 immediately yields a host of familiar examples of functionally complete algebras. Every finite, simple, nonabelian group for instance. A finite ring is functionally complete if and only if it is simple and is not a zero-ring. Any finite, simple member of an arithmetical variety is functionally complete. This would include, in addition to quasiprimal algebras discussed earlier, the r-lattice $\mathbf{MO}_2$ presented on page 190.

Once we leave the domain of Maltsev varieties, no simple characterization of functional completeness is known. We showed in Example 4.8(4) that the two-element distributive lattice is not functionally complete. Contrast that fact with the following example.

Example 7.73. Consider the 2-element implication algebra, $\mathbb{B}_1^{\rightarrow}$, of Exercise 4.45.6. You proved in Exercise 4.75.7 that $\mathbb{B}_1^{\rightarrow}$ does not generate a Maltsev variety (so it is certainly not primal), nor does it have a majority term.

However, $\mathbb{B}_1^{\rightarrow}$ is functionally complete. To see this observe that $x' = x \rightarrow 0$ is the complementation operation on $\{0, 1\}$. Consequently

$$x \vee y = (x \rightarrow y) \rightarrow y \text{ and } x \wedge y = x' \vee y'$$

are polynomials on $\mathbb{B}_1^{\rightarrow}$. Thus $\mathrm{Pol}(\mathbb{B}_1^{\rightarrow})$ contains the Boolean algebra operations, hence $\mathbb{B}_1^{\rightarrow}$ is functionally complete.

Recall that $\mathbb{B}_1^{\rightarrow}$ does generate a congruence distributive variety (Exercise 4.75.7 again) and is therefore simple and nonabelian. Thus this algebra demonstrates the necessity of the "Maltsev" assumption of Theorem 7.72 in a very strong way. Notice that $\{1\}$ is a proper (albeit trivial) subuniverse of $\mathbb{B}_1^{\rightarrow}$. The following very striking theorem shows the degree to which the existence of proper subalgebras dilutes the strength of functional completeness.

Theorem 7.74 (Kaarli, Szendrei, 1992). *Let* $\mathbf{A}$ *be a finite functionally complete algebra with no proper subalgebras. Then* $\mathbf{V}(\mathbf{A})$ *is arithmetical. In fact,* $\mathbf{A}$ *is quasiprimal.*

In this theorem, quasiprimality follows immediately from Theorem 6.10. Let $\mathbf{A}$ be an algebra satisfying the assumption in the statement of the theorem. We shall find a Pixley term for $\mathbf{A}$. The proof requires two lemmas.

Lemma 7.75. *Let* C *be a nonempty subuniverse of* $\mathbf{A}^2$. *Then either* C *is the graph of an automorphism of* $\mathbf{A}$ *or there is an element* u *in* A *such that* $(u, u) \in C$.

Proof. Suppose that C is not an automorphism. Then there are a_1', a_2', b' in A such that (a_1', b'), $(a_2', b') \in C$. Choose n so that $2^n \geq |A|$. Let $f\colon A^n \rightarrow A$ be any function such that $\vec{f}(\{a_1', a_2'\}^n) = A$. Since $\mathbf{A}$ is functionally complete, there is a term t and $a_1, \ldots, a_m \in A$ such that

$$(\forall x_1, \ldots, x_n \in \{a_1', a_2'\})\ f(x_1, \ldots, x_n) = t^{\mathbf{A}}(x_1, \ldots, x_n, a_1, \ldots, a_m).$$

Since $\mathbf{A}$ has no proper subalgebras, $\mathbf{C}$ must be a subdirect square of $\mathbf{A}$. Thus there are $b_1, \ldots, b_m \in A$ with $c_i = (a_i, b_i) \in C$ for $i = 1, \ldots, m$. Let us define

$$g(x_1, \ldots, x_n) = t^{\mathbf{C}}(x_1, \ldots, x_n, c_1, \ldots, c_m) \in \mathrm{Pol}(\mathbf{C})$$

and set $u = t^{\mathbf{A}}(b', b', \ldots, b', b_1, \ldots, b_m) \in A$. From the surjectivity assumption on f, there are $v_1, v_2 \ldots, v_n \in \{a_1', a_2'\}$ such that $f(v_1, v_2, \ldots, v_n) = u$. Since $v_i \in \{a_1', a_2'\}$, our initial assumption on C implies that $(v_i, b') \in C$ for $i = 1, \ldots, n$. Thus computing in $\mathbf{C}$:

$$\begin{aligned} g\big((v_1, b'), \ldots, (v_n, b')\big) &= t^{\mathbf{C}}\big((v_1, b'), \ldots, (v_n, b'), (a_1, b_1), \ldots, (a_m, b_m)\big) = \\ &\big(t^{\mathbf{A}}(v_1, \ldots, v_n, a_1, \ldots, a_m),\ t^{\mathbf{A}}(b', \ldots, b', b_1, \ldots, b_m)\big) = \\ &\big(f(v_1, \ldots, v_n), t^{\mathbf{A}}(b', \ldots, b', b_1, \ldots, b_m)\big) = (u, u). \end{aligned}$$

□

Lemma 7.76. *Let s be an element of* $\mathrm{Clo}_1(\mathbf{A})$ *whose range has minimal cardinality. Then for every pair* $a, b \in \vec{s}(A)$ *there is* $\alpha \in \mathrm{Aut}(\mathbf{A})$ *such that* $\alpha(a) = b$.

Proof. Suppose not. Let $a, b \in \vec{s}(A)$ be a counterexample. Consider the subalgebra $C = \mathrm{Sg}^{\mathbf{A}^2}\{(a,b)\}$. Then C is not an automorphism, so by Lemma 7.75, there is $u \in A$ such that $(u,u) \in C$. Therefore, there is a term $t \in \mathrm{Clo}_1(\mathbf{A})$ such that $t(a) = t(b) = u$. But then $\left|\overrightarrow{ts}(A)\right| < |\vec{s}(A)|$, which is a contradiction. □

Proof of Theorem 7.74. Let $f(x,y,z)$ be any Pixley operation on A. (By this we mean a ternary operation satisfying the equations in Theorem 4.70(c).) Since $\mathbf{A}$ is functionally complete, there is a term t and elements $b_1, \dots, b_m$ in A such that $f(x,y,z) = t^{\mathbf{A}}(x,y,z,b_1,\dots,b_m)$. Choose $s \in \mathrm{Clo}_1(\mathbf{A})$ as in Lemma 7.76 and pick $a \in \vec{s}(A)$. Since $\mathbf{A}$ has no proper subalgebras, for every $i \le m$ there is a unary term r_i such that $r_i^{\mathbf{A}}(a) = b_i$. Let

$$p(x,y,z,w) = t(x,y,z,r_1(w),\dots,r_m(w)).$$

Then $f(x,y,z) = p^{\mathbf{A}}(x,y,z,a)$.

We claim that $q(x,y,z) = p(x,y,z,s(x))$ is a Pixley term for $\mathbf{A}$. q is certainly a term. We verify the identities of Theorem 4.70. Pick $b, c \in A$. By Lemma 7.76, there is $\alpha \in \mathrm{Aut}(\mathbf{A})$ such that $\alpha(a) = s(b)$. Let $b' = \alpha^{-1}(b)$ and $c' = \alpha^{-1}(c)$. Then

$$q(b,b,c) = p(b,b,c,s(b)) = p(b,b,c,\alpha(a)) = \\ \alpha\big(p(b',b',c',a)\big) = \alpha\big(f(b',b',c')\big) = c$$

since f was assumed to be a Pixley operation. The verification of the other two identities is similar. □

The results of this section leave open the question of whether every functionally complete algebra must generate a variety that is either congruence-distributive or congruence-permutable. The answer is no. An example is sketched in Exercise 6.

As a generalization of primality, functional completeness seems to be a relatively weak condition. Besides simplicity, it has no completely general consequences. One can weaken the condition still further by allowing congruences in the following manner.

Definition 7.77. An algebra $\mathbf{A}$ is *affine complete* if $\mathrm{Pol}(\mathbf{A}) = \mathcal{F}(\mathrm{Con}(\mathbf{A}))$. A variety is affine complete if every member is affine complete.

Thus a finite algebra is functionally complete if and only if it is simple and affine complete. Finite affine complete algebras are not hard to find, see for example Exercise 7.

The study of affine complete varieties has proved to be quite rich. Every such variety is congruence-distributive and has no infinite subdirectly irreducible members. Any arithmetical variety generated by a finite algebra with no proper subalgebras is affine complete. A detailed presentation of this topic is given in the monograph by Kaarli and Pixley [KP01].

Exercise Set 7.78.

1. Let $\mathbf{A}$ be a functionally complete algebra of cardinality k. Show that for every $n \geq k$, $k^{k^{(n-k)}} \leq \left|\mathbf{F}_{\mathbf{V}(\mathbf{A})}(n)\right| \leq k^{k^n}$. (Berman)

2. Let $\mathbf{A}$ be finite and functionally complete. Prove that $\mathbf{A}$ is primal if and only if every subalgebra of $\mathbf{A} \times \mathbf{A}$ contains δ_A. (Hint: consider a unary term of minimal range as in Lemma 7.76.) (Kaarli and Pixley [KP01])

3. Prove that a finite algebra is functionally complete if and only if it is simple and has a Pixley polynomial.

4. Let $\mathbf{A}$ be a finite algebra with more than 2 elements.

   (a) Assume the dual-discriminator operation (see Exercise 6.15.10) is a polynomial of $\mathbf{A}$. Suppose that S is a subalgebra of $\mathbf{A}^2$ containing $\delta_A \cup \{(a, b)\}$ for some $a \neq b$. Show that

   $$\begin{gathered}(\forall x \in A)\ \ (a, x) \in S \\ (\forall x \in A)(\forall y \neq a)\ \ (x, y) \in S \\ (\forall x \in A)\ \ (x, a) \in S.\end{gathered}$$

   Conclude that $S = A^2$.

   (b) Prove that $\mathbf{A}$ is functionally complete if and only if the dual-discriminator is a polynomial of $\mathbf{A}$. (Hint: apply Exercise 6.15.10 to $\mathbf{A}^+$.)

5. Consider a 3-person tournament with players $\{a, b, c\}$. Suppose that a beats b, b beats c and c beats a. Define the algebra $\mathbf{T}_3 = \langle \{a, b, c\}, \wedge, \vee \rangle$ in which $x \wedge y$ is the loser and $x \vee y$ the winner of the match between x and y. Furthermore, define $x \wedge x = x \vee x = x$.

   Show that $u(x, y, z) = \big((z \wedge (x \wedge y)) \vee (x \vee y)\big) \wedge \big(z \vee (x \wedge y)\big)$ is a dual-discriminator term for $\mathbf{T}_3$. Conclude that $\mathbf{T}_3$ is functionally complete. (Fried and Pixley, 1979)

6. Let $A = \{a, b, c\}$. Recall that d_A denotes the discriminator on A. Define

operations on A by

$$f(x,y,z,u) = \begin{cases} d_A(x,y,z) & \text{if } u = b \\ a & \text{otherwise} \end{cases}$$

$$g(x) = \begin{cases} c & \text{if } x = b \\ a & \text{if } x \neq b \end{cases} \qquad h(x) = \begin{cases} b & \text{if } x = c \\ a & \text{if } x \neq c \end{cases}$$

Let $\mathbf{A} = \langle A, f, g, h \rangle$. Show that $\mathbf{A}$ is functionally complete but that $\mathsf{V}(\mathbf{A})$ is neither congruence-distributive nor congruence-permutable. (Hint: Let $C_1 = \{(a,a),(a,b),(a,c),(b,a),(c,a)\}$, $C_2 = \{(b,c)\}$ and $B = C_1 \cup C_2$. Then $\vec{f}(B^4), \vec{g}(B), \vec{h}(B) \subseteq C_1$. Thus B is a subuniverse of $\mathbf{A}^2$ and there is a congruence relation, θ, on $\mathbf{B}$ such that $B/\theta = \{C_1, C_2\}$. Therefore $\mathbf{B}/\theta$ is a 2-element set with only a constant operation. Such an algebra can not be a member of a congruence-distributive or congruence-permutable variety.) (Kaarli, 1992)

7. Prove that every finite algebra in an arithmetical variety is affine complete.

Chapter 8

Finite Algebras and Locally Finite Varieties

Throughout this text we have considered numerous results about finite algebras. So it might seem strange to conclude with a chapter devoted to finite algebras. However, our emphasis here is different. In the previous chapters we studied the influence of the global properties of the congruence lattices (distributivity, permutability, etc.) on the behavior of the algebras. In this chapter we shall investigate the interplay between the polynomial operations and the individual congruences (actually, pairs of congruences) of an algebra and what it tells us about structure.

One prototype of this approach is the commutator, which we studied in Sections 7.2 through 7.4. The relation "α centralizes β" does not depend solely on the roles of α and β in the congruence lattice, but in addition involves the particular term operations on the algebra.

Most of the tools needed for this analysis were developed by D. Hobby and R. McKenzie in the late 1980s, and presented in their monograph [HM88]. These techniques are referred to collectively as "tame congruence theory." The theory was refined and extended by several mathematicians, including J. Berman, P. Idziak, K. Kearnes, E. Kiss, A. Szendrei, M. Valeriote and R. Willard.

8.1 Minimal algebras

We take as our starting point Theorem 4.16, which says that for any algebra $\mathbf{A} = \langle A, F \rangle$,

$$\mathrm{Con}(\mathbf{A}) = \mathrm{Con}(\langle A, \mathrm{Pol}_1(\mathbf{A}) \rangle). \tag{8–1}$$

We emphasize that this is not merely a lattice isomorphism, but an equality. Recall that $\mathrm{Pol}(\mathbf{A})$ is the clone on A generated by the operations in F together with all of the constant (nullary) operations on A. $\mathrm{Pol}_1(\mathbf{A})$ consists of the unary members of $\mathrm{Pol}(\mathbf{A})$.

The significance of Equation (8–1) is that unlike a clone, which is a very complicated object, $\mathrm{Pol}_1(\mathbf{A})$ is a familiar structure, namely, a monoid. If one is interested in identifying properties of an algebra that are induced by its congruence lattice, it makes sense to start by investigating its monoid of unary polynomials.

Now the nicest monoids are, of course, groups. What can we say about an algebra **A** if $\mathrm{Pol}_1(\mathbf{A})$ is a group? Well, that's a trick question—there aren't any. This is because $\mathrm{Pol}_1(\mathbf{A})$ contains all of the unary constant operations, which obviously have no inverses. By relaxing the requirement a bit we arrive at the following definition.

Definition 8.1. A finite nontrivial algebra **A** is called *minimal* if every nonconstant member of $\mathrm{Pol}_1(\mathbf{A})$ is invertible.

We warn the reader that the use of the word minimal here is unrelated to that of Section 7.6. The basis for the terminology will become clear as we proceed.

What examples do we know of minimal algebras? First, any two-element algebra is minimal, just on cardinality grounds. As a second construction, let G be any collection of permutations on the finite set A. Then $\langle A, G\rangle$ is minimal since the nonconstant unary polynomials will be precisely the members of the group generated by G.

Thirdly, let **F** be a finite field and **V** a finite-dimensional vector space over **F**. Arguing in a manner similar to Example 4.8, the polynomials of **V** are all operations of the form $a_1x_1 + a_2x_2 + \cdots + a_nx_n + b$ where $a_1, \dots, a_n \in F$ and $b \in V$. Since every nonzero element of **F** is invertible, it follows that **V** is minimal.

In 1984, P. P. Pálfy discovered the remarkable fact that the above two paragraphs constitute a complete inventory of the minimal algebras. In order to prove this, we first establish a couple of simple combinatorial relationships.

Let $f\colon A_1 \times \cdots \times A_n \to A$ be a function. Recall that f *depends on* x_i if there are $a_1, a_2, \dots, a_n, a_i'$ such that

$$f(a_1, \dots, a_{i-1}, a_i, a_{i+1}, \dots, a_n) \neq f(a_1, \dots, a_{i-1}, a_i', a_{i+1}, , \dots, a_n).$$

We say that f is *essentially unary* if it depends on x_i for a unique index i. Similarly, f is essentially k-ary if it depends on exactly k distinct variables. Finally we say that the algebra $\langle A, F\rangle$ is essentially unary if every member of F is essentially unary. Of course we generally use these notions in reference to operations on a set A, but there is no reason they can not be applied to arbitrary mappings.

Lemma 8.2. *Let $f\colon A_1 \times \cdots \times A_n \to A$ with $n > 1$ and assume that f is essentially n-ary. Then there are indices $i < j$ and $c_i \in A_i$, for $i = 1, \dots, n$, such that the binary mapping $f(c_1, \dots, x_i, \dots, x_j, \dots, c_n)$ depends on both x_i and x_j.*

Proof. By induction on n. If $n = 2$ there is nothing to prove. So assume that $n \geq 3$ and the claim holds for all functions of rank less than n. Since f depends on x_3, there are $a_1, \dots, a_n, a_3'$ such that

$$f(a_1, a_2, a_3, \dots, a_n) \neq f(a_1, a_2, a_3', \dots, a_n). \tag{8–2}$$

Consider the $(n-1)$-ary mapping $f(a_1, x_2, \ldots, x_n)$. Inequation (8–2) implies that this function still depends on x_3. If it also depends on x_2 then we can apply the induction hypothesis to complete the proof.

So suppose not. This means that for any $b_3, b_4, \ldots, b_n$, $f(a_1, x_2, b_3, \ldots, b_n)$ is constant. On the other hand, the original function f does depend on x_2. Thus there are $c_1, c_3 \ldots, c_n$ such that $f(c_1, x_2, c_3, \ldots, c_n)$ is not constant. So surely there is c_2 so that $f(a_1, c_2, c_3, \ldots, c_n) \neq f(c_1, c_2, c_3, \ldots, c_n)$. From this we conclude that $f(x_1, x_2, c_3, \ldots, c_n)$ depends on both x_1 and x_2. □

Corollary 8.3. *If the algebra* $\mathbf{A}$ *is not essentially unary then* $\mathrm{Pol}_2(\mathbf{A})$ *contains an operation that depends on both variables.*

Lemma 8.4. *Let* $\mathbf{M}$ *be a minimal algebra of cardinality greater than 2, and suppose that* $f(x, y)$ *is a polynomial of* $\mathbf{M}$ *that depends on both variables. Then* $\langle M, f \rangle$ *is a Latin square.*

Proof. Let us write $x \cdot y$ in place of $f(x, y)$. For every $a \in M$, define $L_a(x) = a \cdot x$ (left multiplication by a). Similarly, $R_a(x) = x \cdot a$. We must show that for every a, the polynomials L_a and R_a are permutations. We shall also use the unary polynomial $g(x) = x \cdot x$.

By assumption, f depends on its second variable. Thus there are elements a, y_1, y_2 in M such that $a \cdot y_1 \neq a \cdot y_2$, which is to say, L_a is not constant. Since $\mathbf{M}$ is minimal, it follows that L_a is a permutation. Similarly, since f depends on its first variable, there is some $d \in M$ such that R_d is a permutation.

Suppose, for the purpose of proof by contradiction, that for some $b \in M$, the map L_b is not a permutation. Then L_b is constant. That is, there is $w \in M$ such that for every x, $L_b(x) = w$. Since L_a is a permutation, there is some $c \in M$ such that $L_a(c) = w$ and surely $a \neq b$. Thus

$$a \cdot c = w = b \cdot c$$

from which it follows that R_c must be constant with value w. But now

$$g(c) = c \cdot c = w = b \cdot b = g(b)$$

so g must have constant value w as well. Therefore

$$a \cdot a = g(a) = w = a \cdot c$$

so $a = c$ since L_a is injective.

Finally, choose $e \in M - \{a, d\}$. Then $e \cdot a = w = g(e) = e \cdot e$, so L_e is the constant w. But then $e \cdot d = w = g(d) = d \cdot d$ which contradicts the assumption that R_d is a permutation. □

This brings us to the promised description of minimal algebras.

Theorem 8.5 (Pálfy [Pál84]). *Let* $\mathbf{M}$ *be a minimal algebra of cardinality greater than 2. Either* $\mathbf{M}$ *is essentially unary or it is polynomially equivalent to a finite-dimensional vector space over a finite field.*

Proof. Assume that $\mathbf{M}$ is not essentially unary. By Corollary 8.3, it has a binary polynomial depending on both variables. Then by Lemma 8.4 and Exercises 4.80.6 and 4.75.1, $\mathbf{M}$ has a Maltsev *polynomial*, q. Let us once again write $\mathbf{M}^+$ for the algebra $\mathbf{M}$ expanded to include a nullary operation for each element of the universe. Thus q is a Maltsev *term* on the algebra $\mathbf{M}^+$. Our objective is to show that $\mathbf{M}^+$ is an Abelian algebra and apply Theorem 7.35. However, this requires a bit of work.

We first show that for any positive integer n, $f \in \mathrm{Pol}_{n+1}(\mathbf{M})$, $\mathbf{a}, \mathbf{b} \in M^n$ and $c, d \in M$

$$f(\mathbf{a}, c) = f(\mathbf{a}, d) \implies f(\mathbf{b}, c) = f(\mathbf{b}, d). \tag{8–3}$$

Suppose first that $n = 1$, so f is binary. If f depends on only one of its variables, the implication is trivially true. On the other hand, if f depends on both variables, we see that $\langle M, f\rangle$ is a Latin square, by Lemma 8.4. In that case, $f(a, c) = f(a, d) \implies c = d \implies f(b, c) = f(b, d)$.

For the general case, we operate variable by variable. Thus

$$\begin{aligned}
f(a_1, a_2, \ldots, a_n, c) = f(a_1, a_2, \ldots, a_n, d) &\implies \\
f(b_1, a_2, \ldots, a_n, c) = f(b_1, a_2, \ldots, a_n, d) &\implies \\
f(b_1, b_2, \ldots, a_n, c) = f(b_1, b_2, \ldots, a_n, d) &\implies \\
\vdots \\
f(b_1, b_2, \ldots, b_n, c) = f(b_1, b_2, \ldots, b_n, d).
\end{aligned}$$

In this derivation, the "$n = 1$ case" is applied to the first implication using the binary polynomial $g_1(x, y) = f(x, a_2, a_3, \ldots, a_n, y)$, to the second implication using $g_2(x, y) = f(b_1, x, a_3, \ldots, a_n, y)$, etc.

Notice that the implication in (8–3) is weaker than the Abelian condition (7–10). However, it turns out to be strong enough to recreate the group operations defined in Theorem 7.34, but with a slightly different proof. Just as before we fix an element 0 in M and define $x + y = q(x, 0, y)$ and $-x = q(0, x, 0)$. Define the polynomials

$$\begin{aligned}
f_1(x, y, z, u) &= q\big(q(x, 0, u),\ 0,\ q(y, u, z)\big) \\
f_2(x, u) &= q(x, u, q(u, x, 0)) \\
f_3(x, y, u) &= q(u, 0, q(x, u, y)).
\end{aligned}$$

Applying (8–3) to $f_1(0, b, 0, b) = f_1(0, b, 0, 0)$ we obtain

$$(a + b) + c = f_1(a, b, c, b) = f_1(a, b, c, 0) = a + (b + c).$$

Similarly, from $f_2(0, a) = f_2(0, 0)$ we deduce $a + (-a) = f_2(a, 0) = f_2(a, a) = 0$ and therefore $f_3(0, 0, b) = f_3(0, 0, 0)$ implies $b + a = f_3(a, b, b) = f_3(a, b, 0) = a + b$. Thus $\langle M, +, -, 0\rangle$ is an Abelian group.

We now strengthen (8–3). For every $n > 0$, $f \in \mathrm{Pol}_{n+1}(\mathbf{M})$, $a \in M$ and $\mathbf{c} \in M^n$ we have

$$f(a, \mathbf{c}) = f(a, \mathbf{0}) + f(0, \mathbf{c}) - f(0, \mathbf{0}) \tag{8–4}$$

where we use $\mathbf{0}$ to denote a sequence of 0's of sufficient length to fill out the function. We verify this relationship by induction on n. Consider the base case, $n = 1$. Let $g(x_1, x_2, y) = f(x_1, y) - f(x_2, y)$. Then $g(0, 0, c) = g(0, 0, 0)$ so by (8–3), $g(a, 0, c) = g(a, 0, 0)$. This is equivalent to equation (8–4).

Assume the equation holds for polynomials of rank less than $n+1$. Define $g(x, y_2, y_3, \ldots, y_n) = f(x, c_1, y_2, \ldots, y_n)$. Applying the induction hypothesis

$$g(a, c_2, \ldots, c_n) = g(a, \mathbf{0}) + g(0, c_2, \ldots, c_n) - g(0, \mathbf{0})$$

from which we obtain

$$\begin{aligned} f(a, \mathbf{c}) = f(a, c_1, \mathbf{0}) + f(0, \mathbf{c}) - f(0, c_1, \mathbf{0}) = \\ \big(f(a, \mathbf{0}) + f(0, c_1, \mathbf{0}) - f(0, \mathbf{0})\big) + f(0, \mathbf{c}) - f(0, c_1, \mathbf{0}) = \\ f(a, \mathbf{0}) + f(0, \mathbf{c}) - f(0, \mathbf{0}) \end{aligned}$$

where we apply the "$n = 1$ case" at the second equality.

Recall that $\text{Clo}(\mathbf{M}^+) = \text{Pol}(\mathbf{M})$. It follows easily from equation (8–4) that $\mathbf{M}^+$ is an Abelian algebra since

$$f(a, \mathbf{c}) = f(a, \mathbf{d}) \implies f(0, \mathbf{c}) = f(0, \mathbf{d}) \implies f(b, \mathbf{c}) = f(b, \mathbf{d}).$$

Finally from Theorem 7.35 we deduce that $\mathbf{M}^+$ is a unital module over a finite ring $\mathbf{R}$. The minimality of $\mathbf{M}$ requires that every nonzero member of $\mathbf{R}$ be invertible. Thus $\mathbf{R}$ is a finite division ring, hence, by Wedderburn's theorem [Her94, Theorem 3.1.1], a finite field.

□

In order to complete our characterization of minimal algebras, we must enumerate the two-element algebras, up to polynomial equivalence. This was first accomplished by Post [Pos41], who described the entire lattice of clones on $\{0, 1\}$. We shall present a self-contained proof that is just complete enough for our current purposes.

For this discussion we shall write 2 to represent $\{0, 1\}$, "+" will denote addition mod 2 and $r(x) = x'$ denotes the complement of $x \in \{0, 1\}$. The symbols 0 and 1 do double-duty denoting both elements and nullary operations. We define seven sets of operations on 2:

$$F_0 = \{0, 1\}, \quad F_1 = \{r, 0, 1\}, \quad F_2 = \{+, 0, 1\}, \quad F_3 = \{\wedge, \vee, r, 0, 1\},$$
$$F_4 = \{\wedge, \vee, 0, 1\}, \quad F_5 = \{\wedge, 0, 1\}, \quad F_6 = \{\vee, 0, 1\}.$$

We write $\mathcal{C}_i = \text{Clo}^2(F_i)$ and $\mathbf{E}_i = \langle 2, F_i \rangle$, for $0 \le i \le 6$.

It is easy to see that these seven clones are ordered as in Figure 8.1, using the observation that $r(x) = x + 1$. It is also not hard to see that these clones are pairwise-distinct. As we have observed on several occasions, $\mathcal{C}_3 = \text{Op}(2)$.

Theorem 8.6. *There are seven distinct clones on $\{0, 1\}$ containing the constant operations. The lattice they form is illustrated in Figure 8.1.*

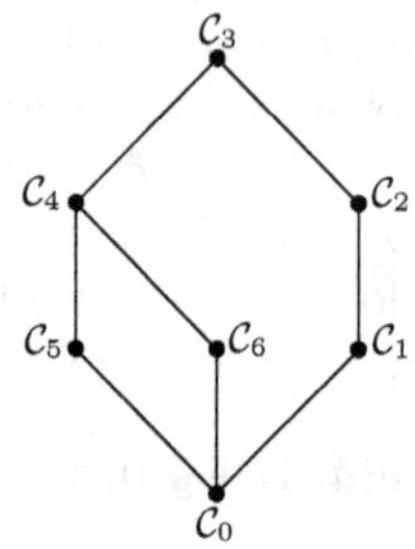

FIGURE 8.1: The lattice of clones on $\{0,1\}$ that contain the constants

Proof. Let $\mathcal{D}$ be a clone on 2 containing 0 and 1. Based on our observations in the previous paragraph, it is enough to show that $\mathcal{D} = \mathcal{C}_i$ for some $i < 7$. Since there are only two nonconstant unary operations on $\{0,1\}$, if $\mathcal{D}$ is essentially unary then $\mathcal{D} \in \{\mathcal{C}_0, \mathcal{C}_1\}$. So let us assume that $\mathcal{D}$ is not essentially unary.

Claim 1. $\mathcal{D}$ contains at least one of the operations "$\wedge$," "$\vee$," or "$+$."

Proof. By Corollary 8.3, $\mathcal{D}$ contains a binary operation f depending on both variables. Let us consider the unary operation $f(x,x)$ and what it says about the multiplication table of f. The fact that f depends on both variables tells us that the table contains at least one nonconstant row and one nonconstant column.

Suppose first that $f(x,x) = x$. Then the multiplication table contains $0,1$ on the diagonal. Since f depends on both variables we must have either $0,0$ or $1,1$ on the off-diagonal entries. In the first case we conclude $f(x,y) = x \wedge y \in \mathcal{D}$ and in the second $f(x,y) = x \vee y \in \mathcal{D}$.

Now suppose that $f(x,x) = x'$. Just as before, the off-diagonal entries must be either $0,0$ or $1,1$. But this time we conclude that $f(x,y)$ is either $(x \vee y)'$ or $(x \wedge y)'$. Either of these operations generates the clone $\mathrm{Op}(2)$, so $\mathcal{D}$ contains all three of $\{\wedge, \vee, +\}$.

Thirdly, suppose that $f(x,x) = 1$. Now there are three possibilities for the off-diagonal entries, namely $0,0$; $0,1$; and $1,0$. The first of these yields the table for the operation $x + y + 1$. Thus $x + y = f(x,y) + 1 \in \mathcal{D}$. The other two cases are the tables for $x \to y$ and $y \to x$. Since $x \vee y = (x \to y) \to y$ we conclude that $\vee \in \mathcal{D}$. (In fact, since $x' = x \to 0$ we can actually conclude that $\mathcal{D} = \mathrm{Op}(2)$.) Since the fourth case, $f(x,x) = 0$ is dual to the third, the proof of Claim 1 is complete.

Claim 2. If $\mathcal{D} \supset \mathcal{C}_2$ then $\mathcal{D} = \mathcal{C}_3$.

Proof. Let $g \in \mathcal{D} - \mathcal{C}_2$. The algebra $\mathbf{A} = \langle 2, +, g \rangle$ lies in a Maltsev variety because of '$+$'. If $\mathbf{A}$ were Abelian, it would be polynomially equivalent to a faithful $\mathbf{R}$-module for some ring $\mathbf{R}$. Since $\langle 2, + \rangle$ has only two endomorphisms, we would have $\mathbf{R} \cong \mathbb{F}_2$ and $\mathbf{A}$ polynomially equivalent to $\langle 2, + \rangle$. But then $g \in \mathrm{Pol}(\langle 2, + \rangle)$, contrary to our assumption.

Thus $\mathbf{A}$ is a simple, nonabelian algebra in a Maltsev variety. By Theorem 7.72, **A** is functionally complete. Since $\mathcal{D}$ contains $\{+, g, 0, 1\}$, we must have $\mathcal{D} = \mathcal{C}_3$.

Claim 3. If $\mathcal{D} \not\subseteq \mathcal{F}(\leq)$ then $r \in \mathcal{D}$.

Proof. Recall that $\mathcal{F}(\leq)$ is the set of isotone operations on 2. Suppose that $g \in \mathcal{D}$ and g is not isotone. Since the only unary operation that is not isotone is r, we may as well assume that g is n-ary with $n > 1$. There must be n-tuples $\mathbf{a}$ and $\mathbf{b}$ with $\mathbf{a} < \mathbf{b}$, $g(\mathbf{a}) = 1$ and $g(\mathbf{b}) = 0$. By reordering the coordinates, we can assume there are indices k and l with $0 \leq l < k \leq n$ such that $\mathbf{a}$ is a string of k zeros followed by $n - k$ ones while $\mathbf{b}$ contains l zeros followed by $n - l$ ones.

Let $h(x) = g(0, \ldots, 0, x \ldots, x, 1 \ldots, 1)$ with l-many zeros, $(k-l)$-many x's and $(n - k)$-many ones. Then $h(0) = g(\mathbf{a}) = 1$ and $h(1) = g(\mathbf{b}) = 0$, so $r = h \in \mathcal{D}$.

Claim 4. If $\mathcal{D} \supset \mathcal{C}_5$ then $\mathcal{D} \supset \mathcal{C}_6$.

Proof. If $\mathcal{D}$ contains an operation that is not isotone, then by Claim 3, $r \in \mathcal{D}$. Consequently, $\mathcal{D} = \mathrm{Op}(2)$ which obviously contains $\mathcal{C}_6$. So we can assume that $\mathcal{D}$ is a clone of order-preserving operations.

Using Exercise 4.10.3 it is easy to see (using the notation of that exercise) that $\mathcal{C}_5 = \{ f_S^n : S \subseteq \{1, 2, \ldots, n\}, n > 0\} \cup \{0, 1\}$. Let $g \in \mathcal{D} - \mathcal{C}_5$ be n-ary, with n as small as possible. Since g is order-preserving, $U = \overleftarrow{g}(1)$ is an upset of 2^n. Since $g \neq f_S^n$ for any S, U is not a principal upset.

Choose two minimal n-tuples, $\mathbf{a}$ and $\mathbf{b}$, of U. Because of our choice of n, $\mathbf{a}$ and $\mathbf{b}$ must disagree in every component. (If they agree in any coordinate, we could leave that coordinate out and get a g of smaller rank.) Let $S = \{ i : a_i = 0 \}$ and define $h(x, y) = g(z_1, \ldots, z_n)$ in which $z_i = x$ for $i \in S$ and $z_i = y$ otherwise. Then $h(1, 1) = 1$ since h is order-preserving, $h(0, 1) = g(\mathbf{a}) = 1$, $h(1, 0) = g(\mathbf{b}) = 1$ and $h(0, 0) = 0$ by the minimality of $\mathbf{a}$ and $\mathbf{b}$. Thus $h(x, y) = x \vee y \in \mathcal{D}$.

To finish the proof of the theorem it is enough to observe that, by Example 4.8(4), $\mathcal{C}_4$ is the clone of all isotone operations. By Claim 3, any properly larger clone must contain r and will therefore be equal to $\mathcal{C}_3$. □

Theorems 8.5 and 8.6 can be combined to give a complete characterization of minimal algebras, up to polynomial equivalence. Note that there is a bit of overlap in the two descriptions. The algebra $\mathbf{E}_2$ is polynomially equivalent to a vector space and, of course, $\mathbf{E}_0$ and $\mathbf{E}_1$ are unary algebras. Also, the algebras $\mathbf{E}_5$ and $\mathbf{E}_6$ are isomorphic if we interchange 0 and 1. Taking these observations into account results in this corollary.

Corollary 8.7. *Every minimal algebra is polynomially equivalent to exactly one of the following.*

(1) *A finite permutation group,*

(2) *A finite-dimensional vector space over a finite field,*

(3) *A two-element Boolean algebra,*

(4) *A two-element lattice,*

(5) *A two-element semilattice.*

Conversely, every nontrivial algebra satisfying one of these five conditions is minimal.

The index number, 1 through 5, appearing in the above corollary is called the *type* of the minimal algebra. Thus, for example, every finite vector space is of type 2. The types 1–5 are also called *unary, affine, Boolean, lattice* and *semilattice* type, respectively.

8.2 Localization and induced algebras

In 1980, P. Pálfy and P. Pudlák explored an unusual technique for inducing structure on a subset of an algebra. It is this construction that McKenzie used as the foundation of tame congruence theory.

Let A be a set, f an n-ary operation on A and U a subset of A. We write $f{\restriction_U}$ in place of the more precise $f{\restriction_{U^n}}$.

Definition 8.8. Let $\mathbf{A}$ be an algebra and U a subset of A.

(1) $\mathrm{Pol}(\mathbf{A})|_U = \{\, f{\restriction_U} : f \in \mathrm{Pol}(\mathbf{A}) \ \&\ \vec{f}(U^n) \subseteq U \,\}$.

(2) $\mathbf{A}|_U = \langle\, U, \mathrm{Pol}(\mathbf{A})|_U \,\rangle$, called *the algebra induced on U by $\mathbf{A}$.*

The algebra $\mathbf{A}|_U$ is the object of which we speak. It is obtained from $\mathbf{A}$, but has no obvious relationship to $\mathbf{A}$. It is not even of the same similarity type as $\mathbf{A}$—in fact it has no similarity type at all. It is a *nonindexed algebra.* By a nonindexed algebra we mean a pair $\langle A, F\rangle$ in which A is a nonempty set and F is a set of operations. Given nonindexed algebras $\langle A, F\rangle$ and $\langle B, G\rangle$, a function $h\colon A \to B$ is a nonindexed homomorphism if there is some way of indexing the sets F and G so that h becomes a homomorphism of the corresponding indexed algebras. We don't try to do too much with this concept, but it should be clear that we can take subalgebras, homomorphic images and powers of nonindexed algebras. On the rare occasion that we wish to generate a variety from $\mathbf{A}|_U$, we fix $\mathrm{Pol}(\mathbf{A})|_U$ itself as the index set and call the resulting algebra $\mathbf{A}\mathrm{I}_U$, *the algebra $\mathbf{A}|_U$ with the normal indexing.*

Definition 8.9. Let $\mathbf{A}$ be an algebra.

(1) $\mathrm{E}(\mathbf{A}) = \{\, e \in \mathrm{Pol}_1(\mathbf{A}) : e \circ e = e \,\}$. The set of *idempotent unary terms of $\mathbf{A}$.*

(2) A *neighborhood* of $\mathbf{A}$ is a subset of the form $\vec{e}(A)$ for some $e \in \mathrm{E}(\mathbf{A})$.

The importance of neighborhoods resides, at least partially, in the fact that if $e \in \mathrm{E}(\mathbf{A})$ and $U = \vec{e}(A)$, then

$$\mathrm{Pol}(\mathbf{A}|_U) = \{\, (e \circ f){\restriction_U} : f \in \mathrm{Pol}(\mathbf{A})\,\}.$$

Pálfy and Pudlák [PP80] discovered that inducing an algebra on a neighborhood retains some of the structure of the congruence lattice.

Theorem 8.10. *Let* $\mathbf{A}$ *be an algebra,* $e \in \mathrm{E}(\mathbf{A})$ *and* $U = \vec{e}(A)$*. There is a complete, surjective, lattice homomorphism* $\mathbf{Con}(\mathbf{A}) \to \mathbf{Con}(\mathbf{A}|_U)$ *given by* $\theta \mapsto \theta{\restriction_U}$.

Proof. Suppose that θ is a congruence on $\mathbf{A}$. Since θ is invariant under $\mathrm{Pol}_1(\mathbf{A})$ we surely have $\theta{\restriction_U}$ invariant under $\mathrm{Pol}_1(\mathbf{A})|_U = \mathrm{Pol}_1(\mathbf{A}|_U)$. Thus, by Theorem 4.16, $\theta{\restriction_U} \in \mathrm{Con}(\mathbf{A}|_U)$.

Let ψ be a congruence on $\mathbf{A}|_U$ and define

$$\widehat{\psi} = \{\, (x, y) \in A^2 : (\forall f \in \mathrm{Pol}_1(\mathbf{A}))\;\; ef(x) \equiv_\psi ef(y)\,\}.$$

$\widehat{\psi}$ is clearly an equivalence relation on A and it is equally easy to see that it is preserved by every member of $\mathrm{Pol}_1(\mathbf{A})$. So another application of Theorem 4.16 shows that $\widehat{\psi}$ is a congruence on $\mathbf{A}$. If $(x, y) \in \psi$ then, since $e \circ f{\restriction_U}$ is a polynomial of $\mathbf{A}|_U$, we get $(x, y) \in \widehat{\psi}$. Thus $\psi \subseteq \widehat{\psi}{\restriction_U}$.

On the other hand, for every $\theta \in \mathrm{Con}(\mathbf{A})$

$$\theta \le \widehat{\psi} \iff \theta{\restriction_U} \le \psi. \tag{8–5}$$

To see this, suppose first that $\theta \le \widehat{\psi}$ and let $(x, y) \in \theta{\restriction_U}$. Then $(x, y) \in \theta \subseteq \widehat{\psi}$ and $x, y \in U$, so $(x, y) = (ex, ey) \in \psi$. Conversely, if $(x, y) \in \theta$, then for every $f \in \mathrm{Pol}_1(\mathbf{A})$, we have $(ef(x), ef(y)) \in \theta$. The fact that $U = \vec{e}(A)$ implies $ef(x), ef(y) \in U$. Thus by assumption $(ef(x), ef(y)) \in \psi$ which is precisely the condition that $(x, y) \in \widehat{\psi}$.

It follows from the previous two paragraphs that $\widehat{\psi}{\restriction_U} = \psi$, thus the restriction map is surjective. Finally to show that restriction preserves arbitrary joins, let $\Theta \subseteq \mathrm{Con}(\mathbf{A})$ and set

$$\psi = \bigvee_{\theta \in \Theta} \theta{\restriction_U}$$

where the computation of ψ is done in $\mathbf{Con}(\mathbf{A}|_U)$. Let $\theta \in \Theta$. Since $\theta{\restriction_U} \le \psi$, it follows from equivalence (8–5) that $\theta \le \widehat{\psi}$. As this holds for every θ, $\bigvee \Theta \le \widehat{\psi}$, so by (8–5) again, $(\bigvee \Theta){\restriction_U} \le \psi$. Since the opposite inclusion is trivially true, we conclude that restriction preserves joins. □

We give a little preview of the connection with minimal algebras. Let $\mathbf{A}$ be a finite simple algebra. Consider the family of subsets

$$\mathcal{S} = \{\, \vec{h}(A) : h \in \mathrm{Pol}_1(\mathbf{A}) \ \& \ |\vec{h}(A)| > 1 \,\}.$$

Let U be a minimal (under inclusion) member of this family. We shall show in Lemma 8.13 that U is a neighborhood of $\mathbf{A}$, that is, $U = \vec{e}(A)$ for some $e \in E(\mathbf{A})$. Thus by the above theorem, the induced algebra $\mathbf{A}|_U$ is simple.

Now consider a nonconstant unary polynomial, f, of $\mathbf{A}|_U$. By definition, $f = g{\restriction}_U$ for some $g \in \mathrm{Pol}_1(\mathbf{A})$ with $\vec{g}(U) \subseteq U$. Define the polynomial $h = ge$ of $\mathbf{A}$. We have $\vec{h}(A) = \vec{g}(U) = \vec{f}(U)$ which is a nontrivial subset of U. By the minimality of U we must have $\vec{h}(A) = U$. Thus f is a permutation of U. This shows that $\mathbf{A}|_U$ is a minimal algebra as defined in the previous section. (And this is the motivation for the term "minimal.")

Furthermore, it will follow from Theorem 8.16 that if U and V are minimal members of $\mathcal{S}$ then $\mathbf{A}|_U \cong \mathbf{A}|_V$. So a finite simple algebra has associated with it a unique minimal algebra.

As an example, consider a finite, simple, nonabelian group, $\mathbf{A}$. We showed in Theorem 7.72 that $\mathbf{A}$ is functionally complete. Therefore every two-element subset $U = \{a, b\}$ is a minimal member of $\mathcal{S}$. Since $\mathrm{Pol}(\mathbf{A}) = \mathrm{Op}(A)$ it is not hard to see that $\mathrm{Pol}(\mathbf{A}|_U) = \mathrm{Op}(U)$. Thus the minimal algebra $\mathbf{A}|_U$ has Boolean type as defined at the end of the last section.

As another example, consider the algebra $\mathbf{B}$ in Theorem 5.28 with $n > 3$. This algebra is simple. To see this, observe first that if $(i, u) \in \theta$ for some $i < n$ and congruence θ, then by multiplying by $i + 1$ we get $(i + 1, u) \in \theta$. So the only congruence with this property is 1_B. Suppose now that $(i, j) \in \theta$ for some $0 \le i < j < n$. Choose k adjacent to one, but not both, of i and j in the graph defining $\mathbf{B}$ (see Figure 5.3). Multiplication by k yields $(u, k) \in \theta$ which, as we just saw, is enough to conclude that $\theta = 1$. Thus $\mathbf{B}$ is simple.

Let $e(x) = 0 \cdot (1 \cdot x)$. It is easy to verify that e is idempotent and $U = \vec{e}(B) = \{1, u\}$. Thus U is a minimal set. What is the type of the minimal algebra? Notice that the operation $x \cdot y$ induces a semilattice on $\mathbf{B}|_U$. This eliminates types 1 and 2. Since any expression containing u must evaluate to u, $\mathbf{B}|_U$ cannot have a lattice structure. Thus $\mathbf{B}|_U$ must be polynomially equivalent to a semilattice.

Restricting our attention to simple algebras is not enough. So now we expand on the ideas introduced above, but in a more general setting. Recall that for congruences $\alpha < \beta$ we write $\alpha \prec \beta$ if $\mathbf{I}[\alpha, \beta] = \{\alpha, \beta\}$. When this occurs, we refer to (α, β) as a *prime quotient.*

Definition 8.11. Let $\mathbf{A}$ be an algebra and $\alpha < \beta$ in $\mathrm{Con}(\mathbf{A})$.

(1) $\mathrm{Sep}_{\mathbf{A}}(\alpha, \beta) = \{\, f \in \mathrm{Pol}_1(\mathbf{A}) : \vec{f}(\beta) \nsubseteq \alpha \,\}$.

(2) $\mathrm{M}_{\mathbf{A}}(\alpha, \beta)$ is the set of minimal members (under inclusion) of

$$\{\, \vec{f}(A) : f \in \mathrm{Sep}_{\mathbf{A}}(\alpha, \beta) \,\}.$$

The members of $\mathrm{M}_{\mathbf{A}}(\alpha,\beta)$ are called (α,β)*-minimal sets of* $\mathbf{A}$.

Notice that the minimal members of $\mathcal{S}$ considered in the simple case above are precisely the members of $\mathrm{M}_{\mathbf{A}}(0,1)$. However, in the general setting a minimal set can have a much more complex structure than in a simple algebra. We shall develop some tools to study minimal sets.

A subset K of $\mathrm{Pol}_1(\mathbf{A})$ is called a *right ideal* if $f \in K$ and $g \in \mathrm{Pol}_1(\mathbf{A})$ implies $f \circ g \in K$.

Lemma 8.12. *Let* $\mathbf{A}$ *be an algebra,* K *a right ideal of* $\mathrm{Pol}_1(\mathbf{A})$*, and* $\alpha < \beta$ *in* $\mathrm{Con}(\mathbf{A})$*. Then*

$$\gamma = \{\, (x,y) \in \beta : (\forall f \in K)\ \ f(x) \equiv_\alpha f(y)\,\} \tag{8–6}$$

is a congruence on $\mathbf{A}$ *with* $\alpha \le \gamma \le \beta$.

Proof. It is easy to see that γ is an equivalence relation on A and $\alpha \subseteq \gamma \subseteq \beta$. To verify that γ is a congruence on $\mathbf{A}$ it is enough to show that for every $g \in \mathrm{Pol}_1(\mathbf{A})$, $g \mid\!\colon \gamma$. But if $(x,y) \in \gamma$ then $f \in K$ implies $fg \in K$, so $\big(fg(x),\, fg(y)\big) \in \alpha$. But this is precisely the condition $(g(x), g(y)) \in \gamma$. □

Let us establish two fundamental properties that we alluded to above.

Theorem 8.13. *Let* $\mathbf{A}$ *be a finite algebra and* $\alpha \prec \beta$ *two congruences on* $\mathbf{A}$*. Every* (α,β)*-minimal set is a neighborhood of* $\mathbf{A}$.

Proof. Let $U \in \mathrm{M}_{\mathbf{A}}(\alpha,\beta)$ and define $K = \big\{\, f \in \mathrm{Pol}_1(\mathbf{A}) : \vec{f}(A) \subseteq U \,\big\}$. Observe that K is a right ideal of the semigroup $\mathrm{Pol}_1(\mathbf{A})$. By Lemma 8.12, the congruence γ defined in (8–6) is a congruence with $\alpha \le \gamma \le \beta$.

By assumption, there is an $f \in \mathrm{Sep}(\alpha,\beta)$ such that $U = \vec{f}(A)$. Choose a pair $(a,b) \in \beta$ such that $(f(a), f(b)) \notin \alpha$. Then $(a,b) \notin \gamma$. Thus $\gamma \ne \beta$. Since $\alpha \prec \beta$ we must have $\gamma = \alpha$. Now since $(f(a), f(b)) \notin \gamma$, there is $g \in K$ such that $(gf(a), gf(b)) \notin \alpha$. Therefore $gf \in \mathrm{Sep}(\alpha,\beta)$ and $\vec{g}(A) \subseteq U$.

We see that $\overrightarrow{gf}(A) = \vec{g}(U) \subseteq U$. By the minimality of U we must have $\vec{g}(U) = U$, that is, g is a permutation of U. Let k be the order of g as a permutation of U. Since $U = \vec{g}(U) \subseteq \vec{g}(A) \subseteq U$ we have $\vec{g}(A) = U$. Let $e = g^k$. Then $\vec{e}(A) = U$ and $e \circ e = e$. Thus U is a neighborhood of $\mathbf{A}$. □

Suppose that $U \in \mathrm{M}_{\mathbf{A}}(\alpha,\beta)$ and $U = \vec{e}(A)$ for some $e \in \mathrm{E}(\mathbf{A})$. It is useful to keep in mind that e is the identity on U. Also, since there are $a, b \in U$ with $(a,b) \in \beta - \alpha$, we see that $(e(a),\, e(b)) = (a,b) \in \beta - \alpha$. Thus $e \in \mathrm{Sep}(\alpha,\beta)$.

Definition 8.14. Let $\mathbf{A}$ be an algebra. Two subsets X and Y of A are called *polynomially isomorphic in* $\mathbf{A}$ if there are $f, g \in \mathrm{Pol}_1(\mathbf{A})$ such that $f{\restriction}_X$ and $g{\restriction}_Y$ are inverse mappings. We write $X \simeq Y$ if there is a polynomial isomorphism from X to Y.

The constructions we have been developing behave very nicely under polynomial isomorphism, as the following lemma attests.

Lemma 8.15. *Let* $\mathbf{A}$ *be an algebra,* $U, V \subseteq A$ *and assume that* $U \simeq V$.

(1) $\mathbf{A}|_U \cong \mathbf{A}|_V$.

(2) *Let* $\alpha < \beta$ *in* $\mathrm{Con}(\mathbf{A})$. *Then* $U \in \mathrm{M}_{\mathbf{A}}(\alpha, \beta)$ *if and only if* $V \in \mathrm{M}_{\mathbf{A}}(\alpha, \beta)$.

Proof. Exercise. □

This lemma holds in complete generality. Most importantly, when $\mathbf{A}$ is finite and α is covered by β, the converse of part (2) holds, as we now show.

Theorem 8.16. *Let* $\mathbf{A}$ *be a finite algebra and* $\alpha \prec \beta$ *be congruences on* $\mathbf{A}$. *Any two* (α, β)*-minimal sets are polynomially isomorphic.*

Proof. Let U_1 and U_2 be (α, β)-minimal sets. By Theorem 8.13, there is e_i in $\mathrm{E}(\mathbf{A})$ such that $\vec{e}_i(A) = U_i$, for $i = 1, 2$. If follows from Definition 8.11 that there is $(a, b) \in \beta\restriction_{U_1} - \alpha\restriction_{U_1}$. We first show that there is a polynomial h with $\vec{h}(A) \subseteq U_2$ and $(h(a), h(b)) \notin \alpha$.

To see this, let $K = \{ e_2 \circ f : f \in \mathrm{Pol}_1(\mathbf{A}) \}$. Then K is a right ideal of $\mathrm{Pol}_1(\mathbf{A})$. Using the same argument as in Lemma 8.13, we determine that

$$\alpha = \{ (x, y) \in \beta : (\forall h \in K)\ \ h(x) \equiv_\alpha h(y)) \}.$$

Since $(a, b) \in \beta - \alpha$ there is $h \in K$ such that $(h(a), h(b)) \in \beta - \alpha$. From the definition of K, $h = e_2 \circ h'$ (for some polynomial h') so $\vec{h}(A) \subseteq \vec{e}_2(A) = U_2$.

We have $U_2 \supseteq \vec{h}(U_1) = \overrightarrow{he_1}(A)$. Since a, b are members of U_1 they are fixed by e_1. So $he_1(a) = h(a) \not\equiv_\alpha h(b) = he_1(b)$. Thus $he_1 \in \mathrm{Sep}(\alpha, \beta)$. Therefore by the minimality of U_2, $\overrightarrow{he}(A) = \vec{h}(U_1) = U_2$.

With a similar argument, there is a polynomial g with $\vec{g}(U_2) = U_1$. Therefore $f = gh\restriction_{U_1}$ is a permutation of the finite set U_1. Let k be the order of f as a permutation. then $(gh)^{k-1} \circ g$ and h are inverse polynomial isomorphisms between U_1 and U_2. □

There are several additional properties of $\mathrm{M}_{\mathbf{A}}(\alpha, \beta)$ that are useful in practice. We list them here.

Theorem 8.17. *Let* $\mathbf{A}$ *be a finite algebra,* $\alpha \prec \beta$ *in* $\mathrm{Con}(\mathbf{A})$, *and* U *an* (α, β)*-minimal set.*

(1) *Assume* $f \in \mathrm{Pol}_1(\mathbf{A})$ *and* $\vec{f}(\beta\restriction_U) \nsubseteq \alpha$. *Then* $\vec{f}(U)$ *is minimal and* $f\restriction_U : U \simeq \vec{f}(U)$.

(2) *If* $(a, b) \in \beta - \alpha$ *then there is* $f \in \mathrm{Pol}_1(\mathbf{A})$ *such that* $\vec{f}(A) = U$ *and* $(f(a), f(b)) \notin \alpha$.

(3) *If* $f \in \mathrm{Sep}_{\mathbf{A}}(\alpha, \beta)$ *then there is* $V \in \mathrm{M}_{\mathbf{A}}(\alpha, \beta)$ *such that* $V \simeq \vec{f}(V)$.

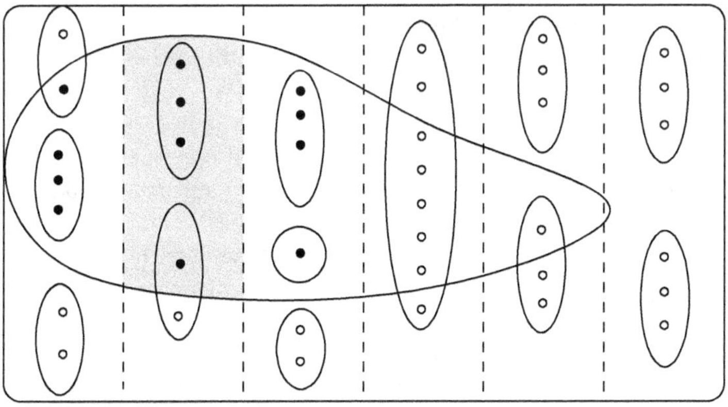

FIGURE 8.2: A minimal set, body, trace and tail

Proof. Since U is minimal, there is $e \in E(\mathbf{A})$ such that $\vec{e}(A) = U$. We start with (2). Let $K = \{\, e \circ f : f \in \mathrm{Pol}_1(\mathbf{A})\,\}$. Then K is a right ideal, so by Lemma 8.12, there is a congruence γ as in (8–6) with $\alpha \leq \gamma \leq \beta$. By assumption, the pair (a, b) is excluded from α, hence $(a, b) \notin \gamma$. Thus $\gamma = \alpha$ and there is $g \in \mathrm{Pol}_1(\mathbf{A})$ such that $\big(eg(a), eg(b)\big) \notin \alpha$. Since $\big(eg(a), eg(b)\big)$ is in β, the polynomial $f = eg \in \mathrm{Sep}(\alpha, \beta)$. Finally, $\vec{f}(A) \subseteq U$. By the minimality of U we must have $\vec{f}(A) = U$.

Now consider (1). By assumption, there is $(a, b) \in \beta{\restriction}_U$ such that $f(a) \not\equiv_\alpha f(b)$. By (2), there is a polynomial g such that $\vec{g}(A) = U$ and $gf(a) \not\equiv_\alpha gf(b)$. Then the polynomial gfe is a member of $\mathrm{Sep}(\alpha, \beta)$ and $\overrightarrow{gfe}(A) \subseteq U$, so by minimality, $\overrightarrow{gf}(U) = \overrightarrow{gfe}(A) = U$. Therefore g is a polynomial isomorphism of $\vec{f}(U)$ with U.

Finally we prove (3). Let $\gamma = \mathrm{Cg}^{\mathbf{A}}(\alpha \cup (\beta{\restriction}_U))$. Clearly $\alpha < \gamma \leq \beta$, so, since $\alpha \prec \beta$ we must have $\gamma = \beta$. Since $f \in \mathrm{Sep}(\alpha, \beta)$, we can apply Theorem 4.17 to obtain $g \in \mathrm{Pol}_1(\mathbf{A})$ and $(a, b) \in \beta{\restriction}_U$ such that $(fg(a), fg(b)) \notin \alpha$. So surely $(g(a), g(b)) \notin \alpha$, which shows that $\vec{g}(\beta{\restriction}_U) \nsubseteq \alpha$. By (1), $V = \vec{g}(U)$ is a minimal set. Also $(g(a), g(b)) \in \beta{\restriction}_V$ which implies $\vec{f}(\beta{\restriction}_V) \nsubseteq \alpha$. Therefore by (1) again, $V \simeq \vec{f}(V)$. □

The structure of a typical (α, β)-minimal set is usually much more complex than in the case of a simple algebra. Figure 8.2 is a schematic of an algebra and two congruences, $\alpha \prec \beta$. The α-classes are indicated by ellipses, the β-classes by the columns in the diagram. Since $\alpha \leq \beta$, each ellipse must be completely contained in a column. The "blob" in the picture is a minimal set.

According to Definition 8.11, if U is a minimal set then $\alpha{\restriction}_U$ is properly contained in $\beta{\restriction}_U$. Thus U must contain a $\beta{\restriction}_U$-class which contains at least two $\alpha{\restriction}_U$-classes. Such a β-class is called a *trace* of U. The minimal set in Figure 8.2 shows 3 traces, one of which is shaded. The union of the traces is called the *body*. The part of the minimal set outside the body is called the *tail*. In the figure, the elements of the body are indicated with solid dots.

We wish to reestablish the relationship, given above for simple algebras, between minimal sets and minimal algebras. In a simple algebra, of course, $0 \prec 1$. Consider an arbitrary finite algebra $\mathbf{A}$ and congruences $\alpha \prec \beta$. How can we formulate a situation in which α acts like 0 and β acts like 1?

For β, we use, instead of an entire minimal set, U, a single trace, N. Note that on $\mathbf{A}|_N$, $\beta\restriction_N = 1$. How do we make α behave like 0? The same way it is always done in algebra, by forming the quotient algebra $\mathbf{A}/\alpha$. Here are the details.

Lemma 8.18. *Let $\mathbf{A}$ be a finite algebra and $\alpha \prec \beta$ congruences on $\mathbf{A}$.*

(1) $U \in \mathrm{M}_{\mathbf{A}}(\alpha, \beta) \implies U/\alpha \in \mathrm{M}_{\mathbf{A}/\alpha}(0, \beta/\alpha)$.

(2) *If N is an (α, β)-trace, then $(\mathbf{A}|_N)/\alpha$ is a minimal algebra.*

Proof. Let U be (α, β)-minimal in $\mathbf{A}$. Then $U = \vec{e}(A)$ for some $e \in \mathrm{E}(\mathbf{A})$. As a polynomial of $\mathbf{A}$, there is a term t and $c_1, \dots, c_n \in A$ such that $e(x) = t^{\mathbf{A}}(x, \mathbf{c})$. Let $g(x) = t^{\mathbf{A}/\alpha}(x, \mathbf{c}/\alpha)$. Then g is a polynomial of $\mathbf{A}/\alpha$ and for every $a \in A$, $e(a)/\alpha = g(a/\alpha)$. It is easy to see that g is idempotent, $g \in \mathrm{Sep}(0, \beta/\alpha)$ and $\vec{g}(A/\alpha) = \vec{e}(A)/\alpha = U/\alpha$.

To see that U/α is minimal, suppose that $V = \vec{h}(A/\alpha) \subseteq U/\alpha$ for some $h \in \mathrm{Sep}(0, \beta/\alpha)$. Using the argument from the previous paragraph in reverse, there is $f \in \mathrm{Pol}_1(\mathbf{A})$ such that $f(x)/\alpha = h(x/\alpha)$. It follows that $f \in \mathrm{Sep}(\alpha, \beta)$ on $\mathbf{A}$. By assumption, there is $(a, b) \in \beta - \alpha$ with $h(a/\alpha) \neq h(b/\alpha)$. The inclusion $V \subseteq U/\alpha$ implies $g \circ h = h$. Therefore

$$ef(a)/\alpha = g\big(f(a)/\alpha\big) = gh(a/\alpha) = h(a/\alpha) \neq h(b/\alpha) = ef(b)/\alpha.$$

Thus $ef \in \mathrm{Sep}(\alpha, \beta)$.

Let $\overline{V} = \overrightarrow{ef}(A) \subseteq U$. By the minimality of U, $\overline{V} = U$. Finally $\overline{V}/\alpha = \overrightarrow{ef}(A)/\alpha = \overrightarrow{gh}(A/\alpha) = V$, so $V = U/\alpha$.

Now we turn to the second item. There is $U \in \mathrm{M}_{\mathbf{A}}(\alpha, \beta)$ with N a trace of U. By (1) U/α is a $(0, \beta/\alpha)$-minimal set of $\mathbf{A}/\alpha$ and N/α is clearly a trace of U/α. By working in the algebra $\mathbf{B} = \mathbf{A}/\alpha$ we can assume that $\alpha = 0$, $\beta \succ 0$, U is a $(0, \beta)$-minimal set, and N is a trace. We must show that $\mathbf{B}|_N$ is a minimal algebra, i.e., if $f \in \mathrm{Pol}_1(\mathbf{B}|_N)$ is not constant, then it is a permutation.

By definition, $f = g_0\restriction_N$ for some $g_0 \in \mathrm{Pol}_1(\mathbf{B})$ such that $\vec{g}_0(N) \subseteq N$. By Theorem 8.13, there is $e \in \mathrm{E}(\mathbf{B})$ such that $\vec{e}(\mathbf{B}) = U$. Let $g = e \circ g_0 \circ e$. Notice that $g\restriction_N = f$ and $\vec{g}(B) = \vec{g}(U) \subseteq U$.

Since f is assumed nonconstant on N, there are $a, b \in N$ with $f(a) \neq f(b)$. Therefore $\vec{g}(\beta) \nsubseteq 0$, i.e., $g \in \mathrm{Sep}(0, \beta)$. Using the minimality of U, $\vec{g}(U) = \vec{g}(B) = U$, so g is a permutation of U. Since $f = g\restriction_N$, f is a permutation of N. □

So when $\alpha \prec \beta$, each (α, β)-trace has an induced minimal algebra structure. This would not be so useful if those structures varied wildly. For fixed α and β they are, in fact, all isomorphic.

Theorem 8.19. *Let* $\mathbf{A}$ *be a finite algebra and* $\alpha \prec \beta$ *congruences. Any two* (α, β)*-traces are polynomially isomorphic.*

Proof. Since any two (α, β)-minimal sets are polynomially isomorphic, it is enough to consider two traces, N and M contained in the same minimal set U. By replacing $\mathbf{A}$ with $\mathbf{A}|_U$, we can assume that $A = U$.

We first show that there is $g \in \mathrm{Sep}(\alpha, \beta)$ such that $\vec{g}(N) \subseteq M$. Since N and M are traces, there are pairs $(a, b) \in (\beta\restriction_N) - \alpha$ and $(c, d) \in (\beta\restriction_M) - \alpha$. Since $\alpha \prec \beta$ in $\mathrm{Con}(\mathbf{A})$, we have $\beta = \mathrm{Cg}^{\mathbf{A}}\big(\alpha \cup \{(a, b)\}\big)$. Now, $(c, d) \in \beta$, so we can apply Theorem 4.17 to obtain (c, d) from the generators of β. Thus there are elements $z_0, \dots, z_n$ and polynomials $f_0, \dots, f_{n-1}$ such that

$$c = z_0, \;\; d = z_n, \;\; \{z_i, z_{i+1}\} = \{f_i(x_i), f_i(y_i)\}, \;\; i < n.$$

In the above, each pair (x_i, y_i) either comes from α, or is equal to (a, b). Therefore all of the z's lie in the same β-class, namely M. If all pairs (x_i, y_i) came from α, then we would have $(c, d) \in \alpha$, which is false. Thus, there is some i such that $(f_i(a), f_i(b)) \in \beta - \alpha$. We can take g to be f_i. Now $x \in N \implies x \equiv_\beta a \implies g(x) \equiv_\beta g(a) \implies g(x) \in M$.

Since $g \in \mathrm{Sep}(\alpha, \beta)$ and $A = U$, we have $\vec{g}(A) \subseteq U = A$. By the minimality of U, $\vec{g}(A) = A$, i.e., g is a permutation of A. The argument in the previous paragraph can be applied again to show that $\overrightarrow{g^{-1}}(M) \subseteq N$. It is then easy to see that g is a polynomial isomorphism of N with M. □

Based on the results of this section, to each prime quotient $\alpha \prec \beta$ on a finite algebra, we can assign a unique minimal algebra, and in particular, a unique *type* from 1 to 5. We denote this type by $\mathrm{typ}(\alpha, \beta)$. It follows from 8.18 and 8.19 that $\mathrm{typ}(\alpha, \beta)$ in $\mathbf{A}$ is the same as $\mathrm{typ}(0, \beta/\alpha)$ in $\mathbf{A}/\alpha$. Thus when computing types, we can often work in a quotient algebra and restrict attention to congruence pairs $(0, \beta)$ in which β is an atom of the congruence lattice.

Example 8.20. Suppose that $\mathbf{A} = \overline{\mathbb{B}}_1$ is the three-element Stone algebra with universe $\{0, e, 1\}$ (see Figure 3.6 on page 68). The congruence lattice of $\mathbf{A}$ is also a 3-element chain. The monolith, μ, identifies e and 1. What are the types of the two prime quotients $(0, \mu)$ and $(\mu, 1)$? First, $\mathbf{A}/\mu$ is exactly the two-element Boolean algebra, which is of course, a minimal algebra of type 3. Thus $\mathrm{typ}(\mu, 1) = 3$.

More interesting is the type of $(0, \mu)$. Observe that $f(x) = e \vee x$ is idempotent and $\vec{f}(A) = \{e, 1\} \in \mathrm{M}_{\mathbf{A}}(0, \mu)$ is both a minimal set and a trace. The two lattice operations of $\mathbf{A}$ are inherited by $\{e, 1\}$ (although the pseudocomplement operation is not), so the trace has at least a lattice structure. Therefore $\mathrm{typ}(0, \mu) \in \{3, 4\}$. Consider the binary relation $\tau = \delta_A \cup \{(e, 1)\}$. It is easy

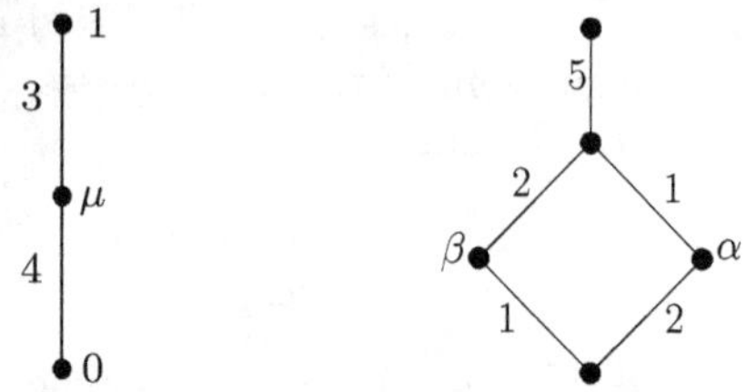

(a) Example 8.20 (b) Example 8.21

FIGURE 8.3: Labeled congruence lattices

to see that the basic operations of **A** as well as the three constant operations all preserve τ. Thus τ is invariant under all of Pol(**A**). If the induced algebra $\mathbf{A}|_{\{e,1\}}$ were Boolean, there would be a unary operation in Pol(**A**) that interchanges e and 1. But such an operation does not preserve τ, since $(e, 1) \in \tau$ while $(1, e) \notin \tau$. So the trace is not Boolean, and therefore $\mathrm{typ}(0, \mu) = 4$.

We can illustrate the types on the prime quotients with a *labeled congruence lattice.* Figure 8.3(a) shows the congruence lattice for the above example. The digits on the edges give the type of the corresponding prime quotient.

Example 8.21. Let $\mathbf{A} = \langle\{0, 1, 2, 3\}, \cdot\rangle$ where the binary operation denotes multiplication modulo 4. The congruence lattice of **A** is shown in Figure 8.3(b). In the figure, α is the congruence that identifies 1 and 3 and nothing else. As a shorthand we can write $\alpha = \{13|0|2\}$. Similarly $\beta = \{02|1|3\}$. Set $\gamma = \alpha \vee \beta$. We shall determine the types of the prime quotients.

We observe that **A** is commutative, associative and satisfies the identity $x^4 \approx x^2$. Therefore, every nonconstant n-ary polynomial is of the form

$$f(x_1, \dots, x_n) = cx_1^{k_1} x_2^{k_2} \cdots x_n^{k_n} \tag{8–7}$$

with $c \in \{1, 2, 3\}$ and every $k_i \leq 3$.

First, it is easy to see that the quotient algebra $\mathbf{A}/\gamma$ is polynomially equivalent to a two-element semilattice. Thus the prime quotient $(\gamma, 1)$ is of semilattice type.

Now we investigate the quotient $(0, \beta)$. The only possible trace is $N_2 = \{0, 2\}$, since that is the only nontrivial β-class. (In fact, the only $(0, \beta)$-minimal set is A itself, with $e_2(x) = x$ the associated idempotent.) Now if $\mathbf{A}|_{N_2}$ is not essentially unary then by Lemma 8.3 there is a binary polynomial $f(x, y)$ such that $f{\restriction}_{N_2}$ depends on both variables. According to (8–7), such an f is of the form $cx^i y^j$ with $1 \leq i, j \leq 3$. But it is easy to see that if both x and y are constrained to lie in $\{0, 2\}$ then f is the constant function 0. Thus $\mathbf{A}|_{N_2}$ is essentially unary, i.e., $\mathrm{typ}(0, \beta) = 1$.

Finally, we turn to $(0, \alpha)$. With $e_3(x) = x^3$ we see that $U_3 = \overrightarrow{e_3}(A) = \{0, 1, 3\}$ and $N_3 = \{1, 3\}$ is the unique trace. The basic operation of **A** induces

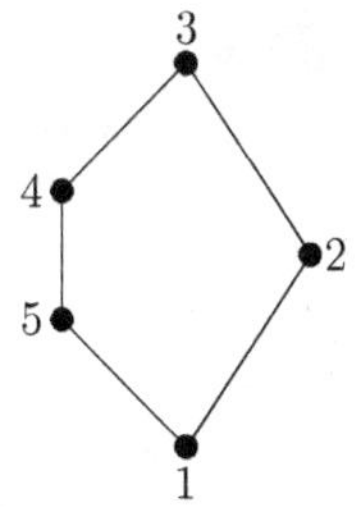

FIGURE 8.4: $\mathcal{T}$: The poset of types

a 2-element group structure on N_3, so $\operatorname{typ}(0,\alpha) \in \{2,3\}$. If $\operatorname{typ}(0,\alpha) = 3$ then $\operatorname{Pol}(\mathbf{A}|_{N_3})$ would contain a semilattice operation $x \wedge y$ on N_3 with, say, $1 \wedge 3 = 1$. However, it is easy to see that the operation cx^iy^j restricted to N_3 can never behave like a semilattice operation. Thus $\operatorname{typ}(0,\alpha) = 2$.

The types of the remaining two prime quotients can be determined using arguments very similar to the ones above. Or, they can be derived from the "transpose property," Theorem 8.38.

Definition 8.22. Let $\mathbf{A}$ be a finite algebra. Then

$$\operatorname{typ}\{\mathbf{A}\} = \{\operatorname{typ}(\alpha,\beta) : \alpha,\beta \in \operatorname{Con}(\mathbf{A}) \text{ and } \alpha \prec \beta\}.$$

If $\mathcal{K}$ is a class of algebras, then $\operatorname{typ}(\mathcal{K}) = \bigcup\{\operatorname{typ}\{\mathbf{A}\} : \mathbf{A} \in \mathcal{K}_{\text{fin}}\}$.

The five types can be ordered as in Figure 8.4. This poset, $\mathcal{T}$, is derived from the ordering on the clones in Figure 8.1 by identifying $\mathcal{C}_0$ and $\mathcal{C}_1$ (the two essentially unary clones) and $\mathcal{C}_5$ and $\mathcal{C}_6$ (the two clones generated by semilattices). Of course, the ordering on $\mathcal{T}$ is a lattice ordering, but the lattice operations do not seem to have any meaning in this context, so we prefer to think of it as a poset.

There are important relationships between properties of a variety and its set of types. Here is a simple example.

Theorem 8.23. *Let $\mathcal{V}$ be a Maltsev variety. Then* $\operatorname{typ}(\mathcal{V}) \subseteq \{2,3\}$.

Proof. Let $\mathbf{A} \in \mathcal{V}_{\text{fin}}$ and (α,β) a prime quotient in $\operatorname{Con}(\mathbf{A})$. By replacing $\mathbf{A}$ with $\mathbf{A}/\alpha$ we can assume that $\alpha = 0$. Let U be a $(0,\beta)$-minimal set, N a trace of U, and $e \in \operatorname{E}(\mathbf{A})$, with $\vec{e}(A) = U$.

Since $\mathcal{V}$ is Maltsev, there is a Maltsev term $q(x,y,z)$ on $\mathcal{V}$. We shall show that $e \circ q$ is a Maltsev polynomial on $\mathbf{A}|_N$. Since the terms and polynomials of $\mathbf{A}|_N$ are the same, the indexed algebra $\mathbf{AI}_N$ lies in a Maltsev variety. Hence it is of either affine or Boolean type.

It is obvious that for every $x,y \in N$, $eq(x,y,y) = eq(y,y,x) = x$. The only nontrivial part of the argument is to verify that $e \circ q$ is, in fact, a polynomial of $\mathbf{A}|_N$. For this we need to check that if $a,b,c \in N$ then $eq(a,b,c) \in N$. But recall that N is a $\beta{\restriction}_U$-class, thus if $a,b,c \in N$ then $a \mathrel{\beta} b \mathrel{\beta} c$. So $eq(a,b,c) \equiv_\beta eq(b,b,c) = c$, and $eq(a,b,c) \in U$, hence $eq(a,b,c) \in N$. ☐

Rather than express this property as an inclusion, Hobby and McKenzie chose to express it as an *omitting types theorem.* Thus, a Maltsev variety omits the types $\{1,4,5\}$. Note that this 3-element set is a downset in the poset $\mathcal{T}$. One of the significant achievements of tame congruence theory is a characterization, for each of the six downsets of $\mathcal{T}$, of those varieties that omit that downset [HM88, Chapter 9].

We conclude this section with a lemma that is useful in finding idempotent operations. It is often considered folklore, and perhaps it is, although Knuth[Knu81] attributes it to R. W. Floyd.

Lemma 8.24. *Let S be a finite set and $f\colon S \to S$ a function. There is a positive integer n such that $f^n = f^{2n}$.*

Proof. There are only finitely many self-maps of S, so there are positive integers k and n such that $f^k = f^{k+n} = f^{k+2n} = \cdots$. Thus we may assume that $f^k = f^{k+n}$ and $n \geq k$. Then $f^n = f^k \circ f^{n-k} = f^{k+n} \circ f^{n-k} = f^{2n}$. □

Exercise Set 8.25.

1. Let $\mathbf{A}$ be a finite algebra and $U \subseteq A$. Show that $\mathrm{Pol}(\mathbf{A}|_U) = \mathrm{Pol}(\mathbf{A})|_U$.

2. $\mathbf{A}$ be an algebra and $\mathbf{B}$ a subalgebra. Show that for every $f \in \mathrm{Pol}(\mathbf{B})$ there is an $\bar{f} \in \mathrm{Pol}(\mathbf{A})$ such that $\bar{f}\restriction_B = f$. (Hint: this is similar to the first paragraph of Lemma 8.18.)

3. Let $\mathbf{A}$ be a finite algebra and $\alpha \prec \beta$ a prime quotient of $\mathrm{Con}(\mathbf{A})$. Let $\theta = \alpha \cup \bigcup\{\, N^2 : N \text{ an } \alpha,\beta\text{-trace}\,\}$. Show that β is the transitive closure of θ, i.e., $\beta = \theta \cup (\theta \circ \theta) \cup (\theta \circ \theta \circ \theta) \cup \cdots$.

4. Let $\mathbf{A}$ be a finite lattice and $\alpha \prec \beta$ a prime quotient in $\mathrm{Con}(\mathbf{A})$. Show that there are $u \prec v$ in $\mathbf{A}$ with $u \mathrel{\beta} v$ and that for any such pairs, $N = \{u, v\}$ is a (α, β)-trace of type 4.

5. Let $\mathbf{B}$ be the algebra of Theorem 5.28 with $n = 3$. Verify that $\mathrm{Con}(\mathbf{B})$ is a 3-element chain with monolith $\mu = \{012|u\}$. Show that $\mathrm{typ}\{\mathbf{B}\} = \{3, 5\}$.

6. Let $\mathbf{A} = \langle\{0, 1, 2, 3, 4\}, \cdot\rangle$ with the operation being multiplication modulo 5. Verify that $\mathrm{Con}(\mathbf{A})$ is a 4-element chain with congruences $\alpha = \{0|14|23\}$ and $\beta = \{0|1234\}$. Determine $\mathrm{typ}\{\mathbf{A}\}$.

7. Prove that if $\mathcal{V}$ is congruence-distributive then $\mathrm{typ}(\mathcal{V})$ omits $\{1, 2, 5\}$. (Hint: use Theorem 4.66 to imitate the argument in 8.23.)

8.3 Centralizers again!

In Chapter 7 we introduced the centralizer congruence and the associated notions of the commutator and of Abelian algebras. At that time we showed that within a Maltsev variety, these constructions had powerful properties. Unfortunately, those properties fail to hold in arbitrary algebras. However, as we show in this section, the centralizer of a congruence does influence the type of the corresponding minimal set.

Recall from Definition 7.16 that for a congruence β on an algebra $\mathbf{A}$, the centralizer of β is the congruence

$$\beta^* = \{ (a,b) \in A^2 : \forall f \in \mathrm{Pol}_{n+1}(\mathbf{A})\ \forall (\mathbf{c},\mathbf{d}) \in \beta, \\ f(a,\mathbf{c}) = f(a,\mathbf{d}) \iff f(b,\mathbf{c}) = f(b,\mathbf{d}) \}.$$

We remark that in the original formulation, the first quantifier is over all *terms* of $\mathbf{A}$, while here we have quantified over *polynomials.* However, using the fact that every polynomial is of the form $t(x_1, x_2, \ldots, x_n, r_1, \ldots, r_m)$ for some term t and $r_1, \ldots, r_m \in A$, it is easy to see that the two definitions are equivalent. In Propositions 7.17 and 7.22 we developed what little we could about the general centralizer before we specialized to Maltsev varieties.

Recall that an algebra $\mathbf{A}$ is Abelian if $1_A^* = 1_A$. We generalize that concept to an arbitrary congruence.

Definition 8.26. Let $\mathbf{A}$ be an algebra, $\alpha \leq \beta$ congruences on $\mathbf{A}$.

(1) β is *Abelian* if $\beta \leq \beta^*$.

(2) β is *Abelian over* α if β/α is Abelian in $\mathbf{A}/\alpha$.

In the language of the commutator, β is Abelian if $[\beta, \beta] = 0$ and β is Abelian over α if $[\beta, \beta] \leq \alpha$. This latter condition can also be expressed as "α has the β,β-term condition," see page 217, except, that in the present context, we might want to call it the β,β-polynomial condition.

We first consider this property in the context of minimal algebras. That essentially unary algebras are Abelian is easy to see, and vector spaces were studied in detail in Section 7.3. Every minimal algebra of one of the remaining three types has two elements and contains a semilattice operation. Observe that with $t(x, y) = x \wedge y$ we have $t(0, 0) = t(0, 1)$ but $t(1, 0) \neq t(1, 1)$. This observation, together with its dual, is enough to show that all three are non-abelian.

To summarize, minimal algebras of types 1 and 2 are Abelian, those of types 3, 4, and 5 are nonabelian. One of the significant results of tame congruence theory is that this distinction carries over to the traces obtained from prime quotients, and impacts the congruence lattice.

Theorem 8.27. *Let* $\mathbf{A}$ *be a finite algebra and* $\alpha \prec \beta$ *a prime quotient of* $\mathrm{Con}(\mathbf{A})$. *Then* β *is Abelian over* α *if and only if* $\mathrm{typ}(\alpha, \beta) \in \{1, 2\}$.

Proof. By switching to the algebra $\mathbf{A}/\alpha$ we can assume that $\alpha = 0$. (See exercise 8.39.1.) So suppose that β is an Abelian congruence and let N be a $(0, \beta)$-trace. It follows easily from the fact that $\mathrm{Pol}(\mathbf{A}|_N) = \mathrm{Pol}(\mathbf{A})|_N$ that $\mathbf{A}|_N$ is a minimal, Abelian algebra, so as we observed just above, its type is either 1 or 2. Hence the same holds for $\mathrm{typ}(0, \beta)$.

Now consider the converse. Assume $\mathrm{typ}(0, \beta) \in \{1, 2\}$. We must show $\beta \leq \beta^*$. Since β is an atom of $\mathrm{Con}(\mathbf{A})$, it is enough to find a pair of distinct elements a, b such that $(a, b) \in \beta$ and $(a, b) \in \beta^*$.

Let U be a $(0, \beta)$-minimal set and N a trace of U. Choose distinct elements a and b in N. We will show that $(a, b) \in \beta^*$. For this, let $f \in \mathrm{Pol}_{n+1}(\mathbf{A})$, $\mathbf{c} \mathrel{\beta} \mathbf{d}$ and suppose that $f(a, \mathbf{c}) = f(a, \mathbf{d})$ but $f(b, \mathbf{c}) \neq f(b, \mathbf{d})$. We shall derive a contradiction.

Since $a, b \in N$, we have $a \mathrel{\beta} b$. Thus

$$f(a, \mathbf{d}) = f(a, \mathbf{c}) \mathrel{\beta} f(b, \mathbf{c}) \mathrel{\beta} f(b, \mathbf{d})$$

so these three elements lie in a single trace, N'. By Theorem 8.19, there is a polynomial isomorphism h from N' to N. By replacing f with $h \circ f$ we can assume that

$$\{a, b, f(a, \mathbf{c}), f(b, \mathbf{c}), f(b, \mathbf{d})\} \subseteq N, \quad f(a, \mathbf{c}) = f(a, \mathbf{d}), \quad f(b, \mathbf{c}) \neq f(b, \mathbf{d}).$$

Write $\mathbf{c} = \langle c_1, \dots, c_n \rangle$ and let $T_i = c_i/\beta$. Note that $d_i \in T_i$ as well. Then

$$x \in N,\ y_1 \in T_1, \dots, y_n \in T_n \implies f(x, y_1, \dots, y_n) \in N \tag{8–8}$$

since $f(x, \mathbf{y}) \mathrel{\beta} f(a, \mathbf{c}) \in N$. By Exercise 8.25.3, there is, for each $i \leq n$, a sequence of overlapping traces $N_{i1}, N_{i2}, \dots, N_{ik}$ connecting c_i to d_i. By allowing repetitions, we can assume that each of these sequences of traces has the same length. That is

$$c_i \in N_{i0},\ d_i \in N_{ik},\ N_{ij} \cap N_{i,j+1} \neq \varnothing, \quad 1 \leq i \leq n,\ 0 \leq j < k.$$

By Theorem 8.19 again, for each i and j there is a polynomial isomorphism g_{ij} from N to N_{ij}. Define, for each $j \leq k$, polynomials f_j by $f_j(x, y_1, \dots, y_n) = f\big(x, g_{1j}(y_1), \dots, g_{nj}(y_n)\big)$. It follows from the definition of f_j and implication (8–8) that $f_j{\restriction}_N \in \mathrm{Pol}(\mathbf{A}|_N)$. Finally, choose elements $c_1', \dots, c_n'$ and $d_1', \dots, d_n'$ such that $g_{i0}(c_i') = c_i$ and $g_{ik}(d_i') = d_i$.

We have yet to use our assumption, that $\mathrm{typ}(0, \beta) \in \{1, 2\}$. We consider the two cases separately. Suppose first that the type is 2. Then the minimal algebra $\mathbf{A}|_N$ is a vector space over a finite field, F. The polynomials are the F-affine operations on N. A similar argument was given in Example 4.8. Thus, for every $j \leq k$, there are scalars $r_j, s_{j1}, \dots, s_{jn} \in F$ and $t_j \in N$ such that $f_j \restriction_N (x, y_1, \dots, y_n) = r_j x + s_{j1} y_1 + \cdots + s_{jn} y_n + t_j$.

Claim. For every $j < k$, $r_j = r_{j+1}$.

Proof. Since N_{ij} and $N_{i,j+1}$ overlap, we can find $u_i, v_i \in N$, for $i \le n$, such that $g_{ij}(u_i) = g_{i,j+1}(v_i)$. Then

$$\begin{aligned} r_j x + \sum\nolimits_i s_{ji}u_i + t_j = f_j(x, u_1, \ldots, u_n) = \\ f(x, g_{1j}(u_1), \ldots, g_{nj}(u_n)) = f(x, g_{1,j+1}(v_1), \ldots, g_{n,j+1}(v_n)) = \\ f_{j+1}(x, v_1, \ldots, v_n) = r_{j+1}x + \sum\nolimits_i s_{ji}v_i + t_{j+1}. \end{aligned}$$

Hence $(r_j - r_{j+1})x$ does not depend on x. This is only possible (in a vector space) if $r_j = r_{j+1}$.

From the claim it follows that $r_0 = r_k$. We now compute

$$r_0 a + \sum\nolimits_i s_{0i}c'_i + t_0 = f(a, \mathbf{c}) = f(a, \mathbf{d}) = r_k a + \sum\nolimits_i s_{ki}d'_i + t_k$$

from which we conclude that $\sum_i s_{0i}c'_i + t_0 = \sum_i s_{ki}d'_i + t_k$. But this implies

$$f(b, \mathbf{c}) = r_0 b + \sum\nolimits_i s_{0i}c'_i + t_0 = r_k b + \sum\nolimits_i s_{ki}d'_i + t_k = f(b, \mathbf{d})$$

which is a contradiction.

Now we turn to the possibility that $\operatorname{typ}(0, \beta) = 1$. This means that every polynomial of $\mathbf{A}|_N$ is essentially unary. We argue in a manner much like the claim in the previous case.

Claim. For every $j < k$, $f_j \restriction_N (x, y_1, \ldots, y_n)$ depends on x if and only if $f_{j+1}\restriction_N$ depends on x.

Proof. If $f_j\restriction_N$ depends only on its first variable, there are elements u, u' such that for every $\mathbf{v}$, $f_j(u, \mathbf{v}) \neq f_j(u', \mathbf{v})$. Choose the vector $\mathbf{v}$ so that $g_{ij}(v_i) \in N_{ij} \cap N_{i,j+1}$. Then there is another vector $\mathbf{v}'$ such that $g_{i,j+1}(v'_i) = g_{ij}(v_i)$. Consequently

$$f_{j+1}(u, \mathbf{v}') = f_j(u, \mathbf{v}) \neq f_j(u', \mathbf{v}) = f_{j+1}(u', \mathbf{v}').$$

Since f_{j+1} can only depend on one variable, it must be the first. The argument for the converse is similar.

Now if $f_0\restriction_N$ depends on its first variable, then by the claim, the same is true for every $f_j\restriction_N$. For all i and j, let u_{ij} lie in $N_{ij} \cap N_{i,j+1}$ and choose u'_{ij} such that $g_{ij}(u'_{ij}) = u_{ij}$. Then from our first-variable assumption

$$\begin{aligned} f(b, \mathbf{c}) = f_0(b, \mathbf{c}') = f_0(b, \mathbf{u}'_0) = f(b, \mathbf{u}_0) = f_1(b, \mathbf{u}'_0) = f_1(b, \mathbf{u}'_1) = \cdots \\ = f_k(b, \mathbf{u}'_{k-1}) = f_k(b, \mathbf{d}') = f(b, \mathbf{d}) \end{aligned}$$

where $\mathbf{u}_j = \langle u_{1j}, \ldots, u_{nj} \rangle$. This contradicts our initial assumption.

Finally, if any of the f_j's fail to depend on x, then none of them depend on x. In that case

$$\begin{aligned} f(b, \mathbf{c}) = f_0(b, \mathbf{c}') = f_0(a, \mathbf{c}') = f(a, \mathbf{c}) = \\ f(a, \mathbf{d}) = f_0(a, \mathbf{d}') = f_0(b, \mathbf{d}') = f(b, \mathbf{d}) \end{aligned}$$

again, a contradiction. □

Here is a simple application. Let **A** be the group $\mathbf{S}_5$. Then Con(**A**) is a 3-element chain: $0 < \mu < 1$ in which μ corresponds to the normal subgroup A_5. Since **A** lies in a Maltsev variety, by Theorem 8.23, the types of the two prime quotients must each be either 2 or 3. Because $\mathbf{A}/\mu$ is a two-element Abelian group, its type is 2. Equivalently, $\operatorname{typ}(\mu, 1) = 2$. We can see the same thing by observing that $[1, 1] \leq \mu$ (since for any $x, y \in S_5$, the element $xyx^{-1}y^{-1}$ is an even permutation) and applying the theorem.

For the other prime quotient, $0 \prec \mu$, $\mathbf{A}_5$ is nonabelian, so we surely do not have $[\mu, \mu] = 0$, thus μ is nonabelian. Consequently, by Theorem 8.27, $\operatorname{typ}(0, \mu) \neq 2$, so it must be 3.

As another example, look back at Example 8.21 from the previous section. The congruence 1 is not Abelian over γ since $0 \cdot 1 = 0 \cdot 3$ while $1 \cdot 1 \neq 1 \cdot 3$. Of course, this only shows that $\operatorname{typ}(\gamma, 1) \in \{3, 4, 5\}$. Additional arguments are needed to determine the type exactly. On the other hand, using (8–7) it is not hard to convince oneself that γ is itself Abelian. It follows that all four of the prime quotients below γ must have type 1 or 2. But which is which? What we need is a condition that separates types 1 and 2.

Definition 8.28. Let **A** be an algebra, $\alpha \leq \beta$ congruences on **A**.

(1) We say β is *strongly Abelian over* α if

$$(\forall f \in \operatorname{Pol}_{n+1}(\mathbf{A}))\ (\forall (a, b) \in \beta)\ (\forall \mathbf{c}\ \beta\ \mathbf{d}\ \beta\ \mathbf{e}) :$$
$$f(a, \mathbf{c}) \equiv_\alpha f(b, \mathbf{d}) \implies f(a, \mathbf{e}) \equiv_\alpha f(b, \mathbf{e}).$$

(2) **A** is *strongly Abelian* if 1_A is strongly Abelian over 0_A.

To get a hint of the meaning of this condition, consider two prototypical examples of Abelian algebras. It is easy to see that every essentially unary algebra is strongly Abelian. On the other hand, a nontrivial Abelian group is not strongly Abelian since for any $a \neq 0$, $0 - 0 = a - a$ but $0 - 0 \neq a - 0$.

It is not hard to see that β strongly Abelian over α implies β is Abelian over α. Thus a minimal algebra is type 1 if and only if it is strongly Abelian.

Theorem 8.29. *Let* **A** *be a finite algebra,* $\alpha \prec \beta$ *a prime quotient in* Con(**A**). *Then* β *is strongly Abelian over* α *if and only if* $\operatorname{typ}(\alpha, \beta) = 1$.

Proof. As before, we can assume that $\alpha = 0$ (Exercise 8.39.1). If β is strongly Abelian over α, then for any $(0, \beta)$-trace, N, the algebra $\mathbf{A}|_N$ is minimal and strongly Abelian, hence has type 1.

We turn to the converse. Assume that $\operatorname{typ}(0, \beta) = 1$. We build our way up to the desired conclusion.

Claim 1. Let $f \in \operatorname{Pol}_2(\mathbf{A})$ and let N, N_1 and N_2 be $(0, \beta)$-traces. If $\vec{f}(N_1 \times N_2) \subseteq N$ then $f{\restriction}_{N_1 \times N_2}$ is essentially unary.

Proof. Since $N_1 \simeq N \simeq N_2$, there are polynomial isomorphisms $h_1 \colon N \to N_1$

and $h_2\colon N \to N_2$. Then $f(h_1(x_1), h_2(x_2))\restriction_N \in \mathrm{Pol}_2(\mathbf{A}|_N)$. Since the minimal algebra $\mathbf{A}|_N$ is of type 1, it is essentially unary. Since h_1 and h_2 are invertible, it follows that f too is essentially unary.

Claim 2. Let f be an n-ary polynomial, $T_1, \ldots, T_n$ be β-classes, and N a $(0, \beta)$-trace. Write $T = T_1 \times \cdots \times T_n$. If $\vec{f}(T) \subseteq N$ then $f\restriction_T$ is essentially unary.

Proof. Suppose not. Since the assertion is obviously true for $n = 1$, we must have $n \geq 2$. If $n > 2$ then, by Lemma 8.2, there is a binary counterexample. Thus we can assume that $n = 2$ and f depends on both variables.

This means there are $a_1 \in T_1$ and $a_2 \in T_2$ such that each of the unary polynomials $f(x, a_2)$ and $f(a_1, x)$ are nonconstant. It follows from this that both T_1 and T_2 have cardinality greater than 1. Therefore, by Exercise 8.25.3, there are traces N_1 and N_2 such that $f_1(x) = f(x, a_2)$ is nonconstant on N_1 and $f_2(x) = f(a_1, x)$ is nonconstant on N_2.

Let b be arbitrary in T_2. By Exercise 8.25.3 again, there are traces $M_1, M_2, \ldots, M_k$ and elements $u_1, \ldots, u_{k-1}$, with $a_2 \in M_1$, $u_i \in M_i \cap M_{i+1}$, for $i = 1, \ldots, k-1$, and $b \in M_k$. By applying Claim 1 successively to $N_1 \times M_i$ for $i = 1, \ldots, k$, we deduce that

$$f(x, a_2) = f(x, u_1) = f(x, u_2) = \cdots = f(x, u_{k-1}) = f(x, b).$$

Similarly, for any $c \in T_1$, $f(a_1, y) = f(c, y)$. But this means that for any $(c, b) \in N_1 \times N_2$,

$$f_1(c) = f(c, a_2) = f(c, b) = f(a_1, b) = f_2(b)$$

which implies that f_1 is constant on N_1, a contradiction. This proves Claim 2.

Our objective is to prove that β is strongly Abelian. For this, let f be a member of $\mathrm{Pol}_{n+1}(\mathbf{A})$, $(a, b) \in \beta$ and $\mathbf{c}\ \beta\ \mathbf{d}\ \beta\ \mathbf{e}$. Suppose that $f(a, \mathbf{e}) \neq f(b, \mathbf{e})$. We shall show that $f(a, \mathbf{c}) \neq f(b, \mathbf{d})$.

Let $T_0 = a/\beta$ and $T_i = c_i/\beta$, for $i = 1, \ldots, n$. Note that $f(a, \mathbf{e})$ and $f(b, \mathbf{e})$ are distinct but β-equivalent. By Theorem 8.17(2), there is $U \in \mathrm{M}(0, \beta)$ and $h \in \mathrm{Pol}_1(\mathbf{A})$ such that $\vec{h}(A) = U$ and $hf(a, \mathbf{e}) \neq hf(b, \mathbf{e})$. Let $g = h \circ f$. There is a trace N containing $g(a, \mathbf{e})$ and $g(b, \mathbf{e})$. Thus

$$\vec{g}(T_0 \times T_1 \times \cdots \times T_n) \subseteq N.$$

By Claim 2, $g\restriction_T$ depends on only one variable. Since $g(a, \mathbf{e}) \neq g(b, \mathbf{e})$, it must be the first variable. But then

$$g(a, \mathbf{c}) = g(a, \mathbf{e}) \neq g(b, \mathbf{e}) = g(b, \mathbf{d})$$

as desired. □

Returning once again to Example 8.21, we wish to determine whether the prime quotients $(0, \alpha)$ and $(0, \beta)$ have type 1 or 2. Recall that $(1, 3) \in \alpha$.

Taking $a = c = 1$, $b = d = 3$ and $e = 1$ we see that $a \cdot c = b \cdot d$ while $a \cdot e \neq b \cdot e$, showing that α is not strongly Abelian, so $\operatorname{typ}(0, \alpha) \neq 1$. The congruence β, on the other hand, is strongly Abelian. Since the argument is largely the same as the one we gave on page 260, we omit it.

Theorems 8.27 and 8.29 allow us to distinguish prime quotients of type 1 from type 2 from types $\{3, 4, 5\}$. It is significant that we can recognize types 1 and 2 as global properties of the congruences within the entire algebra. On the other hand, the structure of minimal sets of types 1 and 2 can be quite complex.

Contrast the above observations with the situation for the nonabelian types. recall that a minimal algebra of type 3, 4, or 5 must have two elements and contain a semilattice operation in its clone of polynomials. While those properties are not reflected in the global structure of the algebra, they do impact the structure of a typical minimal set. Since we will be concerned with algebras of the form $\mathbf{A}|_U$ for some minimal set U, it is helpful to make the following definition.

Definition 8.30. Let $\mathbf{C}$ be a finite algebra and $\alpha < \beta$ in $\operatorname{Con}(\mathbf{C})$. We say that $\mathbf{C}$ is (α, β)-*minimal* if $\mathrm{M}_{\mathbf{C}}(\alpha, \beta) = \{C\}$.

The point here is that if $\mathbf{A}$ is a finite algebra, $U \in \mathrm{M}_{\mathbf{A}}(\alpha, \beta)$, and $U = \vec{e}(A)$ for some $e \in \mathrm{E}(\mathbf{A})$, then $\mathbf{A}|_U$ is an $(\alpha{\restriction_U}, \beta{\restriction_U})$-minimal algebra. In particular this holds for any prime quotient $\alpha \prec \beta$.

Suppose that $\mathbf{A}$ is (α, β)-minimal for some prime quotient (α, β) of nonabelian type, and let N be a trace. So $(\mathbf{A}/\alpha)|_{N/\alpha}$ is a minimal algebra of type 3, 4, or 5. Therefore $N/\alpha = \{0/\alpha, 1/\alpha\}$ and there is a polynomial that behaves like the meet operation on N/α. That is,

there is $f \in \operatorname{Pol}_2(\mathbf{A})|_N$ and $f(x, y)/\alpha = (x/\alpha) \wedge (y/\alpha)$ for $x, y \in \{0, 1\}$.

Let $g_0(x) = f(x, 0)$ and $g_1(x) = f(x, 1)$. Since $g_1(0) \not\equiv_\alpha g_1(1)$, the minimality of $\mathbf{A}$ requires that g_1 be a permutation of A. On the other hand, since A is finite and $g_0(0) \equiv_\alpha g_0(1)$, it is impossible for g_0 to be a permutation. This observation gives rise to the following definition.

Definition 8.31. Let $\mathbf{A}$ be a finite algebra, $t(x, \mathbf{y})$ be a term, and let $\mathbf{c}, \mathbf{d}$ be members of A^n. The polynomials $f(x) = t(x, \mathbf{c})$ and $g(x) = t(x, \mathbf{d})$ are called *twins.* If $U \in \mathrm{M}_{\mathbf{A}}(\alpha, \beta)$ and if $\mathbf{c}$ and $\mathbf{d}$ come from the body of U, then f and g are called (α, β)-*body twins.*

Returning to the discussion, in a minimal algebra of nonabelian type, the polynomials g_0 and g_1 defined above form a pair of (α, β)-body twins, only one of which is a permutation. We shall now prove the converse, and, along the way, derive some additional information.

Lemma 8.32. *Let $\mathbf{A}$ be an (α, β)-minimal algebra and let f_0 and f_1 be (α, β)-body twins. Suppose that f_1 is a permutation while f_0 is not. Then*

(1) *The body, B, of* $\mathbf{A}$ *contains a single trace,* N*;*

(2) $N/\alpha = \{c/\alpha, d/\alpha\}$ *for some elements* $c, d \in B$*;*

(3) (α, β) *is a nonabelian prime quotient.*

Proof. It is easy to see that if $\mathbf{A}$ is (α, β)-minimal then $\mathbf{A}/\alpha$ is $(0, \beta/\alpha)$-minimal. Thus we assume that $\alpha = 0$. This means that for any unary polynomial, f,

$$\text{either } f \text{ is a permutation or } \big((x, y) \in \beta \implies f(x) = f(y)\big). \tag{8–9}$$

By assumption, there is a term t and vectors $\mathbf{c}$ and $\mathbf{d}$ from B such that $f_1(x) = t(x, \mathbf{c})$ and $f_0(x) = t(x, \mathbf{d})$. Consider the sequence of polynomials

$$t(x, c_1, c_2, c_3, \dots, c_n),\ t(x, d_1, c_2, c_3, \dots, c_n),\ t(x, d_1, d_2, c_3, \dots, c_n), \dots, \\ t(x, d_1, d_2, d_3 \dots, d_n).$$

There must be two adjacent polynomials in this sequence in which one is a permutation and the other is not. Thus we can find a binary polynomial $p(x, y)$ and elements $c, d \in B$ such that $p(x, c)$ and $p(x, d)$ form a pair of $(0, \beta)$-body twins, in which only $p(x, c)$ is a permutation.

By holding y fixed and applying Lemma 8.24 to $f(x) = p(x, y)$, there is a polynomial $h(x, y)$ such that $h(h(x, y), y) = h(x, y)$. Thus, for any b, the unary polynomial $h(x, b)$ is idempotent. We set $g_1(x) = h(x, c)$ and $g_0(x) = h(x, d)$. Then g_1 is an idempotent permutation, that is, the identity map, while g_0 is idempotent, but not a permutation.

Claim 1. If $b \neq c$ then $h(x, b)$ is not a permutation.

Proof. Suppose the contrary. Then $h(a, b) = a = h(a, c)$, for every $a \in A$. Then for every a, the polynomial $h(a, x)$ is not a permutation. By (8–9), $(y, z) \in \beta \implies h(a, y) = h(a, z)$. Since $h(x, c) \neq h(x, d)$, we conclude that $(c, d) \notin \beta$.

By assumption, c and d come from the body, which means that neither c/β nor d/β are singletons. Pick $c' \neq c$ with $c' \mathrel{\beta} c$ and $d' \neq d$, $d' \mathrel{\beta} d$. Then

$$h(d, d) = h(d', d) = h(d', d').$$

The first equality follows from (8–9), since $h(x, d)$ is not a permutation, the second from the previous paragraph. From this we conclude that $h(x, x)$ is not a permutation. Therefore using the fact that $h(x, c) = x$ and (8–9) again,

$$c = h(c, c) = h(c', c') = h(c', c) = c'$$

a contradiction, proving Claim 1.

Claim 2. B has cardinality 2.

Proof. Suppose not. Then there are elements $a, b \in B$ such that $|\{a, b, c\}| = 3$ and $(a, b) \in \beta$. By Claim 1, neither $h(x, a)$ nor $h(x, b)$ is a permutation.

Next observe that

$$(r,s)\in\beta \ \&\ h(r,x) \text{ not a permutation } \Longrightarrow \qquad h(s,x) \text{ not a permutation.} \tag{8–10}$$

To see this, suppose that $(r,s)\in\beta$ and $h(r,x)$ is not a permutation. Then from the previous paragraph and (8–9), $h(s,a) = h(r,a) = h(r,b) = h(s,b)$ so we conclude that $h(s,x)$ also fails to be a permutation.

Let $N = c/\beta$. Since N is a trace, there must be $c' \in N - \{c\}$. From our earlier computations, $h(c,c) = c$, so $\vec{h}(N^2) \subseteq N$. By Claim 1, $h(x,c')$ is constant on N, say $h(x,c') = c''$ for all $x \in N$. So $h(c'',c) = c'' = h(c'',c')$, showing that $h(c'',x)$ is not a permutation. Then from (8–10), for every $e \in N$, $h(e,x)$ is not a permutation, so it is constant on N. Finally

$$c = h(c,c) = h(c,c') = c'' = h(c',c') = h(c',c) = c'$$

a contradiction. This proves Claim 2.

Claim 2, together with the assumption that $\alpha = 0$ yields the first two conclusions of the theorem. We see that $B = N = \{c,d\}$ and $0 \prec \beta$. Finally, since $h(c,d) = h(d,d)$ (from (8–9)) while $h(c,c) = c \neq d = h(d,c)$, we deduce that $(0,\beta)$ is nonabelian. □

Thus a quotient (α,β) being prime and nonabelian is equivalent to the existence of a pair of body twins, and this in turn implies that the body contains a single trace. We can extract a bit of information about the α-classes of the trace as well.

Theorem 8.33. *Let* $\mathbf{A}$ *be a finite* (α,β)*-minimal algebra in which* $\alpha \prec \beta$*, and suppose that* β *is nonabelian over* α*. Let* N *be the unique* (α,β)*-trace. Then* $N/\alpha = \{O, I\}$ *and* $|I| = 1$*. If* $\operatorname{typ}(\alpha,\beta) \in \{3,4\}$*, then* $|O| = 1$ *as well.*

Proof. Since β is nonabelian over α, $\operatorname{typ}(\alpha,\beta) \in \{3,4,5\}$ by Theorem 8.27. As we discussed earlier, $\mathbf{A}$ has a pair of body twins, so by Theorem 8.32, $\mathbf{A}$ has a unique trace N. Because of the type, N has exactly two α-classes, call them O and I, and $(\mathbf{A}|_N)/\alpha$ contains a semilattice operation under which we can assume $O < I$. Let $g \in \operatorname{Pol}_2(\mathbf{A})$ induce that semilattice operation on N/α.

Let $0 \in O$ and $1 \in I$ and define $d(x) = g(x,x)$. Since $d(0) \in O$ and $d(1) \in I$, we surely have $d \in \operatorname{Sep}(\alpha,\beta)$. By the minimality of $\mathbf{A}$, this means that d is a permutation of A. Since A is finite, there is a positive integer n such that $d^n(x) = x$.

Let $p(x,y) = d^{n-1}(g(x,y))$. Then $p(x,x) = x$ and p still induces a semilattice operation on N/α. By applying Lemma 8.24 to each variable of p we can assume that $p(p(x,y),y) = p(x,p(x,y)) = p(x,y)$.

Let $a \in I$. Then $p(a,a) \not\equiv_\alpha p(0,a)$, so the unary polynomial $p(x,a)$ is a permutation. But since this map is also idempotent, it must be the identity. Similarly $p(a,x)$ is the identity. Therefore $a = p(1,a) = 1$ since both a and

1 are members of I. Thus $I = \{1\}$. Let us note for future reference that we have also shown that $\langle\{0,1\}, p\rangle$ is a semilattice.

Finally, if $\mathrm{typ}(\alpha, \beta) \in \{3, 4\}$ then we have a join operation as well as a meet. Applying the same argument as above, we conclude that $|O| = 1$. □

Let us reiterate that in the above argument, we constructed a binary polynomial p and showed that $\langle\{0,1\}, p\rangle$ is a semilattice. We give a name to this object.

Definition 8.34. Let $\mathbf{A}$ be an algebra, $\alpha < \beta$ in $\mathrm{Con}(\mathbf{A})$ and $(a, b) \in \beta - \alpha$.

(1) (a, b) is an (α, β)*-1 snag* if, for some $f \in \mathrm{Pol}_2(\mathbf{A})$, $f(a, b) = f(b, a) = a$ and $f(b, b) = b$.

(2) (a, b) is an (α, β)*-2 snag* if, for some $f \in \mathrm{Pol}_2(\mathbf{A})$, $f(a, a) = f(a, b) = f(b, a) = a$ and $f(b, b) = b$.

Clearly every 2-snag is a 1-snag. In the previous theorem, $(0, 1)$ is a 2-snag. Snags can be used to distinguish types.

Proposition 8.35. *Let* $\mathbf{A}$ *be a finite algebra and* (α, β) *a prime quotient of* $\mathrm{Con}(\mathbf{A})$.

(1) (α, β) *is Abelian if and only if there are no* (α, β)*-2 snags.*

(2) (α, β) *is strongly Abelian if and only if there are no* (α, β)*-1 snags.*

Proof. Theorem 8.33 constructs a 2-snag if (α, β) is nonabelian. Conversely, a 2-snag violates the Abelian condition. That is enough to prove (1). Similarly, a 1-snag prevents β from being strongly Abelian over α. Since every 2-snag is a 1-snag, to complete the proof, we must show that if $\mathrm{typ}(\alpha, \beta) = 2$ then there is an (α, β)-1 snag.

Let U be an (α, β)-minimal set, B the (α, β)-body of U, and N a trace contained in B. There is a polynomial $p(x, y)$ in $\mathrm{Pol}_2(\mathbf{A})$ inducing a group operation on N/α. Let $b, c \in N$ with $(b, c) \notin \alpha$. Then for any $a \in N$

$$p^N(a, b) = p^N(a, c) \implies p^{N/\alpha}(a/\alpha, b/\alpha) = p^{N/\alpha}(a/\alpha, c/\alpha) \implies (b, c) \in \alpha.$$

Since the conclusion is false, the unary polynomial $p(a, x)$ is a member of $\mathrm{Sep}(\alpha, \beta)$, so it is a permutation of U. Since (α, β) is Abelian, Lemma 8.32 (applied to the (α, β)-minimal algebra $\mathbf{A}|_U$) shows that any body twin of $p(a, x)$ is also a permutation. Thus $p(b, x)$ is a permutation of U for every $b \in B$. The same holds for the polynomial $p(x, b)$, for every $b \in B$.

Thus $\langle B, p\rangle$ is a quasigroup. Consequently, there is a ternary polynomial q inducing a Maltsev operation on B. Pick any pair $(a, b) \in (\beta - \alpha)\restriction_B$. Then $f(x, y) = q(x, b, y)$ is a 1-snag. □

Let us recapitulate, for later use, the constructions that have appeared in the last few propositions.

Corollary 8.36. *Let* **A** *be a finite* (α, β)*-minimal algebra in which* $\alpha \prec \beta$. *Let* B *and* N *denote the body and a trace of* **A** *respectively.*

(1) *If* $\text{typ}(\alpha, \beta) = 5$ *then* $B = N$, $N/\alpha = \{O, I\}$ *and* $I = \{1\}$. *For any* $0 \in O$ *there is a polynomial* f *so that* $\langle \{0, 1\}, f{\restriction}_{\{0,1\}} \rangle$ *is a meet-semilattice.*

(2) *If* $\text{typ}(\alpha, \beta) = 4$ *then* $B = N = \{0, 1\}$ *and* $\mathbf{A}|_N$ *is polynomially equivalent to a 2-element lattice.*

(3) *If* $\text{typ}(\alpha, \beta) = 3$ *then* $B = N = \{0, 1\}$ *and* $\mathbf{A}|_N$ *is polynomially equivalent to a 2-element Boolean algebra.*

(4) *If* $\text{typ}(\alpha, \beta) = 2$ *then there is an idempotent polynomial* q *such that* $q{\restriction}_B$ *is a Maltsev operation.*

Proof. The first three parts follow easily from Lemma 8.32 and Theorem 8.33. Let (α, β) be Abelian. In Proposition 8.35 we constructed a polynomial q_0 which induces a Maltsev operation on B. Let $g(x) = q_0(x, x, x)$. Since q_0 is idempotent on B, we must have $g \in \text{Sep}(\alpha, \beta)$. By minimality, g is a permutation of A. Thus $q(x, y, z) = g^{-1} \circ q_0(x, y, z)$ is the desired idempotent polynomial inducing a Maltsev operation on B. □

We can utilize these last few results to prove that if two prime quotients are transposes, then they share the same type. Recall that the intervals $\mathbf{I}[\alpha_0, \beta_0]$ and $\mathbf{I}[\alpha_1, \beta_1]$ are called *transposes*, and we write $\mathbf{I}[\alpha_0, \beta_0] \nearrow \mathbf{I}[\alpha_1, \beta_1]$ if $\beta_0 \wedge \alpha_1 = \alpha_0$ and $\beta_0 \vee \alpha_1 = \beta_1$.

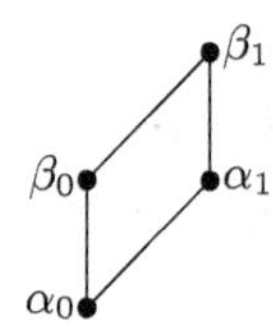

Lemma 8.37. *Let* U *be a neighborhood of the finite algebra* **A**, θ *a congruence on* **A**, *and* N *a* $\theta{\restriction}_U$*-class. Then the mapping* $\psi \mapsto \psi{\restriction}_N$ *is a surjective lattice homomorphism from* $\mathbf{I}[0, \theta]$ *to* $\mathbf{Con}(\mathbf{A}|_N)$.

Proof. The map in question is the composition of the restriction maps

$$\mathbf{I}[0, \theta] \xrightarrow{\ \restriction_U\ } \mathbf{I}[0, \theta{\restriction}_U\,] \xrightarrow{\ \restriction_N\ } \mathbf{Con}(\mathbf{A}|_N).$$

We know from Theorem 8.10 that the first of these is a surjective lattice homomorphism. Thus we may as well assume that $U = A$. The restriction map always preserves meets. That restriction preserves joins of congruences in $\mathbf{I}[0, \theta]$ follows easily from Corollary 2.19. Finally, to see that the map is surjective, let $\psi \in \text{Con}(\mathbf{A}|_N)$ and define

$$\widehat{\psi} = \{\, (x, y) \in \theta : (\forall f \in \text{Pol}_1(\mathbf{A})) : \\ \{f(x), f(y)\} \cap N \neq \varnothing \implies f(x) \equiv_\psi f(y) \,\}.$$

It is straightforward to verify that $\widehat{\psi}$ is a congruence on **A** and that $\widehat{\psi}{\restriction}_N \leq \psi$. For the reverse inclusion, suppose that $(x, y) \in \psi$, $f \in \text{Pol}_1(\mathbf{A})$ and $f(x)$ is in N. Then since N is a θ-class, $\vec{f}(N) \subseteq N$, so $f{\restriction}_N \in \text{Pol}_1(\mathbf{A}|_N)$. Therefore $f(x) \equiv_\psi f(y)$. □

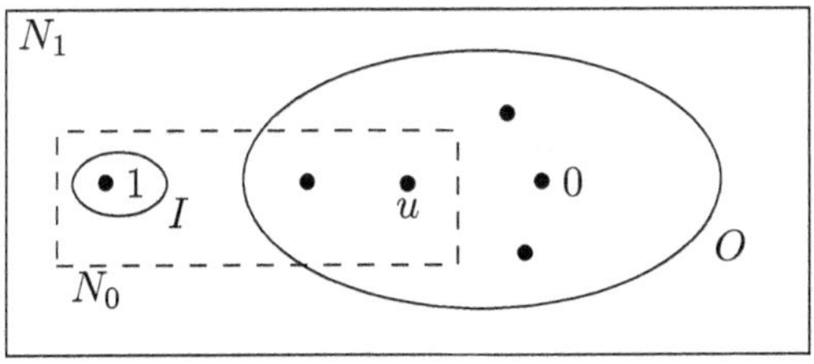

FIGURE 8.5: The traces in Theorem 8.38

Theorem 8.38. *Let (α_0, β_0) and (α_1, β_1) be prime quotients in a finite algebra* $\mathbf{A}$ *and assume that* $\mathbf{I}[\alpha_0, \beta_0] \nearrow \mathbf{I}[\alpha_1, \beta_1]$. *Then* $\mathrm{M}_{\mathbf{A}}(\alpha_0, \beta_0) = \mathrm{M}_{\mathbf{A}}(\alpha_1, \beta_1)$ *and* $\mathrm{typ}(\alpha_0, \beta_0) = \mathrm{typ}(\alpha_1, \beta_1)$.

Proof. The equality of the minimal sets is left as Exercise 8.39.3. Let us address the equality of the types. Fix a set U in $\mathrm{M}(\alpha_0, \beta_0)$.

We first consider the case that $\mathrm{typ}(\alpha_1, \beta_1) \in \{3, 4, 5\}$. According to Theorem 8.33, U has a unique (α_1, β_1)-minimal trace, N_1, which is a union of two α_1-classes, O and $I = \{1\}$. Let 0 be a member of O. By Lemma 8.37, we must have $(\alpha_1 \restriction_{N_1}) \vee (\beta_0 \restriction_{N_1}) = \beta_1 \restriction_{N_1}$. Since $(1, 0) \in \beta_1 - \alpha_1$, there is $u \in N_1$ such that $(u, 1) \in \beta_0 - \alpha_0$ (use Corollary 2.19). Let N_0 be the (α_0, β_0)-trace containing $\{u, 1\}$. Since $\beta_0 \leq \beta_1$, we see that $N_0 = 1/\beta_0 \subseteq 1/\beta_1 = N_1$. (See Figure 8.5.)

Suppose that N' is any (α_0, β_0)-trace excluding 1. As a nontrivial β_0-class, it must be contained in a nontrivial β_1-class. But the only such nontrivial class is N_1. Since $N_1 - \{1\} = O$, we conclude that N' lies in a single α_1-class. By Lemma 8.37 again, $(\alpha_1 \restriction_{N_1}) \wedge (\beta_0 \restriction_{N_1}) = \alpha_0 \restriction_{N_1}$, so N' lies in a single α_0-class, which is impossible for a trace. Thus N_0 is the unique (α_0, β_0)-trace.

We conclude that both N_1/α_1 and N_0/α_0 are two-element algebras. There is a polynomial that induces the meet operation on N_1/α_1. That same polynomial induces the meet on N_0/α_0, so $\mathrm{typ}(\alpha_0, \beta_0) \in \{3, 4, 5\}$ as well. If $\mathrm{typ}(\alpha_1, \beta_1) \neq 5$ then N_1 is a two-element set, so the same must be true of N_0. Thus $\mathbf{A}|_{N_1} = \mathbf{A}|_{N_0}$ and they obviously have the same type. On the other hand, if $\mathrm{typ}(\alpha_0, \beta_0) \neq 5$ then there is a polynomial inducing a join operation on N_0. That polynomial also induces the join on N_1, so $\mathrm{typ}(\alpha_1, \beta_1) \neq 5$. Thus the two types are equal.

Now we turn to the case that (α_1, β_1) is Abelian. According to Proposition 7.45, $[\beta_0, \beta_0] \leq \beta_0$ and $[\beta_0, \beta_0] \leq [\beta_1, \beta_1] \leq \alpha_1$, hence $[\beta_0, \beta_0] \leq \alpha_0$, that is, (α_0, β_0) is also Abelian. An argument similar to that of Proposition 7.45 shows that if (α_1, β_1) is strongly Abelian, then so is (α_0, β_0).

All that remains is to show that $\mathrm{typ}(\alpha_1, \beta_1) = 2 \implies \mathrm{typ}(\alpha_0, \beta_0) \neq 1$. The argument in Proposition 8.35(2) shows that any pair $(b, c) \in \beta_1 - \alpha_1$ forms a 1-snag. By choosing (b, c) in $\beta_0 - \alpha_1$ (which must be possible since $\alpha_1 \not\subseteq \beta_0$), the same polynomial yields an (α_0, β_0)-1 snag. So $\mathrm{typ}(\alpha_0, \beta_0) \neq 1$. □

Exercise Set 8.39.

1. Let $\mathbf{A}$ be an algebra, $\delta \leq \alpha \leq \beta$ congruences on $\mathbf{A}$. Show that β is Abelian over α in $\mathbf{A}$ if and only if β/δ is Abelian over α/δ in $\mathbf{A}/\delta$. Do the same with "Abelian" replaced by "strongly Abelian."

2. Let $\mathbf{A}$ be a finite algebra, $\alpha < \beta$ in $\mathrm{Con}(\mathbf{A})$, $U \in \mathrm{M}_{\mathbf{A}}(\alpha, \beta)$, and $U = \vec{e}(A)$ for some $e \in \mathrm{E}(\mathbf{A})$. Then $\mathbf{A}|_U$ is an $(\alpha{\restriction_U}, \beta{\restriction_U})$-minimal algebra.

3. Prove that if $\alpha_i < \beta_i$ are congruences on a finite algebra and $\mathbf{I}[\alpha_0, \beta_0] \nearrow \mathbf{I}[\alpha_1, \beta_1]$, then $\mathrm{M}(\alpha_0, \beta_0) = \mathrm{M}(\alpha_1, \beta_1)$.

4. Let $\mathbf{A}$ be a finite algebra and $0_A = \alpha_0 \prec \alpha_1 \prec \cdots \prec \alpha_n = 1_A$ be a chain of congruence covers. Prove that $\mathbf{A}$ has no $(0_A, 1_A)$-2 snags if and only if, for every $i < n$, (α_i, α_{i+1}) is Abelian.

8.4 Applications

Tame congruence theory provides numerous insights into the structure of a finite algebra, far more than we can cover in this text. We have, however, developed enough of the theory to demonstrate its power. In this final section, we give several examples.

Recall from Section 5.2 that if $\mathcal{K}$ is a finite set of finite algebras generating a congruence-distributive variety $\mathcal{V}$, and if $\mathbf{A}$ is a subdirectly irreducible member of $\mathcal{V}$, then $\mathbf{A} \in \mathsf{HS}(\mathcal{K})$. Here is a related theorem that does not require any global assumption on $\mathcal{V}$.

Theorem 8.40. *Let $\mathcal{K}$ be a finite set of finite algebras and assume that $\mathbf{A}$ is finite, simple, and of Boolean or lattice type. If $\mathbf{A} \in \mathsf{V}(\mathcal{K})$ then $\mathbf{A} \in \mathsf{HS}(\mathcal{K})$.*

In this theorem, by the type of $\mathbf{A}$, we mean $\mathrm{typ}(0_A, 1_A)$ which is defined since $\mathbf{A}$ is simple. We require a lemma.

Lemma 8.41. *Let $\mathbf{A}$ be a finite algebra and (α, β) be a prime congruence quotient of type 3 or 4. There is a smallest congruence δ such that $\alpha \vee \delta = \beta$.*

Proof. Let N be an (α, β)-trace. According to Theorem 8.33, $N = \{0, 1\}$ with $(0, 1) \notin \alpha$. Let $\delta = \mathrm{Cg}^{\mathbf{A}}(0, 1)$. Since $(0, 1) \in \beta$ we have $\delta \leq \beta$. Thus, $\alpha \leq \alpha \vee \delta \leq \beta$. By assumption, $\alpha \prec \beta$ and $\alpha \neq \alpha \vee \delta$ since $(0, 1) \notin \alpha$. Therefore $\alpha \vee \delta = \beta$.

On the other hand, suppose γ is any congruence such that $\alpha \vee \gamma = \beta$. By Lemma 8.37, $(\alpha{\restriction_N}) \vee (\gamma{\restriction_N}) = \beta{\restriction_N}$. Since $N = \{0, 1\}$, this is only possible if $(0, 1) \in \gamma$. Thus $\delta \leq \gamma$. □

Proof of Theorem 8.40. Let $\mathbf{A} \in \mathsf{V}(\mathcal{K})$. Since $\mathsf{V} = \mathsf{HSP}$, $\mathbf{A} \cong \mathbf{C}/\alpha$ for some finite algebra $\mathbf{C}$ such that $\mathbf{C} \leq \mathbf{C}_1 \times \cdots \times \mathbf{C}_m$ and each $\mathbf{C}_i \in \mathcal{K}$. (Note that we do not assume that this representation of $\mathbf{C}$ is subdirect.) Since $\mathbf{A}$ is simple, $\alpha \prec 1$ in $\mathrm{Con}(\mathbf{C})$ and $\mathrm{typ}(\alpha, 1_C) = \mathrm{typ}(0_A, 1_A) \in \{3, 4\}$.

Let η_i be the projection kernel $\mathbf{C} \to \mathbf{C}_i$, for $i \leq m$, and let δ be the least congruence such that $\alpha \vee \delta = 1$ (which exists by Lemma 8.41). Certainly $\delta \neq 0$. Since $\bigwedge_i \eta_i = 0$, there is $j \leq m$ such that $\delta \nleq \eta_j$. By the minimality of δ, $\alpha \vee \eta_j \neq 1$. But $\alpha \prec 1$, so we must have $\eta_j \leq \alpha$. Therefore

$$\mathbf{A} \cong \mathbf{C}/\alpha \in \mathsf{H}(\mathbf{C}/\eta_j) \subseteq \mathsf{HS}(\mathcal{K}). \qquad \square$$

Inspired by group theory, an algebra $\mathbf{A}$ is called *solvable* if there is a finite sequence $0_A = \beta_0 < \beta_1 < \cdots < \beta_n = 1_A$ in $\mathrm{Con}(\mathbf{A})$, such that β_{k+1} is Abelian over β_k, for all $k < n$. Similarly, $\mathbf{A}$ is called *strongly solvable* if each β_{k+1} is strongly Abelian over β_k.

Suppose that $\mathbf{A}$ is a finite algebra. By Exercise 8.39.4, $\mathbf{A}$ is solvable if and only if $\mathbf{A}$ has no $(0_A, 1_A)$-2 snags. Theorem 8.27 implies that the absence of 2-snags is equivalent to $\mathrm{typ}\,\mathbf{A} \subseteq \{1, 2\}$. Analogous conditions hold for strong solvability. Finally, a locally finite algebra is called *locally solvable* (or *locally strongly solvable*) if every finite subalgebra is solvable (strongly solvable).

Proposition 8.42. *Let $\mathcal{V}$ be a locally finite variety. The class of locally solvable members of $\mathcal{V}$ forms a subvariety, as does the class of locally strongly solvable members.*

Proof. It is easy to see that the class of locally solvable algebras is closed under both subalgebra and arbitrary products. We must check closure under homomorphic images.

Let $\mathbf{A}$ be locally solvable and let $\mathbf{B} = \mathbf{A}/\theta$. Suppose that $\mathbf{B}$ fails to be locally solvable. Then there is a finite subalgebra $\mathbf{C}_0$ of $\mathbf{B}$, which is not solvable. This means there is a nonabelian prime quotient (α_0, β_0) in $\mathrm{Con}(\mathbf{C}_0)$.

By local finiteness, there is a finite subalgebra $\mathbf{C}$ of $\mathbf{A}$, such that $\mathbf{C}/(\theta{\restriction}_C) \cong \mathbf{C}_0$. Therefore, by the correspondence theorem, there are congruences α and β such that $\theta{\restriction}_C \leq \alpha \prec \beta \leq 1$ in $\mathrm{Con}(\mathbf{C})$ and β is nonabelian over α (by Exercise 8.39.1). Therefore $\mathrm{typ}\{\mathbf{C}\} \nsubseteq \{1, 2\}$. From our comments above, $\mathbf{C}$ is not solvable, contradicting the local solvability of $\mathbf{A}$.

A similar argument works for strong solvability. $\square$

Suppose that $\mathbf{A}$ and $\mathbf{B}$ are finite subdirectly irreducible algebras in a congruence-distributive variety. It follows from Corollary 5.11 that if $\mathsf{V}(\mathbf{A}) = \mathsf{V}(\mathbf{B})$ then $\mathbf{A} \cong \mathbf{B}$. A similar result is proved in [FM81] for finite simple algebras in a congruence-modular variety. Without a global assumption on $\mathcal{V}$, a relationship like this fails to hold. We can show, however, that the type of the generating algebra is preserved.

Theorem 8.43. *Let $\mathbf{A}$ and $\mathbf{B}$ be finite simple algebras and assume that $\mathsf{V}(\mathbf{A}) = \mathsf{V}(\mathbf{B})$. Then $\mathrm{typ}(\mathbf{A}) = \mathrm{typ}(\mathbf{B})$. If that type is 3 or 4, then $\mathbf{A} \cong \mathbf{B}$.*

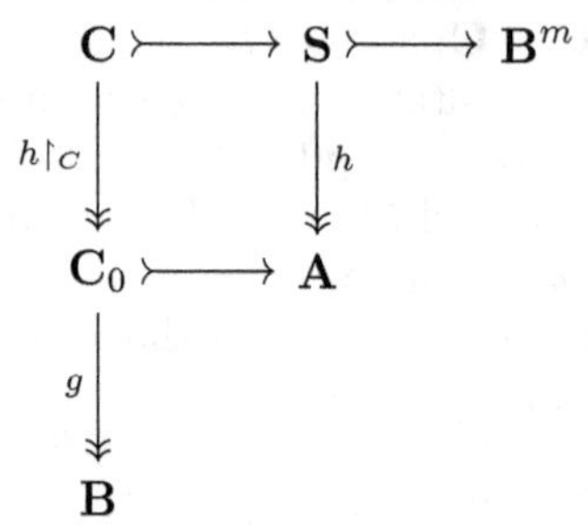

FIGURE 8.6: Diagram for Theorem 8.43

Proof. Let $\mathcal{V} = \mathsf{V}(\mathbf{A}) = \mathsf{V}(\mathbf{B})$. If $\text{typ}(\mathbf{A}) = 1$ then $\mathbf{A}$ is strongly solvable, so by Proposition 8.42, $\mathcal{V}$ is locally strongly solvable. Since $\mathbf{B} \in \mathcal{V}$ we must have $\text{typ}(\mathbf{B}) = 1$. If $\text{typ}(\mathbf{A}) = 2$ then $\mathcal{V}$ is still locally solvable, so $\text{typ}(\mathbf{B}) \in \{1, 2\}$. But if $\text{typ}(\mathbf{B}) = 1$ then arguing in reverse, $\text{typ}(\mathbf{A}) = 1$ which is false.

If the types of $\mathbf{A}$ and $\mathbf{B}$ are both in $\{3, 4\}$ then applying Theorem 8.40, $\mathbf{A} \in \mathsf{HS}(\mathbf{B})$ and $\mathbf{B} \in \mathsf{HS}(\mathbf{A})$. By finiteness $\mathbf{A} \cong \mathbf{B}$, so obviously $\text{typ}(\mathbf{A}) = \text{typ}(\mathbf{B})$.

The only combination left to rule out is $\text{typ}(\mathbf{A}) = 5$ and $\text{typ}(\mathbf{B}) \in \{3, 4\}$. So suppose that $\mathbf{A}$ and $\mathbf{B}$ have those types and let us derive a contradiction. Since $\mathbf{A} \in \mathsf{HSP}(\mathbf{B})$, there is an integer m, an algebra $\mathbf{S} \leq \mathbf{B}^m$, and a surjective homomorphism $h \colon \mathbf{S} \to \mathbf{A}$. Take m to be as small as possible, and let $\theta = \ker h$. Since $\mathbf{A}$ is simple, θ must be a coatom of $\text{Con}(\mathbf{S})$. See Figure 8.6.

On the other hand, since $\mathbf{B}$ is of Boolean or lattice type, by Theorem 8.40, $\mathbf{B} \in \mathsf{HS}(\mathbf{A})$. Therefore, there is a subalgebra $\mathbf{C}_0$ of $\mathbf{A}$ and a surjective homomorphism $g \colon \mathbf{C}_0 \to \mathbf{B}$. Let $C = \overleftarrow{h}(C_0)$. This is a subuniverse of $\mathbf{S}$. Let $\delta = \ker(g \circ h{\restriction}_C)$. Since $\mathbf{B}$ is simple, δ is a coatom of $\text{Con}(\mathbf{C})$. Also, $\delta \geq \theta{\restriction}_C$.

For $i \leq m$ let $p_i \colon \mathbf{S} \to \mathbf{B}$ be the projection map, and write $\eta_i = \ker(p_i)$. Using the same argument as in Theorem 8.40, there is an index $j \leq m$ such that $\eta_j{\restriction}_C \leq \delta$. Let us write η in place of η_j. We have

$$\mathbf{C}/(\eta{\restriction}_C) \leq \mathbf{S}/\eta \leq \mathbf{B} \text{ and } \mathbf{C}/\delta \cong \mathbf{B}.$$

Since all of these algebras are finite, it follows on cardinality grounds that $\delta = \eta{\restriction}_C$ and all three of the above embeddings are isomorphisms. In particular,

$$\text{for every } b \in B \text{ there is } c \in C \text{ such that } p_j(c) = b. \tag{8–11}$$

Let $\eta' = \bigwedge_{i \neq j} \eta_i$. By the minimality of m we must have $\eta' \not\leq \theta$. Pick a pair $(a, b) \in \eta' - \theta$. By Theorem 8.17(2),

$$(\forall U \in \text{M}_{\mathbf{S}}(\theta, 1))\ (\exists f \in \text{Pol}_1(\mathbf{S}))\ \vec{f}(S) = U\ \&\ (f(a), f(b)) \notin \theta. \tag{8–12}$$

Let $U_1 \in \mathrm{M}_{\mathbf{C}}(\eta{\restriction}_C, 1)$. Consider

$$\mathcal{G} = \{\, f \in \mathrm{E}(\mathbf{S}) : \; f{\restriction}_C \in \mathrm{E}(\mathbf{C}), \text{ and } \vec{f}(C) = U_1 \,\}.$$

Claim. $\mathcal{G}$ is nonempty.

Proof. By Theorem 8.13, there is $r \in \mathrm{E}(\mathbf{C})$ such that $\vec{r}(C) = U_1$. By Exercise 8.25.2 there is $\bar{r} \in \mathrm{Pol}_1(\mathbf{S})$ such that $\bar{r}{\restriction}_C = r$. By Lemma 8.24, there is an integer k such that $\bar{r}^{2k} = \bar{r}^k$. Since $\bar{r}^k{\restriction}_C = r^k = r$, we see that $\bar{r}^k \in \mathcal{G}$.

Since $\mathcal{G}$ is nonempty, we can pick an $e \in \mathcal{G}$ that minimizes $\vec{e}(S)$ under inclusion. Because $\vec{e}(C) = U_1$ is $(\eta{\restriction}_C, 1)$-minimal, we see that $\vec{e}(1_C) \not\subseteq \eta{\restriction}_C$. On the other hand, $\theta{\restriction}_C \leq \delta = \eta{\restriction}_C$, so we conclude $\vec{e}(1_S) \not\subseteq \theta$. Therefore, by another application of 8.17, there is $U \in \mathrm{M}_{\mathbf{S}}(\theta, 1)$ such that $\vec{e}(S) \supseteq U$.

Using (8–12) we can find a pair $(u, v) \in (\eta'{\restriction}_U) - \theta$. By (8–11), there is $u' \in C$ such that $p_j(u') = p_j(u)$, i.e., $(u, u') \in \eta$. Therefore $u = e(u) \;\eta\; e(u')$. Setting $u_1 = e(u')$ and applying the same argument to v we obtain

$$\big(\exists(u,v) \in (\eta'{\restriction}_U) - \theta\big)\ \big(\exists u_1, v_1 \in U_1\big)\ \ (u, u_1) \in \eta \ \ \&\ \ (v, v_1) \in \eta.$$

Let N_1 be a trace of U_1. Since $\mathrm{typ}(\eta{\restriction}_C, 1) = \mathrm{typ}(\mathbf{B}) \in \{3, 4\}$, we know from 8.33 that $|N_1| = 2$. We can write $N_1 = \{0, 1\}$.

Since $(u, v) \in \eta' - \theta$ we surely have $u \neq v$, so $(u, v) \notin \eta$ (since $\eta \wedge \eta' = 0$). It follows that $(u_1, v_1) \notin \eta$. Moreover, since N_1 is the unique trace of U_1, we can assume $u_1 = 0$ and $v_1 = 1$, and therefore both $(u, 0)$ and $(v, 1)$ are members of η.

Claim. The clone of $\mathbf{S}|_{\{u,v\}}$ contains lattice operations.

Assume for the moment that the claim holds. Let N be a trace of U. Since $\mathrm{typ}(\theta, 1) = \mathrm{typ}(\mathbf{A}) = 5$, N consists of two θ-classes, namely u/θ and v/θ. It follows from the claim that $(\mathbf{S}|_N)/\theta$ is a minimal algebra of type 3 or 4. This implies $\mathrm{typ}(\theta, 1) \in \{3, 4\}$, which is a contradiction, proving the theorem.

It remains to prove the claim. Since $\mathrm{typ}(\eta{\restriction}_C, 1_C) \in \{3, 4\}$, there is a binary polynomial q_1 on $\mathbf{C}$ which acts like a join operation on N_1. By Exercise 8.25.2 there is $q \in \mathrm{Pol}_2(\mathbf{S})$ such that $q{\restriction}_C = q_1$. Take $p(x) = eq(x, x)$ and choose n such that $p^{2n} = p^n$ (Lemma 8.24). Then $p^n \in \mathcal{G}$, so the minimality of e implies that $\overrightarrow{p^n}(S) = \vec{e}(S) \supseteq U$.

Finally, define $r(x, y) = p^{n-1}eq(x, y)$. Then r still behaves like the join operation on N_1. Also, $r(u, u) = p^n(u) = u$, $r(v, v) = v$ and

$$r(u, v) \equiv_{\eta'} r(v, v) = v$$
$$r(u, v) \equiv_{\eta} r(0, 1) = 1 \equiv_{\eta} v.$$

But $\eta \wedge \eta' = 0_S$, so these imply $r(u, v) = v$. Similarly, $r(v, u) = v$. Thus r is a join operation on $\{u, v\}$. A similar argument produces a meet operation. □

f_1	0	1	2
0	0	0	0
1	2	1	0
2	1	0	2

f_2	0	1	2
0	0	0	0
1	0	1	2
2	1	1	2

FIGURE 8.7: Operation tables for Example 8.44

Example 8.44. Consider the algebras $\mathbf{A}_i = \langle \{0,1,2\}, f_i \rangle$, for $i = 1, 2$, in which the operation tables for f_1 and f_2 are given in Figure 8.7. Both algebras are simple, so Theorem 8.43 applies. However, $\mathbf{A}_1$ has type 3 while $\mathbf{A}_2$ has type 4. Thus the theorem assures us that they do not generate the same variety. In fact, it is not hard to check that $\mathbf{A}_2 \models y \cdot x \approx y \cdot (y \cdot x)$ while $\mathbf{A}_1$ fails to satisfy this identity. (Take $x = 0$, $y = 1$.) We can also apply Theorem 8.40 to conclude that $\mathbf{A}_2 \notin \mathbf{V}(\mathbf{A}_1)$ as well. The identity $(x \cdot y) \cdot x \approx (y \cdot x) \cdot y$ holds in $\mathbf{A}_1$ but fails in $\mathbf{A}_2$, with $x = 0$, $y = 2$. This example comes from [BB96b].

Theorems 8.40 and 8.43 certainly suggest a close connection between the types 3 and 4 and congruence-distributivity. In fact Exercise 8.25.7 asserts that if $\mathcal{V}$ is congruence-distributive then $\operatorname{typ}\{\mathcal{V}\} \subseteq \{3, 4\}$. Put another way, $\mathcal{V}$ omits types $\{1, 2, 5\}$. However, the converse of this is not true, even for locally finite varieties.

On $\{0, 1\}$ define the following operations and algebras:

$$f_1(x) = x, \quad f_2(x) = x', \quad g_1(x) = x', \quad g_2(x) = 1$$
$$\mathbf{P}_i = \langle \{0,1\}, \wedge, f_i, g_i, 0, 1 \rangle, \quad \text{for } i = 1, 2.$$

Let $\mathbf{P} = \mathbf{P}_1 \times \mathbf{P}_2$. The variety generated by $\mathbf{P}$ is known as *Polin's variety,* [DF80, Pol77]. It is not hard to see that $\operatorname{Con}(\mathbf{P})$ is

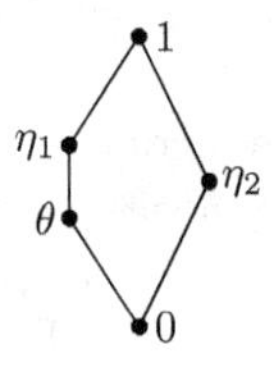

In this picture, η_1 and η_2 are the usual projection congruences and θ is the congruence identifying $(0,0)$ with $(0,1)$. Obviously $\mathbf{V}(\mathbf{P})$ is not congruence-distributive. Although we shall not prove it here, it is a fact that $\operatorname{typ}\{\mathbf{V}(\mathbf{P})\} = \{3\}$. Thus, there must be some other consequence of congruence-distributivity that fails to hold in this variety.

The answer lies in the tails. Although type 3 implies that the trace is unique and has a distributive congruence lattice, it does not follow that the entire minimal set is congruence-distributive. With a bit of computation one can show that the only (θ, η_1)-minimal set is P itself. Of course the trace of this minimal set has two elements. The remaining two elements lie in the tail. The congruence lattice of the minimal set is still $\mathbf{N}_5$.

Let us say that a finite algebra $\mathbf{A}$ has *empty tails* if, for every $\alpha \prec \beta$ in $\mathrm{Con}(\mathbf{A})$, the tail of every (α, β)-minimal set is empty. (Equivalently, every minimal set is equal to its body.)

Proposition 8.45. *Every congruence-distributive variety has empty tails.*

Proof. Let $p_0, p_1, \ldots, p_n$ be Jónsson terms (Theorem 4.66) for a congruence-distributive variety, $\mathcal{V}$. Let $\mathbf{A}$ be a finite algebra in $\mathcal{V}$ and (α, β) a prime congruence quotient of $\mathbf{A}$. As we have already observed, $\mathrm{typ}(\alpha, \beta) \in \{3, 4\}$. By replacing $\mathbf{A}$ by $\mathbf{A}/\alpha$, we can assume that $\alpha = 0$. Let U be a minimal set. By Theorem 8.33, $N = \{0, 1\}$ is the unique trace of U.

Since U is a neighborhood, there is $e \in \mathrm{E}(\mathbf{A})$ such that $\vec{e}(A) = U$. By replacing each p_i with ep_i we can assume that both U and N are closed under $p_0, \ldots, p_n$. Note that the p_i's will still satisfy the conditions in Theorem 4.66(c).

Suppose, by way of contradiction, that c is an element of the tail of U, i.e., $c \in U$ and $c \notin N$. For $i \leq n$, let $f_i(x) = p_i(0, x, 1)$. Then

$$f_i(c) = p_i(0, c, 1) \equiv_\beta p_i(0, c, 0) = 0$$

by 4.66(c)(i). Thus $f_i(c) \in N$. Since f_i must map the finite set N to itself and it maps c into N, it must be the case that f_i is not a permutation of U. Since U is $(0, \beta)$-minimal, it must be that f_i is constant on U.

We argue by induction that for all $i \leq n$, $f_i(x) = 0$ on U. Since each $f_i{\restriction}_U$ is constant, it is enough to show this for a single value of x. For this we use the identities in 4.66(c). First, $f_0(x) = p_0(0, x, 1) = 0$ by (ii). Then, if i is even,

$$0 = f_i(0) = p_i(0, 0, 1) = p_{i+1}(0, 0, 1) = f_{i+1}(0)$$

by (iv), while if i is odd

$$0 = f_i(1) = p_i(0, 1, 1) = p_{i+1}(0, 1, 1) = f_{i+1}(1)$$

using (v).

But now, using (iii), $0 = f_n(x) = 1$, a contradiction. □

As we shall now show, these two conditions: $\mathrm{typ}(\mathcal{V}) \subseteq \{3, 4\}$ and empty tails, are both necessary and sufficient for congruence distributivity. Note that under the assumption that the types are all 3 and 4, the condition "empty tails" is equivalent to "every minimal set has cardinality 2." We require one preliminary observation.

Proposition 8.46. *Let $\mathbf{A}$ be a finite algebra and let $(\alpha_1, \beta_1), \ldots, (\alpha_n, \beta_n)$ be a list of all of the prime congruence quotients in $\mathbf{A}$. For each $i \leq n$ choose $U_i \in \mathrm{M}_A(\alpha_i, \beta_i)$. There is a natural subdirect representation*

$$\mathbf{Con}(\mathbf{A}) \rightarrowtail \prod_{i=1}^{n} \mathbf{Con}(\mathbf{A}|_{U_i}) \ .$$

Proof. By Theorems 8.10 and 8.13 the restriction maps r_i from $\mathbf{Con}(\mathbf{A})$ to $\mathbf{Con}(\mathbf{A}|_{U_i})$ are surjective lattice homomorphisms. Let $\Theta = \bigcap_i \ker(r_i)$. If $\Theta \neq 0$ then (since $\mathrm{Con}(\mathbf{A})$ is finite) there are congruences $\alpha \prec \beta$ such that $(\alpha, \beta) \in \Theta$. But by assumption, there is some $i \leq n$ such that $(\alpha, \beta) = (\alpha_i, \beta_i)$. Then $\alpha\restriction_{U_i} \neq \beta\restriction_{U_i}$, which contradicts $(\alpha, \beta) \in \ker(r_i)$. □

Theorem 8.47. *Let $\mathcal{V}$ be locally finite. $\mathcal{V}$ is congruence-distributive if and only if* $\mathrm{typ}\{\mathcal{V}\} \subseteq \{3, 4\}$ *and $\mathcal{V}$ has empty tails.*

Proof. Exercise 8.25.7 and Proposition 8.45 show the necessity of the two conditions. So assume that $\mathrm{typ}\{\mathcal{V}\} \subseteq \{3,4\}$ and that $\mathcal{V}$ has empty tails. Let $\mathbf{A} = \mathbf{F}_{\mathcal{V}}(3)$. By local finiteness, A is finite. By Theorem 4.66 it is enough to show that $\mathbf{Con}(\mathbf{A})$ is distributive. The above proposition reduces this to checking that $\mathbf{Con}(\mathbf{A}|_U)$ is distributive, where (α, β) is a prime quotient and $U \in \mathrm{M}_{\mathbf{A}}(\alpha, \beta)$.

But since the type of (α, β) is assumed to be either lattice or Boolean, U has a unique trace, $N = \{0, 1\}$, and since $\mathcal{V}$ has empty tails, $U = N$. Since $\mathbf{A}|_N$ is polynomially equivalent to either a two-element lattice or Boolean algebra, it is congruence-distributive. □

Theorem 8.47 demonstrates the deep connection between the "classical" notion of congruence-distributivity and the recent discoveries of tame congruence theory. Hobby and McKenzie [HM88] derive a similar result for congruence-modularity, namely that $\mathcal{V}$ omit types 1 and 5 and has empty tails. Valeriote and Willard [VW91] characterize locally finite Maltsev varieties with a somewhat analogous condition.

Given a finite algebra $\mathbf{A}$, is it possible to verify that the generated variety satisfies the conditions in Theorem 8.47? In general, the answer is no. A recent deep result of Freese and Valeriote [FV09] shows that there is no polynomial-time algorithm that accepts a finite algebra as input and determines whether it generates a congruence-distributive variety.

By contrast, the same paper contains a proof that if $\mathbf{A}$ is a finite *idempotent* algebra, then the conditions in Theorem 8.47 can be checked rather easily. In fact, if $\mathbf{A}$ is finite and idempotent, and if $\mathsf{S}(\mathbf{A}^2)$ has only lattice and Boolean types and has empty tails, then $\mathsf{V}(\mathbf{A})$ is congruence-distributive. Since the argument is a nice application of the concepts developed in this chapter, we present the proof.

Recall that $\mathcal{T}$ is the poset of types, as pictured in Figure 8.4. We shall write $i \leq_{\mathcal{T}} j$ to indicate that i lies below j in this poset. In Section 7.6 we defined an algebra to be *strictly simple* if it is finite, simple, and has no proper nontrivial subalgebras.

Lemma 8.48. *Let $\mathcal{V}$ be a locally finite, idempotent variety and suppose that $i \in \mathrm{typ}\{\mathcal{V}\}$. Then $\mathcal{V}$ contains a strictly simple algebra $\mathbf{B}$ with* $\mathrm{typ}(\mathbf{B}) \leq_{\mathcal{T}} i$.

Proof. Every locally finite variety contains a strictly simple algebra (see the discussion on page 234). Thus, if $i = 3$, the claim is immediate. So let us assume that $i \neq 3$.

The class $\mathcal{S} = \{\mathbf{A} \in \mathcal{V}_{\text{fin}} : (\exists j \in \text{typ}\{\mathbf{A}\})\ j \leq_{\mathcal{T}} i\}$ is nonempty by assumption. Let $\mathbf{B}$ be a member of $\mathcal{S}$ of smallest cardinality. By the minimality of $\mathbf{B}$, there is a congruence $\beta \succ 0$ such that $\text{typ}(0, \beta) = j \leq_{\mathcal{T}} i$. Let C be a nontrivial β-class of $\mathbf{B}$. Since $\mathcal{V}$ is idempotent, C is a subuniverse.

If $j = 2$ then β is Abelian, hence

$$[1_C, 1_C] = [\beta{\restriction}_C, \beta{\restriction}_C] \leq [\beta, \beta]{\restriction}_C = 0_C$$

so $\mathbf{C}$ is a finite Abelian algebra. Choose a nontrivial algebra $\mathbf{D}$ in $\mathsf{HS}(\mathbf{C})$ of minimal cardinality. Then $\mathbf{D}$ is strictly simple and Abelian. Hence $\text{typ}(\mathbf{D}) \leq_{\mathcal{T}} 2 \leq_{\mathcal{T}} i$. $\mathbf{D}$ is the strictly simple algebra required by the lemma. A similar argument works for $j = 1$ using the strong Abelian property.

For the remainder of the proof we can assume that $j \in \{4, 5\}$. Let $N = \{0, 1\}$ be a $(0, \beta)$-trace of $\mathbf{B}$ and $C = \text{Sg}^{\mathbf{B}}(\{0, 1\})$. Define $\gamma = \text{Cg}^{\mathbf{C}}(0, 1)$ and choose $\alpha \in \text{Con}(\mathbf{C})$ with $\alpha \prec \gamma$. Note that $(0, 1) \notin \alpha$. We shall show that $\text{typ}(\alpha, \gamma) \leq_{\mathcal{T}} j$.

Let $U \in \text{M}_{\mathbf{C}}(\alpha, \gamma)$. By 8.17(2), there is a unary polynomial f_0 of $\mathbf{C}$ such that $\vec{f_0}(C) = U$ and $f_0(0) \not\equiv_\alpha f_0(1)$. Let M be a trace of U containing $f_0(0)$ and $f_0(1)$. Using Exercise 8.25.2 there is $f \in \text{Pol}_1(\mathbf{B})$ with $f{\restriction}_C = f_0$. Then by 8.17(1), $N' = \vec{f}(N)$ is a $(0, \beta)$-trace and $N' \subseteq M$.

Since β is nonabelian, $(f(0), f(1))$ is a $(0, \beta)$-2 snag, hence also an (α, γ)-2 snag. By Proposition 8.35, $\text{typ}(\alpha, \gamma) \neq 2$.

Suppose $\text{typ}(\alpha, \gamma) = 3$. Then by Corollary 8.36, $|M| = 2$, so $M = N' = \{f(0), f(1)\}$. The algebra $\mathbf{C}|_M$ is Boolean, so it has a polynomial g_0 that acts like complementation, i.e., $g_0 f(0) = f(1)$ and $g_0 f(1) = f(0)$. As in the previous paragraph, there is $g \in \text{Pol}(\mathbf{B})$ such that $(g{\restriction}_C){\restriction}_{N'} = g_0$. Then g induces a complementation on $\mathbf{B}|_{N'}$ which contradicts $\text{typ}(0, \beta) \in \{4, 5\}$.

From these two paragraphs, we conclude that if $j = 4$ then $\text{typ}(\alpha, \gamma) \leq_{\mathcal{T}} j$. So suppose that $j = 5$ and $\text{typ}(\alpha, \gamma) = 4$. Then $M = N'$ has cardinality 2. Therefore $\mathbf{C}|_{N'}$ has lattice operations while $\mathbf{B}|_{N'}$ does not, which is impossible.

To summarize, we have shown that $\text{typ}(\alpha, \gamma) \leq_{\mathcal{T}} j$, $j \in \{4, 5\}$, and $\mathbf{C}$ is a subalgebra of $\mathbf{B}$. By the minimality of $\mathbf{B}$, we must have $B = C$, which is to say, $\mathbf{B}$ is generated by any one of its traces. Since $B = C = \text{Sg}(\{0, 1\}) \subseteq 0/\beta \subseteq B$ we conclude that $\beta = 1_B$, i.e., $\mathbf{B}$ is simple.

Finally, we will show that $\mathbf{B}$ has no proper, nontrivial subalgebras. For this it is enough to show that if $a \neq b$ in B then $\text{Sg}^{\mathbf{B}}(\{a, b\}) = B$. By Exercise 8.25.3, there is an element c such that $\{a, c\}$ is a trace. Since every trace generates $\mathbf{B}$, there is a binary term t such that $t(a, c) = b$. By idempotence, $t(a, a) = a$. Hence, the binary polynomial $t(a, x)$ maps the trace $\{a, c\}$ to the trace $\{a, b\}$, so $\{a, b\}$ generates $\mathbf{B}$. □

Theorem 8.49. *Let* $\mathbf{A}$ *be a finite idempotent algebra and* $\mathcal{D}$ *a downset of* $\mathcal{T}$. *If* $\mathsf{S}(\mathbf{A})$ *omits* $\mathcal{D}$ *then* $\mathsf{V}(\mathbf{A})$ *omits* $\mathcal{D}$.

Proof. Assume $\text{typ}\{\mathbf{V}(\mathbf{A})\} \cap \mathcal{D} \neq \varnothing$. By Lemma 8.48, there is a strictly simple algebra $\mathbf{S}$ in $\mathbf{V}(\mathbf{A})$ with $\text{typ}(\mathbf{S}) \in \mathcal{D}$. Since $\mathbf{V}(\mathbf{A}) = \mathbf{HSP}(\mathbf{A})$ is locally finite, there is an integer n, a finite subalgebra, $\mathbf{B}$, of $\mathbf{A}^n$, and a surjective homomorphism $h\colon \mathbf{B} \to \mathbf{S}$. Choose $\mathbf{B}$ to minimize n. We shall show that $n = 1$.

Let $\theta = \ker(h)$. Note that $\theta \prec 1_B$ since $\mathbf{S}$ is simple. So $\text{typ}(\theta, 1) = \text{typ}(\mathbf{S}) \in \mathcal{D}$. Let $p_1\colon \mathbf{B} \to \mathbf{A}$ be the projection to the first coordinate and let A' be the image of p_1 in $\mathbf{A}$. Note that $|A'| > 1$ by the minimality of n.

For every $a \in A'$ define $B_a = \overleftarrow{p}_1(a)$. Idempotence implies that each B_a is a subalgebra of $\mathbf{B}$. Thus $\vec{h}(B_a)$ is a subalgebra of $\mathbf{S}$. But $\mathbf{S}$ is strictly simple, so either $|\vec{h}(B_a)| = 1$ or $\vec{h}(B_a) = S$. The latter case, since $\mathbf{B}_a \leq \mathbf{A}^{n-1}$, contradicts the minimality of n.

Thus, for every $a \in A'$, $|\vec{h}(B_a)| = 1$. This means that $\ker(p_1) \leq \theta$. But then $\mathbf{S} \cong \mathbf{B}/\theta \in \mathbf{HS}(\mathbf{A})$ which implies that $n = 1$. □

By taking $\mathcal{D} = \{1, 2, 5\}$ we see that verifying the first condition of Theorem 8.47 for $\mathbf{V}(\mathbf{A})$ requires that we check only the prime quotients in $\mathbf{S}(\mathbf{A})$. We now show that we can do the same with the second condition, although it requires the prime quotients in $\mathbf{S}(\mathbf{A}^2)$.

Theorem 8.50. *Let* $\mathbf{A}$ *be a finite idempotent algebra.* $\mathbf{V}(\mathbf{A})$ *is congruence-distributive if and only if* $\text{typ}\{\mathbf{S}(\mathbf{A})\} \subseteq \{3, 4\}$ *and* $\mathbf{S}(\mathbf{A}^2)$ *has empty tails.*

Proof. We have already discussed the left-to-right direction. So let us assume the two conditions in the statement of the theorem. Theorem 8.49 implies that $\mathbf{V}(\mathbf{A})$ omits types 1, 2, and 5. In order to apply Theorem 8.47 we must show that $\mathbf{V}(\mathbf{A})$ has empty tails.

Let us assume that there is a finite algebra $\mathbf{D}$ in $\mathbf{V}(\mathbf{A})$ and $\bar{\alpha} \prec \bar{\beta}$ in $\text{Con}(\mathbf{D})$ such that an $(\bar{\alpha}, \bar{\beta})$-minimal set has a tail. We shall derive a contradiction.

By local finiteness, there is a positive integer n, a finite algebra $\mathbf{B} \leq \mathbf{A}^n$ and a congruence $\theta \in \text{Con}(\mathbf{B})$ such that $\mathbf{D} \cong \mathbf{B}/\theta$. By the correspondence theorem, there are $\alpha, \beta \in \text{Con}(\mathbf{B})$ such that $\theta \leq \alpha \prec \beta$ and every (α, β)-minimal set has a tail. Thus we can forget about $\mathbf{D}$ and assume that there is a finite subpower of $\mathbf{A}$ with nonempty tails.

Choose the minimal possible n, and for that n choose $\mathbf{B}$ of minimal cardinality. Now choose a minimal congruence β on $\mathbf{B}$ such that there is a congruence $\alpha \prec \beta$ so that the (α, β)-minimal sets have tails. Let U be such a minimal set and N a trace of U. Since $\text{typ}(\alpha, \beta) \in \{3, 4\}$, N is the unique trace and $N = \{0, 1\}$.

Claim 1. $\beta = \text{Cg}^{\mathbf{B}}(0, 1)$.

Proof. We first show that β is join irreducible. If not, then $\beta = \alpha \vee \beta_1$ for some congruence $\beta_1 < \beta$. Choose β_1 as small as possible. Let $\alpha_1 = \alpha \wedge \beta_1$. Then the minimality of β_1 and the fact that $\alpha \prec \beta$ implies that $\alpha_1 \prec \beta_1$. By Exercise 8.39.3, $\text{M}(\alpha, \beta) = \text{M}(\alpha_1, \beta_1)$, so (α_1, β_1) has a minimal set with a tail. This contradicts the minimality of β.

So β is join-irreducible, with lower cover α. If $\mathrm{Cg}(0,1) < \beta$ then we would have $\mathrm{Cg}(0,1) \leq \alpha$. But this is false since $(0,1) \notin \alpha$. This proves Claim 1.

By assumption, U has a tail, i.e., $U - N$ is nonempty.

Claim 2. For any $t \in U - N$, B is generated by $\{0,1,t\}$.

Proof. Let $C = \mathrm{Sg}^{\mathbf{B}}(\{0,1,t\})$. Suppose that $C \subset B$. Let $\alpha' = \alpha{\restriction}_C$ and $\beta' = \beta{\restriction}_C$. Then $\alpha' < \beta'$ and $(0,1) \in \beta' - \alpha'$. There is a pair of congruences δ and γ such that $\alpha' \leq \delta \prec \gamma \leq \beta'$ and $(0,1) \in \gamma - \delta$.

Let $V \in \mathrm{M}_{\mathbf{C}}(\delta,\gamma)$. By the minimality of $\mathbf{B}$, V must have an empty tail. Since $\mathrm{typ}(\delta,\gamma) \in \{3,4\}$, $|V| = 2$.

V is a neighborhood of $\mathbf{C}$, so by Theorem 8.17 and Exercise 8.25.2 there is $f \in \mathrm{Pol}_1(\mathbf{B})$ such that $\vec{f}(C) = V$ and $f(0) \not\equiv_\delta f(1)$. From this we deduce that

$$V = \{f(0), f(1)\}, \quad f(t) \in V, \quad f(0) \not\equiv_\alpha f(1).$$

Therefore $f \in \mathrm{Sep}_{\mathbf{B}}(\alpha,\beta)$. By Theorem 8.17 there is a minimal set $W = \vec{f}(U)$ and $W \simeq U$. Then $\vec{f}(N)$ will be the trace of W. But t is in the tail of U while $f(t) \in \vec{f}(N)$ is in the body of W. This is impossible since f is a polynomial isomorphism. This contradiction shows that $C = B$, proving the claim.

Since $\mathbf{B} \leq \mathbf{A}^n$, there are projection maps $p_i \colon \mathbf{B} \to \mathbf{A}$, for $i = 1,\ldots,n$. Let $\eta_i = \ker(p_i)$. For a *proper* subset I of $\{1,2,\ldots,n\}$, let $\rho_I = \bigcap_{i\in I} \eta_i$. Note that since I is a proper subset, the minimality of n requires that $\rho_I > 0$ in $\mathrm{Con}(\mathbf{B})$.

Claim 3. If $\beta \not\leq \rho_I$ then $\beta \wedge \rho_I = \alpha \wedge \rho_I$ and $\beta \vee \rho_I = \alpha \vee \rho_I$.

Proof. Let us write ρ in place of ρ_I. By Claim 1, $(0,1) \notin \rho$. Therefore $\beta \wedge \rho < \beta$, so $\beta \wedge \rho \leq \alpha$. Hence $\beta \wedge \rho = \alpha \wedge \rho$.

Suppose that $\alpha \vee \rho < \beta \vee \rho$. Then $\alpha \vee \rho \not\geq \beta$. Choose a congruence $\delta \in \mathbf{I}[\alpha \vee \rho, \beta \vee \rho]$ maximal with the property that $\delta \not\geq \beta$. Let γ be an upper cover of δ. By the maximality of δ, $\gamma \geq \beta$ so $\delta \vee \beta = \gamma$. Also, $\alpha \leq \delta \wedge \beta < \beta$, so $\delta \wedge \beta = \alpha$.

This shows that $\mathbf{I}[\alpha,\beta] \nearrow \mathbf{I}[\delta,\gamma]$. By Exercise 8.39.3, $\mathrm{M}(\alpha,\beta) = \mathrm{M}(\delta,\gamma)$. Therefore there is a (δ,γ)-minimal set of cardinality greater than 2, which is to say, it has a tail. But then the algebra $\mathbf{B}/\rho$ has a $(\delta/\rho,\gamma/\rho)$-minimal set with a tail. Since $\mathbf{B}/\rho \leq \mathbf{A}^{n-1}$, this contradicts the minimality of n.

Claim 4. Either $\beta \leq \rho_I$ or $\alpha \vee \rho_I = 1$.

Proof. Let us again write ρ in place of ρ_I. Assume that $\beta \not\leq \rho$. By Claim 3, $\alpha \wedge \rho = \beta \wedge \rho$ and $\alpha \vee \rho = \beta \vee \rho$. This says that $\mathrm{Con}(\mathbf{B})$ contains a pentagon:

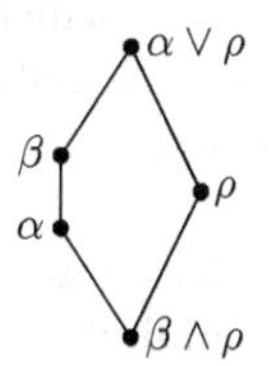

Let $C = 0/(\alpha \vee \rho)$. Since $\mathbf{B}$ is idempotent, C is a subuniverse. Note that since $\alpha \vee \rho \geq \beta$ and $(0,1) \in \beta$, we have $1 \in C$.

Let $M = C \cap U$. Since C is a congruence class in B, M is a congruence class in U. Therefore by Lemma 8.37, the restriction map is a surjective lattice homomorphism from $\mathbf{I}[0_B, \alpha \vee \rho]$ to $\mathbf{I}[0_M, (\alpha \vee \rho)\restriction_M]$. Since $0, 1 \in M$ we have $\alpha\restriction_M \prec \beta\restriction_M$. Therefore the subdirect indecomposability of $\mathbf{N}_5$ implies that $\mathrm{Con}(\mathbf{B}|_M)$ must contain a pentagon, so surely $|M| > 2$.

But $M = C \cap U$, so M must contain some element, t, from the tail of U. Therefore $\{0, 1, t\} \subseteq C$, so by Claim 2, $C = B$. But then $\alpha \vee \rho = 1$, proving the claim.

Claim 5. Assume that $n > 1$. There are $i, j \leq n$ such that $\alpha \vee \eta_i < 1$ and $\alpha \vee \eta_j = 1$.

Proof. Suppose that for every $i \leq n$, $\alpha \vee \eta_i = 1$. Choose a congruence γ such that $\alpha \leq \gamma \prec 1$. Since $\mathrm{typ}(\gamma, 1) \in \{3, 4\}$, Lemma 8.41 yields a smallest congruence δ such that $\gamma \vee \delta = 1$.

Then, for every $i \leq n$, $\alpha \leq \gamma \implies 1 = \alpha \vee \eta_i = \gamma \vee \eta_i$. Hence $\delta \leq \eta_i$ by the minimality of δ. But then $\delta \leq \bigwedge_i \eta_i = 0$, which means that $1 = \gamma \vee \delta = \gamma$, which is false.

On the other hand, suppose that for all $i \leq n$, $\alpha \vee \eta_i < 1$. By Claim 4, (with $I = \{i\}$) $\beta \leq \eta_i$ for all i, so just as above, $\beta = 0$, which is also false.

Finally, we can complete the proof by showing that $n \leq 2$, for then, $\mathbf{B}$ is a member of $\mathsf{S}(\mathbf{A}^2)$, which contradicts the assumption that $\mathsf{S}(\mathbf{A}^2)$ has no tails. Suppose that $n > 2$. By Claim 5, there are i and j such that $\alpha \vee \eta_i < 1$ and $\alpha \vee \eta_j = 1$. Let $\rho = \rho_{\{i,j\}} = \eta_i \wedge \eta_j$. Note that $\{i, j\}$ is a proper subset since $n > 2$. By Claim 4, either $\beta \leq \rho$ or $\alpha \vee \rho = 1$.

But $\alpha \vee \rho \leq \alpha \vee \eta_i < 1$ and if $\beta \leq \rho$ then $\alpha < \beta \leq \rho \leq \eta_j$, so $\alpha \vee \eta_j = \eta_j \neq 1$. This contradiction concludes the proof of the theorem.

□

As an example, consider a finite lattice $\mathbf{A}$. We showed in Exercise 8.25.4 that $\mathrm{typ}(\mathbf{A}) = \{4\}$ and that every minimal set has cardinality 2. It follows easily from Theorem 8.50 that $\mathbf{A}$ generates a congruence-distributive variety. (This should come as no surprise to the reader.)

Example 8.51. Let us return to the algebras in Example 8.44. Both algebras are idempotent, so we can apply Theorem 8.50. The algebra $\mathbf{A}_2$ has type 4, which is good news. However, $\{0, 1\}$ forms a subalgebra isomorphic to a 2-element semilattice, so that has type 5. Furthermore, $\{1, 2\}$ is a subalgebra isomorphic to a right-zero semigroup, which has type 1. Thus the variety generated by $\mathbf{A}_2$ is not congruence-distributive.

The algebra $\mathbf{A}_1$ is a different story. It has type 3 and no proper, nontrivial subalgebras. By Theorem 8.49, this is enough to conclude that $\mathrm{typ}(\mathsf{V}(\mathbf{A})) = \{3\}$. (Take $\mathcal{D} = \{1, 2, 4, 5\}$.) In order to apply Theorem 8.50, we must check that no subalgebra of $(\mathbf{A}_1)^2$ has a minimal set with a tail. Let us first observe that the polynomial $g(x) = x \cdot 1$ is idempotent and $\vec{g}(A_1) = \{0,1\}$ which must be a minimal set. Thus $\mathbf{A}_1$ itself has empty tails.

Since $\mathbf{A}_1$ is idempotent, $(\mathbf{A}_1)^2$ has quite a few subalgebras. There are of course nine 1-element subalgebras, but these are of no interest. There are also six subalgebras of the form $\overleftarrow{p}_i(a)$, for $a \in A_1$ and $i = 1, 2$. All of these are isomorphic to $\mathbf{A}_1$, so as we just observed, they have empty tails. There are also two automorphisms, namely, the identity and a mapping that transposes 1 and 2. As subalgebras, they are also isomorphic to $\mathbf{A}_1$. Finally, there is a subalgebra $B = \overleftarrow{p}_1(0) \cup \overleftarrow{p}_2(0)$, of order 5. $\mathbf{B}$ is a subdirect product of two copies of $\mathbf{A}_1$, and its only proper, nontrivial congruences are the projection kernels. Since $\mathbf{B}/\eta_i \cong \mathbf{A}_1$, we see that $(\eta_i, 1_B)$ has empty tails. Then using Exercise 8.39.3, we determine that $(0, \eta_i)$ also has empty tails.

Thus the theorem shows that $\mathsf{V}(\mathbf{A}_1)$ is congruence-distributive. Therefore, Theorem 4.66 guarantees that this variety has Jónsson terms. Using a computer [Fre11], we were able to find the following terms:

$$p_0(x,y,z) = x, \quad p_1(x,y,z) = (x(yx)) \cdot ((x(zx)) \cdot (z(yx))),$$
$$p_2(x,y,z) = (z(yx)) \cdot ((yz)x), \quad p_3(x,y,z) = z.$$

For our final application, we explore the outer reaches of "structure." As we have remarked before, structure is not a straightforward concept. One useful approach to the question is through Maltsev conditions.

Even properties that are not directly characterizable by Maltsev conditions often have them as consequences. Every directly representable variety is congruence-permutable, which is a Maltsev condition. Discriminator varieties are both arithmetical and congruence-regular. Both of these properties are characterized by Maltsev conditions.

Recall that we can formalize the notion of Maltsev condition through that of interpretation. Let us write $\mathcal{V} \leq_I \mathcal{W}$ if there is an interpretation of the variety $\mathcal{V}$ in $\mathcal{W}$. (See Definition 4.76 on page 131 for the definition of interpretation. In fact, the reader might want to review Section 4.8 before continuing.) Note that "$\leq_I$" is reflexive and transitive, but it is not anti-symmetric. For example there are interpretations of rings into Abelian groups and vice versa, but the two varieties are not even term-equivalent.

For a fixed, finitely based variety $\mathcal{W}$, the *strong Maltsev class* defined by $\mathcal{W}$ is the class of all varieties $\mathcal{V}$ such that $\mathcal{W} \leq_I \mathcal{V}$. The class of congruence-permutable varieties is an example of a strong Maltsev class.

A *Maltsev class* is defined by an infinite sequence $\mathcal{W}_0 \geq_I \mathcal{W}_1 \geq_I \mathcal{W}_2 \geq_I \cdots$ of finitely based varieties. $\mathcal{V}$ belongs to the Maltsev class if, for some $i \in \omega$, $\mathcal{V} \geq_I \mathcal{W}_i$. Congruence-distributivity is an example of this phenomenon. Finally, a *weak Maltsev class* is the intersection of a countable family of Maltsev classes.

For this example, we shall be restricting our attention to *idempotent* Maltsev conditions. That is, the defining varieties (the $\mathcal{W}$'s in the above paragraphs) will be required to be idempotent. This does not seem to be a severe restriction. In practice, most Maltsev conditions that have arisen in practice are idempotent.

If we want to understand minimal structure, we should first identify those varieties that have no structure at all. One example is clear: a set has no structure. To be more precise, an algebra with no basic operations has no identifiable structure. But we can cast our net a little wider. For example, a left-zero semigroup has a basic binary operation, but that operation is nothing but a projection, so a left-zero semigroup is term-equivalent to a set.

Definition 8.52. Let $\mathcal{S}et$ denote the variety of sets — that is, algebras with no basic operations. A variety, $\mathcal{W}$, is *passive* if and only if $\mathcal{W} \leq_I \mathcal{S}et$, otherwise it is *active*.

The word "passive" shouldn't be seen as having an overly negative connotation. It follows from our example above that the variety of semigroups is passive. Nevertheless, plenty of people spend their entire lives studying semigroups. We merely want to suggest a certain absence of structure. For instance, since $\mathcal{S}et$ is not congruence-distributive, we can immediately conclude that the variety of semigroups fails to be congruence-distributive. More generally, a passive variety can satisfy no nontrivial Maltsev condition.

In [Tay77] W. Taylor proved a converse to this observation, under the assumption of idempotence.

Definition 8.53. Let $\mathcal{W}$ be a variety, $n > 1$, and t an n-ary term of $\mathcal{W}$. We call t a *Taylor term* for $\mathcal{W}$ if $\mathcal{W}$ satisfies identities

$$\begin{aligned} t(x_{11}, x_{12}, \ldots, x_{1n}) &\approx t(y_{11}, y_{12}, \ldots, y_{1n}) \\ t(x_{21}, x_{22}, \ldots, x_{2n}) &\approx t(y_{21}, y_{22}, \ldots, y_{2n}) \\ &\vdots \\ t(x_{n1}, x_{n2}, \ldots, x_{nn}) &\approx t(y_{n1}, y_{n2}, \ldots, y_{nn}) \\ t(x, x, \ldots, x) &\approx x \end{aligned}$$

in which every x_{ij} and y_{ij} is a variable and $x_{ii} \neq y_{ii}$ for $1 \leq i, j \leq n$.

Every Maltsev term (for congruence-permutability) and majority term is a Taylor term. In fact, every near-unanimity term is Taylor. A binary term is Taylor if and only if it is idempotent and commutative. From these observations we see that almost every variety we have discussed in this text, including semilattices, has a Taylor term.

Theorem 8.54. *Let $\mathcal{W}$ be an idempotent variety. Then $\mathcal{W}$ is active if and only if $\mathcal{W}$ has a Taylor term.*

Before presenting the proof of Taylor's theorem, we isolate a construction, first presented by Padmanabhan and Quackenbush in [PQ73]. Let f be an n-ary, and g a k-ary operation. Define the kn-ary operation h by

$$h(x_{11}, \ldots, x_{nk}) = f\big(g(x_{11}, \ldots, x_{1k}), g(x_{21}, \ldots, x_{2k}), \ldots, g(x_{n1}, \ldots, x_{nk})\big).$$

If both f and g are idempotent then they can be obtained from h by repeating variables. In fact

$$\begin{aligned} f(y_1, \ldots, y_n) &= h(y_1, y_1, \ldots, y_1, y_2, y_2, \ldots, y_2, \ldots, y_n) \\ g(y_1, \ldots, y_k) &= h(y_1, y_2, \ldots, y_k, y_1, y_2, \ldots, y_k, \ldots, y_k). \end{aligned}$$

Let us denote this construction by $h = f \star g$. Note that if f and g are idempotent, then so is $f \star g$. By the way, when an operation f can be obtained from h by repeating and permuting variables, we call f a *polymer* of h.

Proof of Theorem 8.54. In the variety of sets, every term operation is a projection. On a nontrivial set, no projection operation satisfies the axioms of a Taylor term. Hence if $\mathcal{W}$ has a Taylor term then $\mathcal{W} \not\leq_I \mathcal{S}et$.

Now assume that $\mathcal{W}$ is active. We shall derive a Taylor term. Let Σ be an equational base for $\mathcal{W}$, and let F be the set of basic operation symbols. Suppose that $f \in F$ is m-ary. Let $\gamma(f)$ denote the first-order sentence

$$(\forall \mathbf{x}) \ \ (f(\mathbf{x}) \approx x_1) \curlyvee (f(\mathbf{x}) \approx x_2) \curlyvee \cdots \curlyvee (f(\mathbf{x}) \approx x_m)$$

where $\mathbf{x} = \langle x_1, x_2, \ldots, x_m \rangle$. This sentence says that f represents a projection operation. Notice that if D were an interpretation into $\mathcal{S}et$, then for any set S, S^D would be a model of $\gamma(f)$. But by assumption, no such interpretation from $\mathcal{W}$ exists. Therefore the set $\Sigma \cup \{ \gamma(f) : f \in F \}$ must fail to have a model. By the compactness theorem, there are finite sets Σ_0 of Σ and F_0 of F such that the operations appearing in Σ_0 come from F_0 and $\Sigma_0 \cup \{ \gamma(f) : f \in F_0 \}$ has no model.

Let $\mathcal{W}_0 = \mathrm{Mod}(\Sigma_0)$. Note that $\mathcal{W}_0$ is a finitely based variety of finite type and $\mathcal{W}_0 \not\leq_I \mathcal{S}et$. Since F_0 is finite, we can add finitely many additional identities to Σ_0 and ensure that $\mathcal{W}_0$ is idempotent.

An identity $s \approx t$ is called *height 1* if both s and t are terms of height 1. (s has height 1 if it is of the form $f(x_1, \ldots, x_m)$ for some basic operation f.) Our objective is to define a variety, $\mathcal{W}_2$, term-equivalent to $\mathcal{W}_0$, that has a base consisting of height 1 identities. We do this in two steps.

For the first step we use a simple device to reduce the height of each term. For each proper subterm of height 2 appearing in Σ_0, add to F_0 a new operation symbol. Replace the appearance of that subterm with the new symbol and add an identity equating the subterm with the new symbol.

An example should make this clear. Suppose that $F_0 = \{f_1, \ldots, f_5\}$ and that Σ_0 contains the identity

$$f_1(f_2(x, f_3(y, z), x) \approx f_4(f_5(x, y, z)).$$

Since the left-hand side has a proper subterm of height 2, we add a new operation symbol f_6 and replace this identity with the pair of identities

$$\begin{aligned} f_2(x, f_3(y, z)) &\approx f_6(x, y, z) \\ f_1(f_6(x, y, z), x) &\approx f_4(f_5(x, y, z)). \end{aligned}$$

By repeating this process, we can produce a set of identities in which every term has height at most 2. Let Σ_1 be the resulting set of identities in the operation symbols F_1 and $\mathcal{W}_1 = \mathrm{Mod}(\Sigma_1)$. Note that $\mathcal{W}_1$ is term-equivalent to $\mathcal{W}_0$.

Now, to squash a height 2 term down to height 1, we use the trick described just before the proof. By applying the "$\star$" operation repeatedly, we can find a single term h (of very large rank, m) such that every basic operation of F_1 is a polymer of h.

Finally, define $\bar{h} = h \star h$. Then every term of height at most 2 is a polymer of $\bar{h}$. In this way, we can convert Σ_1 into a set of height 1 identities in the single operation symbol $\bar{h}$. Let Σ_2 be that set of identities, together with the identity

$$\begin{aligned} &\bar{h}(x_1, x_1, \dots, x_1, x_2, x_2, \dots, x_2, \dots, x_m) \approx \\ &\qquad \bar{h}(x_1, x_2, \dots, x_m, x_1, x_2 \dots, x_m, \dots). \end{aligned} \tag{8–13}$$

The variety $\mathcal{W}_2 = \mathrm{Mod}(\Sigma_2)$ fulfills the goal we set for ourselves earlier: $\mathcal{W}_2$ is idempotent, defined by height 1 identities, and $\mathcal{W}_2 \equiv_t \mathcal{W}_1 \equiv_t \mathcal{W}_0$.

Add to Σ_2 the identities

$$\begin{aligned} \bar{h}(x_{11}, \dots, x_{mm}) &\approx h\big(h(x_{11}, \dots, x_{1m}), \dots, h(x_{m1}, \dots, x_{mm})\big) \\ h(x, x, \dots, x) &\approx x. \end{aligned} \tag{8–14}$$

Call the resulting set Σ_3. Note that $\mathcal{W}_3 = \mathrm{Mod}(\Sigma_3)$ is still term-equivalent to $\mathcal{W}_0$.

We claim that $\bar{h}$ is a Taylor term for $\mathcal{W}_0$. Suppose not. It follows easily from (8–14) that $\bar{h}$ is idempotent. Say, the kth identity of Definition 8.53 fails. Since the only requirement is that the variable in position k be different on each side of the identity, it must be that in every member of Σ_2, the kth variables on each side agree. This applies, in particular, to (8–13). By dividing k by m, we can find i, j with $1 \le i, j \le m$ and $k = im + j$. We deduce that in (8–13), we must have $x_i = x_j$, so $i = j$.

But now we can interpret $\mathcal{W}_3$ into $\mathcal{S}et$ by mapping $h(x_1, \dots, x_m)$ to x_i and $\bar{h}(x_1, \dots, x_{m^2})$ to x_k. This contradicts the relationship

$$\mathcal{W}_3 \equiv_t W_0 \not\leq_I \mathcal{S}et. \qquad \square$$

The point of Theorem 8.54 is that the class of varieties possessing a Taylor term is the largest proper, idempotent Maltsev class. Any larger class would contain a passive variety $\mathcal{W}$. If $\mathcal{V}$ is any variety at all, then $\mathcal{S}et \le_I \mathcal{V}$, hence $\mathcal{W} \le_I \mathcal{V}$. Therefore $\mathcal{V}$ lies in this larger Maltsev class, so the class consists of all varieties.

It is time to explain the relationship to tame congruence theory. Suppose that $\mathcal{W}$ is a passive variety. Then there is an interpretation D of $\mathcal{W}$ into $\mathcal{S}et$. Let 2 denote a two-element set. Then 2^D is a two-element member of $\mathcal{W}$ in which every polynomial is either a projection or constant. Thus 2^D

is a minimal algebra of type 1. To summarize, every passive variety admits type 1. We now proceed to demonstrate that the converse holds for locally finite varieties.

Theorem 8.55. *Let $\mathcal{V}$ be a locally finite variety. If $1 \in \text{typ}\{\mathcal{V}\}$ then $\mathcal{V}$ has no Taylor term.*

Proof. By assumption, there is a finite algebra, $\mathbf{A}$, in $\mathcal{V}$ and an atom, β, of $\text{Con}(\mathbf{A})$ such that $\text{typ}(0, \beta) = 1$. Let U be a $(0, \beta)$-minimal set and N a trace of U. By Theorem 8.13, there is $e \in \text{E}(\mathbf{A})$ such that $U = \vec{e}(A)$.

Suppose that t is a Taylor term for $\mathcal{V}$, and define $f = e \circ t^{\mathbf{A}}\restriction_N$. Since t is idempotent, f is a polynomial of $\mathbf{A}|_N$, and therefore, f is a Taylor term for $\mathbf{M} = \mathbf{A}\text{I}_N$. However, as a minimal algebra of type 1, the only idempotent terms of $\mathbf{M}$ are the projections. Thus $\mathbf{M}$ cannot support a Taylor term. □

Putting these arguments together, we obtain the following characterization.

Corollary 8.56. *Let $\mathcal{V}$ be a locally finite, idempotent variety. The following are equivalent.*

(1) *$\mathcal{V}$ omits type 1;*

(2) *$\mathcal{V}$ is active;*

(3) *$\mathcal{V}$ has a Taylor term.*

Proof. The equivalence of (2) and (3) comes from Theorem 8.54, while (3) implies (1) is Theorem 8.55. The paragraph preceding 8.55 argues that (1) implies (2). □

Some final remarks. In the literature, the definition of a Taylor term has the additional requirement that in the identities of Definition 8.53, every x_{ij} and y_{ij} come from $\{x, y\}$. Let us, for a moment, call such a term a "super Taylor term." Obviously, every super Taylor term is a Taylor term, by its very definition. On the other hand, if t is a Taylor term, then by specializing all of the variables to x and y, we determine that t is, in fact, super. Thus "Taylor" and "super Taylor" coincide.

Corollary 8.56 can be significantly strengthened in several ways. First, the idempotence assumption is only used in the proof of (2) implies (3). It is possible to do away with that assumption completely. But in fact, much more is true. The existence of a Taylor term is a Maltsev condition, but not a *strong* Maltsev condition (since there is no bound on the rank of the term). Recent work (see [KMM09] and also [Sig10]) shows that every active variety $\mathcal{V}$ has a term $t(x, y, z, u)$ satisfying the identities

$$t(x, x, x, x) \approx x \text{ and } t(x, y, z, y) \approx t(y, z, x, x).$$

Thus the class of active varieties is strong Maltsev.

Finally, local finiteness is used in Corollary 8.56 only to prove that (3) implies (1). Kearnes and Kiss [KK] prove the more general assertion that if $\mathcal{V}$ is an arbitrary variety with a Taylor term, then no algebra in $\mathcal{V}$ has a nontrivial, strongly Abelian congruence.

Bibliography

[AS68] A. A. Akataev and D. M. Smirnov. Lattices of subvarieties of algebra varieties. *Algebra i Logika*, 7(1):5–25, 1968.

[Bak77] K. A. Baker. Finite equational bases for finite algebras in a congruence-distributive equational class. *Advances in Math.*, 24(3):207–243, 1977.

[BB75] J. T. Baldwin and J. Berman. The number of subdirectly irreducible algebras in a variety. *Algebra Universalis*, 5(3):379–389, 1975.

[BB96a] C. Bergman and J. Berman. Morita equivalence of almost-primal clones. *J. Pure Appl. Algebra*, 108:175–201, 1996.

[BB96b] J. Berman and S. Burris. A computer study of 3-element groupoids. In *Logic and algebra (Pontignano, 1994)*, volume 180 of *Lecture Notes in Pure and Appl. Math.*, pages 379–429. Dekker, New York, 1996.

[Ber80] J. Berman. A proof of Lyndon's finite basis theorem. *Discrete Math.*, 29:229–233, 1980.

[BF48] G. Birkhoff and O. Frink. Representations of lattices by sets. *Trans. Amer. Math. Soc.*, 64:299–316, 1948.

[Bir35] G. Birkhoff. On the structure of abstract algebras. *Proc. Cambr. Philos. Soc.*, 31:433–454, 1935.

[BP75] K. A. Baker and A. F. Pixley. Polynomial interpolation and the Chinese remainder theorem for algebraic systems. *Math. Z.*, 143:165–174, 1975.

[BS81] S. Burris and H. P. Sankappanavar. *A Course in Universal Algebra.* Springer-Verlag, New York, 1981. Available from `http://www.math.uwaterloo.ca/~snburris/htdocs/ualg.html`.

[Bur95] S. Burris. Computers and universal algebra: some directions. *Algebra Universalis*, 34:61–71, 1995.

[BW02] K. A. Baker and J. Wang. Definable principal subcongruences. *Algebra Universalis*, 47(2):145–151, 2002.

[CK76] D. M. Clark and P. H. Krauss. Para primal algebras. *Algebra Universalis*, 6:165–192, 1976.

[Col66] R. R. Colby. On indecomposable modules over rings with minimum condition. *Pacific J. Math.*, 19:23–33, 1966.

[Csá70] B. Csákány. Characterizations of regular varieties. *Acta Sci. Math. (Szeged)*, 31:187–189, 1970.

[Day69] A. Day. A characterization of modularity for congruence lattices of algebras. *Canad. Math. Bull.*, 12:167–173, 1969.

[DF80] A. Day and R. Freese. A characterization of identities implying congruence modularity. I. *Canad. J. Math.*, 32(5):1140–1167, 1980.

[DP02] B. A. Davey and H. A. Priestley. *Introduction to lattices and order*. Cambridge University Press, New York, second edition, 2002.

[FM81] R. Freese and R. McKenzie. Residually small varieties with modular congruence lattices. *Trans. Amer. Math. Soc.*, 264:419–430, 1981.

[FM87] R. Freese and R. McKenzie. *Commutator Theory for Congruence Modular Varieties*. Number 125 in London Mathematical Society Lecture Notes. Cambridge University Press, Cambridge, 1987.

[FP79] E. Fried and A. F. Pixley. The dual discriminator function in universal algebra. *Acta Sci. Math. (Szeged)*, 41(1-2):83–100, 1979.

[Fre90] R. Freese. On the two kinds of probability in algebra. *Algebra Universalis*, 27(1):70–79, 1990.

[Fre11] R. Freese. UACalc: a universal algebra calculator. Available from `http://www.uacalc.org`, 2011.

[FV09] R. Freese and M. A. Valeriote. On the complexity of some Maltsev conditions. *Internat. J. Algebra Comput.*, 19(1):41–77, 2009.

[Grä79] G. Grätzer. *Universal Algebra*. Springer-Verlag, New York, second edition, 1979.

[GS63] G. Grätzer and E. T. Schmidt. Characterizations of congruence lattices of abstract algebras. *Acta Sci. Math. (Szeged)*, 24:34–59, 1963.

[Gum80] H. P. Gumm. An easy way to the commutator in modular varieties. *Arch. Math. (Basel)*, 34(3):220–228, 1980.

[Gum83] H. P. Gumm. Geometrical methods in congruence modular algebras. *Mem. Amer. Math. Soc.*, 45(286):viii+79, 1983.

[Her79] C. Herrmann. Affine algebras in congruence modular varieties. *Acta Sci. Math. (Szeged)*, 41(1-2):119–125, 1979.

[Her94] I. N. Herstein. *Noncommutative rings*, volume 15 of *Carus Mathematical Monographs*. Mathematical Association of America, Washington, DC, 1994. Reprint of the 1968 original, with an afterword by Lance W. Small.

[HH79] J. Hagemann and C. Herrmann. A concrete ideal multiplication for algebraic systems and its relation to congruence distributivity. *Arch. Math. (Basel)*, 32(3):234–245, 1979.

[Hin05] P. G. Hinman. *Fundamentals of mathematical logic*. A. K. Peters Ltd., Wellesley, MA, 2005.

[HM88] D. Hobby and R. McKenzie. *The Structure of Finite Algebras*, volume 76 of *Contemporary Mathematics*. American Math. Soc., Providence, R. I., 1988. Available from `http://www.ams.org/online_bks/conm76/`.

[HŚ60] A. Hulanicki and S. Świerczkowski. Number of algebras with a given set of elements. *Bull. Acad. Polon. Sci. Sér. Sci. Math. Astronom. Phys.*, 8:283–284, 1960.

[JM] M. Jackson and G. McNulty. The equational complexity of Lyndon's algebra. *Algebra Universalis*. to appear.

[JM59] Ju. I. Janov and A. A. Mučnik. Existence of k-valued closed classes without a finite basis. *Dokl. Akad. Nauk SSSR*, 127:44–46, 1959.

[Jón67] B. Jónsson. Algebras whose congruence lattices are distributive. *Math. Scand.*, 21:110–121, 1967.

[Jón68] B. Jónsson. Equational classes of lattices. *Math. Scand.*, 22:187–196 (1969), 1968.

[Jón79] B. Jónsson. Congruence varieties. 1979. Appendix 3 in [Grä79].

[JT61] B. Jónsson and A. Tarski. On two properties of free algebras. *Math. Scand.*, 9:95–101, 1961.

[Kaa92] K. Kaarli. On varieties generated by functionally complete algebras. *Algebra Universalis*, 29(4):495–502, 1992.

[Kis85] E. Kiss. Injectivity and related concepts in modular varieties: I–II. *Bull. Aust. Math. Soc.*, 32:35–53, 1985.

[KK] K. Kearnes and E. Kiss. *The Shape of Congruence Lattices*. American Math. Soc. to appear.

[KMM09] K. Kearnes, P. Marković, and R. McKenzie. A note on Mal'cev conditions for omitting type 1 in locally finite varieties. Preprint, 2009.

[Knu81] D. E. Knuth. *The art of computer programming. Vol. 2.* Addison-Wesley Publishing Co., Reading, Mass., second edition, 1981. Seminumerical algorithms, Addison-Wesley Series in Computer Science and Information Processing.

[KP01] K. Kaarli and A. F. Pixley. *Polynomial completeness in algebraic systems.* Chapman & Hall/CRC, Boca Raton, FL, 2001.

[Kru73] R. L. Kruse. Identities satisfied by a finite ring. *J. Algebra*, 26:298–318, 1973.

[KS97] K. Kearnes and Á. Szendrei. A characterization of minimal locally finite varieties. *Trans. Amer. Math. Soc.*, 349(5):1749–1768, 1997.

[Lak71] H. Lakser. The structure of pseudocomplemented distributive lattices. I. Subdirect decomposition. *Trans. Amer. Math. Soc.*, 156:335–342, 1971.

[Lee70] K. B. Lee. Equational classes of distributive pseudo-complemented lattices. *Canad. J. Math.*, 22:881–891, 1970.

[Ło55a] J. Łoś. On the extending of models. I. *Fund. Math.*, 42:38–54, 1955.

[Ło55b] J. Łoś. Quelques remarques, théorèmes et problèmes sur les classes définissables d'algèbres. In *Mathematical interpretation of formal systems*, pages 98–113. North-Holland Publishing Co., Amsterdam, 1955.

[L′v73] I. V. L′vov. Varieties of associative rings. I, II. *Algebra i Logika*, 12:269–297, 363; ibid. 12 (1973), 667–688, 735, 1973.

[Lyn51] R. C. Lyndon. Identities in two-valued calculi. *Trans. Amer. Math. Soc.*, 71:457–465, 1951.

[Lyn54] R. C. Lyndon. Identities in finite algebras. *Proc. Amer. Math. Soc.*, 5:8–9, 1954.

[Lyn59] R. C. Lyndon. Properties preserved under homomorphism. *Pacific J. Math.*, 9:143–154, 1959.

[Mad06] R. D. Maddux. *Relation algebras*, volume 150 of *Studies in Logic and the Foundations of Mathematics.* Elsevier B. V., Amsterdam, 2006.

[Mal36] A. I. Maltsev. Untersuchungen aus dem gebiete der mathematischen logik. *Matematicheskii Sbornik*, 1:323–336, 1936.

[Mal54] A.I. Maltsev. On the general theory of algebraic systems. *Mat. Sbornik*, 77:3–20, 1954.

[McK70] R. McKenzie. Equational bases for lattice theories. *Math. Scand.*, 27:24–38, 1970.

[McK72] R. McKenzie. Equational bases and nonmodular lattice varieties. *Trans. Amer. Math. Soc.*, 174:1–43, 1972.

[McK78] R. McKenzie. Para primal varieties: A study of finite axiomatizability and definable principal congruences in locally finite varieties. *Algebra Universalis*, 8(3):336–348, 1978.

[McK82] R. McKenzie. Narrowness implies uniformity. *Alg. Univ.*, 15:67–85, 1982.

[McK96a] R. McKenzie. The residual bounds of finite algebras. *Internat. J. Algebra Comput.*, 6(1):1–28, 1996.

[McK96b] R. McKenzie. Tarski's finite basis problem is undecidable. *Internat. J. Algebra Comput.*, 6(1):49–104, 1996.

[Men10] E. Mendelson. *Introduction to mathematical logic.* Discrete Mathematics and its Applications (Boca Raton). CRC Press, Boca Raton, FL, fifth edition, 2010.

[Mit78] A. Mitschke. Near unanimity identities and congruence distributivity in equational classes. *Algebra Universalis*, 8(1):29–32, 1978.

[MMT87] R. McKenzie, G. McNulty, and W. Taylor. *Algebras, Lattices, Varieties*, volume I. Wadsworth & Brooks/Cole, Belmont, CA, 1987.

[MR65] W. D. Maurer and J. L. Rhodes. A property of finite simple non-abelian groups. *Proc. Amer. Math. Soc.*, 16:552–554, 1965.

[MS72] S. O. Macdonald and A. P. Street. On laws in linear groups. *Quart. J. Math. Oxford Ser. (2)*, 23:1–12, 1972.

[Mur65] V. L. Murskiĭ. The existence in the three-valued logic of a closed class with a finite basis having no finite complete system of identities. *Dokl. Akad. Nauk SSSR*, 163:815–818, 1965.

[Mur75] V. L. Murskiĭ. The existence of a finite basis of identities, and other properties of "almost all" finite algebras. *Problemy Kibernet.*, (30):43–56, 1975.

[Mur79] V. L. Murskiĭ. The number of k-element algebras with a binary operation which do not have a finite basis of identities. *Problemy Kibernet.*, (35):5–27, 208, 1979.

[MW70] G. Michler and R. Wille. Die primitiven Klassen arithmetischer Ringe. *Math. Z.*, 113:369–372, 1970.

[MW73] A. Mitschke and H. Werner. On groupoids representable by vector spaces over finite fields. *Arch. Math. (Basel)*, 24:14–20, 1973.

[Neu67] H. Neumann. *Varieties of Groups*. Springer–Verlag, Berlin, 1967.

[Ol′70] A. Ju. Ol′šanskiĭ. The finite basis problem for identities in groups. *Izv. Akad. Nauk SSSR Ser. Mat.*, 34:376–384, 1970.

[OP64] S. Oates and M. B. Powell. Identical relations in finite groups. *J. Algebra*, 1:11–39, 1964.

[Pál84] P. P. Pálfy. Unary polynomials in algebras. I. *Algebra Universalis*, 18(3):262–273, 1984.

[Par80] R. E. Park. A four-element algebra whose identities are not finitely based. *Algebra Universalis*, 11(2):255–260, 1980.

[Per69] P. Perkins. Bases for equational theories of semigroups. *J. Algebra*, 11:298–314, 1969.

[Pie82] R. S. Pierce. *Associative algebras*, volume 88 of *Graduate Texts in Mathematics*. Springer-Verlag, New York, 1982. Studies in the History of Modern Science, 9.

[Pig72] D. Pigozzi. On some operations on classes of algebras. *Algebra Universalis*, 2:346–353, 1972.

[Pix63] A. F. Pixley. Distributivity and permutability of congruence relations in equational classes of algebras. *Proc. Amer. Math. Soc.*, 14:105–109, 1963.

[Pol76] S. V. Polin. Identities of finite algebras. *Sibirsk. Mat. Ž.*, 17(6):1356–1366, 1439, 1976.

[Pol77] S. V. Polin. Identities in congruence lattices of universal algebras. *Mat. Zametki*, 22(3):443–451, 1977.

[Pos41] E. Post. *Two-valued iterative systems of mathematical logic*. Princeton University Press, 1941.

[PP80] P. P. Pálfy and P. Pudlák. Congruence lattices of finite algebras and intervals in subgroup lattices of finite groups. *Algebra Universalis*, 11(1):22–27, 1980.

[PQ73] R. Padmanabhan and R. W. Quackenbush. Equational theories of algebras with distributive congruences. *Proc. Amer. Math. Soc.*, 41:373–377, 1973.

[PRS95] A. Pilitowska, A. Romanowska, and J. D. H. Smith. Affine spaces and algebras of subalgebras. *Algebra Universalis*, 34(4):527–540, 1995.

[Qua72] R. W. Quackenbush. Equational classes generated by finite algebras. *Algebra Universalis*, 1:265–266, 1971/72.

[Rob55] H. Robbins. A remark on Stirling's formula. *Amer. Math. Monthly*, 62:26–29, 1955.

[Sco56] D. Scott. Equationally complete extensions of finite algebras. *Proc. Amer. Math. Soc.*, 59:35–38, 1956.

[Sie45] W. Sierpiński. Sur les fonctions de plusieurs variables. *Fund. Math.*, 33:169–173, 1945.

[Sig10] M. H. Siggers. A strong Mal'cev condition for locally finite varieties omitting the unary type. *Algebra Universalis*, 64(1–2):15–20, 2010.

[Sio61] F. M. Sioson. Free-algebraic characterizations of primal and independent algebras. *Proc. Amer. Math. Soc.*, 12:435–439, 1961.

[Smi76] J. D. H. Smith. *Maltsev Varieties*, volume 554 of *Lecture Notes in Mathematics*. Springer-Verlag, 1976.

[Sze92] Á. Szendrei. Strictly simple algebras and minimal varieties. In A. Romanowska and J. D. H. Smith, editors, *Universal Algebra and Quasigroup Theory*, pages 209–239. Heldermann Verlag, Berlin, 1992.

[Tar54] A. Tarski. Contributions to the theory of models. I. *Nederl. Akad. Wetensch. Proc. Ser. A.*, 57:572–581 = Indagationes Math. 16, 572–581 (1954), 1954.

[Tay73] W. Taylor. Characterizing Mal'cev conditions. *Algebra Universalis*, 3:351–397, 1973.

[Tay77] W. Taylor. Varieties obeying homotopy laws. *Canad. J. Math.*, 29(3):498–527, 1977.

[VL70] M. R. Vaughan-Lee. Uncountably many varieties of groups. *Bull. London Math. Soc.*, 2:280–286, 1970.

[VW91] M. A. Valeriote and R. Willard. A characterization of congruence permutable locally finite varieties. *J. Algebra*, 140(2):362–369, 1991.

[Wer74] H. Werner. Congruences on products of algebras and functionally complete algebras. *Algebra Universalis*, 4:99–105, 1974.

Index of Notation

References to the initial definitions of most of the notation used in this book are given in the following table. Most of the symbols are listed alphabetically. At the bottom of the table are those symbols that resist alphabetization.

Symbol	Page	Description
$\mathcal{A}_n$	111	The variety of Abelian groups of exponent n
$\mathcal{Alg}_\rho$	96	The class of all algebras of similarity type ρ
$\mathrm{Aut}(\mathbf{A})$	8	The set of automorphisms of $\mathbf{A}$
$\mathbb{B}_n$	67	The n-atom Boolean algebra
$\overline{\mathbf{B}}$	67	The Boolean algebra $\mathbf{B}$ with a new unit attached
$\mathrm{Cg}^{\mathbf{A}}(\theta)$	17	The congruence on $\mathbf{A}$ generated by θ
$\mathrm{Clo}^A(F)$	80	The clone on A generated by F
$\mathrm{Clo}(\mathbf{A})$	82	The clone of term operations on $\mathbf{A}$
$\mathrm{Clo}_n^A(F)$	81	The n-ary members of $\mathrm{Clo}^A(F)$
$\mathrm{Con}(\mathbf{A})$	17	The set of congruences on $\mathbf{A}$
$\mathrm{Con}_{\mathrm{fi}}(\mathbf{A})$	117	The set of fully invariant congruences of $\mathbf{A}$
$\mathcal{Cr}_n$	60	The class of commutative rings satisfying $x^n \approx x$
δ_B	169	The diagonal relation on B
$\mathrm{Dn}(\mathbf{P})$	24	The set of downsets of $\mathbf{P}$
$\mathrm{E}(\mathbf{A})$	252	The set of idempotent unary terms of $\mathbf{A}$
$\mathrm{End}(\mathbf{A})$	117	The set of endomorphisms of $\mathbf{A}$
$\mathrm{Eq}(A)$	14	The set of equivalence relations on A
$\mathcal{F}(\Theta)$	90	The set of all operations preserving every member of Θ
$\mathbf{F}_{\mathcal{K}}(X)$	99	The free $\mathcal{K}$-algebra over X
$\mathbb{F}_n$	77	The finite field of order n
$\mathsf{H}(\mathcal{K})$	9	Homomorphic images of members of $\mathcal{K}$
$\mathrm{Hom}(\mathbf{A}, \mathbf{B})$	13	The set of homomorphisms from $\mathbf{A}$ to $\mathbf{B}$
ι_A	8	The identity map on A
$\mathbf{I}[a, b]$	23	The interval subposet from a to b
$\mathrm{Id}(\mathcal{K})$	106	All identities true of all members of $\mathcal{K}$
$\mathrm{Id}_n(\mathcal{K})$	157	All n-variable identities that hold in $\mathcal{K}$
$\mathrm{Idl}(\mathbf{R})$	33	The set of ideals of a ring

Symbol	**Page**	**Description**
$\mathrm{Sub}(\mathbf{A})$	8	The set of subuniverses of the algebra $\mathbf{A}$
$\mathrm{Sub}_1(\mathbf{A})$	175	The set of trivial subuniverses of $\mathbf{A}$
sup	22	Least upper bound
$\mathcal{T}$	261	The poset of tame-congruence types
$T_\rho(X)$	95	The set of terms of type ρ over X
$\mathrm{Th}(\mathcal{K})$	153	All first-order sentences true in every member of $\mathcal{K}$
$\mathrm{typ}(\alpha,\beta)$	259	The type of β over α
$\mathrm{typ}\{\mathbf{A}\}$	261	The type set of $\mathbf{A}$
$\mathrm{Up}(\mathbf{P})$	24	The set of upsets of $\mathbf{P}$
$\mathsf{V}(\mathcal{K})$	72	The variety generated by $\mathcal{K}$
X_ω, X_n	99	$X_\omega=\{x_1,x_2,\dots\}$, $X_n=\{x_1,\dots,x_n\}$
$\mathbb{Z}$, $\mathbb{Q}$, $\mathbb{R}$	3	The sets of integers, rational and real numbers
$\mathbb{Z}(p^\infty)$	59	Quasi-cyclic group
A^B	3	All functions from B to A
ω	3	The natural numbers $\{0,1,2,3,\dots\}$
$Y\subseteq_\omega X$	3	Y is a finite subset of X
$\mathbf{A}\cong\mathbf{B}$	8	$\mathbf{A}$ is isomorphic to $\mathbf{B}$
$\prod S_i$	8	Direct product of the S_i's
$f^{\mathbf{A}}$	7	Operation symbol f evaluated in algebra $\mathbf{A}$
$x\,\theta\,y$	14	$(x,y)\in\theta$
$\theta\circ\psi$	14	Relative product of θ and ψ
$\theta^{\smile}$	14	The converse of the binary relation θ
$x\equiv_\theta y$	14	$(x,y)\in\theta$
a/θ	15	The equivalence class of a modulo θ
$a\lll X$	22	a is a lower bound of X
0_L	22	Smallest element of the lattice $\mathbf{L}$
1_L	22	Largest element of the lattice $\mathbf{L}$
0_A	14	Smallest equivalence relation on the set A
1_A	14	Largest equivalence relation on the set A
$\mathbf{P}^\partial$	21	The dual of the poset $\mathbf{P}$
$a\prec b$	23	a is covered by b
$x\wedge y$	22	Greatest lower bound of $\{x,y\}$
$x\vee y$	22	Least upper bound of $\{x,y\}$
$\bigwedge X$	30	Greatest lower bound of X
$\bigvee X$	30	Least upper bound of X
$X^{\blacktriangleleft}$, $X^{\blacktriangleright}$	39	The polars of X under a Galois connection
$(a]$	41	The principal ideal generated by a
$\mathbf{2}$	42	The set $\{0,1\}$, often with a lattice or Boolean algebra structure

Index

www.ingramcontent.com/pod-product-compliance
Lightning Source LLC
Chambersburg PA
CBHW060330220326
41598CB00023B/2667
* 9 7 8 1 0 3 2 9 2 2 1 8 8 *